APPLIED CALCULUS

FOR BUSINESS AND ECONOMICS, LIFE SCIENCES, AND SOCIAL SCIENCES

APPLIED CALCULUS

FOR BUSINESS AND ECONOMICS, LIFE SCIENCES, AND SOCIAL SCIENCES

RAYMOND A. BARNETT
Merritt College

MICHAEL R. ZIEGLER
Marquette University

DELLEN PUBLISHING COMPANY
San Francisco and Santa Clara, California

© Copyright 1982 by Dellen Publishing Company, 3600 Pruneridge Avenue, Santa Clara, California 95051

Printed in the United States of America
10 9 8 7 6 5 4 3 2 1

LIBRARY OF CONGRESS CATALOGING IN PUBLICATION DATA

Barnett, Raymond A.
  Applied calculus for business and economics, life
sciences, and social sciences.

  Includes index.
  1. Calculus.   I. Ziegler, Michael R.   II. Title.
QA303.B2827   515   81-17453
ISBN 0–89517–036–1   AACR2

CONTENTS

v

CHAPTER **8** DIFFERENTIAL EQUATIONS 367

CHAPTER **9** TAYLOR POLYNOMIALS AND SERIES; L'HÔPITAL'S RULE 413

CHAPTER **10** NUMERICAL TECHNIQUES 463

CHAPTER **11** PROBABILITY AND CALCULUS 513

APPENDIX **A** REVIEW: SETS AND ALGEBRA 573

TABLES   **623**

ANSWERS   **633**

INDEX   **671**

PREFACE

Many colleges and universities now offer calculus courses that emphasize topics that are most useful to students in business and economics, the life sciences, and social sciences. Because of this trend, the authors have reviewed course outlines and college catalogs from a large number of colleges and universities, and on the basis of this survey, selected the topics, applications, and emphasis found in this text.

The book is designed for students who have had 1½–2 years of high school algebra or its equivalent. However, because much of this material is forgotten through lack of use, Appendix A and Chapter 1 review basic topics from intermediate algebra which can be treated in a systematic way or can be referred to as needed. If Chapter 7, involving trigonometric functions is covered, then it will be useful to have had a course in trigonometry sometime in the past.

IMPORTANT FEATURES

EMPHASIS
Emphasis is on computational skills, ideas, and problem solving rather than mathematical theory. Most derivations and proofs are omitted except where their inclusion adds significant insight into a particular concept. General concepts and results are usually presented only after particular cases have been discussed.

EXERCISE SETS
The exercise sets are designed so that an average or below-average student will experience success, and a very capable student will be challenged. They are mostly divided into A (routine, easy mechanics), B (more difficult mechanics), and C (difficult mechanics and some theoretical) levels.

APPLICATIONS   There are sufficient applications included in this book to convince even the most skeptical student that mathematics is really useful (see the Applications Index on page xiii). The majority of the applications are included at the end of exercise sets and are generally divided into business and economics, life science, and social science groupings. An instructor with students from all three disciplines can let them choose applications from their own fields of interest, or if most students are from one of the three areas, then special emphasis can be placed there. Most of the applications are simplified versions of actual real-world problems taken from professional journals and professional books associated with the given subjects. No specialized experience is required to solve any of the applications included in this book.

STUDENT AND INSTRUCTOR AIDS

STUDENT AIDS   Dotted **"think boxes"** are used to enclose steps that are usually performed mentally (see Section 2-7).

Examples and developments are often **annotated** to help students through critical stages (see Section 2-7).

A **second color** is used functionally to indicate key steps (see Section 1-3).

Boldface type is used to introduce new terms and important comments.

Answers to odd-numbered problems are included in the back of the book.

Chapter review sections include a review of all important terms and symbols, a comprehensive review exercise, and a practice test. Answers to all review exercises and practice test problems are included in the back of the book.

A **solutions manual** by Suman Shah is available at a nominal cost through a book store. The manual includes detailed solutions to all odd-numbered problems, all chapter review exercises, and all practice test problems.

INSTRUCTOR AIDS   A uniquely designed **test battery** is included in the instructor's manual, which can be obtained from the publisher without charge. The test battery includes quizzes and chapter tests (two forms of each). All tests have easy-to-grade solution keys and sample student-solution sheets. The format is 8½ by 11 inches for ease of reproduction.

Answers to even-numbered problems, which are not included in the text, are given in the instructor's manual.

A **solutions manual** by Suman Shah (see student aids) is available to instructors without cost from the publisher.

ACKNOWLEDGMENTS

In addition to the authors many others are involved in the successful publication of a book. We wish to thank personally:

Susan Boren, University of Tennessee at Martin

Sandra Gossum, University of Tennessee at Martin

Freida Holly, Metropolitan State College

Stanley Lukawecki, Clemson University

John Plachy, Metropolitan State College

Walter Roth, University of North Carolina at Charlotte

Wesley Sanders, Sam Houston State College

Arthur Sparks, University of Tennessee at Martin

Martha Stewart, University of North Carolina at Charlotte

James Strain, Midwestern State College

Michael Vose, Austin Community College

Scott Wright, Loyola Marymount University

Dennis Zill, Loyola Marymount University

We also wish to thank:

Janet Bollow for an outstanding book design

John Drooyan for the many sensitive and beautiful photographs seen throughout the book

Charles Burke for carefully checking all examples and problems (a tedious but extremely important job)

Phyllis Niklas for her thorough and expert editing and her ability to guide the book smoothly through all production details

Don Dellen, the publisher, who continues to provide all the support services and encouragement an author could hope for

APPLICATIONS INDEX

Most of the applications in this text are simplified versions of actual real-world problems taken from professional journals and professional books associated with the given subjects. No specialized background outside of that provided in the text is necessary to solve any of the applications.

CHAPTER 1
REVIEW: GRAPHS AND FUNCTIONS

CONTENTS

1 REVIEW: GRAPHS AND FUNCTIONS

In Appendix A we review basic algebraic operations. Here, we will review graphs of equations in two variables, the general concept of relation and function, and some special functions, including linear, quadratic, exponential, and logarithmic functions.

1-1 CARTESIAN COORDINATE SYSTEM

Recall that a **Cartesian (rectangular) coordinate system** in a plane is formed by taking two mutually perpendicular real number lines intersecting at their origins (**coordinate axes**), one horizontal and one vertical,

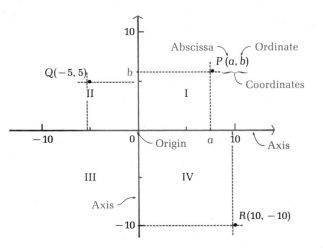

FIGURE 1 The Cartesian coordinate system.

and then assigning unique **ordered pairs** of numbers (**coordinates**) to each point P in the plane (Fig. 1). The first coordinate (**abscissa**) is the distance of P from the vertical axis, and the second coordinate (**ordinate**) is the distance of P from the horizontal axis. In Figure 1 the coordinates of point P are (a, b). By reversing the process, each ordered pair of real numbers can be associated with a unique point in a plane. The coordinate axes divide a plane into four parts (**quadrants**) numbered from I to IV in a counterclockwise direction.

1-2   STRAIGHT LINES; SLOPE

In this section we will review some of the properties of straight lines and their equations. The following problems are basic:

1. Given a linear equation, draw its graph.
2. Given certain information about a straight line in a coordinate system, find its equation.

GRAPHING LINEAR EQUATIONS

A **solution** of an equation in two variables is an ordered pair of real numbers that satisfy the equation. For example, $(0, -3)$ is a solution of $3x - 4y = 12$. The **solution set** of an equation in two variables is the set of all solutions of the equation. When we say that we **graph an equation** in two variables, we mean that we graph its solution set on a rectangular coordinate system.

The graph of any equation of the form

$$Ax + By = C \qquad \text{Standard form} \qquad (1)$$

where A, B, and C are constants (A and B not both zero), is a straight line. Every straight line in a Cartesian coordinate system is the graph of an equation of this type. Also, the graph of any equation of the form

$$y = mx + b \qquad (2)$$

where m and b are constants, is a straight line. Form (2) is simply a special case of (1). To graph either (1) or (2) we plot any two points of their solution set and use a straightedge to draw the line through these two points. The points where the line crosses the axes—called the **intercepts**—are often the easiest to find when dealing with form (1). To find the **y intercept**, we let $x = 0$ and solve for y; to find the **x intercept**, we let $y = 0$ and solve for x. It is sometimes wise to find a third point as a check.

Example 1   (A)   The graph of $3x - 4y = 12$ is

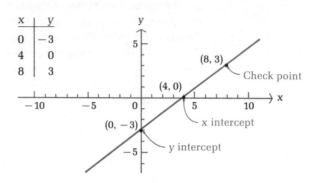

x	y
0	−3
4	0
8	3

(B)   The graph of $y = 2x - 1$ is

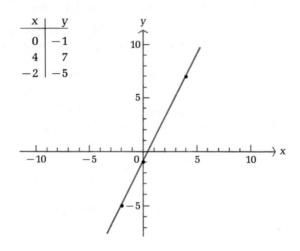

x	y
0	−1
4	7
−2	−5

Problem 1   Graph.

(A) $4x - 3y = 12$ (B) $y = \dfrac{x}{2} + 2$

Answers   (A)   (B)

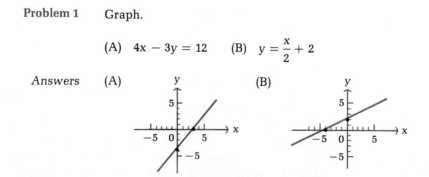

SLOPE It is very useful to have a numerical measure of the "steepness" of a line. The concept of slope is widely used for this purpose. The **slope** of a line through the two points (x_1, y_1) and (x_2, y_2) is given by the following formula:

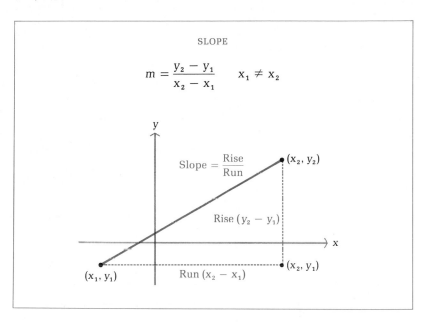

SLOPE

$$m = \frac{y_2 - y_1}{x_2 - x_1} \qquad x_1 \neq x_2$$

The slope of a vertical line is not defined. (Why?)

Example 2 Find the slope of the line through $(-2, 5)$ and $(4, -7)$.

Solution Let $(x_1, y_1) = (-2, 5)$ and $(x_2, y_2) = (4, -7)$. Then

$$m = \frac{y_2 - y_1}{x_2 - x_1} = \frac{-7 - 5}{4 - (-2)} = \frac{-12}{6} = -2$$

Note that we also could have let $(x_1, y_1) = (4, -7)$ and $(x_2, y_2) = (-2, 5)$, since this simply reverses the sign in both the numerator and the denominator and the slope does not change:

$$m = \frac{5 - (-7)}{-2 - 4} = \frac{12}{-6} = -2$$

Problem 2 Find the slope of the line through $(3, -6)$ and $(-2, 4)$.

Answer -2

In general, the slope of a line may be positive, negative, 0, or not defined. Each of these cases is interpreted geometrically in Table 1.

TABLE 1 Going from left to right

LINE	SLOPE	EXAMPLE
Rising	Positive	
Falling	Negative	
Horizontal	0	
Vertical	Not defined	

SLOPE–INTERCEPT FORM The constants m and b in the equation

$$y = mx + b \qquad (3)$$

have special geometric significance.

If we let $x = 0$, then $y = b$, and we observe that the graph of (3) crosses the y axis at $(0, b)$. The constant b is called the **y intercept.** For example, the y intercept of the graph of $y = -4x - 1$ is -1.

To determine the geometric significance of m, let us choose two points (x_1, y_1) and (x_2, y_2) on the line $y = mx + b$. Since these points lie on the line, their coordinates must satisfy the equation $y = mx + b$. Thus,

$$y_1 = mx_1 + b \qquad \text{and} \qquad y_2 = mx_2 + b$$

Solving both for b, we obtain

$$b = y_1 - mx_1 \qquad \text{and} \qquad b = y_2 - mx_2$$

Hence,

$$y_1 - mx_1 = y_2 - mx_2$$

Solving this equation for m (a good exercise for the reader) we obtain

$$m = \frac{y_2 - y_1}{x_2 - x_1}$$

Thus, m is the slope of the line given by $y = mx + b$. Now we know why

$$y = mx + b$$

is called the **slope–intercept form** of the equation of a line.

Example 3   (A)   Find the slope and y intercept, and graph $y = -\frac{2}{3}x - 3$.
(B)   Write the equation of the line with slope $\frac{2}{3}$ and y intercept -2.

Solutions   (A)   Slope $= m = -\frac{2}{3}$ (B) $m = \frac{2}{3}$ and $b = -2$; thus,
y intercept $= b = -3$ $y = \frac{2}{3}x - 2$

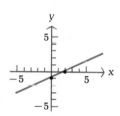

Problem 3   Write the equation of the line with slope $\frac{1}{2}$ and y intercept -1. Graph.

Answer $y = \frac{1}{2}x - 1$

POINT–SLOPE FORM   If a line has slope m and passes through the fixed point (x_1, y_1) and if (x, y)
is any other point on the line (Fig. 2), then

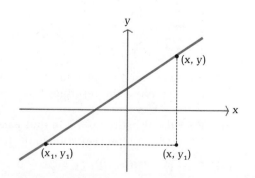

FIGURE 2

$$\frac{y - y_1}{x - x_1} = m$$

That is,

$$y - y_1 = m(x - x_1)$$

This form is called the **point–slope form** of the equation of a line. It is important to note that x and y in the equation are variables and that x_1, y_1, and m are constants. The point–slope form is extremely useful, since it enables us to find an equation for a line if we know its slope and a point that it passes through, or if we simply only know the coordinates of two points on the line.

Example 4 (A) Find the equation of a line that has slope $\frac{1}{2}$ and passes through $(-4, 3)$. Write the final answer in the form $Ax + By = C$.

(B) Write the equation of a line that passes through the two points $(-3, 2)$ and $(-4, 5)$. Write the resulting equation in the form $y = mx + b$.

Solutions (A) $y - y_1 = m(x - x_1)$

Let $m = \frac{1}{2}$ and $(x_1, y_1) = (-4, 3)$. Then

$$y - 3 = \tfrac{1}{2}(x + 4) \qquad \text{Multiply by 2}$$
$$2y - 6 = x + 4$$
$$-x + 2y = 10 \quad \text{or} \quad x - 2y = -10$$

(B) First, find the slope of the line using the slope formula:

$$m = \frac{y_2 - y_1}{x_2 - x_1} = \frac{5 - 2}{-4 - (-3)} = \frac{3}{-1} = -3$$

Now use

$$y - y_1 = m(x - x_1)$$

with $m = -3$ and $(x_1, y_1) = (-3, 2)$:

$$y - 2 = -3(x + 3)$$
$$y - 2 = -3x - 9$$
$$y = -3x - 7$$

Problem 4 (A) Find the equation of a line that has slope $\frac{2}{3}$ and passes through $(6, -2)$. Write the resulting equation in the form $Ax + By = C$, $A > 0$.

(B) Find the equation of a line that passes through $(2, -3)$ and $(4, 3)$. Write the resulting equation in the form $y = mx + b$.

Answers (A) $2x - 3y = 18$ (B) $y = 3x - 9$

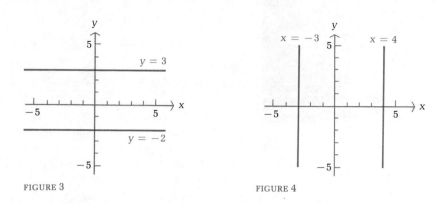

FIGURE 3   FIGURE 4

VERTICAL AND HORIZONTAL LINES

If a line is horizontal (slope 0), then all points on the line have the same y value, while x can assume any value (Fig. 3). Thus, its equation is of the form

$$0x + y = C$$

or simply

$$y = C \qquad \text{Horizontal line}$$

If a line is vertical, then its slope is not defined. However, all x values are alike, while y can take on any value (Fig. 4). Thus, its equation is of the form

$$x + 0y = C$$

or simply

$$x = C \qquad \text{Vertical line}$$

Example 5

The equation of a horizontal line through $(-2, 3)$ is $y = 3$, and the equation of a vertical line through the same point is $x = -2$.

Problem 5

Find the equations of the horizontal and vertical lines through $(4, -5)$.

Answer $y = -5, x = 4$

PARALLEL LINES

If two nonvertical lines are parallel, then they have the same slope; if two lines have the same slope, they are parallel.

Example 6

Find the slope of $y = mx + 7$ so that it is parallel to $3x - 2y = 4$.

Solution

To find the slope of $3x - 2y = 4$, write the equation in the form $y = mx + b$ and identify m:

$$3x - 2y = 4$$
$$-2y = -3x + 4$$
$$y = \tfrac{3}{2}x - 2$$

Thus, $m = \tfrac{3}{2}$, and

$$y = \tfrac{3}{2}x + 7$$

is parallel to $3x - 2y = 4$ (since they have the same slope).

Problem 6 Find the slope of $y = mx - 3$ so that it is parallel to $2x + 3y = 6$.

Answer $m = -\tfrac{2}{3}$

APPLICATIONS We will now see how equations of lines occur naturally in certain applications.

Example 7 The management of a company that manufactures roller skates has fixed costs (costs at zero output) of $300 per day and total costs of $4,300 per day at an output of 100 pairs of skates per day. Assume that cost C is linearly related to output.

(A) Find the slope of the line joining the points associated with outputs of 0 and 100; that is, the line passing through (0, 300) and (100, 4,300).

(B) Find the equation of the line relating output to cost. Write the final answer in the form $C = mx + b$.

(C) Graph the cost equation from part B for $0 \le x \le 200$.

Solutions (A) $m = \dfrac{y_2 - y_1}{x_2 - x_1} = \dfrac{4,300 - 300}{100 - 0} = \dfrac{4,000}{100} = 40$

(B) We must find the equation of the line that passes through (0, 300) with slope 40. We use the point–slope formula:

$$C - C_1 = m(x - x_1)$$
$$C - 300 = 40(x - 0)$$
$$C = 40x + 300$$

(C)

x	C
0	300
100	4,300
200	8,300

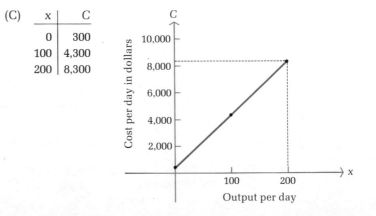

Problem 7   Answer parts A and B in Example 7 for fixed costs of $250 per day and total costs of $3,450 per day at an output of 80 pairs of skates per day.

Answers   (A) $m = 40$ (B) $C = 40x + 250$

EXERCISE 1-2

A   *Graph in a rectangular coordinate system.*

1. $y = 2x - 3$

2. $y = \dfrac{x}{2} + 1$

3. $2x + 3y = 12$

4. $8x - 3y = 24$

Find the slope and y intercept of the graph of each equation.

5. $y = 2x - 3$

6. $y = \dfrac{x}{2} + 1$

7. $y = -\frac{2}{3}x + 2$

8. $y = \frac{3}{4}x - 2$

Write the equation of a line with the indicated slope and y intercept.

9. Slope $= -2$
    y intercept $- 4$

10. Slope $= -\frac{2}{3}$
    y intercept $= -2$

11. Slope $= -\frac{3}{5}$
    y intercept $= 3$

12. Slope $= 1$
    y intercept $= -2$

B   *Graph in a rectangular coordinate system.*

13. $y = -\frac{2}{3}x - 2$

14. $y = -\frac{3}{2}x + 1$

15. $3x - 2y = 10$

16. $5x - 6y = 15$

17. $x = 3$ and $y = -2$

18. $x = -3$ and $y = 2$

Find the slope of the graph of each equation. (First write the equation in the form $y = mx + b$.)

19. $3x + y = 5$

20. $2x - y = -3$

21. $2x + 3y = 12$

22. $3x - 2y = 10$

Write the equation of the line through each indicated point with the indicated slope. Transform the equation into the form $y = mx + b$.

23. $m = -3$, $(4, -1)$.

24. $m = -2$, $(-3, 2)$

25. $m = \frac{2}{3}$, $(-6, -5)$

26. $m = \frac{1}{2}$, $(-4, 3)$

Find the slope of the line that passes through the given points.

27. $(1, 3)$ and $(7, 5)$

28. $(2, 1)$ and $(10, 5)$

29. $(-5, -2)$ and $(5, -4)$

30. $(3, 7)$ and $(-6, 4)$

Write the equation of the line through each indicated pair of points. Write the final answer in the form $Ax + By = C$, $A > 0$.

31. (1, 3) and (7, 5) **32.** (2, 1) and (10, 5)
33. (−5, −2) and (5, −4) **34.** (3, 7) and (−6, 4)

Write the equations of the vertical and horizontal lines through each point.

35. (3, −5) **36.** (−2, 7) **37.** (−1, −3) **38.** (6, −4)

Find the equation of the line, given the information in each problem. Write the final answer in the form $y = mx + b$.

39. Line passes through (−2, 5) with slope −½.
40. Line passes through (3, −1) with slope −⅔.
41. Line passes through (−2, 2) parallel to $y = -\frac{1}{2}x + 5$.
42. Line passes through (−4, −3) parallel to $y = 2x - 3$.
43. Line passes through (−2, −1) parallel to $x - 2y = 4$.
44. Line passes through (−3, 2) parallel to $2x + 3y = -6$.

C **45.** Graph $y = mx - 2$ for $m = 2$, $m = \frac{1}{2}$, $m = 0$, $m = -\frac{1}{2}$, and $m = -2$, all on the same coordinate system.
46. Graph $y = -\frac{1}{2}x + b$ for $b = -4$, $b = 0$, and $b = 4$, all on the same coordinate system.

Write the equation of the line through the indicated points. Be careful!

47. (2, 7) and (2, −3) **48.** (−2, 3) and (−2, −1)
49. (2, 3) and (−5, 3) **50.** (−3, −3) and (0, −3)

APPLICATIONS

BUSINESS & ECONOMICS

51. *Simple interest.* If $P (the principal) is invested at an interest rate of r, then the amount A that is due after t years is given by

$$A = Prt + P$$

If $100 is invested at 6% (r = 0.06), then $A = 6t + 100$, $t \geq 0$.
(A) What will $100 amount to after 5 years? After 20 years?
(B) Graph the equation for $0 \leq t \leq 20$.
(C) What is the slope of the graph? (The slope indicates the increase in the amount A for each additional year of investment.)

52. *Cost equation.* The management of a company manufacturing surfboards has fixed costs (zero output) of $200 per day and total costs of $1,400 per day at a daily output of twenty boards.

(A) Assuming the total cost per day (C) is linearly related to the total output per day (x), write an equation relating these two quantities. [*Hint:* Find the equation of a line that passes through (0, 200) and (20, 1,400).]

(B) What are the total costs for an output of twelve boards per day?

(C) Graph the equation for $0 \leq x \leq 20$.

[*Note:* The slope of the line found in part A is the increase in total cost for each additional unit produced and is called the *marginal cost*. More will be said about the concept of marginal cost later.]

53. *Demand equation.* A manufacturing company is interested in introducing a new power mower. Its market research department gave the management the demand-price forecast listed in the table.

PRICE	ESTIMATED DEMAND
$ 70	7,800
$120	4,800
$160	2,100
$200	0

(A) Plot these points, letting d represent the number of mowers people are willing to buy (demand) at a price of p each.

(B) Note that the points in part A lie along a straight line. Find the equation of that line.

[*Note:* The slope of the line found in part B indicates the decrease in demand for each $1 increase in price.]

54. *Depreciation.* Office equipment was purchased for $20,000 and is assumed to have a scrap value of $2,000 after 10 years. If its value is depreciated linearly (for tax purposes) from $20,000 to $2,000:

(A) Find the linear equation that relates value (V) in dollars to time (t) in years.

(B) What would be the value of the equipment after 6 years?

(C) Graph the equation for $0 \leq t \leq 10$.

[*Note:* The slope found in part A indicates the decrease in value per year.]

LIFE SCIENCES 55. *Nutrition.* In a nutrition experiment, a biologist wants to prepare a special diet for the experimental animals. Two food mixes, A and B, are available. If mix A contains 20% protein and mix B contains 10%

protein, what combinations of each mix will provide exactly 20 grams of protein? Let x be the amount of A used and let y be the amount of B used. Then write a linear equation relating x, y, and 20. Graph this equation for $x \geq 0$ and $y \geq 0$.

56. *Ecology.* As one descends into the depths of the ocean, pressure increases linearly. The pressure is 15 pounds per square inch on the surface and 30 pounds per square inch 33 feet below the surface.

(A) If p is pressure in pounds and d is the depth below the surface in feet, write an equation that expresses p in terms of d. [*Hint:* Find the equation of a line that passes through (0, 15) and (33, 30).]

(B) What is the pressure at 12,540 feet (the average depth of the ocean)?

(C) Graph the equation for $0 \leq d \leq 12{,}540$.

[*Note:* The slope found in part A indicates the change in pressure for each additional foot of depth.)

SOCIAL SCIENCES

57. *Psychology.* In an experiment on motivation, J. S. Brown trained a group of rats to run down a narrow passage in a cage to receive food in a goal box. He then connected the rats, using a harness, to an overhead wire that was attached to a spring scale. A rat was placed at different distances d (in centimeters) from the goal box, and the pull p (in grams) of the rat toward the food was measured. Brown found that the relationship between these two variables was very close to being linear and could be approximated by the equation

$$p = -\tfrac{1}{5}d + 70 \qquad 30 \leq d \leq 175$$

(See J. S. Brown, *Journal of Comparative Physiology and Psychology*, 1948, 41: 450–465.)

(A) What was the pull when $d = 30$? When $d = 175$?

(B) Graph the equation.

(C) What is the slope of the line?

1-3 RELATIONS AND FUNCTIONS

INTRODUCTION

The relation–function concept is one of the most important concepts in mathematics. The idea of correspondence plays a central role in its formulation. You have already had experiences with correspondences in everyday life. For example:

To each item on the shelf in a grocery store there corresponds a price.

To each name in a telephone book there corresponds one or more telephone number(s).

To each square there corresponds an area.
To each number there corresponds its cube.
To each student there corresponds a grade-point average.

One of the most important aspects of any science (managerial, life, social, physical, etc.) is the establishment of correspondences among various types of phenomena. Once a correspondence is known, predictions can be made. A cost analyst would like to predict costs for various levels of output in a manufacturing process; a medical researcher would like to know the correspondence between heart disease and overweightness; a psychologist would like to predict the level of performance after a given number of repetitions of a task; and so on.

What do all of the above examples of correspondence have in common? Each deals with the matching of elements from a first set, called the **domain** of the correspondence, with elements in a second set, called the **range** of the correspondence. Let us consider two of the above examples in more detail. Suppose from a student record office we select five names with their corresponding grade-point averages (GPAs) and telephone numbers. The information is summarized in Tables 2 and 3.

TABLE 2 Correspondence I	
DOMAIN	**RANGE**
Name	*GPA*
Jones, Robert ——→	2.4
Jones, Ruth ——→	2.9
Jones, Sally ——→	3.8
Jones, Samuel ——→	3.4
Jones, Sandra ╱	

TABLE 3 Correspondence II	
DOMAIN	**RANGE**
Name	*Telephone number*
Jones, Robert ——→	841-2315
	841-2403
Jones, Ruth ——→	838-5106
Jones, Sally ╱	
Jones, Samuel ——→	715-0176
Jones, Sandra ——→	732-1934

RELATIONS AND FUNCTIONS

Correspondences I and II are examples of relations; correspondence I is an example of a function. These two important terms, *relation* and *functions*, are defined below:

DEFINITION OF RELATION AND FUNCTION

A **relation** is a correspondence between a first set called the **domain** and a second set called the **range** such that to each element in the domain there corresponds *one or more* elements in the range.

A **function** is a relation with the added restriction that to each domain value there corresponds *one and only one* range value. (All functions are relations, but some relations are not functions.)

From these definitions, we can see that correspondence I is a relation, and it is also a function, since to each domain value (name) there corresponds one and only one range value (GPA). On the other hand, correspondence II is not a function, since in one case two range values (telephone numbers) correspond to one domain value (name)—Robert Jones has two telephone numbers.

COMMON WAYS OF SPECIFYING RELATIONS AND FUNCTIONS

The arrow method of specifying relations and functions illustrated in Tables 2 and 3 is convenient for an introduction to the subject—using it we can easily identify which relations are functions. However, in actual practice, relations and functions are more generally specified as shown in Table 4.

When using graphs, it is common practice to associate domain values with the horizontal axis and range values with the vertical axis. Thus, the first coordinate (abscissa) of a point on a graph is a domain value and the second coordinate (ordinate) is a range value. The graph in Table 4 does not specify a function since more than one range value corresponds to a given domain value.

Example 8

Given the set $F = \{(0, 0), (1, -1), (1, 1), (4, -2), (4, 2)\}$:
(A) Write this relation using arrows as in Tables 2 and 3. Indicate the domain and range.

TABLE 4 Common ways of specifying relations and functions

METHOD	ILLUSTRATION	EXAMPLE	
Equation	$y = x^2 + x$	$x = 2$ corresponds to	$y = 6$
		$x = 1$ corresponds to	$y = 2$
Table	$\begin{array}{c\|c} p & C \\ \hline 2 & 14 \\ 4 & 18 \\ 6 & 22 \end{array}$	$p = 4$ corresponds to	$C = 18$
		$p = 6$ corresponds to	$C = 22$
Set of ordered pairs of elements	$\{(2, 14), (4, 18), (6, 22)\}$	6 corresponds to 22	
		2 corresponds to 14	
Graph		$x = 4$ corresponds to	$y = \pm 2$
		$x = 1$ corresponds to	$y = \pm 1$

(B) Graph the set in a rectangular coordinate system.

(C) Is the relation a function? Explain.

Solutions (A)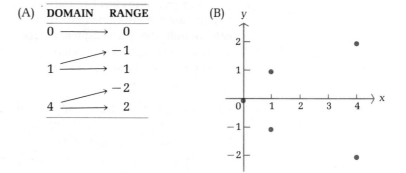

(C) The relation is not a function, since more than one range value corresponds to a given domain value. (For F to be a function, no two ordered pairs in F can have the same first coordinate.)

Problem 8 Repeat Example 8 for the set $F = \{(-2, 4), (-1, 1), (0, 0), (1, 1), (2, 4)\}$.

Answers (A)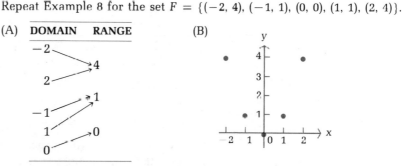

(C) The relation is a function, since each domain value corresponds to exactly one range value.

RELATIONS AND FUNCTIONS SPECIFIED BY EQUATIONS

Let us now focus on relations and functions specified by equations. Consider the equation

$$y = 3 - x^2$$

For each **input** x we obtain one **output** y. For example:

If $x = 2$, then $y = 3 - 2^2 = 3 - 4 = -1$
If $x = -3$, then $y = 3 - (-3)^2 = 3 - 9 = -6$

The input values are domain values and the output values are range values. The equation $y = 3 - x^2$ assigns each domain value x exactly one range value y (hence, the equation specifies a function). Any variable that assumes domain values is called an **independent variable**; any variable that assumes range values is called a **dependent variable.** In the above example, x is independent and y is dependent (the value of y depends on the value assigned to x).

Most equations in two variables specify relations, but when does an equation specify a function? **If for each value of the independent variable (input) there corresponds exactly one value of the dependent variable (output), then the equation specifies a function. If there is more than one output for at least one input, then the equation does not specify a function.**

Example 9

Assuming x is an independent variable, which of the equations below specifies a function? For the equation that does not specify a function, find a domain value (value of x) that corresponds to more than one range value (value of y).

$$x^2 + y^2 = 4 \qquad 2x + y = 3$$

Solution

The equation $2x + y = 3$ specifies a function, since for each value of x, there corresponds exactly one value of y. For example, if $x = 2$, then

$$2(2) + y = 3$$
$$4 + y = 3$$
$$y = -1 \qquad \text{And no other value}$$

On the other hand, the equation $x^2 + y^2 = 4$ does not specify a function, since, for example, if $x = 0$, then

$$0^2 + y^2 = 4$$
$$y^2 = 4$$
$$y = \pm \sqrt{4} = \pm 2$$

Thus, two values of y (the dependent variable) correspond to a given value of x (the independent variable).

Problem 9

Repeat Example 9 for the equations $y - x^2 = 0$ and $y^2 - x = 0$.

Answer

$y - x^2 = 0$ specifies a function; $y^2 - x = 0$ does not specify a function, since, for example, if $x = 4$, then $y = \pm 2$

FUNCTION NOTATION

We have just seen that a function involves two sets of elements—a domain and a range—and a rule of correspondence that enables us to assign each element in the domain to exactly one element in the range. We can use letters to denote names for numbers, and in essentially the same way, we will now use different letters to denote names for functions. For example, f and g may be used to name the two functions

$$f: \quad y = 2x + 1 \qquad g: \quad y = x^2 + 2x - 3$$

If x represents an element in the domain of a function f, then we will often use the symbol $f(x)$ in place of y to designate the number in the

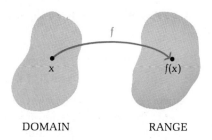

FIGURE 5   DOMAIN   RANGE

range of the function f to which x is paired (Fig. 5). It is important not to think of $f(x)$ as the product of f and x. The symbol $f(x)$ is read "f of x," or "the value of f at x." The variable x is an independent variable; both y and $f(x)$ are dependent variables.

This new function notation is extremely important, and its correct use should be mastered as early as possible. For example, in place of the more formal representation of the functions f and g above, we can now write

$$f(x) = 2x + 1 \qquad g(x) = x^2 + 2x - 3$$

The symbols $f(x)$ and $g(x)$ have some advantages over the variable y in certain situations. For example, if we write $f(3)$ and $g(5)$, then each symbol indicates in a concise way that these are range values of particular functions associated with particular domain values. Let us find $f(3)$ and $g(5)$.

To find $f(3)$, we replace x by 3 wherever x occurs in $f(x) = 2x + 1$ and evaluate the right side:

$$f(3) = 2(3) + 1$$
$$= 6 + 1$$
$$= 7$$

Thus,

$f(3) = 7$ The function f assigns the range value 7 to the domain value 3; the ordered pair (3, 7) belongs to f

To find $g(5)$, we replace x by 5 wherever x occurs in $g(x) = x^2 + 2x - 3$ and evaluate the right side:

$$g(5) = 5^2 + 2(5) - 3$$
$$= 25 + 10 - 3$$
$$= 32$$

Thus,

$g(5) = 32$ The function g assigns the range value 32 to the domain value 5; the ordered pair (5, 32) belongs to g

It is very important to understand and remember the definition of $f(x)$:

THE $f(x)$ SYMBOL

For any element x in the domain of the function f, the symbol **$f(x)$** represents the element in the range of f corresponding to x in the domain of f. If x is an input value, then $f(x)$ is an output value; or, symbolically, $f: x \to f(x)$. The ordered pair $(x, f(x))$ belongs to the function f.

Figure 6, which illustrates a "function machine," may give you additional insight into the nature of function and the symbol $f(x)$. We can think of a function machine as a device that produces exactly one output (range) value for each input (domain) value. (If more than one output value were produced for an input value, then the machine would be a "relation machine" instead of a function machine.)

For the function $f(x) = 2x + 1$, the machine takes each domain value (input), multiplies it by 2, then adds 1 to the result to produce the range value (output). Different rules inside the machine result in different functions.

Unless stated to the contrary, we shall adhere to the following convention regarding domains and ranges for relations and functions specified by equations:

FIGURE 6   Function machine—
exactly one output for
each input.

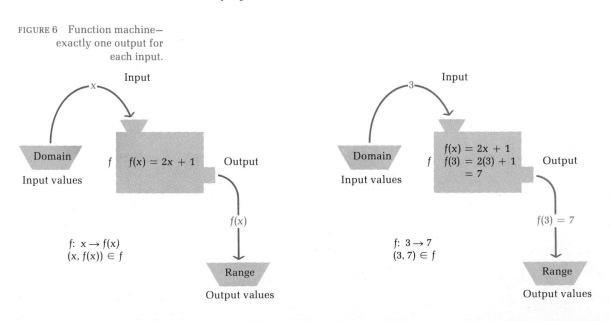

AGREEMENT ON DOMAINS AND RANGES

If a relation or function is specified by an equation and the domain is not indicated, then we shall assume that the domain is the set of all real number replacements of the independent variable (inputs) that produce real values for the dependent variable (outputs). The range is the set of all outputs corresponding to input values.

Example 10   If

$$f(x) = \frac{12}{x - 2} \qquad g(x) = 1 - x^2 \qquad h(x) = \sqrt{x - 1}$$

then:

(A) $f(6) = \dfrac{12}{6 - 2} = \dfrac{12}{4} = 3$

(B) $g(-2) = 1 - (-2)^2 = 1 - 4 = -3$

(C) $f(0) + g(1) - h(10) = \dfrac{12}{0 - 2} + (1 - 1^2) - \sqrt{10 - 1}$

$$= \frac{12}{-2} + 0 - \sqrt{9}$$

$$= -6 - 3 = -9$$

(D)   The domain of h is the set of all x such that

$$x - 1 \geq 0$$
$$x \geq 1$$

Problem 10   Use the functions f, g, and h in Example 10 to find:
(A) $f(-2)$ (B) $g(-1)$ (C) $f(3)/h(5)$ (D)   Domain of f

Answers   (A) -3 (B) 0 (C) 6 (D)   All real numbers except $x = 2$

Example 11   For $f(x) = 2x - 3$, find:

(A) $f(a)$ (B) $f(a + h)$ (C) $\dfrac{f(a + h) - f(a)}{h}$

Solutions   (A) $f(a) = 2a - 3$

(B) $f(a + h) = 2(a + h) - 3 = 2a + 2h - 3$

(C) $\dfrac{f(a + h) - f(a)}{h} = \dfrac{[2(a + h) - 3] - [2a - 3]}{h}$

$$= \frac{2a + 2h - 3 - 2a + 3}{h} = \frac{2h}{h} = 2$$

Problem 11   Repeat Example 11 for $f(x) = 3x - 2$.

Answers   (A) $3a - 2$ (B) $3a + 3h - 2$ (C)   3

COMPOSITE FUNCTIONS   Consider the function h given by

$$h(x) = \sqrt{x - 1}$$

This function can be thought of as a combination of two simpler functions f and g, where

$$f(u) = \sqrt{u} \quad \text{and} \quad u = g(x) = x - 1$$

That is,

$$h(x) = f[g(x)]$$

The function h is called a **composite function.** It is defined for all values of the range of g that are in the domain of f (see Fig. 7).

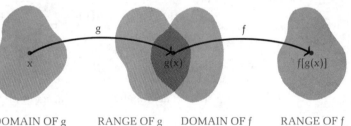

FIGURE 7   Composite function.

DOMAIN OF g RANGE OF g DOMAIN OF f RANGE OF f

Example 12   If $f(x) = \sqrt{x}$ and $g(x) = x - 1$, then:
(A) $f[g(5)] = f[5 - 1] = f(4) = \sqrt{4} = 2$
(B) $g[f(9)] = g[\sqrt{9}] = g(3) = 3 - 1 = 2$
(C) $f[g(a)] = f(a - 1) = \sqrt{a - 1}$
(D) $f[g(x)] = f(x - 1) = \sqrt{x - 1}$

Problem 12   For $f(x) = |x|$ and $g(x) = x + 2$, find:*
(A) $f[g(-3)]$ (B) $g[f(-2)]$ (C) $f[g(a)]$ (D) $f[g(x)]$

Answers   (A)   1   (B)   4   (C) $|a + 2|$ (D) $|x + 2|$

*Recall that $|x|$ is called the **absolute value** of x. It is defined by

$$|x| = \begin{cases} x & \text{if } x > 0 \\ 0 & \text{if } x = 0 \\ -x & \text{if } x < 0 \end{cases}$$

EXERCISE 1-3

A   *Indicate whether each relation is a function.*

1.

DOMAIN	RANGE
3 ⟶ 0	
5 ⟶ 1	
7 ⟶ 2	

2.

DOMAIN	RANGE
−1 ⟶ 5	
−2 ⟶ 7	
−3 ⟶ 9	

3.

DOMAIN	RANGE
3	5
	6
4	7
5	8

4.

DOMAIN	RANGE
8 ⟶ 0	
9 ⟶ 1	
⟶ 2	
10 ⟶ 3	

5.

DOMAIN	RANGE
3	
6	5
9	
12	6

6.

DOMAIN	RANGE
−2	
−1	
0	6
1	

The relations in Problems 7–12 are specified by graphs. Indicate whether each relation is a function.

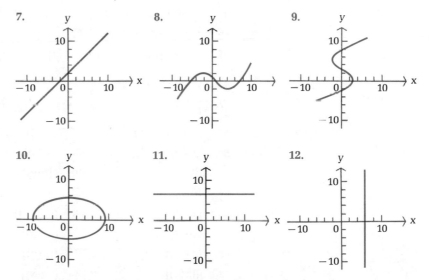

The equations in Problems 13–24 specify relations. Which equations specify functions? For each equation that does not specify a function, find a value of x that corresponds to more than one value of y (x is independent and y is dependent).

13. $y = 3x - 1$ **14.** $y = \dfrac{x}{2} - 1$ **15.** $y = x^2 - 3x + 1$

16. $y = x^3$ **17.** $y^2 = x$ **18.** $x^2 + y^2 = 25$

19. $x = y^2 - y$ 20. $x = (y - 1)(y + 2)$ 21. $y = x^4 - 3x^2$

22. $2x - 3y = 5$ 23. $y = \dfrac{x + 1}{x - 1}$ 24. $y = \dfrac{x^2}{1 - x}$

If $f(x) = 3x - 2$ and $g(x) = x - x^2$, find each of the following:

25. $f(2)$ 26. $f(1)$ 27. $f(-1)$

28. $f(-2)$ 29. $g(3)$ 30. $g(1)$

31. $f(0)$ 32. $f(\frac{1}{3})$ 33. $g(-3)$

34. $g(-2)$ 35. $f(1) + g(2)$ 36. $g(1) + f(2)$

37. $g(2) - f(2)$ 38. $f(3) - g(3)$ 39. $g(3) \cdot f(0)$

40. $g(0) \cdot f(-2)$ 41. $\dfrac{g(-2)}{f(-2)}$ 42. $\dfrac{g(-3)}{f(2)}$

B   State the domain and range for each relation, and indicate whether the relation is a function.

43. $F = \{(1, 1), (2, 1), (3, 2), (3, 3)\}$

44. $f = \{(2, 4), (4, 2), (2, 0), (4, -2)\}$

45. $G = \{(-1, -2), (0, -1), (1, 0), (2, 1), (3, 2), (4, 1)\}$

46. $g = \{(-2, 0), (0, 2), (2, 0)\}$

47. $y = 6 - 2x, \quad x \in \{0, 1, 2, 3\}$ 48. $y = \dfrac{x}{2} - 4, \quad x \in \{0, 1, 2, 3, 4\}$

49. $y^2 = x, \quad x \in \{0, 1, 4\}$ 50. $y = x^2, \quad x \in \{-2, 0, 2\}$

If $f(x) = 2x + 1, g(x) = x^2 - x$, and $h(x) = \sqrt{x}$, find each of the following:

51. $f(3) + g(-2)$ 52. $g(-1) - f(1)$ 53. $h(9) - g(-2)$

54. $g(-2) - h(4)$ 55. $f[h(4)]$ 56. $h[f(4)]$

57. $h[g(2)]$ 58. $g[h(9)]$ 59. $g(e)$

60. $f(a)$ 61. $h(u)$ 62. $g(t)$

63. $g(2 + h)$ 64. $f(2 + h)$ 65. $f(a + h)$

66. $g(a + h)$ 67. $h[f(a)]$ 68. $f[h(a)]$

69. $h[f(x)]$ 70. $f[h(x)]$ 71. $h[g(a)]$

72. $g[h(a)]$ 73. $h[g(x)]$ 74. $g[h(x)]$

75. $\dfrac{f(2 + h) - f(2)}{h}$ 76. $\dfrac{f(a + h) - f(a)}{h}$ 77. $\dfrac{g(2 + h) - g(2)}{h}$

78. $\dfrac{g(a + h) - g(a)}{h}$

Find the domain of each function.

79. $f(x) = \sqrt{x}$ 80. $f(x) = \dfrac{1}{\sqrt{x}}$

81. $f(x) = \dfrac{x - 3}{(x - 5)(x + 3)}$

82. $f(x) = \dfrac{x + 1}{x - 2}$

83. $f(x) = \sqrt{x - 1}$

84. $f(x) = \sqrt{x + 1}$

C Find the domain of each function.

85. $f(x) = \dfrac{1}{x^2 - x - 6}$

86. $f(x) = \sqrt{x^2 - 1}$

87. If

$$f(x) = \begin{cases} x^2 & \text{when } x < 1 \\ 2x & \text{when } x \geq 1 \end{cases}$$

find:

(A) $f(-1)$ (B) $f(0)$ (C) $f(1)$ (D) $f(3)$

88. If

$$f(x) = \begin{cases} -x & \text{when } x \leq 0 \\ x & \text{when } x > 0 \end{cases}$$

find:

(A) $f(-3)$ (B) $f(-1)$ (C) $f(0)$ (D) $f(5)$

APPLICATIONS

Each of the statements in Problems 89–94 can be described by a function. Write an equation that specifies each function.

BUSINESS & ECONOMICS

89. *Cost function.* The cost $C(x)$ of x records at \$4 per record. (The cost depends on the number of records purchased.)

90. *Cost function.* The cost $C(x)$ of manufacturing x pairs of skis if fixed costs are \$400 per day and the variable costs are \$70 per pair of skis manufactured. (The cost per day depends on the number of skis manufactured per day.)

LIFE SCIENCES

91. *Temperature conversion.* The temperature in Celsius degrees $C(F)$ can be found from the temperature in Fahrenheit degrees F by subtracting 32 from the Fahrenheit temperature and multiplying the difference by $\frac{5}{9}$.

92. *Ecology.* The pressure $P(d)$ in the ocean in pounds per square inch depends on the depth d. To find the pressure, divide the depth by 33, add 1 to the quotient, and multiply the result by 15.

SOCIAL SCIENCES

93. *Psychology.* For all 12-year-old children, IQ depends on the mental age as determined by certain standardized tests. To find an IQ, divide a mental age (MA) by 12 and multiply the quotient by 100.

94. *Politics.* The percentage of seats y won by a given party in a two-party election depends on the percentage of the two-party votes x received by the given party. The percentage of seats y can be approximated for $0.4 \le x \le 0.6$ by multiplying x by 2.5 and subtracting 0.7 from the product.

1-4   GRAPHING LINEAR AND QUADRATIC FUNCTIONS

In this section we will review the graphs of several special types of functions that are frequently encountered.

LINEAR FUNCTIONS

Any nonvertical line in a rectangular coordinate system defines a linear function. That is to say, any function f defined by an equation of the form

$$f(x) = ax + b$$

where a and b are constants, is called a **linear function.** We know from Section 1-2 that the graph of this equation is a straight line (nonvertical) with slope a and y intercept b.

Example 13

Graph the linear function defined by

$$f(x) = -\frac{x}{2} + 3$$

and indicate its slope and y intercept.

Solution

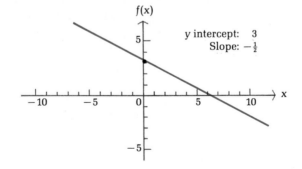

y intercept: 3
Slope: $-\frac{1}{2}$

Problem 13

Graph the linear function defined by

$$f(x) = \frac{x}{3} + 1$$

and indicate its slope and y intercept.

Answer

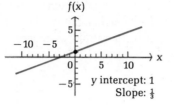

y intercept: 1
Slope: $\frac{1}{3}$

QUADRATIC FUNCTIONS

Any function defined by an equation of the form

$$f(x) = ax^2 + bx + c \qquad a \neq 0$$

where a, b, and c are constants and x is a variable, is called a **quadratic function.**

Let us start by graphing two simple quadratic functions:

$$f(x) = x^2 \qquad \text{and} \qquad g(x) = -x^2$$

We evaluate these functions for integer values from their domains, find corresponding range values, then plot the resulting ordered pairs and join these points with a smooth curve. The first two steps are usually done mentally or on scratch paper.

GRAPHING $f(x) = x^2$

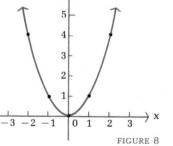

FIGURE 8

DOMAIN VALUES	RANGE VALUES	ELEMENTS OF f
x	$y - f(x)$	$(x, f(x))$
-2	$y = f(-2) = (-2)^2 = 4$	$(-2, 4)$
-1	$y = f(-1) = (-1)^2 - 1$	$(-1, 1)$
0	$y = f(0) = 0^2 = 0$	$(0, 0)$
1	$y = f(1) = 1^2 = 1$	$(1, 1)$
2	$y = f(2) = 2^2 = 4$	$(2, 4)$

GRAPHING $g(x) = -x^2$

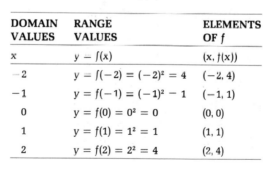

FIGURE 9

DOMAIN VALUES	RANGE VALUES	ELEMENTS OF g
x	$y = g(x)$	$(x, g(x))$
-2	$y = g(-2) = -(-2)^2 = -4$	$(-2, -4)$
-1	$y = g(-1) = -(-1)^2 = -1$	$(-1, -1)$
0	$y = g(0) = -0^2 = 0$	$(0, 0)$
1	$y = g(1) = -1^2 = -1$	$(1, -1)$
2	$y = g(2) = -2^2 = -4$	$(2, -4)$

Both of the curves shown in Figures 8 and 9 are called **parabolas.** It is shown in a course in analytic geometry that the graph of any quadratic function is also a parabola. In general:

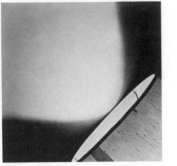

GRAPH OF $f(x) = ax^2 + bx + c, \quad a \neq 0$

The graph of a quadratic function f is a parabola that has its **axis** (line of symmetry) parallel to the vertical axis. It opens upward if $a > 0$ and downward if $a < 0$. The intersection point of the axis and parabola is called the **vertex.**

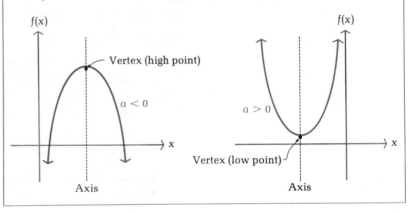

GRAPHING QUADRATIC FUNCTIONS BY COMPLETING THE SQUARE

In addition to the point-by-point method of graphing quadratic functions described above, let us consider another approach that will give us added insight into these functions. (A brief review of completing the square, which is discussed in Section A-7, may prove useful first.) We illustrate the method through an example, and then generalize the results.

Consider the quadratic function given by

$$f(x) = 2x^2 - 8x + 5$$

If we can find the vertex of the graph, then the rest of the graph can be sketched with relatively few points. In addition, we will then have found the maximum or minimum value of the function. We start by transforming the equation into the form

$$f(x) = A(x - B)^2 + C \qquad A, B, C \text{ constants}$$

by completing the square:

$f(x) = 2x^2 - 8x + 5$ Factor the coefficient of x^2 out of the first two terms

$f(x) = 2(x^2 - 4x) + 5$

$\quad\; = 2(x^2 - 4x + ?) + 5$ Complete the square within parentheses

$\quad\; = 2(x^2 - 4x + 4) + 5 - 8$ We added 4 to complete the square inside the parentheses; but because of the 2 on the outside we have actually added 8, so we must subtract 8

$\quad\; = 2(x - 2)^2 - 3$ The transformation is complete

Thus,

$$f(x) = \underbrace{2(x - 2)^2}_{\substack{\text{Never negative} \\ \text{(Why?)}}} - 3$$

When $x = 2$, the first term on the right vanishes, and we add 0 to -3. For *any* other value of x we will add a positive number to -3, thus making $f(x)$ larger. Therefore, $f(2) = -3$ is the minimum value of $f(x)$ for *all* x. A very important result!

The point $(2, -3)$ is the lowest point on the parabola and is also the vertex. The vertical line $x = 2$ is the axis of the parabola. We plot the vertex and the axis and a couple of points on either side of the axis to complete the graph (Fig. 11).

x	f(x)
2	−3
1	−1
3	−1
0	5
4	5

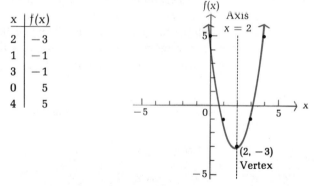

FIGURE 11

Note the important results we have obtained with this approach. We have found:

1. Axis of the parabola
2. Vertex of the parabola
3. Minimum value of $f(x)$
4. Graph of $y = f(x)$

By proceeding in essentially the same way with the general quadratic function given by

$$f(x) = ax^2 + bx + c \qquad a \neq 0$$

we can obtain the following general results:

QUADRATIC FUNCTION $f(x) = ax^2 + bx + c, \quad a \neq 0$

1. Axis (of symmetry) of the parabola:

$$x = -\frac{b}{2a}$$

2. Maximum or minimum value of $f(x)$:

$$f\left(-\frac{b}{2a}\right) \qquad \begin{array}{l} \text{Minimum if } a > 0 \\ \text{Maximum if } a < 0 \end{array}$$

3. Vertex of the parabola:

$$\left(-\frac{b}{2a}, f\left(-\frac{b}{2a}\right)\right)$$

To graph a quadratic function using the method of completing the square, we can either actually complete the square as in the earlier example or use the properties listed in the box—some people can more readily remember a formula, others a process. We will use the boxed properties in the next example.

Example 14   Graph by finding axis of symmetry, maximum or minimum of $f(x)$, and vertex:

$$f(x) = 12x - 2x^2 \qquad 0 \leq x \leq 6$$

Solution   *Axis of symmetry:*

$$x = -\frac{b}{2a} = -\frac{12}{2(-2)} = 3$$

Maximum value of $f(x)$ (since $a = -2 < 0$):

$$f(3) = 12(3) - 2(3)^2 = 18$$

Vertex: **(3, 18)**

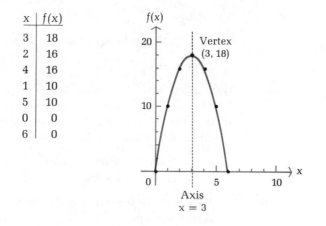

x	f(x)
3	18
2	16
4	16
1	10
5	10
0	0
6	0

Problem 14   Graph as in Example 14.

$$f(x) = x^2 - 2x - 3 \qquad -1 \le x \le 4$$

Answer   Minimum: $f(1) = -4$

APPLICATION: The market research department of a company recommended to manage-
MARKET RESEARCH   ment that the company manufacture and market a promising new product. After extensive surveys, the research department backed up the
recommendation with the **demand equation**

$$x = f(p) = 6{,}000 - 30p \tag{1}$$

where x is the number of units that retailers are likely to buy per month
at \$p per unit. Notice that as the price goes up, the **number of units goes
down.** From the financial department, the following **cost equation** was
obtained:

$$C = g(x) = 72{,}000 + 60x \tag{2}$$

where \$72,000 is the fixed cost (tooling and overhead) and \$60 is the variable cost per unit (materials, labor, marketing, transportation, storage,
etc.). The **revenue equation** (the amount of money, R, received by the
company for selling x units at \$p per unit) is

$$R = xp \tag{3}$$

And, finally, the **profit equation** is

$$P = R - C \tag{4}$$

where P is profit, R is revenue, and C is cost.

We notice that the cost equation (2) expresses C as a function of x and the demand equation (1) expresses x as a function of p. Substituting (1) into (2), we obtain cost C as a linear function of price p:

$$\begin{aligned} C &= 72{,}000 + 60(6{,}000 - 30p) \\ &= 432{,}000 - 1{,}800p \end{aligned} \qquad \text{Linear function} \tag{5}$$

Similarly, substituting (1) into (3), we obtain revenue R as a quadratic function of price p:

$$\begin{aligned} R &= (6{,}000 - 30p)p \\ &= 6{,}000p - 30p^2 \end{aligned} \qquad \text{Quadratic function} \tag{6}$$

Now let us graph equations (5) and (6) in the same coordinate system. We obtain Figure 12. Notice how much information is contained in this graph. Let us compute the **break-even points;** that is, the prices at which cost equals revenue (the points of intersection of the two graphs above). Find p so that

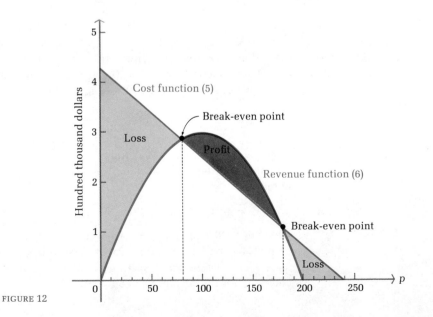

FIGURE 12

$$C - R$$

$$432{,}000 - 1{,}800p = 6{,}000p - 30p^2$$

$$30p^2 - 7{,}800p + 432{,}000 = 0$$

$$p^2 - 260p + 14{,}400 = 0$$

$$p = \frac{260 \pm \sqrt{260^2 - 4(14{,}400)}}{2}$$

$$= \frac{260 \pm 100}{2}$$

$$= \$80, \quad \$180$$

Thus, at a price of $80 or $180 per unit the company will break even. Between these two prices it is predicted that the company will make a profit.

At what price will a **maximum profit** occur? To find out, we write

$$P = R - C$$
$$= (6{,}000p - 30p^2) - (432{,}000 - 1{,}800p)$$
$$= -30p^2 + 7{,}800p - 432{,}000$$

Since this is a quadratic funciton, the maximum profit occurs at

$$p = -\frac{b}{2a} = -\frac{7{,}800}{2(-30)} = \$130$$

Note that this is not the price at which the maximum revenue occurs. The latter occurs at $p = \$100$, as shown in Figure 12.

EXERCISE 1-4

A Graph each linear function, and indicate its slope and y intercept.

1. $f(x) = 2x - 4$

2. $g(x) = \dfrac{x}{2}$

3. $h(x) = 4 - 2x$

4. $f(x) = -\dfrac{x}{2} + 3$

5. $g(x) = -\tfrac{2}{3}x + 4$

6. $f(x) = 3$

B Graph each quadratic function, and include the axis of symmetry, vertex, and maximum or minimum value.

7. $f(x) = x^2 + 8x + 16$

8. $h(x) = x^2 - 2x - 3$

9. $f(u) = u^2 - 2u + 4$

10. $f(x) = x^2 - 10x + 25$

11. $h(x) = 2 + 4x - x^2$

12. $g(x) = -x^2 - 6x - 4$

13. $f(x) = 6x - x^2$

14. $G(x) = 16x - 2x^2$

15. $F(s) = s^2 - 4$

16. $g(t) = t^2 + 4$

17. $F(x) = 4 - x^2$

18. $G(x) = 9 - x^2$

Find the maximum or minimum value of each function. Do not graph.

19. $f(x) = 4x^2 - 16x + 9$ 20. $h(x) = 3 + 18x - 3x^2$
21. $f(t) = 2t(t - 24)$ 22. $g(x) = 3x(x + 12)$
23. $A(x) = x(50 - x)$ 24. $A(x) = x(100 - 2x)$

C *Graph each quadratic function, and include the axis of symmetry, vertex, and maximum or minimum value.*

25. $f(x) = x^2 - 7x + 10$ 26. $g(t) = t^2 - 5t + 2$
27. $g(t) = 4 + 3t - t^2$ 28. $h(x) = 2 - 5x - x^2$

APPLICATIONS

BUSINESS & ECONOMICS

29. *Cost equation.* The cost equation for a particular company to produce stereos is found to be

$$C = g(n) = 96{,}000 + 80n$$

where $96,000 represents fixed costs (tooling and overhead) and $80 is the variable cost per unit (material, labor, etc.). Graph this function for $0 \le n \le 1{,}000$.

30. *Demand equation.* After extensive surveys the research department in a stereo company produced the demand equation

$$n = f(p) = 8{,}000 - 40p \qquad 100 \le p \le 200$$

where n is the number of units that retailers are likely to purchase per week at a price of $$p$ per unit. Graph the function for the indicated domain.

31. Suppose that in the market research example in this section the demand equation (1) is changed to $x = 9{,}000 - 30p$ and the cost equation (2) is changed to $C = 90{,}000 + 30x$.
    (A) Express cost C as a linear function of price p.
    (B) Express revenue R as a quadratic function of price p.
    (C) Graph the cost and revenue functions found in parts A and B in the same coordinate system, and identify the regions of profit and loss.
    (D) Find the break-even points; that is, find the prices to the nearest dollar at which $R = C$. (A hand calculator might prove useful here.)
    (E) Find the price that produces the maximum revenue.

LIFE SCIENCES

32. *Medicine.* The velocity (v) of blood, in centimeters per second, at x centimeters from the center of a given artery (see the figure) is given by

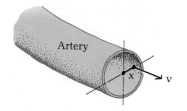

Artery

$$v = f(x) = 1.28 - 20{,}000x^2 \qquad 0 \le x \le 8 \times 10^{-3}$$

Graph this quadratic function for the indicated values of x.

33. *Air pollution.* On an average summer day in a large city, the pollu-ion index at 8:00 AM is 20 parts per million and it increases linearly 15 parts per million for each hour until 3:00 PM. Let $P(x)$ be the amount of pollutants in the air x hours after 8:00 AM.
   (A) Express $P(x)$ as a linear function of x.
   (B) What is the air pollution index at 1:00 PM?
   (C) Graph the function P for $0 \le x \le 7$.
   (D) What is the slope of the graph? (The slope is the increase in pollution for each hour increase in time.)

SOCIAL SCIENCES   34. *Psychology—sensory perception.* One of the oldest studies in psy-chology concerns the following question: Given a certain level of stimulation (light, sound, weight lifting, electric shock, and so on), how much should the stimulation be increased for a person to notice the difference? In the middle of the nineteenth century, E. H. Weber (a German physiologist) formulated a law that still carries his name: If Δs is the change in stimulus that will just be noticeable at a stim-ulus level s, then the ratio of Δs to s is a constant. Thus,

$$\frac{\Delta s}{s} = k$$

Hence, the amount of change that will be noticed is a linear function of the stimulus level, and we note that the greater the stimulus the more it takes to notice a difference. In an experiment on weight lift-ing, the constant k for a given individual was found to be $\frac{1}{30}$.
   (A) Find Δs (the just noticeable difference) at the 30 pound level. At the 90 pound level.
   (B) Graph $\Delta s = s/30$ for $0 \le s \le 120$.
   (C) What is the slope of the graph?

1-5   EXPONENTIAL FUNCTIONS

Until now we have considered mostly **algebraic functions**—that is, func-tions that can be defined using the algebraic operations of addition, sub-traction, multiplication, division, powers, and roots. In no case has a variable been an exponent. In this and the next section we will con-sider two new kinds of functions that use variable exponents in their definitions.

To start, note that

$$f(x) = 2^x \qquad \text{and} \qquad g(x) = x^2$$

are not the same function. The function g is a quadratic function, which we have just discussed, and the function f is a new function, called an **exponential function.** In general, an exponential function is a function defined by the equation

$$f(x) = b^x \qquad b > 0, \quad b \neq 1$$

where b is a constant, called the **base,** and the exponent is a variable. The domain of f is the set of all real numbers.

If asked to graph an exponential function such as

$$f(x) = 2^x$$

most students would not hesitate. They would likely make up a table by assigning integers to x, plot the resulting ordered pairs of numbers, and then join the plotted points with a smooth curve (see Fig. 13). What has been overlooked? The exponent form 2^x has not been defined for *all* real numbers x.

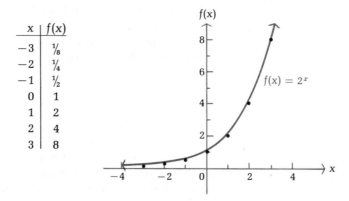

x	f(x)
-3	$\frac{1}{8}$
-2	$\frac{1}{4}$
-1	$\frac{1}{2}$
0	1
1	2
2	4
3	8

FIGURE 13

We know what all the expressions 2^3, 2^{-3}, $2^{2/3}$, $2^{-3/5}$, $2^{1.5}$, and $2^{-2.35}$ mean (that is, 2^r, where r is any rational number—see Section A-3), but what does

$$2^{\sqrt{2}}$$

mean? We cannot answer this question now. The definition of 2^x for x irrational requires a concept considered in calculus. There we can show that 2^x can be approximated as closely as we like by using rational number approximations for x. In particular, since $\sqrt{2} = 1.414213\ldots$, the sequence

$$2^{1.4}, \ 2^{1.41}, \ 2^{1.414}, \ \ldots$$

approximates $2^{\sqrt{2}}$, and as we move to the right the approximation improves. In addition, after irrational exponents are defined in calculus, it can be shown that all five laws of exponents continue to hold for these new exponents.

Here, we assume these results for irrational exponents and we also assume that if we plot $f(x) = 2^x$ for irrational x, the points will lie on the curve we obtained in Figure 13.

Graphs of $f(x) = b^x$ are of two basic types, depending on whether the base b is between zero and one or b is greater than one. The graphs of

$$y = 2^x \qquad \text{and} \qquad y = (\tfrac{1}{2})^x = 2^{-x}$$

are typical of these two cases (see Fig. 14).

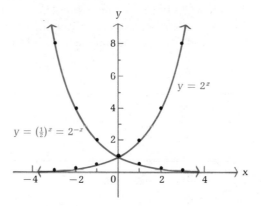

FIGURE 14

A great variety of growth phenomena can be described by exponential functions, which is the reason these functions are often referred to as **growth functions**. They are used to describe the growth of money at compound interest; population growth of people, animals, and bacteria; radioactive decay (negative growth); and the growth of learning a skill such as typing or swimming relative to practice.

Even though it was convenient to use the bases 2 and $\tfrac{1}{2}$ for the introduction of exponential functions, a certain irrational number, denoted by e, is the base that is used most frequently—both for theoretical and practical reasons. In fact, the use of the function

$$f(x) = e^x$$

is so widespread that it is often referred to as *the* exponential function. The reasons for the use of the base e are made clear in calculus. There it is shown that e is approximated by $(1 + 1/n)^n$ to any decimal accuracy desired by taking n (an integer) sufficiently large. To five decimal places, we find that

$$e \approx 2.71828$$

Similarly, for a given x, e^x can be approximated to any decimal accuracy desired by using $(1 + 1/n)^{nx}$ for sufficiently large n. Because of the widespread use of e^x and e^{-x}, tables for their evaluation are readily available. In fact, most business and scientfic hand calculators have the capability of evaluating e^x in two or three simple steps. A short table (Table I) for e^x and e^{-x} is included in the back of the book for those of you who are not using hand calculators.

Example 15   Graph $y = 10e^{0.5x}$, $-4 \leq x \leq 4$, using Table I or a hand calculator.

Solution   The kind of tabulation shown here is useful for this purpose. Note that the third column is obtained directly from the second column by using either Table I or a calculator. The fourth column is obtained from the third column simply by multiplying each entry by 10. Now we graph the ordered pairs (x, y) from the first and fourth columns.

x	$0.5x$	$e^{0.5x}$	$y = 10e^{0.5x}$
0	0	1.00	10.0
1	0.5	1.65	16.5
-1	-0.5	0.61	6.1
2	1.0	2.72	27.2
-2	-1.0	0.37	3.7
3	1.5	4.48	44.8
-3	-1.5	0.22	2.2
4	2.0	7.39	73.9
-4	-2.0	0.14	1.4

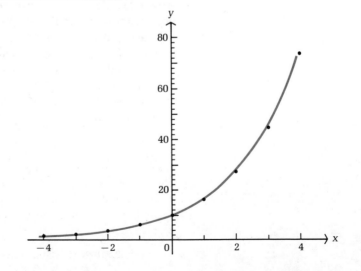

Problem 15   Graph $y = 10e^{-0.5x}$, $-4 \leq x \leq 4$, using Table I or a hand calculator.

Answer

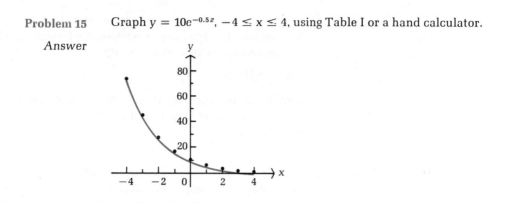

EXERCISE 1-5

A   *Graph each equation for $-3 \leq x \leq 3$. Plot points using integers for x, and then join the points with a smooth curve.*

1. $y = 3^x$

2. $y = 10 \cdot 2^x$
    [Note: $10 \cdot 2^x \neq 20^x$]

3. $y = (\frac{1}{3})^x = 3^{-x}$

4. $y = 10 \cdot (\frac{1}{2})^x = 10 \cdot 2^{-x}$

5. $y = 10 \cdot 3^x$

6. $y = 10 \cdot (\frac{1}{3})^x = 10 \cdot 3^{-x}$

B   *Graph each equation for $-3 \leq x \leq 3$. Use Table I or a calculator if the base is e.*

7. $y = 10 \cdot 2^{2x}$

8. $y = 10 \cdot 2^{-3x}$

9. $y = e^x$

10. $y = e^{-x}$

11. $y = 10e^{0.2x}$

12. $y = 100e^{0.1x}$

13. $y = 100e^{-0.1x}$

14. $y = 10e^{-0.2x}$

C   15. Graph $y = e^{-x^2}$ for $x = -1.5, -1.0, -0.5, 0, 0.5, 1.0, 1.5$, and then join these points with a smooth curve. Use Table I or a calculator. (This is a very important curve in probability and statistics.)

16. Graph $y = y_0 2^x$, where y_0 is the value of y when $x = 0$. (Express the vertical scale in terms of y_0.)

17. Graph $y = 2^x$ and $x = 2^y$ on the same coordinate system.

18. Graph $y = 10^x$ and $x = 10^y$ on the same coordinate system.

APPLICATIONS

BUSINESS & ECONOMICS   19. *Exponential growth.*   If we start with 2¢ and double up each day, after n days we would have 2^n¢. Graph $f(n) = 2^n$ for $1 \leq n \leq 10$. (Label the vertical scale so that the graph will not go off the paper.)

20. *Compound interest.* If a certain amount of money P (the principal) is invested at $100r\%$ interest compounded annually, the amount of money (A) after t years is given by

$$A = P(1 + r)^t$$

Graph this equation for $P = \$100$, $r = 0.10$, and $0 \le t \le 6$. How much money would a person have after 10 years if no interest were withdrawn?

LIFE SCIENCES

21. *Bacteria growth.* A single cholera bacterium divides every $\frac{1}{2}$ hour to produce two complete cholera bacteria. If we start with 100 bacteria, in t hours (assuming adequate food supply) we will have

$$A = 100 \cdot 2^{2t}$$

bacteria. Graph this equation for $0 \le t \le 5$.

22. *Ecology.* The atmospheric pressure (P, in pounds per square inch) may be calculated approximately from the formula

$$P = 14.7e^{-0.21h}$$

where h is the altitude above sea level in miles. Graph this equation for $0 \le h \le 12$.

SOCIAL SCIENCES

23. *Learning curves.* The performance record of a particular person learning to type is given approximately by

$$N = 100(1 - e^{-0.1t})$$

where N is the number of words per minute and t is the number of weeks of instruction. Graph this equation for $0 \le t \le 40$.

24. *Small group analysis.* After a lengthy investigation, sociologists Stephan and Mischler found that if the members of a discussion group of ten were ranked according to the number of times each participated, then the number of times, $N(k)$, the kth-ranked person participated was given approximately by

$$N(k) = N_1e^{-0.11(k-1)} \qquad 1 \le k \le 10$$

where N_1 was the number of times the top-ranked person participated in the disussion. Graph the equation assuming $N_1 = 100$. [For a general discussion of this phenomenon, see J. S. Coleman, *Introduction to Mathematical Sociology* (London: The Free Press of Glencoe, 1964), pp. 28–31.]

1-6   LOGARITHMIC FUNCTIONS

Now we are ready to consider logarithmic functions. These functions are closely related to exponential functions.

DEFINITION OF LOGARITHMIC FUNCTIONS

If we start with an exponential function f defined by

$$y = 2^x \tag{1}$$

and interchange the variables, we obtain an equation that defines a new relation g defined by

$$x = 2^y \tag{2}$$

Any ordered pair of numbers that belongs to f will belong to g if we interchange the order of the components. For example, (3, 8) satisfies equation (1) and (8, 3) satisfies equation (2). Thus, the domain of f becomes the range of g and the range of f becomes the domain of g. Graphing f and g on the same coordinate system (Fig. 15), we see that g is also a function. We call this new function the **logarithmic function with base 2,** and write

$$y = \log_2 x \qquad \text{if and only if} \qquad x = 2^y$$

Note that in Figure 15 if we fold the paper along the dashed line $y = x$, the two graphs match exactly.

In general, we define the logarithmic functions with base b as follows:

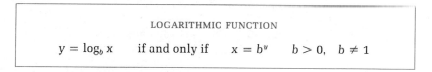

LOGARITHMIC FUNCTION

$$y = \log_b x \qquad \text{if and only if} \qquad x = b^y \qquad b > 0, \quad b \neq 1$$

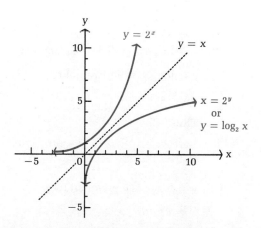

FIGURE 15

In words, **the logarithm of a number x to a base b ($b > 0$, $b \neq 1$) is the exponent to which b must be raised to equal x**. It is important to remember that $y = \log_b x$ and $x = b^y$ describe the same function, while $y = b^x$ is the related exponential function. Look at Figure 15 again

Since the domain of an exponential function includes all real numbers and its range is the set of positive real numbers, the **domain** of a logarithmic function is the set of all positive real numbers and its **range** is the set of all real numbers. Remember that the logarithm of zero or a negative number is not defined.

Example 16   Change from logarithmic form to exponential form.
(A) $\log_5 25 = 2$ is equivalent to $25 = 5^2$
(B) $\log_9 3 = \frac{1}{2}$ is equivalent to $3 = 9^{1/2}$
(C) $\log_2 \frac{1}{4} = -2$ is equivalent to $\frac{1}{4} = 2^{-2}$

Problem 16   Change to an equivalent exponential form.
(A) $\log_3 9 = 2$ (B) $\log_4 2 = \frac{1}{2}$ (C) $\log_3 \frac{1}{9} = -2$

Answers   (A) $9 = 3^2$ (B) $2 = 4^{1/2}$ (C) $\frac{1}{9} = 3^{-2}$

Example 17   Change from exponential form to logarithmic form.
(A) $64 = 4^3$ is equivalent to $\log_4 64 = 3$
(B) $6 = \sqrt{36}$ is equivalent to $\log_{36} 6 = \frac{1}{2}$
(C) $\frac{1}{8} = 2^{-3}$ is equivalent to $\log_2 \frac{1}{8} = -3$

Problem 17   Change to an equivalent logarithmic form.
(A) $49 = 7^2$ (B) $3 = \sqrt{9}$ (C) $\frac{1}{3} = 3^{-1}$

Answers   (A) $\log_7 49 = 2$ (B) $\log_9 3 = \frac{1}{2}$ (C) $\log_3 \frac{1}{3} = -1$

Example 18   Find y, b, or x.
(A) $y = \log_4 16$ (B) $\log_2 x = -3$
(C) $y = \log_8 4$ (D) $\log_b 100 = 2$

Solutions   (A) $y = \log_4 16$ is equivalent to   (B) $\log_2 x = -3$ is equivalent to

$16 = 4^y$ $x = 2^{-3}$

Thus,   Thus,

$y = 2$ $x = \dfrac{1}{2^3} = \dfrac{1}{8}$

(C) $y = \log_8 4$ is equivalent to

$4 = 8^y$ or $2^2 = 2^{3y}$

Thus,

$$3y = 2$$
$$y = \tfrac{2}{3}$$

(D) $\log_b 100 = 2$ is equivalent to

$$100 = b^2$$

Thus,

$$b = 10 \qquad \text{Recall that } b \text{ cannot be negative}$$

Problem 18   Find y, b, or x.
(A) $y = \log_9 27$ (B) $\log_3 x = -1$ (C) $\log_b 1{,}000 = 3$

Answers   (A) $y = \tfrac{3}{2}$ (B) $x = \tfrac{1}{3}$ (C) $b = 10$

**PROPERTIES OF
LOGARITHMIC
FUNCTIONS**

Logarithmic functions have several very useful properties that follow directly from their definitions. These properties will enable us to convert multiplication problems into addition problems, division problems into subtraction problems, and power and root problems into multiplication problems. We will also be able to solve exponential equations such as $2 = 1.06^n$.

LOGARITHMIC PROPERTIES

$$(b > 0, \quad b \neq 1, \quad M > 0, \quad N > 0)$$

1. $\log_b b^x = x$

2. $\log_b MN = \log_b M + \log_b N$

3. $\log_b \dfrac{M}{N} = \log_b M - \log_b N$

4. $\log_b M^p = p \log_b M$

5. $\log_b M = \log_b N$ if and only if $M = N$

6. $\log_b 1 = 0$

The first property follows directly from the definition of a logarithmic function. Here, we will **sketch a proof** for property 2. The other properties are established in a **similar way**. Let

$$u = \log_b M \qquad \text{and} \qquad v = \log_b N$$

Or, in equivalent exponential form,

$$M = b^u \qquad \text{and} \qquad N = b^v$$

Now, see if you can provide reasons for each of the following steps:

$$\log_b MN = \log_b b^u b^v = \log_b b^{u+v} = u + v = \log_b M + \log_b N$$

Example 19

(A) $\log_b \dfrac{wx}{yz}$ $\boxed{\begin{aligned} &= \log_b wx - \log_b yz \\[6pt] &= \log_b w + \log_b x - (\log_b y + \log_b z) \end{aligned}}$

$$= \log_b w + \log_b x - \log_b y - \log_b z$$

(B) $\log_b (wx)^{3/5}$ $\boxed{= \tfrac{3}{5} \log_b wx}$

$$= \frac{3}{5}(\log_b w + \log_b x)$$

Problem 19

Write in simpler logarithmic forms, as in Example 19.

(A) $\log_b \dfrac{R}{ST}$ (B) $\log_b \left(\dfrac{R}{S}\right)^{2/3}$

Answers   (A) $\log_b R - \log_b S - \log_b T$ (B) $\tfrac{2}{3}(\log_b R - \log_b S)$

The following example and problem, though somewhat artificial, will give you additional practice in using basic logarithmic properties.

Example 20   Find x so that

$$\tfrac{3}{2} \log_b 4 - \tfrac{2}{3} \log_b 8 + \log_b 2 = \log_b x$$

Solution $\tfrac{3}{2} \log_b 4 - \tfrac{2}{3} \log_b 8 + \log_b 2 = \log_b x$

$\log_b 4^{3/2} - \log_b 8^{2/3} + \log_b 2 = \log_b x$ ⟶ Property 4

$\log_b 8 - \log_b 4 + \log_b 2 = \log_b x$

$\log_b \dfrac{8 \cdot 2}{4} = \log_b x$ ⟶ Properties 2 and 3

$\log_b 4 = \log_b x$

$x = 4$ ⟶ Property 5

Problem 20   Find x so that

$$3 \log_b 2 + \tfrac{1}{2} \log_b 25 - \log_b 20 = \log_b x$$

Answer $x = 2$

EXERCISE 1-6

A   *Rewrite in exponential form.*

1. $\log_3 27 = 3$ 2. $\log_2 32 = 5$ 3. $\log_{10} 1 = 0$
4. $\log_e 1 = 0$ 5. $\log_4 8 = \tfrac{3}{2}$ 6. $\log_9 27 = \tfrac{3}{2}$

Rewrite in logarithmic form.

7. $49 = 7^2$

8. $36 = 6^2$

9. $8 = 4^{3/2}$

10. $9 = 27^{2/3}$

11. $A = b^u$

12. $M = b^x$

Find each of the following:

13. $\log_{10} 10^3$

14. $\log_{10} 10^{-5}$

15. $\log_2 2^{-3}$

16. $\log_3 3^5$

17. $\log_{10} 1{,}000$

18. $\log_6 36$

Write in terms of simpler logarithmic forms as in Example 19.

19. $\log_b \dfrac{P}{Q}$

20. $\log_b FG$

21. $\log_b L^5$

22. $\log_b w^{15}$

23. $\log_b \dfrac{p}{qrs}$

24. $\log_b PQR$

B *Find x, y, or b.*

25. $\log_3 x = 2$

26. $\log_2 x = 2$

27. $\log_7 49 = y$

28. $\log_3 27 = y$

29. $\log_b 10^{-4} = -4$

30. $\log_b c^{-2} = -2$

31. $\log_4 x = \frac{1}{2}$

32. $\log_{25} x = \frac{1}{2}$

33. $\log_{1/3} 9 = y$

34. $\log_{49} \frac{1}{7} = y$

35. $\log_b 1{,}000 = \frac{3}{2}$

36. $\log_b 4 = \frac{2}{3}$

Write in terms of simpler logarithmic forms going as far as you can with logarithmic properties (see Example 19).

37. $\log_b \dfrac{x^5}{y^3}$

38. $\log_b x^2 y^3$

39. $\log_b \sqrt[3]{N}$

40. $\log_b \sqrt[5]{Q}$

41. $\log_b x^2 \sqrt[3]{y}$

42. $\log_b \sqrt[3]{\dfrac{x^2}{y}}$

43. $\log_b(50 \cdot 2^{-0.2t})$

44. $\log_b(100 \cdot 1.06^t)$

45. $\log_b P(1 + r)^t$

46. $\log_e Ae^{-0.3t}$

47. $\log_e 100e^{-0.01t}$

48. $\log_{10}(67 \cdot 10^{-0.12x})$

Find x.

49. $\log_b x = \frac{2}{3} \log_b 8 + \frac{1}{2} \log_b 9 - \log_b 6$

50. $\log_b x = \frac{2}{3} \log_b 27 + 2 \log_b 2 - \log_b 3$

51. $\log_b x = \frac{3}{2} \log_b 4 - \frac{2}{3} \log_b 8 + 2 \log_b 2$

52. $\log_b x = 3 \log_b 2 + \frac{1}{2} \log_b 25 - \log_b 20$

C 53. Find the logarithm of 1 for any permissible base.

54. Why is 1 not a suitable logarithmic base? [*Hint:* Try to find $\log_1 8$.]

55. Write $\log_{10} y - \log_{10} c = 0.8x$ in an exponential form that is free of logarithms.

56. Write $\log_e x - \log_e 25 = 0.2t$ in an exponential form that is free of logarithms.

1-7	CHAPTER REVIEW

IMPORTANT TERMS
AND SYMBOLS

1-1 *Cartesian coordinate system.* rectangular coordinate system, Cartesian coordinate system, coordinate axes, ordered pair, coordinates, abscissa, ordinate, quadrants, (a, b)

1-2 *Straight lines; slope.* solution of an equation in two variables, solution set, graph of an equation, x intercept, y intercept, slope, equation of a line: slope–intercept form, equation of a line: point–slope form, horizontal lines, vertical lines, parallel lines, $y = mx + b$, $y - y_1 = m(x - x_1)$, $y = C$, $x = C$

1-3 *Relations and functions.* correspondence, relation, function, domain, range, independent variable, dependent variable, input, output, function notation, composite function, $f(x)$, $f[g(x)]$, $f: x \rightarrow f(x)$

1-4 *Graphing linear and quadratic functions.* linear function, quadratic function, parabola, axis of a parabola, vertex of a parabola, maximum, minimum, demand equation, cost equation, revenue equation, profit equation, break-even point, $f(x) = ax + b$, $f(x) = ax^2 + bx + c, a \neq 0$

1-5 *Exponential functions.* exponential function, base, growth function, $y = b^x, b > 0, b \neq 1$

1-6 *Logarithmic functions.* base, $y = \log_b x$ if and only if $x = b^y$, $b > 0$, $b \neq 1$

EXERCISE 1-7	CHAPTER REVIEW

Work through all the problems in this chapter review and check your answers in the back of the book. (Answers to all review problems are there.) Where weaknesses show up, review appropriate sections in the text. When you are satisfied that you know the material, take the practice test following this review.

A 1. Graph

$$y = \frac{x}{2} - 2$$

in a rectangular coordinate system. Indicate the slope and the y intercept.

2. Write the equation of a line that passes through (4, 3) with slope $\frac{1}{2}$. Write the final answer in the form $y = mx + b$.

3. Graph $x - y = 2$ in a rectangular coordinate system. Indicate the slope.

4. For $f(x) = 2x - 1$ and $g(x) = x^2 - 2x$, find $f(-2) + g(-1)$.

5. Graph the linear function f given by the equation

    $$f(x) = \tfrac{2}{3} x - 1$$

    Indicate the slope of the graph.

6. Evaluate for $x = -3$:
    (A) $f(x) = \sqrt{x^2 - 2x + 1}$ (B) $g(x) = 2x^{-2}$

7. Graph $y = 2^x$ and $y = 2^{-x}$, $-3 \le x \le 3$, on the same coordinate system.

8. (A) Write $\log_3 27 = 3$ in exponential form.
    (B) Write $16 = 2^4$ in logarithmic form.

9. $\log_{10} 10^{-3} = ?$

10. Write the following in terms of simpler logarithms:

    $$\log_b \frac{wx}{y}$$

11. Which of the equations below specify functions (x is an independent variable)?
    (A) $2x + y = 6$ (B) $y^2 = x + 1$

B 12. Find the equation of a line that passes through $(-2, 3)$ and $(6, -1)$. Write the answer in the form $Ax + By = C, A > 0$. What is the slope of the line?

13. Graph $3x - y = 9$ in a rectangular coordinate system. What is the slope of the graph?

14. Write the equations of the vertical line and the horizontal line that pass through $(-5, 2)$. Graph both equations on the same coordinate system.

15. For $f(x) = 2x - 1$ and $g(x) = x^2 - 2x$, find $g[f(2)]$.

16. Graph $g(x) = 8x - 2x^2$, $x \ge 0$, in a rectangular coordinate system. Indicate the vertex and axis.

17. Graph $y = 10 \cdot 2^{3x}$ and $y = 10 \cdot 2^{-3x}$, $-2 \le x \le 2$, on the same coordinate system.

18. Graph $y = 100e^{-0.1x}$, $0 \le x \le 10$, using Table I or a calculator.

19. (A) Find b: $\log_b 9 = 2$ (B) Find x: $\log_4 x = -\tfrac{3}{2}$

20. Write the following in terms of simpler logarithmic forms:

    $$\log_b \frac{\sqrt[3]{x}}{uv}$$

21. Write the following in terms of simpler logarithmic forms:
    $\log_b(100 \cdot 1.06^t)$

C 22. Write the equation of the line that passes through the points $(4, -3)$ and $(4, 5)$.

23. Find the equation of a line that passes through $(-3, 6)$ and is parallel to $2x - 3y = 6$. Write the answer in the form $y = mx + b$.

24. Find the domain of the function f specified by each equation.

    (A) $f(x) = \dfrac{5}{x - 3}$ (B) $f(x) = \sqrt{x - 1}$

25. For $f(x) = x^2 - 2$, find:

    (A) $\dfrac{f(3 + h) - f(3)}{h}$ (B) $\dfrac{f(x + h) - f(x)}{h}$

26. Graph

    $$f(x) = \frac{|x|}{x} + 1$$

    in a rectangular coordinate system.

27. Graph $A = A_0 e^{0.2x}$, $0 \le x \le 10$, using Table I.

28. Write $\ln y - \ln c = -0.2x$ in an exponential form free of logarithms.

29. A sporting goods store sells a tennis racket that cost $30 for $48 and a pair of jogging shoes that cost $20 for $32.
    (A) If the markup policy of the store for items that cost over $10 is assumed to be linear and is reflected in the pricing of these two items, write an equation that relates retail price R to cost C.
    (B) What should be the retail price of a pair of skis that cost $105.

PRACTICE TEST: **CHAPTER 1**

---

1. Graph $3x + 6y = 18$ in a rectangular coordinate system. Indicate the slope, x intercept, and y intercept.

2. Find the equation of a line that passes through $(-2, 5)$ and $(2, -1)$. Write the answer in the form $y = mx + b$.

3. Write an equation of a line that passes through $(2, -3)$ and is parallel to $2x - 4y = 5$. Write the final answer in the form $Ax + By = C$, $A > 0$.

4. Graph (in the same coordinate system) the vertical and horizontal lines that pass through $(2, -3)$. Indicate the equation of each line.

5. For $f(x) = 2x - x^2$ and $g(x) = |x|$, find:
   (A) $f(-2) - g(-3)$ (B) $f(a + 1)$

6. For $f(x) = \sqrt{x}$ and $g(x) = x^2 + 2x$, find:

   (A) $f[g(2)]$ (B) $g[f(a)]$

7. Write in terms of simpler logarithmic forms: $\log_b \dfrac{5\sqrt{w}}{x^3 y}$

8. Find x, y, or b:

   (A) $y = \log_4 2$ (B) $\log_{1/2} x = -2$ (C) $\log_b 8 = 3$

9. Graph $y = 10e^{0.1x}$, $0 \le x \le 10$, using Table I or a calculator.

10. An electronic computer was purchased by a company for $20,000 and is assumed to have a salvage value of $2,000 after 10 years (for tax purposes). If its value is depreciated linearly from $20,000 to $2,000:

   (A) Find the linear equation that relates value V in dollars to time t in years.

   (B) What would be the value of the computer after 6 years?-

CHAPTER 2
THE DERIVATIVE

CONTENTS

2 THE DERIVATIVE

2-1 INTRODUCTION

How do algebra and calculus differ? The two words, *static* and *dynamic* probably come as close as any in expressing the difference between the two disciplines. In algebra, we solve equations for a particular value of a variable—a static notion. In calculus, we are interested in how a change in one variable affects another variable—a dynamic notion.

Figure 1 illustrates three basic problems in calculus. It may surprise you to learn that all three problems—as different as they appear—are mathematically related. The solutions to these problems and the discovery of their relationship required the creation of a new kind of mathematics. Isaac Newton (1642–1727) of England and Gottfried Wilhelm von Leibniz (1646–1716) of Germany simultaneously and independently developed this new mathematics, called **the calculus**—it was an idea whose time had come.

FIGURE 1

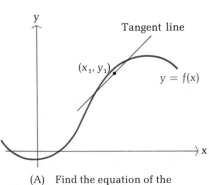

(A) Find the equation of the tangent line at (x_1, y_1) given $y = f(x)$

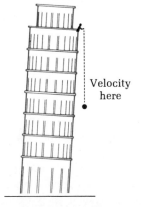

(B) Find the instantaneous velocity of a falling object

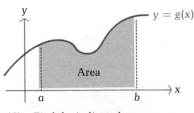

(C) Find the indicated area bounded by $y = g(x)$, $x = a$, $x = b$, and the x axis

In addition to solving the problems described in Figure 1, calculus will enable us to solve many important problems. Until fairly recently, calculus was primarily used in the physical sciences, but now, people in many other disciplines are finding it a useful tool.

2-2   LIMITS AND CONTINUITY

Basic to the study of calculus are the concepts of *limit* and *continuity*. These concepts help us describe, in a precise way, the behavior of a range value $f(x)$ when a domain value x is close to a particular value c. Our approach will be informal, concentrating on concept development and understanding rather than formal details.

CONTINUITY

You already have had experience with the word *continuous* in ordinary English, where it means "unbroken" or "connected." As a technical mathematical word applied to functions, *continuity* has a similar meaning. If we look at the function graphs in Figure 2, we see that some of the graphs are broken. A function whose graph is broken (disconnected) at a certain point is said to be **discontinuous** at that point; if the graph is not broken at a point, then the graph is said to be **continuous** at that point. A function is **continuous over an interval** if it is continuous (not

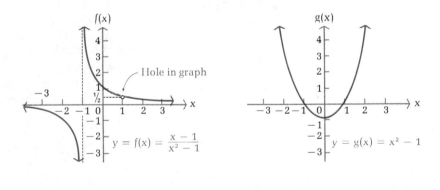

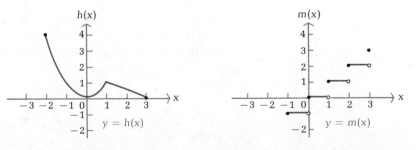

FIGURE 2 Continuous and discontinuous functions. A solid dot indicates the point is on the graph; a hollow dot indicates the point is not on the graph.

broken) at each value on the interval. Stated in another way, we say that a function is continuous over an interval if the graph can be drawn over the interval without taking our pen or pencil off the paper.

Example 1   Refer to Figure 2.
(A)   Function f is discontinuous at $x = -1$ and at $x = 1$, but is continuous on any interval not containing either $x = -1$ or $x = 1$.
(B)   Function g is continuous for all real numbers.

Problem 1   Refer to Figure 2.
(A)   Where is function h discontinuous on the interval $-2 \leq x \leq 3$?
(B)   On which of the following intervals is the function m continuous?

$$0 \leq x \leq 2 \qquad 1 \leq x \leq 2 \qquad 1.1 \leq x \leq 1.5$$

Answers   (A)   Nowhere   (B) $1.1 \leq x \leq 1.5$

Why are we interested in continuous functions? There are many processes that are performed in calculus that are only generally valid on intervals over which the involved functions are continuous (we will have more to say about this later). In addition, continuous functions are often used to model real-world phenomena, since they are usually easier to work with than other types of functions.

LIMITS   Let us look at function f in Figure 2 from a different point of view. We just observed that function f is discontinuous at $x = -1$ and at $x = 1$. What happens to $f(x)$ when x is close to but not equal to either -1 or 1? From the graph of f, we see that when x approaches -1 (from either side of -1), $f(x)$ does not appear to approach any fixed number. On the other hand, when x approaches the other point of discontinuity, $+1$ (from either side), it appears that $f(x)$ approaches $\frac{1}{2}$. To state the latter, we write

$$f(x) \to \tfrac{1}{2} \qquad \text{as} \qquad x \to 1$$

or more concisely,

$$\lim_{x \to 1} f(x) = \tfrac{1}{2}$$

which is read, "the limit of $f(x)$ as x approaches 1 is $\frac{1}{2}$." Note that $f(x)$ never takes on the value of $\frac{1}{2}$ for any x near 1; $\frac{1}{2}$ is the number that $f(x)$ *approaches* as x approaches 1, and it is this number ($\frac{1}{2}$) that we call the *limit*, even though the function is not even defined at $x = 1$.

In the case where $x \to -1$, we write

$$\lim_{x \to -1} f(x) \quad \text{does not exist}$$

since $f(x)$ does not approach any fixed number as x approaches -1 (from either side).

We now state an informal definition of **the limit of a function f as x approaches a number c.** A precise definition will not be needed for our discussion, but one is given in the footnote below.*

LIMIT (INFORMAL DEFINITION)

We write

$$\lim_{x \to c} f(x) = L$$

if the functional value $f(x)$ is close to the single real number L whenever x is close to but not equal to c.

IMPORTANT OBSERVATIONS

1. If $f(x)$ does not approach one number L when x approaches c from the left and from the right, the limit as x approaches c does not exist.
2. The function f does not have to be defined at $x = c$ (but it can be) in order for a limit to exist when x approaches c.

Example 2   Use Figure 2 and the definition of limit to find $\lim_{x \to 0} f(x)$.

Solution   When x is close to 0, then $f(x)$ is close to 1; that is,

$$f(x) \to 1 \quad \text{as} \quad x \to 0$$

Or, in terms of limit notation, we write

$$\lim_{x \to 0} f(x) = 1$$

Problem 2   Use Figure 2 and the definition of limit to find:
(A) $\lim_{x \to 1} h(x)$ (B) $\lim_{x \to 2} m(x)$

Answers   (A) 1   (B) Does not exist

LIMITS AND CONTINUITY   We now see how the limit concept can be used to make the notion of continuity precise. Let us summarize our observations about the behavior of $f(x)$ in Figure 2 at $x = 0$, a point of continuity, and at $x = 1$, a point of discontinuity.

Behavior of $f(x)$ at $x = 0$, a point of continuity:

$$\lim_{x \to 0} f(x) = 1 = f(0)$$

When x is close to 0, $f(x)$ is close to $f(0)$.

*To make the informal definition of limit precise, the use of the word *close* must be made more precise. This is done as follows: We write $\lim_{x \to c} f(x) = L$ if for each $e > 0$, there exists a $d > 0$ such that $|f(x) - L| < e$ whenever $0 < |x - c| < d$. This definition is used to establish particular limits and to prove many useful properties of limits that will be helpful to us in finding particular limits.

Behavior of f(x) at x = 1, a point of discontinuity:

$$\lim_{x \to 1} f(x) = \tfrac{1}{2} \neq f(1) \qquad f(1) \text{ is not defined}$$

Thus, we see that if $\lim_{x \to c} f(x) \neq f(c)$, then f is not continuous at $x = c$. We now introduce a formal definition of continuity in terms of a limit.

CONTINUITY

A function f is **continuous at the point x = c** if

1. $f(c)$ exists (is defined)
2. $\lim\limits_{x \to c} f(x) = f(c)$

A function is **continuous on an interval** I if it is continuous at each point on I.

This definition states that for a function f to be continuous at $x = c$, f must be defined at $x = c$; and when x is close to c, $f(x)$ must stay close to $f(c)$. To use the definition to test for continuity of a function f at $x = c$, we compute $f(c)$ and $\lim_{x \to c} f(x)$. If both exist (are some number, including zero) and if they are the same, then the function f is continuous at $x = c$.

From the definition of continuity, it can be shown that the following properties hold:

CONTINUITY PROPERTIES

1. $f(x) = k$ (a constant function) is continuous for all real x
2. $f(x) = x^n$, n a natural number, is continuous for all real x
3. Polynomial functions* are continuous for all real numbers.

For continuous functions $y = f(x)$ and $y = g(x)$:

4. $f(x) \pm g(x)$ and $f(x) \cdot g(x)$ are continuous

5. $\dfrac{f(x)}{g(x)}$ is continuous, except for those values of x such that $g(x) = 0$

6. $\sqrt[n]{f(x)}$ is continuous (x is restricted to avoid even roots of negative numbers)

---

*The function P is a polynomial function if $P(x)$ can be written in the form

$$P(x) = a_n x^n + a_{n-1} x^{n-1} + \cdots + a_1 x + a_0, \qquad a_n \neq 0$$

where the coefficients are real constants.

Each of the following equations defines a continuous function for all real numbers, but with the indicated restrictions:

$$f(x) = 3x^2 - 2x + 5 \qquad \text{A polynomial function}$$

$$f(x) = \frac{x^2 - 2}{x + 3} \qquad x \neq -3$$

$$f(x) = \sqrt{x - 1} \qquad x \geq 1$$

$$f(x) = (x^2 - 1)(x + 2) - (3x^2 - 2x + 5)$$

Example 3   Discuss the continuity of

$$f(x) = \frac{x + 3}{x^2 - x - 6}$$

Solution $$\frac{x + 3}{x^2 - x - 6} = \frac{x + 3}{(x - 3)(x + 2)}$$

Since $f(x)$ is the quotient of two polynomials, f is continuous for all real x, except at $x = 3$ and $x = -2$, where the denominator is zero (continuity properties 3 and 5).

Problem 3   Discuss the continuity of

$$f(x) = \frac{6}{x^2 - 3x - 10}$$

Answer   Discontinuous only at $x = 5$ and $x = -2$.

LIMITS AT POINTS OF CONTINUITY   We now consider the problem of finding limits of functions when we do not have a graph to refer to (which is most of the time). If we know a function is continuous at $x = c$, then to find $\lim_{x \to c} f(x)$ we simply evaluate $f(x)$ at $x = c$ and write

$$\lim_{x \to c} f(x) = f(c)$$

Example 4   (A) $\lim\limits_{x \to 2} (2x^2 - 3x + 1) = ?$ (B) $\lim\limits_{x \to 3} \dfrac{x^2 - 5}{x + 2} = ?$

Solutions   (A)   Since polynomial functions are continuous for all real values (continuity property 3), we write

$$\lim_{x \to 2} (2x^2 - 3x + 1) = 2(2^2) - 3(2) + 1 = 3$$

(B)   Since quotients of polynomial functions (rational functions) are continuous for all real numbers, except for those values that make the denominator equal to zero (continuity properties 3 and 5), we can write

$$\lim_{x \to 3} \frac{x^2 - 5}{x + 2} = \frac{3^2 - 5}{3 + 2} = \frac{4}{5}$$

Problem 4   (A) $\lim_{x \to -2} (x^3 - 2x^2 + 1) = ?$ (B) $\lim_{x \to 2} \frac{x - 2}{x^2 + 4} = ?$

Answers   (A) -15 (B)   0

LIMITS AT POINTS OF DISCONTINUITY

As we mentioned at the beginning of this section, if a function f is discontinuous at $x = c$, then $\lim_{x \to c} f(x)$ may or may not exist. The following properties of limits, which we state without proof, are of use in finding limits (if they exist) at points of discontinuity, or at points where it is not certain whether a function is or is not continuous.

LIMIT PROPERTIES

If k and c are constants, n is a positive integer, and

$$\lim_{x \to c} f(x) = A \qquad \lim_{x \to c} g(x) = B$$

then:

1. $\lim_{x \to c} k = k$

2. $\lim_{x \to c} kf(x) = k \lim_{x \to c} f(x) = kA$

3. $\lim_{x \to c} [f(x) \pm g(x)] = \lim_{x \to c} f(x) \pm \lim_{x \to c} g(x) = A \pm B$

4. $\lim_{x \to c} [f(x) \cdot g(x)] = [\lim_{x \to c} f(x)][\lim_{x \to c} g(x)] = AB$

5. $\lim_{x \to c} \dfrac{f(x)}{g(x)} = \dfrac{\lim_{x \to c} f(x)}{\lim_{x \to c} g(x)} = \dfrac{A}{B} \qquad B \neq 0$

6. $\lim_{x \to c} \sqrt[n]{f(x)} = \sqrt[n]{\lim_{x \to c} f(x)} = \sqrt[n]{A}$
     (x is restricted to avoid even roots of negative numbers)

Example 5   Given

$$f(x) = \frac{x - 1}{x^2 - 1}$$

answer the following questions without referring to Figure 1:
(A)   Where is f discontinuous?
(B) $\lim_{x \to 1} f(x) = ?$
(C) $\lim_{x \to -1} f(x) = ?$

Solutions   (A)   The polynomial $x^2 - 1$ is 0 when $x = -1$ or 1; therefore, f is (by continuity property 5) discontinuous at $x = -1$ and $x = 1$.

(B) $\lim\limits_{x \to 1} \dfrac{x - 1}{x^2 - 1}$

We cannot use limit property 5, since $\lim_{x \to 1}(x^2 - 1) = 0$, so let us factor the denominator.

$= \lim\limits_{x \to 1} \dfrac{\cancel{(x - 1)}^{1}}{\cancel{(x - 1)}_{1}(x + 1)}$

We are interested in what $(x - 1)/(x - 1)(x + 1)$ approaches as x approaches (but does not equal) 1. The factor $(x - 1)$ cancels for any value of $x \neq 1$, and we can now use limit property 5.

$= \lim\limits_{x \to 1} \dfrac{1}{x + 1}$

$= \dfrac{1}{2}$

(C) $\lim\limits_{x \to -1} \dfrac{x - 1}{x^2 - 1}$

Proceed as in part B.

$= \lim\limits_{x \to -1} \dfrac{\cancel{(x - 1)}^{1}}{\cancel{(x - 1)}_{1}(x + 1)}$

$= \lim\limits_{x \to -1} \dfrac{1}{x + 1}$

Does not exist

What does $1/(x + 1)$ approach as x approaches (but does not equal) -1? The denominator has a limit of zero, so limit property 5 cannot be used. As the denominator approaches zero, the fraction $1/(x + 1)$ can be made as large in absolute value as you like; hence, it does not approach any fixed number. The limit does not exist.

Compare the results we obtained in Example 5 with those obtained earlier in this section, where we introduced the concept of limit using Figure 2.

Problem 5   Given

$$f(x) = \dfrac{2 + x}{4 - x^2}$$

(A)   Where is f discontinuous?
(B) $\lim\limits_{x \to -2} f(x) = ?$
(C) $\lim\limits_{x \to 2} f(x) = ?$

Answers   (A) $x = \pm 2$ (B) $\frac{1}{4}$ (C)   Does not exist

Example 6   Given $f(x) = x^2 + 1$, find

$$\lim\limits_{h \to 0} \dfrac{f(1 + h) - f(1)}{h}$$

Solution

$$\frac{f(1 + h) - f(1)}{h} = \frac{[(1 + h)^2 + 1] - [1^2 + 1]}{h} \qquad \text{Note:} \quad h \neq 0$$

$$= \frac{1 + 2h + h^2 + 1 - 2}{h}$$

$$= \frac{2h + h^2}{h} = \frac{\overset{1}{\cancel{h}}(2 + h)}{\underset{1}{\cancel{h}}} = 2 + h$$

Thus,

$$\lim_{h \to 0} \frac{f(1 + h) - f(1)}{h} = \lim_{h \to 0} (2 + h) = 2$$

Problem 6   Given $g(x) = x^2 - 2$, find

$$\lim_{h \to 0} \frac{g(2 + h) - g(2)}{h}$$

Answer   4

EXERCISE 2-2

A   *Problems 1–6 refer to the following function graph:*

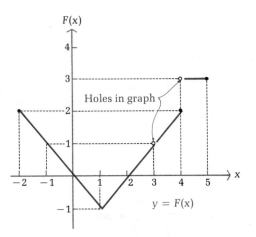

1. Where is the function F discontinuous?
2. Where is the function F continuous?
3. (A)   Is F continuous at $x = -1$?
   (B) $F(-1) = ?$
   (C) $\lim_{x \to -1} F(x) = ?$

4. (A) Is F continuous at x = 3?
   (B) F(3) = ?
   (C) $\lim_{x \to 3} F(x) = ?$

5. (A) Is F continuous at x = 1?
   (B) F(1) = ?
   (C) $\lim_{x \to 1} F(x) = ?$

6. (A) Is F continuous at x = 4?
   (B) F(4) = ?
   (C) $\lim_{x \to 4} F(x) = ?$

Find points of discontinuity (if they exist) for each function.

7. $f(x) = 3x^2 - 2x + 1$

8. $g(x) = 3 - 2x^2$

9. $f(x) = \dfrac{2}{x - 3}$

10. $g(x) = \dfrac{x + 5}{x}$

11. $F(x) = \dfrac{2x - 1}{(x + 3)(x - 3)}$

12. $G(x) = \dfrac{3x + 5}{(x + 4)(x - 1)}$

13. $f(x) = \dfrac{1}{x^2 - x - 6}$

14. $g(x) = \dfrac{3}{x^2 - 3x - 10}$

B　*Find points of discontinuity (if they exist) for each function.*

15. $h(x) = \dfrac{x^2 - x - 6}{x^2 - 9}$

16. $m(x) = \dfrac{x^2 - 2x}{x^2 + 2x - 8}$

Find each limit.

17. $\lim_{x \to 5} (2x^2 - 3)$

18. $\lim_{x \to 2} (x^2 - 8x + 2)$

19. $\lim_{x \to 4} (x^2 - 3x)$

20. $\lim_{x \to 1} (3x^2 - 9x - 2)$

21. $\lim_{x \to 2} \dfrac{5x}{2 + x^2}$

22. $\lim_{x \to 10} \dfrac{2x + 5}{3x - 5}$

23. $\lim_{x \to 0} \dfrac{x^2 - 3x}{x}$

24. $\lim_{x \to 0} \dfrac{2x - 3x^2}{x}$

25. $\lim_{x \to 3} \dfrac{x^2 - x - 6}{x^2 - 9}$

26. $\lim_{x \to 2} \dfrac{x^2 + 2x - 8}{x^2 - 2x}$

27. $\lim_{h \to 0} \dfrac{f(2 + h) - f(2)}{h}$
    for $f(x) = 3x + 1$

28. $\lim_{h \to 0} \dfrac{f(3 + h) - f(3)}{h}$
    for $f(x) = 5x - 1$

29. $\lim_{h \to 0} \dfrac{f(3 + h) - f(3)}{h}$
    for $f(x) = x^2 + 1$

30. $\lim_{h \to 0} \dfrac{f(2 + h) - f(2)}{h}$
    for $f(x) = x^2 - 5$

C *Find each limit.*

31. $\lim\limits_{x \to 2} \sqrt{x^2 + 2x}$

32. $\lim\limits_{x \to 4} \sqrt[3]{x^2 - 3x}$

33. $\lim\limits_{x \to -3} \dfrac{x^2 - x - 6}{x^2 - 9}$

34. $\lim\limits_{x \to 0} \dfrac{x^2 + 2x - 8}{x^2 - 2x}$

APPLICATIONS

BUSINESS & ECONOMICS

35. *Cost function.* A sky diving facility at an airport charges sky divers as follows: A mandatory extra safety chute is \$1. The plane cost is \$5 for the first 5,000 feet, or any fraction thereof; then \$2 per 1,000 feet, or any fraction thereof, for additional altitudes above 5,000 feet. The maximum altitude is 10,000 feet. The costs are graphed in the figure.

$C(x)$

Total cost in dollars

15

10

5

0 .... 5 .... 10 ....→ x

Thousands of feet

(A) Where is the function C discontinuous?

(B) $\lim\limits_{x \to 3} C(x) = ?,\quad C(3) = ?$

(C) $\lim\limits_{x \to 5} C(x) = ?,\quad C(5) = ?$

(D) $\lim\limits_{x \to 7.5} C(x) = ?,\quad C(7.5) = ?$

(E) Is C continuous at $x = 7.5$?

LIFE SCIENCES

36. *Animal supply.* A medical laboratory raises its own rabbits. The number of rabbits $N(t)$ available at any time t depends on the number of births and deaths. When a birth or death occurs, the function N generally has a discontinuity, as shown in the figure.

(A) Where is the function N discontinuous?

(B) $\lim\limits_{t \to t_5} N(t) = ?,\quad N(t_5) = ?$

(C) $\lim\limits_{t \to t_3} N(t) = ?,\quad N(t_3) = ?$

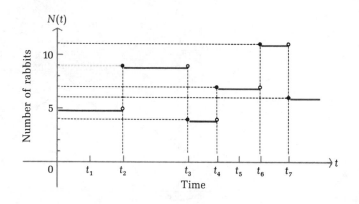

SOCIAL SCIENCES

37. *Learning.* The graph shown here might represent the history of a particular person learning the material on limits and continuity in this book. At time t_2, the student's mind goes blank during a quiz. At time t_4, the instructor explains a concept particularly well, and suddenly, a big jump in understanding takes place.

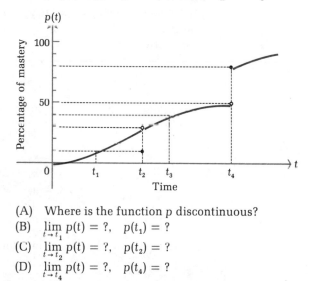

(A) Where is the function p discontinuous?

(B) $\lim\limits_{t \to t_1} p(t) = ?, \quad p(t_1) = ?$

(C) $\lim\limits_{t \to t_2} p(t) = ?, \quad p(t_2) = ?$

(D) $\lim\limits_{t \to t_4} p(t) = ?, \quad p(t_4) = ?$

2-3 INCREMENTS, TANGENT LINES, AND RATES OF CHANGE

We will now use the concept of limit to solve two of the three basic problems of calculus stated at the beginning of this chapter. The parts of Figure 1 that we will concentrate on are repeated in Figure 3 (next page).

INCREMENTS Before pursuing these problems, we digress for a moment to introduce *increment* notation. If we are given a function defined by $y = f(x)$ and the

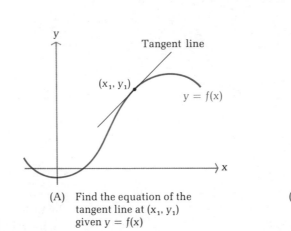

(A) Find the equation of the tangent line at (x_1, y_1) given $y = f(x)$

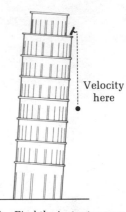

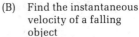

Velocity here

(B) Find the instantaneous velocity of a falling object

FIGURE 3

independent variable x changes from x_1 to x_2, then the dependent variable y will change from $y_1 = f(x_1)$ to $y_2 = f(x_2)$ (see Fig. 4). Mathematically, the change in x and the corresponding change in y, called **increments in x and y,** respectively, are denoted by Δx and Δy (read "delta x" and "delta y").

INCREMENTS

For $y = f(x)$

(see Fig. 4)

$$\Delta x = x_2 - x_1$$
$$x_2 = x_1 + \Delta x$$
$$\Delta y = y_2 - y_1$$
$$= f(x_2) - f(x_1)$$
$$= f(x_1 + \Delta x) - f(x_1)$$

Δy represents the change in y corresponding to a Δx change in x

---

Example 7   Given the function

$$f(x) = \frac{x^2}{2}$$

(A) Find Δx, Δy, and $\Delta y/\Delta x$ for $x_1 = 1$ and $x_2 = 2$.

(B) Find

$$\frac{f(x_1 + \Delta x) - f(x_1)}{\Delta x}$$

for $x_1 = 1$ and $\Delta x = 2$.

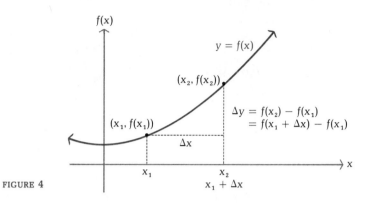

$f(x)$

$y = f(x)$

$(x_2, f(x_2))$

$\Delta y = f(x_2) - f(x_1)$
$= f(x_1 + \Delta x) - f(x_1)$

$(x_1, f(x_1))$

Δx

x_1 x_2 $x_1 + \Delta x$ x

FIGURE 4

Solutions (A) $\Delta x = x_2 - x_1 = 2 - 1 = 1$

$\Delta y = f(x_2) - f(x_1)$

$= f(2) - f(1) = \dfrac{4}{2} - \dfrac{1}{2} = \dfrac{3}{2}$

$\dfrac{\Delta y}{\Delta x} = \dfrac{f(x_2) - f(x_1)}{x_2 - x_1} = \dfrac{3/2}{1} = \dfrac{3}{2}$

(B) $\dfrac{f(x_1 + \Delta x) - f(x_1)}{\Delta x} = \dfrac{f(1 + 2) - f(1)}{2}$

$= \dfrac{f(3) - f(1)}{2} = \dfrac{(9/2) - (1/2)}{2} = \dfrac{4}{2} = 2$

Problem 7 Given the function $f(x) = x^2 + 1$:
(A) Find Δx, Δy, and $\Delta y/\Delta x$ for $x_1 = 2$ and $x_2 = 3$.
(B) Find

$$\dfrac{f(x_1 + \Delta x) - f(x_1)}{\Delta x}$$

for $x_1 = 1$ and $\Delta x = 2$.

Answers (A) $\Delta x = 1$, $\Delta y = 5$, $\Delta y/\Delta x = 5$ (B) 4

TANGENT LINE From plane geometry, we know that a tangent to a circle is a line that passes through one point on the circle, but how do we define and find a tangent line to a graph of a function at a point? The concept of slope of a straight line (see Section 1-2) will play a central role in the process. If we pass a straight line through two points on the graph of $y = f(x)$, as in Figure 5 (next page), we obtain a secant line. Given the coordinates of the two points, we can find the slope of the secant line using the point–slope formula from Section 1-2:

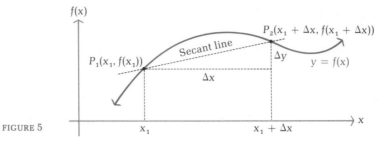

FIGURE 5

$$\text{Secant line slope} = \frac{y_2 - y_1}{x_2 - x_1}$$

$$= \frac{f(x_1 + \Delta x) - f(x_1)}{\Delta x}$$

$$= \frac{\Delta y}{\Delta x}$$

As we let Δx tend to zero, P_2 will approach P_1, and it appears that the secant lines will approach a limiting position and the secant slopes will approach a limiting value (see Fig. 6). If they do, then we will call the line that the secant lines approach the *tangent line to the curve* at $(x_1, f(x_1))$, and the limiting slope will be the slope of the tangent line. This leads to the following definition of a tangent line:

TANGENT LINE

Given the graph of $y = f(x)$, then the **tangent line** at $(x_1, f(x_1))$ is the line that passes through this point with slope

$$\text{Tangent line slope} = \lim_{\Delta x \to 0} \frac{f(x_1 + \Delta x) - f(x_1)}{\Delta x} \qquad (1)$$

if the limit exists. The slope of the tangent line is also referred to as the **slope of the curve** at $(x_1, f(x_1))$.

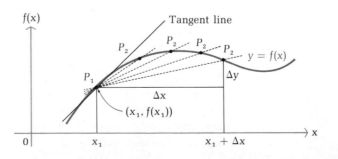

FIGURE 6 Dotted lines are secant lines for smaller and smaller Δx.

Example 8   Given $f(x) = x^2$, find the equation of the tangent line at $x = 1$. Sketch the graph of f, the tangent line at $(1, f(1))$, and the secant line passing through $(1, f(1))$ and $(2, f(2))$.

Solution   First, we find the slope of the tangent line using equation (1).

$$\frac{f(1 + \Delta x) - f(1)}{\Delta x} = \frac{(1 + \Delta x)^2 - 1^2}{\Delta x}$$

We are computing the slope of a secant line passing through $(1, f(1))$ and $(1 + \Delta x, f(1 + \Delta x))$— see Figure 5

$$= \frac{1 + 2\Delta x + \Delta x^2 - 1}{\Delta x}$$

$$= \frac{2\Delta x + \Delta x^2}{\Delta x}$$

$$= \frac{\Delta x(2 + \Delta x)}{\Delta x} = 2 + \Delta x \qquad \Delta x \neq 0$$

$$\text{Tangent line slope} = \lim_{\Delta x \to 0} \frac{f(1 + \Delta x) - f(1)}{\Delta x}$$

$$= \lim_{\Delta x \to 0} (2 + \Delta x) = 2$$

Now, to find the tangent line **equation,** we use the point–slope formula and substitute our known values:

$$y - y_1 = m(x - x_1)$$
$$x_1 = 1$$
$$y_1 = f(x_1) - f(1) = 1^2 = 1$$
$$m = 2$$

So,

$$y - 1 = 2(x - 1)$$
$$y - 1 = 2x - 2$$
$$y = 2x - 1 \qquad \text{Tangent line equation}$$

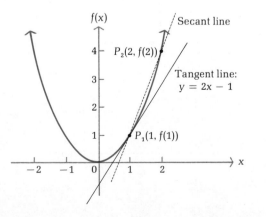

Problem 8   Find the equation of the tangent line for the graph of $f(x) = x^2$ at $x = 2$. Write the answer in the form $y = mx + b$.

Answer $y = 4x - 4$

AVERAGE AND INSTANTANEOUS RATES OF CHANGE

We now show how increments and limits can be used to analyze rate problems. In the process, we will solve the second basic calculus problem we stated at the beginning of the chapter.

Example 9   A small steel ball dropped from a tower will fall a distance of y feet in x seconds, as given approximately by the formula (from physics) $y = f(x) = 16x^2$. Let us determine the ball's position on a coordinate line at various times (Fig. 7). Our ultimate objective is to find the ball's *velocity* at a given instant, say, at the end of 2 seconds.

(A)   Find x_2 and Δy for $x_1 = 2$ and $\Delta x = 1$.

(B)   Find the average velocity for the time change in part A.

(C)   Find an expression for the average velocity from $x = 2$ to $x = 2 + \Delta x$, where Δx represents a small but arbitrary change in time and $\Delta x \neq 0$ (see Fig. 7).

(D)   Find $\lim\limits_{\Delta x \to 0} (\Delta y / \Delta x)$ using $\Delta y / \Delta x$ from part C.

Solutions   (A) $x_2 = x_1 + \Delta x = 2 + 1 = 3$

$$\begin{aligned}
\Delta y &= f(x_1 + \Delta x) - f(x_1) \\
&= f(3) - f(2) \\
&= 16(3^2) - 16(2^2) \\
&= 144 - 64 = 80 \text{ ft}
\end{aligned}$$

Distance fallen from end of 2 seconds to end of 3 seconds (see Fig. 7)

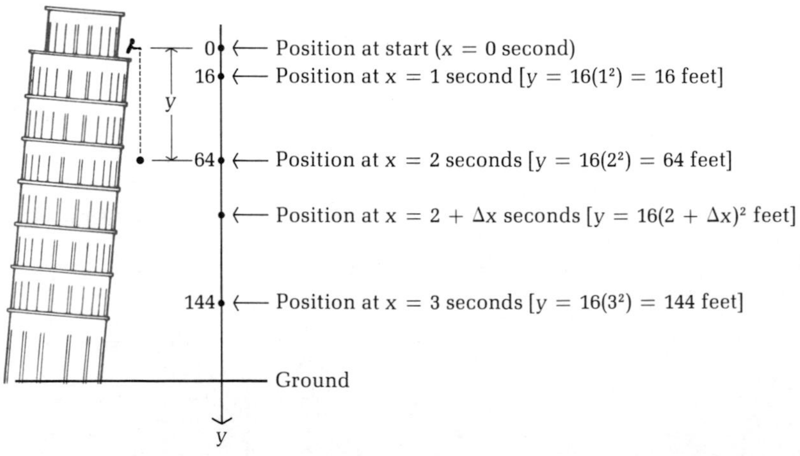

0 ← Position at start ($x = 0$ second)

16 ← Position at $x = 1$ second [$y = 16(1^2) = 16$ feet]

64 ← Position at $x = 2$ seconds [$y = 16(2^2) = 64$ feet]

← Position at $x = 2 + \Delta x$ seconds [$y = 16(2 + \Delta x)^2$ feet]

144 ← Position at $x = 3$ seconds [$y = 16(3^2) = 144$ feet]

Ground

FIGURE 7

Note: Positive y direction is down.

(B)   Recall the formula $d = rt$, which can be written in the form

$$r = \frac{d}{t} = \frac{\text{Total distance}}{\text{Elapsed time}} = \text{Average rate}$$

For example, if a person drives from San Francisco to Los Angeles—a distance of about 420 miles—in 10 hours, then the average rate is

$$r = \frac{d}{t} = \frac{420}{10} = 42 \text{ miles per hour}$$

Sometimes the person will be traveling faster and sometimes slower, but the *average rate* is 42 miles per hour. In our present problem, it is clear from Figure 7 that the ball is *accelerating* (falling faster and faster), but we can compute an average rate, or velocity, just as we did for the trip from San Francisco to Los Angeles:

$$\textbf{Average velocity} = \frac{\text{Total distance}}{\text{Elapsed time}}$$

$$= \frac{\Delta y}{\Delta x} = \frac{f(3) - f(2)}{1} = \frac{80}{1} = 80 \text{ feet per second}$$

Thus, the average velocity from the end of 2 seconds to the end of 3 seconds is 80 feet per second.

(C)   Average velocity $= \dfrac{\Delta y}{\Delta x} = \dfrac{f(2 + \Delta x) - f(2)}{\Delta x}$ $\Delta x \neq 0$

$$= \frac{16(2 + \Delta x)^2 - 16(2^2)}{\Delta x}$$

$$= \frac{16(4 + 4\Delta x + \Delta x^2) - 64}{\Delta x}$$

$$= \frac{64 + 64\Delta x + 16\Delta x^2 - 64}{\Delta x}$$

$$= \frac{64\Delta x + 16\Delta x^2}{\Delta x} = \frac{\Delta x(64 + 16\Delta x)}{\Delta x} = 64 + 16\Delta x$$

Note that if $\Delta x = 1$, the average velocity is 80 feet per second; if $\Delta x = 0.5$, then the average velocity is 72 feet per second; if $\Delta x = 0.01$, then the average velocity is 64.16 feet per second; and so on. The smaller Δx gets, the closer the average velocity gets to 64 feet per second.

(D) $\displaystyle \lim_{x \to 0} \frac{\Delta y}{\Delta x} = \lim_{\Delta x \to 0} \frac{f(2 + \Delta x) - f(2)}{\Delta x}$

$$= \lim_{\Delta x \to 0} (64 + 16\Delta x)$$

$$= 64 \text{ feet per second}$$

We call 64 feet per second the **instantaneous velocity** at $x = 2$ seconds, and we have solved the second basic problem stated at the beginning of this chapter!

The discussion in Example 9 leads to the following general definitions of average rate and instantaneous rate:

AVERAGE AND INSTANTANEOUS RATES

For $y = f(x)$

$$\textbf{Average rate} = \frac{\Delta y}{\Delta x} = \frac{f(x_2) - f(x_1)}{x_2 - x_1} = \frac{f(x_1 + \Delta x) - f(x_1)}{\Delta x}$$

$$\textbf{Instantaneous rate} = \lim_{\Delta x \to 0} \frac{\Delta y}{\Delta x} = \lim_{\Delta x \to 0} \frac{f(x_1 + \Delta x) - f(x_1)}{\Delta x}$$

if the limit exists

Problem 9   For the falling steel ball in Example 9, find:
(A)   The average velocity from $x = 1$ to $x = 2$
(B)   The average velocity from $x = 1$ to $x = 1 + \Delta x$
(C)   The instantaneous velocity at $x = 1$

Answers   (A)   48 feet per second   (B) $32 + 16\Delta x$ (C)   32 feet per second

Now we will consider a slightly different type of rate problem, but we will use the same approach as in Example 9.

Example 10   Suppose a produce grower is willing to supply crates of oranges according to the supply function illustrated in Figure 8 [$S(x) = 100x^2$]. At \$2 per crate, the supplier would be willing to supply $S(2) = 100(2^2) = 400$ crates of oranges; at \$4 per crate, the supplier would be willing to suply $S(4) = 100(4^2) = 1{,}600$ crates. As the price goes up, the supplier is willing to supply more oranges, just as we would expect.
(A)   What is the average rate of change in supply from \$2 per crate to \$4 per crate?
(B)   What is the average rate of change in supply from \$2 per crate to \$(2 + \Delta x) per crate?
(C)   What value does $\Delta S/\Delta x$ in part B approach as Δx tends to zero?

Solutions   (A) $\dfrac{\Delta S}{\Delta x} = \dfrac{S(x_2) - S(x_1)}{x_2 - x_1}$

$$= \frac{S(4) - S(2)}{4 - 2}$$

$$= \frac{1{,}600 - 400}{2} = \frac{1{,}200}{2} = 600 \text{ crates per dollar}$$

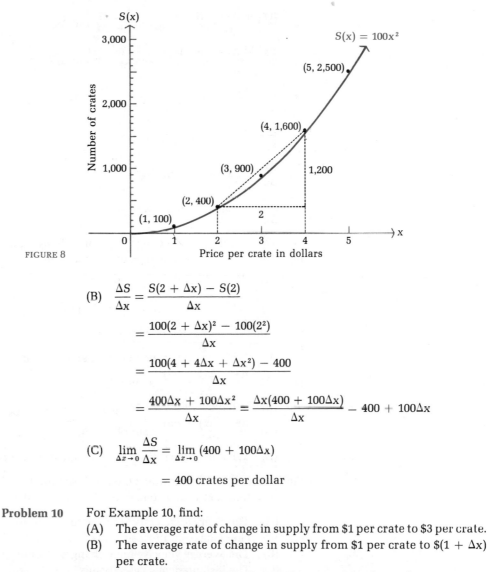

FIGURE 8

$$(B) \quad \frac{\Delta S}{\Delta x} = \frac{S(2 + \Delta x) - S(2)}{\Delta x}$$

$$= \frac{100(2 + \Delta x)^2 - 100(2^2)}{\Delta x}$$

$$= \frac{100(4 + 4\Delta x + \Delta x^2) - 400}{\Delta x}$$

$$= \frac{400\Delta x + 100\Delta x^2}{\Delta x} = \frac{\Delta x(400 + 100\Delta x)}{\Delta x} - 400 + 100\Delta x$$

$$(C) \quad \lim_{\Delta x \to 0} \frac{\Delta S}{\Delta x} = \lim_{\Delta x \to 0} (400 + 100\Delta x)$$

$$= 400 \text{ crates per dollar}$$

Problem 10   For Example 10, find:
(A)   The average rate of change in supply from $1 per crate to $3 per crate.
(B)   The average rate of change in supply from $1 per crate to $(1 + \Delta x)$ per crate.
(C)   What value does $\Delta S/\Delta x$ in part B approach as Δx tends to zero?

Answers   (A)   400 crates per dollar   (B) $200 + 100\Delta x$
(C)   200 crates per dollar

EXERCISE 2-3

In Problems 1–14 find the indicated quantities for $y = f(x) = 3x^2$.

A   1. Δx, Δy,   and $\Delta y/\Delta x$,   given $x_1 = 1$ and $x_2 = 4$
2. Δx, Δy,   and $\Delta y/\Delta x$,   given $x_1 = 2$ and $x_2 = 5$

3. $\dfrac{f(x_1 + \Delta x) - f(x_1)}{\Delta x}$, given $x_1 = 1$ and $\Delta x = 2$

4. $\dfrac{f(x_1 + \Delta x) - f(x_1)}{\Delta x}$, given $x_1 = 2$ and $\Delta x = 1$

5. $\dfrac{y_2 - y_1}{x_2 - x_1}$, given $x_1 = 1$ and $x_2 = 3$

6. $\dfrac{y_2 - y_1}{x_2 - x_1}$, given $x_1 = 2$ and $x_2 = 3$

7. $\dfrac{\Delta y}{\Delta x}$, given $x_1 = 1$ and $x_2 = 3$

8. $\dfrac{\Delta y}{\Delta x}$, given $x_1 = 2$ and $x_2 = 3$

B 9. The average rate of change of y, for x changing from 1 to 4
   10. The average rate of change of y, for x changing from 2 to 5

11. (A) $\dfrac{f(2 + \Delta x) - f(2)}{\Delta x}$ (simplify)

    (B) What does the ratio in part A approach as Δx approaches zero?

12. (A) $\dfrac{f(3 + \Delta x) - f(3)}{\Delta x}$ (simplify)

    (B) What does the ratio in part A approach as Δx approaches zero?

13. (A) $\dfrac{f(4 + \Delta x) - f(4)}{\Delta x}$ (simplify)

    (B) What does the ratio in part A approach as Δx approaches zero?

14. (A) $\dfrac{f(5 + \Delta x) - f(5)}{\Delta x}$ (simplify)

    (B) What does the ratio in part A approach as Δx approaches zero?

Suppose an object moves along the y axis so that its location is $y = f(x) = x^2 + x$ at time x (y is in meters and x is in seconds). Find:

15. (A) The average velocity (the average rate of change of y) for x changing from 1 to 3 seconds
    (B) The average velocity for x changing from 1 to $(1 + \Delta x)$ seconds
    (C) The instantaneous velocity at $x = 1$

16. (A) The average velocity (the average rate of change of y) for x changing from 2 to 4 seconds
    (B) The average velocity for x changing from 2 to $(2 + \Delta x)$ seconds
    (C) The instantaneous velocity at $x = 2$

In Problems 17 and 18 for the graph of $y = f(x) = x^2 + x$, find:

17. (A) The slope of the secant line joining $(1, f(1))$ and $(3, f(3))$
    (B) The slope of the secant line joining $(1, f(1))$ and
        $(1 + \Delta x, f(1 + \Delta x))$

(C)   The slope of the tangent line at $(1, f(1))$

(D)   The equation of the tangent line at $(1, f(1))$

18.   (A)   The slope of the secant line joining $(2, f(2))$ and $(4, f(4))$

(B)   The slope of the secant line joining $(2, f(2))$ and
$(2 + \Delta x, f(2 + \Delta x))$

(C)   The slope of the tangent line at $(2, f(2))$

(D)   The equation of the tangent line at $(2, f(2))$

C   19.   If an object moves on the x axis so that it is at $x = f(t) = t^2 - t$ at
time t (t measured in seconds and x measured in meters), find the
instantaneous velocity of the object at $t = 2$.

20.   Find the equation of the tangent line for the graph of $y = x^2 - x$
at $x = 2$.

APPLICATIONS

BUSINESS & ECONOMICS

21.   *Income.*   The per capita income in the United States from 1969 to
1973 is given approximately in the table. Find the average rate of
change of per capita income for a time change from:

(A)   1969 to 1971   (B)   1971 to 1973

YEAR	1969	1970	1971	1972	1973
INCOME	$3,700	$3,900	$4,100	$4,500	$5,000

22.   *Demand function.*   Suppose in a given grocery store people are
willing to buy $D(x)$ pounds of chocolate candy per day at $x per
pound, as given by the demand function

$$D(x) = 100 - x^2 \qquad \$1 \le x \le \$10$$

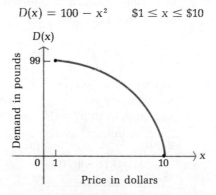

Note that as price goes up, demand goes down (see the figure).

(A)   Find the average rate of change in demand for a price change
from $2 to $5; that is, find $\Delta y / \Delta x$ for $x_1 = 2$ and $x_2 = 5$.

(B)   Simplify:

$$\frac{D(2 + \Delta x) - D(2)}{\Delta x}$$

(C)   What does the ratio in part B approach as Δx approaches zero? [This is called "the instantaneous rate of change of $D(x)$ with respect to x at x = 2."]

LIFE SCIENCES

23.   *Medicine.*   The area of a small (healing) wound in square millimeters, where time is measured in days, is given in the table.

AREA	400	360	180	120	90	72	60
DAYS	0	1	2	3	4	5	6

Find the average rate of change of area for the time change from:
(A)   0 to 2 days   (B)   4 to 6 days

24.   *Weight–height.*   A formula relating the approximate weight of an average person and his or her height is

$W(h) = 0.0005h^3$

where $W(h)$ is in pounds and h is in inches.
(A)   Find the average rate of change of weight for a height change from 60 to 70 inches.
(B)   Simplify:

$$\frac{W(60 + \Delta h) - W(60)}{\Delta h}$$

(C)   What does the ratio in part B approach as Δx approaches zero? [This is called "the instantaneous rate of change of $W(h)$ with respect to h at h = 60."]

SOCIAL SCIENCES

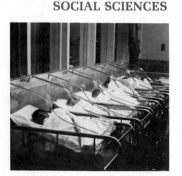

25.   *Illegitimate births.*   The approximate numbers of illegitimate births per 1,000 live births in the United States from 1940 to 1970 are given in the table. Find the average rate of change of illegitimate births per 1,000 live births for the time change from:
(A)   1940 to 1945   (B)   1965 to 1970

YEAR	1940	1945	1950	1955	1960	1965	1970
ILLEGITIMATE BIRTHS *Per 1,000 live births*	38	41	40	47	54	80	120

26.   *Learning.*   A certain person learning to type has an achievement record given approximately by the function

$$N(t) = 60\left(1 - \frac{2}{t}\right) \qquad 3 \le t \le 10$$

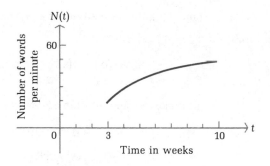

where $N(t)$ is in number of words per minute and t is in weeks. Find the average rate of change of the number of words per minute for the change in time from:

(A) 4 to 6 weeks   (B) 8 to 10 weeks

2-4   DEFINITION OF THE DERIVATIVE

In the last section we found that the special limit

$$\lim_{\Delta x \to 0} \frac{f(x_1 + \Delta x) - f(x_1)}{\Delta x} \tag{1}$$

if it exists, gives us the slope of the tangent line to the graph of $y = f(x)$ at $(x_1, f(x_1))$. It also gives us the instantaneous rate of change of y per unit change in x at $x = x_1$. Formula (1) is of such basic importance to calculus and to the applications of calculus that we will give it a name and study it in detail. To keep formula (1) simple and general, we will drop the subscript on x_1 and think of the ratio

$$\frac{f(x + \Delta x) - f(x)}{\Delta x}$$

as a function of Δx, with x held fixed as we let Δx tend to zero. We are now ready to define one of the basic concepts in calculus, the *derivative*:

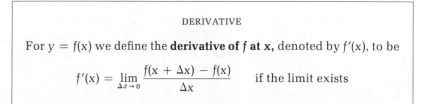

DERIVATIVE

For $y = f(x)$ we define the **derivative of f at x,** denoted by $f'(x)$, to be

$$f'(x) = \lim_{\Delta x \to 0} \frac{f(x + \Delta x) - f(x)}{\Delta x} \qquad \text{if the limit exists}$$

Thus, taking the derivative of a function f at x creates a new function f' that gives, among other things, the instantaneous rate of change of $y = f(x)$ and the slope of the tangent line to the graph of $y = f(x)$ for each x.

Example 11   Find $f'(x)$, the derivative of f at x, for $f(x) = 4x - x^2$.

Solution   To find $f'(x)$, we find

$$\lim_{\Delta x \to 0} \frac{f(x + \Delta x) - f(x)}{\Delta x}$$

To make the computation easier, we introduce a two-step process:

Step 1.   *Find $[f(x + \Delta x) - f(x)]/\Delta x$ and simplify.*

$$\frac{f(x + \Delta x) - f(x)}{\Delta x} = \frac{[4(x + \Delta x) - (x + \Delta x)^2] - [4x - x^2]}{\Delta x}$$

$$= \frac{4x + 4\Delta x - x^2 - 2x\Delta x - \Delta x^2 - 4x + x^2}{\Delta x}$$

$$= \frac{4\Delta x - 2x\Delta x - \Delta x^2}{\Delta x}$$

$$= \frac{\Delta x}{\Delta x}(4 - 2x - \Delta x)$$

$$= 4 - 2x - \Delta x \qquad \Delta x \neq 0$$

Step 2.   *Find the limit of the result of step 1.*

$$\lim_{\Delta x \to 0} \frac{f(x + \Delta x) - f(x)}{\Delta x} = \lim_{\Delta x \to 0} (4 - 2x - \Delta x)$$

$$= 4 - 2x$$

Thus, $f'(x) = 4 - 2x$

Problem 11   Find $f'(x)$, the derivative of f at x, for $f(x) = 8x - 2x^2$.

Answer $f'(x) = 8 - 4x$

TANGENT LINES AND SLOPE FUNCTION   In the last section we defined the slope of the tangent line to the graph of $y = f(x)$ at $(x_1, f(x_1))$ to be

$$\lim_{\Delta x \to 0} \frac{f(x_1 + \Delta x) - f(x_1)}{\Delta x}$$

if the limit exists. This, of course, is $f'(x_1)$, the derivative of f at $x = x_1$. To find the equation of a tangent line to the graph of $y = f(x)$ at $(x_1, f(x_1))$, we use the point–slope form for the equation of a line, $y - y_1 = m(x - x_1)$, and the fact that $m = f'(x_1)$ and $y_1 = f(x_1)$ to obtain:

TANGENT LINE

The equation of the tangent line to the graph of $y = f(x)$ at $x = x_1$, is

$$y - f(x_1) = f'(x_1)(x - x_1) \qquad \text{if } f'(x_1) \text{ exists}$$

Example 12   In Example 11 we started with the function specified by $f(x) = 4x - x^2$ and found the derivative of f at x to be $f'(x) = 4 - 2x$. This derived function will give us the slope of a tangent line at any point on the graph of $y = f(x)$; hence, it can be called a **slope function.** Thus,

$$m = f'(x) = 4 - 2x$$

We will use this slope function in the following problems.

(A)   Find the slopes of the tangent lines at $x = 1, 2,$ and 3.

(B)   Find the equations of the tangent lines at $x = 1, 2,$ and 3.

(C)   Sketch the tangent lines on the graph of $y = 4x - x^2$ at $x = 1, 2,$ and 3.

Solutions   (A) $m_1 = f'(1) = 4 - 2(1) = 2$
$m_2 = f'(2) = 4 - 2(2) = 0$
$m_3 = f'(3) = 4 - 2(3) = -2$

(B)   TANGENT LINE AT $x = 1$ TANGENT LINE AT $x = 2$

$y - f(1) = f'(1)(x - 1)$ $y - f(2) = f'(2)(x - 2)$
$y - 3 = 2(x - 1)$ $y - 4 = 0(x - 2)$
$y - 3 = 2x - 2$ $y = 4$
$y = 2x + 1$

TANGENT LINE AT $x = 3$

$y - f(3) = f'(3)(x - 3)$
$y - 3 = -2(x - 3)$
$y - 3 = -2x + 6$
$y = -2x + 9$

(C)

OBSERVATIONS

1.   Slope is positive when curve is rising.
2.   Slope is zero at high point.
3.   Slope is negative when curve is falling.

The observations in part C of Example 12 will be very useful to us in the next chapter, when we consider the use of the derivative in graphing and the solution of maxima–minima problems.

Problem 12   Repeat Example 12 for $y = f(x) = 8x - 2x^2$ and $f'(x) = 8 - 4x$ found in Problem 11.

Answers   (A) $m_1 = 4, m_2 = 0, m_3 = -4$

(B) $y = 4x + 2, y = 8, y = -4x + 18$

(C)

y

Slope = 0

Slope = −4

5

Slope = 4

0   5   → x

NONEXISTENCE OF THE DERIVATIVE

The existence of a derivative at $x = a$ depends on the existence of a limit at $x = a$; that is, on the existence of

$$f'(a) = \lim_{\Delta x \to 0} \frac{f(a + \Delta x) - f(a)}{\Delta x}$$

If the limit does not exist at $x = a$, we say that the function f is **nondifferentiable at $x = a$ or $f'(a)$ does not exist.** Geometrically, a tangent line may not exist at a point, or a point may have a vertical tangent line (recall, the slope of a vertical line is not defined). Figure 9 illustrates several situations for which the derivative does not exist. Also, a derivative cannot exist at $x = a$ if the function is not defined at $x = a$. Looking at it from the other way around, we can prove that **if a derivative exists at a point, then the function must be continuous at that point.**

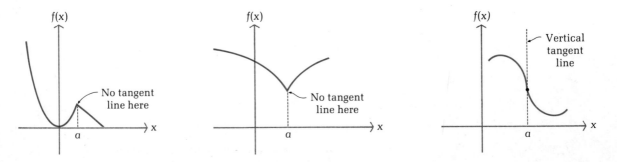

FIGURE 9   Slope does not exist at each indicated point.

INSTANTANEOUS RATE OF CHANGE

From the definition of instantaneous rate of change of $f(x)$ at x given in Section 2-3, we see that it is simply the derivative of f at x, that is, $f'(x)$.

Example 13

Return to the example concerning the falling steel ball in Section 2-3. Find a function that will give the instantaneous velocity, v, of the ball at any time x. Find the velocity at $x = 2, 3,$ and 5 seconds.

Solution

Recall that the distance y (in feet) that the ball falls in x seconds is given by

$$y = f(x) = 16x^2$$

The instantaneous velocity function is $v = f'(x)$; thus,

$$v = f'(x) = \lim_{\Delta x \to 0} \frac{f(x + \Delta x) - f(x)}{\Delta x}$$

Find $f'(x)$ using the two-step process described in Example 11:

Step 1. *Find $[f(x + \Delta x) - f(x)]/\Delta x$ and simplify.*

$$\frac{f(x + \Delta x) - f(x)}{\Delta x} = \frac{[16(x + \Delta x)^2] - [16x^2]}{\Delta x}$$

$$= \frac{16x^2 + 32x\Delta x + 16\Delta x^2 - 16x^2}{\Delta x}$$

$$= \frac{32x\Delta x + 16\Delta x^2}{\Delta x}$$

$$= \frac{\Delta x}{\Delta x}(32x + 16\Delta x) = 32x + 16\Delta x \qquad \Delta x \neq 0$$

Step 2. *Find the limit of the result of step 1.*

$$\lim_{\Delta x \to 0} \frac{f(x + \Delta x) - f(x)}{\Delta x} = \lim_{\Delta x \to 0} (32x + 16\Delta x)$$

$$= 32x$$

Thus,

$$v = f'(x) = 32x$$

The instantaneous velocities at $x = 2, 3,$ and 5 seconds are

$$f'(2) = 32(2) = 64 \text{ feet per second}$$
$$f'(3) = 32(3) = 96 \text{ feet per second}$$
$$f'(5) = 32(5) = 160 \text{ feet per second}$$

Problem 13

A steel ball falls so that its distance y (in feet) at time x (in seconds) is given by $y = f(x) = 16x^2 - 4x$.

(A) Find a function that will give the instantaneous velocity v at time x.

(B) Find the velocity at $x = 2, 4,$ and 6 seconds.

Answers (A) $v = f'(x) = 32x - 4$
(B) $f'(2) = 60$ feet per second, $f'(4) = 124$ feet per second, $f'(6) = 188$ feet per second

MARGINAL COST

In business and economics one is often interested in the rate at which something is taking place. A manufacturer, for example, is not only interested in the total cost $C(x)$ at certain production levels x, but is also interested in the rate of change of costs at various production levels.

In economics the word **marginal** refers to a rate of change, that is, to a derivative. Thus, if

$C(x) =$ Total cost of producing x units during some unit of time

then

$C'(x) =$ Marginal cost
$=$ Rate of change in cost per unit change in production at an output level of x units

Example 14 Suppose the total cost $C(x)$ in thousands of dollars for manufacturing x thousand popular record albums is given by
$$C(x) = 2 + 8x - x^2 \qquad 0 \le x \le 3$$

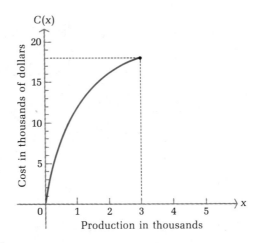

Find:
(A) The marginal cost at x
(B) The marginal cost at $x = 1, 2,$ and 3 thousand levels of production

Solutions (A) Marginal cost at x is

$$C'(x) = \lim_{\Delta x \to 0} \frac{C(x + \Delta x) - C(x)}{\Delta x}$$

which we find using the two-step process discussed in Example 11 (steps omitted here).

Marginal cost $= C'(x) = 8 - 2x$

(B)   Marginal costs at a production level of $x = 1, 2,$ and 3 thousand albums are

$$C'(1) = 8 - 2(1) = 6 \qquad \text{\$6,000 per 1,000 increase in production}$$
$$C'(2) = 8 - 2(2) = 4 \qquad \text{\$4,000 per 1,000 increase in production}$$
$$C'(3) = 8 - 2(3) = 2 \qquad \text{\$2,000 per 1,000 increase in production}$$

Notice that, as production goes up, the marginal cost goes down, as we might expect.

Problem 14   Repeat Example 14 with the cost function $C(x) = 3 + 10x - x^2, 0 \le x \le 4$.

Answers   (A)   Marginal cost $= C'(x) = 10 - 2x$
(B)   Marginal costs at a production level of 1,000, 2,000, and 3,000 albums are $C'(1) = \$8$ thousand, $C'(2) = \$6$ thousand, and $C'(3) = \$4$ thousand.

SUMMARY   The concept of the derivative is a very powerful mathematical idea, and its applications are many and varied. In the next three sections we will develop derivative formulas and general properties of derivatives that will enable us to find the derivatives of many functions without having to go through the (two-step) limiting process each time.

EXERCISE 2-4

A   *Find $f'(x)$ for each of the indicated functions; then find $f'(1), f'(2),$ and $f'(3)$.*

1. $f(x) = 2x - 3$ 2. $f(x) = 4x + 3$

B   3. $f(x) = 6x - x^2$ 4. $f(x) = 8x - x^2$

5.   If an object moves along a line so that it is at $y = f(x) = 4x^2 - 2x$ at time x (in seconds), find the instantaneous velocity function $v = f'(x)$, and find the velocity at times 1, 3, and 5 seconds (y is measured in feet).

6.   Repeat Problem 5 with $f(x) = 8x^2 - 4x$.

7.   Given $y = f(x) = x^2, \quad -3 \le x \le 3$:
(A)   Find the slope function $m = f'(x)$.
(B)   Find the slope of the tangent line to the graph of $y = x^2$ at $x = -2, 0,$ and 2.
(C)   Find the equations of the tangents at $x = -2, 0,$ and 2.
(D)   Sketch the tangent lines on the graph at $x = -2, 0,$ and 2.

8.   Repeat Problem 7 for $y = f(x) = x^2 + 1, \quad -3 \le x \le 3$.

C   **9.** For $f(x) = x^3 + 2x$, find:
            (A) $f'(x)$ (B) $f'(1)$ and $f'(3)$

      10. For $f(x) = x^2 - 3x^3$, find:
            (A) $f'(x)$ (B) $f'(1)$ and $f'(2)$

APPLICATIONS

BUSINESS & ECONOMICS

11. *Marginal cost.* The total cost $C(x)$ in hundreds of dollars for manufacturing x hundred transistor radios is given by

$$C(x) = 3 + 10x - x^2 \qquad 0 \le x \le 4$$

(A) Find the marginal cost at x.

(B) Find the marginal cost at $x = 1$, 3, and 4 hundred levels of production.

LIFE SCIENCES

12. *Negative growth.* A colony of bacteria was treated with a poison, and the number of survivors $N(t)$, in thousands, after t hours was found to be given approximately by

$$N(t) = t^2 - 8t + 16 \qquad 0 \le t \le 4$$

(A) Find $N'(t)$.

(B) Find the rate of change of the colony at $t = 1, 2$, and 3.

SOCIAL SCIENCES

13. *Learning.* A private foreign language school found that the average person learned $N(t)$ basic phrases in t continuous hours, as given approximately by

$$N(t) = 14t - t^2 \qquad 0 \le t \le 7$$

(A) Find $N'(t)$.

(B) Find the rate of learning at $t = 1, 3$, and 6 hours.

2-5   DERIVATIVES OF CONSTANTS, POWER FORMS, SUMS, AND DIFFERENCES

In the last section we defined the derivative of f at x as

$$f'(x) = \lim_{\Delta x \to 0} \frac{f(x + \Delta x) - f(x)}{\Delta x}$$

(if the limit exists) and we used this definition and a two-step process to find the derivatives of a number of functions. In this and the next two sections we will develop some rules based on this definition that will enable us to determine the derivatives of a rather large class of functions without having to go through the two-step process each time.

Before starting on these rules, we list some symbols that are widely used to represent derivatives:

DERIVATIVE NOTATION

Given $y = f(x)$, then

$$f'(x) \qquad y' \qquad \frac{dy}{dx} \qquad D_x f(x)$$

all represent the derivative of f at x.

Each of these symbols for derivatives has its particular advantage in certain situations. All of them will become familiar to you after a little experience.

DERIVATIVE OF A CONSTANT

Suppose

$$f(x) = C \qquad C \text{ a constant}$$

Geometrically, the graph of $f(x) = C$ is a horizontal straight line with slope zero; hence, we would expect $D_x C = 0$

Then

$$f'(x) = \lim_{\Delta x \to 0} \frac{f(x + \Delta x) - f(x)}{\Delta x}$$

$$= \lim_{\Delta x \to 0} \frac{C - C}{\Delta x}$$

$$= \lim_{\Delta x \to 0} \frac{0}{\Delta x}$$

$$= \lim_{\Delta x \to 0} 0 = 0$$

And we conclude that **the derivative of any constant is zero.**

DERIVATIVE OF A CONSTANT

If $y = f(x) = C$, then

$$f'(x) = 0$$

Also, $y' = 0$, $dy/dx = 0$, and $D_x C = 0$.

[*Note:* When we write $D_x C = 0$, we mean $D_x f(x) = 0$, where $f(x) = C$.]

Example 15   (A)   If $f(x) = 3$, then $f'(x) = 0$.   (B)   If $y = -1.4$, then $y' = 0$.
(C)   If $y = \pi$, then $dy/dx = 0$.   (D) $D_x 23 = 0$

Problem 15   Find:
(A) $f'(x)$ for $f(x) = -24$ (B) y' for $y = 12$
(C) dy/dx for $y = -\sqrt{7}$ (D) $D_x(-\pi)$

Answers   All are zero.

POWER RULE   Suppose

$$f(x) = x^n \qquad n \text{ a positive integer}$$

To find $f'(x)$, we start with the definition and try to find a derivative formula for this power function:

$$f'(x) = \lim_{\Delta x \to 0} \frac{f(x + \Delta x) - f(x)}{\Delta x}$$

$$= \lim_{\Delta x \to 0} \frac{(x + \Delta x)^n - x^n}{\Delta x}$$

To find this limit we take advantage of the binomial formula:

$$(a + b)^n = a^n + na^{n-1}b + \frac{n(n-1)}{2}a^{n-2}b^2 + \cdots + b^n$$

Thus,

$$(x + \Delta x)^n = x^n + nx^{n-1}\Delta x + \frac{n(n-1)}{2}x^{n-2}\Delta x^2 + \cdots + \Delta x^n$$

and

$$\frac{(x + \Delta x)^n - x^n}{\Delta x} = nx^{n-1} + \frac{n(n-1)}{2}x^{n-2}\Delta x + \cdots + \Delta x^{n-1}$$

[Again, remember that since Δx represents one quantity, $\Delta x^n = (\Delta x)^n$ and not $\Delta(x^n)$.] In the right member of the last equation, all the terms after the first contain a Δx; so, holding x fixed, all terms except the first tend to zero as Δx tends to zero. Thus,

$$f'(x) = \lim_{\Delta x \to 0} \left[nx^{n-1} + \frac{n(n-1)}{2}x^{n-2}\Delta x + \cdots + \Delta x^{n-1} \right]$$

$$= nx^{n-1}$$

and we conclude that

$$D_x x^n = nx^{n-1}$$

It can be shown that this formula holds for *any* real number n. We will assume this general result for the remainder of this book.

POWER RULE

If $y = f(x) = x^n$, n a real number, then

$$f'(x) = nx^{n-1}$$

Example 16

(A) If $f(x) = x^5$, then $f'(x) = 5x^{5-1} = 5x^4$.

(B) If $y = \dfrac{1}{x^3} = x^{-3}$, then $y' = -3x^{-3-1} = -3x^{-4}$ or $\dfrac{-3}{x^4}$.

(C) If $y = x^{5/3}$, then $\dfrac{dy}{dx} = \tfrac{5}{3}x^{(5/3)-1} = \tfrac{5}{3}x^{2/3}$.

(D) $D_x\sqrt{x} = D_x x^{1/2} = \tfrac{1}{2}x^{(1/2)-1} = \tfrac{1}{2}x^{-1/2} = \dfrac{1}{2\sqrt{x}}$

Problem 16

Find:

(A) $f'(x)$ for $f(x) = x^3$ (B) y' for $y = x^{3/2}$

(C) $\dfrac{dy}{dx}$ for $y = \dfrac{1}{x^2}$ or x^{-2} (D) $D_x\dfrac{1}{\sqrt{x}}$

Answers

(A) $3x^2$ (B) $\tfrac{3}{2}x^{1/2}$ (C) $-2x^{-3}$ (D) $-\tfrac{1}{2}x^{-3/2}$ or $\dfrac{-1}{2\sqrt{x^3}}$

DERIVATIVE OF A CONSTANT TIMES A FUNCTION

Suppose

$$f(x) = ku(x) \qquad \text{with } k \text{ a constant, and } u'(x) \text{ exists}$$

Then

$$f'(x) = \lim_{\Delta x \to 0} \frac{f(x + \Delta x) - f(x)}{\Delta x}$$

$$= \lim_{\Delta x \to 0} \frac{ku(x + \Delta x) - ku(x)}{\Delta x}$$

$$= \lim_{\Delta x \to 0} k\frac{u(x + \Delta x) - u(x)}{\Delta x}$$

$$= k\lim_{\Delta x \to 0} \frac{u(x + \Delta x) - u(x)}{\Delta x}$$

$$= ku'(x)$$

Thus, **the derivative of a constant times a differentiable function is the constant times the derivative of the function.**

CONSTANT TIMES A FUNCTION RULE

If $y = f(x) = ku(x)$, then

$$f'(x) = ku'(x)$$

Also, $y' = ku'$, $dy/dx = k\,du/dx$, and $D_x ku(x) = kD_x u(x)$.

Example 17

(A) If $f(x) = 3x^2$, then $f'(x) = 3 \cdot 2x^{2-1} = 6x$.

(B) If $y = \dfrac{1}{2x^4} = \dfrac{1}{2}x^{-4}$, then $y' = \dfrac{1}{2}(-4x^{-4-1}) = -2x^{-5}$ or $\dfrac{-2}{x^5}$.

(C) If $y = 8x^{3/2}$, then $\dfrac{dy}{dx} = 8 \cdot \dfrac{3}{2}x^{(3/2)-1} = 12x^{1/2}$ or $12\sqrt{x}$.

(D) $D_x \dfrac{4}{\sqrt{x^3}} = D_x \dfrac{4}{x^{3/2}} = D_x 4x^{-3/2} = 4\left[-\dfrac{3}{2}x^{(-3/2)-1}\right]$

$$= -6x^{-5/2} \text{ or } -\dfrac{6}{\sqrt{x^5}}$$

Problem 17

Find:

(A) $f'(x)$ for $f(x) = 4x^5$ (B) y' for $y = \dfrac{1}{3x^3}$

(C) $\dfrac{dy}{dx}$ for $y = 6x^{1/3}$ (D) $D_x \dfrac{9}{\sqrt[3]{x}}$

Answers   (A) $20x^4$ (B) $-x^{-4}$ (C) $2x^{-2/3}$ (D) $-3x^{-4/3}$ or $\dfrac{-3}{\sqrt[3]{x^4}}$

**DERIVATIVE
OF A SUM
AND DIFFERENCE**

Suppose $f(x) = u(x) + v(x)$ and $u'(x)$ and $v'(x)$ exist. Then

$$f'(x) = \lim_{\Delta x \to 0} \frac{f(x + \Delta x) - f(x)}{\Delta x}$$

$$= \lim_{\Delta x \to 0} \frac{[u(x + \Delta x) + v(x + \Delta x)] - [u(x) + v(x)]}{\Delta x}$$

$$= \lim_{\Delta x \to 0} \frac{u(x + \Delta x) - u(x) + v(x + \Delta x) - v(x)}{\Delta x}$$

$$= \lim_{\Delta x \to 0}\left[\frac{u(x + \Delta x) - u(x)}{\Delta x} + \frac{v(x + \Delta x) - v(x)}{\Delta x}\right]$$

$$= \lim_{\Delta x \to 0} \frac{u(x + \Delta x) - u(x)}{\Delta x} + \lim_{\Delta x \to 0} \frac{v(x + \Delta x) - v(x)}{\Delta x}$$

$$= u'(x) + v'(x)$$

So we see that **the derivative of the sum of two differentiable functions is the sum of the derivatives.** Similarly, we can show that **the derivative of the difference of two differentiable functions is the difference of the derivatives.** Together, we then have the sum and difference rule for differentiation.

SUM AND DIFFERENCE RULE

If $y = f(x) = u(x) \pm v(x)$, then

$$f'(x) = u'(x) \pm v'(x)$$

      With this and the other rules stated above, we will be able to compute the derivatives of all polynomials and a variety of other functions.

Example 18

(A) If $f(x) = 3x^2 + 2x$, then $f'(x) = (3x^2)' + (2x)' = 6x + 2$.

(B) If $y = 4 + 2x^3 - 3x^{-1}$, then $y' = (4)' + (2x^3)' - (3x^{-1})' = 6x^2 + 3x^{-2}$.

(C) If $y = \sqrt[3]{x} - 3x$, then $\dfrac{dy}{dx} = \dfrac{d}{dx}x^{1/3} - \dfrac{d}{dx}3x = \dfrac{1}{3}x^{-2/3} - 3$.

(D) $D_x\left(\dfrac{8}{\sqrt[4]{x}} + x^7\right) = D_x 8x^{-1/4} + D_x x^7 = -2x^{-5/4} + 7x^6$

Problem 18

Find:

(A) $f'(x)$ for $f(x) = 3x^4 - 2x^3 + x^2 - 5x + 7$

(B) y' for $y = 3 - 7x^{-2}$

(C) $\dfrac{dy}{dx}$ for $y = 5x^3 - \sqrt[4]{x}$

(D) $D_x\left(\dfrac{6}{\sqrt[3]{x}} + \dfrac{2}{x^3}\right)$

Answers   (A) $12x^3 - 6x^2 + 2x - 5$ (B) $14x^{-3}$ (C) $15x^2 - \frac{1}{4}x^{-3/4}$

(D) $-2x^{-4/3} - 6x^{-4}$

APPLICATIONS

Example 19

The distance y in feet that a steel ball falls in x seconds is given by

$$y = f(x) = 16x^2$$

Find the instantaneous velocity function $v = f'(x)$. Find the velocity at $x = 1$ and 6 seconds.

Solution $v = f'(x) = 16(2x^{2-1}) = 32x$
                            $f'(1) = 32(1) = 32$ feet per second
                            $f'(6) = 32(6) = 192$ feet per second

Problem 19   A steel ball falls so that its distance y in feet after x seconds is given by

$$y = f(x) = 16x^2 - 4x$$

(A) Find the instantaneous velocity function.
(B) Find the velocity at $x = 2$ and 5 seconds.

Answers   (A) $v = 32x - 4$
           (B) $f'(2) = 60$ feet per second; $f'(5) = 156$ feet per second

Example 20   (A) Find the slope function $m = f'(x)$ for $y = f(x) = 4x - x^2$.
                  (B) Find the slope of the tangent to the graph of $y = 4x - x^2$ at $x = 1$,
                       2, and 3.

Solutions   (A) $m = f'(x) = (4x)' - (x^2)' = 4 - 2x$
             (B) $m_1 = f'(1) = 4 - 2(1) = 2$
                  $m_2 = f'(2) = 4 - 2(2) = 0$
                  $m_3 = f'(3) = 4 - 2(3) = -2$

Problem 20   Repeat Example 20 for $y = f(x) = 8x - 2x^2$.

Answers   (A) $m = 8 - 4x$ (B) $m_1 = 4; m_2 = 0; m_3 = -4$

Example 21   The total cost $C(x)$ in thousands of dollars for manufacturing x thousand
                  record albums is given by

$$C(x) = 2 + 8x - x^2 \qquad 0 \le x \le 3$$

(A) The marginal cost at a production level of x is

$$C'(x) = (2)' + (8x)' - (x^2)' = 8 - 2x$$

(B) The marginal cost at $x = 1$ is

$$C'(1) = 8 - 2(1) = 6 \qquad \$6,000 \text{ per } 1,000 \text{ increase in production}$$

(C) The marginal cost at $x = 3$ is

$$C'(3) = 8 - 2(3) = 2 \qquad \$2,000 \text{ per } 1,000 \text{ increase in production}$$

Problem 21   Repeat Example 21 with the cost function $C(x) = 3 + 10x - x^2, 0 \le x \le 4$.

Answers   (A) Marginal cost $= C'(x) = 10 - 2x$
           (B) $C'(1) = 8$ $\$8,000$ per 1,000 increase in production
           (C) $C'(3) = 4$ $\$4,000$ per 1,000 increase in production

EXERCISE 2-5

Find each of the following:

A 1. $f'(x)$ for $f(x) = 12$ 2. $\dfrac{dy}{dx}$ for $y = -\sqrt{3}$

3. $D_x 23$ 4. y' for $y = \pi$

5. $\dfrac{dy}{dx}$ for $y = x^{12}$ 6. $D_x x^5$

7. $f'(x)$ for $f(x) = x$ 8. y' for $y = x^7$

9. $f'(x)$ for $f(x) = 2x^4$ 10. $\dfrac{dy}{dx}$ for $y = -3x$

11. $D_x(\tfrac{1}{3}x^6)$ 12. y' for $y = \tfrac{1}{2}x^4$

B 13. $D_x(2x^{-5})$ 14. y' for $y = -4x^{-1}$

15. $f'(x)$ for $f(x) = -3x^{1/3}$ 16. $\dfrac{dy}{dx}$ for $y = -8x^{1/4}$

17. $\dfrac{dy}{dx}$ for $y = 3x^5 - 2x^3 + 5$ 18. $f'(x)$ for $y = 2x^2 - 0x + 5$

19. $D_x(3x^{-4} + 2x^{-2})$ 20. y' for $y = 2x^{-3} - 4x^{-1}$

21. $\dfrac{dy}{dx}$ for $y = \dfrac{3}{x^2}$ 22. $f'(x)$ for $y = \dfrac{1}{x^4}$

23. $D_x \dfrac{1}{\sqrt[3]{x}}$ 24. y' for $y - \dfrac{10}{\sqrt[5]{x}}$

25. $\dfrac{dy}{dx}$ for $y = \dfrac{12}{\sqrt{x}} - 3x^{-2} + x$

26. $f'(x)$ for $f(x) = 2x^{-3} - \dfrac{6}{\sqrt[3]{x^2}} + 7$

27. Given the equation $y = f(x) = 6x - x^2$, find:
(A)   The slope function $m = f'(x)$.
(B)   The slope of the tangent to the graph at $x = 2$ and at $x = 4$.
(C)   The value(s) of x such that the slope is zero.

28. Repeat Problem 27 for $y = f(x) = 2x^2 + 8x$.

29. Repeat Problem 27 for $y = f(x) = \tfrac{1}{3}x^3 - 3x^2 + 2$.

30. Repeat Problem 27 for $y = f(x) = 2x^3 - 3x^2 - 5$.

31. If an object moves along the y axis (marked in feet) so that its position at time x in seconds is given by $y = f(x) = 176x - 16x^2$, find:
(A)   The instantaneous velocity function $v = f'(x)$
(B)   The velocity at $x = 0, 3$, and 6 seconds
(C)   The time(s) when $v = 0$

32. Repeat Problem 31 for $y = f(x) = 80x - 10x^2$.

33. Repeat Problem 31 for $y = f(x) = 10 + 40x - 5x^2$.

34. Repeat Problem 31 for $y = f(x) = -20 + 120x - 15x^2$.

C   35. $D_x \dfrac{x^4 - 3x^3 + 5}{x^2}$

36. y' for $y = \dfrac{2x^5 - 4x^3 + 2x}{x^3}$

37. $\dfrac{dy}{dx}$ for $y = \dfrac{\sqrt{x} - 6}{\sqrt{x^3}}$

38. $f'(x)$ for $f(x) = \dfrac{\sqrt[3]{x} + 3}{\sqrt[3]{x^2}}$

APPLICATIONS

BUSINESS & ECONOMICS

39. *Advertising.* Using past records it is estimated that a company will sell $N(x)$ units of a product after spending $\$x$ thousand on advertising, as given by

$$N(x) = 60x - x^2 \qquad 5 \le x \le 30$$

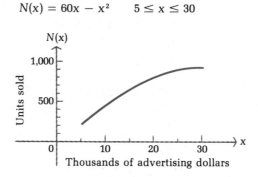

(A) Find $N'(x)$, the rate of change of sales per unit change in money spent on advertising at the $\$x$ thousand level.

(B) Find $N'(10)$ and $N'(20)$ and interpret.

40. *Marginal average cost.* (This topic is treated in detail in Section 2-8.) Economists often work with average costs—cost per unit output—rather than total costs. We would expect higher average costs, because of plant inefficiency, at low output levels and also at output levels near plant capacity. Therefore, we would expect the graph of an average cost function to be U-shaped. Suppose that for a given firm the total cost of producing x thousand units is given by

$$C(x) = x^3 - 6x^2 + 12x$$

Then the average cost $\overline{C}(x)$ is given by

$$\overline{C}(x) = \frac{C(x)}{x} = x^2 - 6x + 12$$

(A) Find the marginal average cost $\overline{C}'(x)$.

(B) Find the marginal average cost at $x = 2, 3,$ and $4,$ and interpret.

LIFE SCIENCES

41. *Medicine.* A person x inches tall has a pulse rate of y beats per minute, as given approximately by

$$y = 590x^{-1/2} \qquad 30 \leq x \leq 75$$

What is the instantaneous rate of change of pulse rate at the:
(A) 36 inch level? (B) 64 inch level?

42. *Ecology.* A coal-burning electrical generating plant emits sulfur dioxide into the surrounding air. The concentration $C(x)$ in parts per million is given approximately by

$$C(x) = \frac{0.1}{x^2}$$

where x is the distance from the plant in miles. Find the (instantaneous) rate of change of concentration at:
(A) $x - 1$ mile (B) $x - 2$ miles

SOCIAL SCIENCES

43. *Learning.* If a person learns y items in x hours, as given by

$$y = 50\sqrt{x} \qquad 0 \leq x \leq 9$$

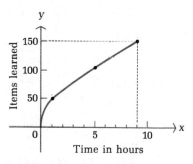

find the rate of learning at the end of:
(A) 1 hour (B) 9 hours

44. *Learning.* If a person learns y items in x hours, as given by

$$y = 21 \sqrt[3]{x^2} \qquad 0 \leq x \leq 8$$

find the rate of learning at the end of:
(A) 1 hour (B) 8 hours

2-6   DERIVATIVES OF PRODUCTS AND QUOTIENTS

The derivative rules discussed in the last section added substantially to our ability to compute and apply derivatives to many practical problems. In this and the next section we will add a few more rules that will increase this ability even further.

DERIVATIVES OF PRODUCTS

There are cases where a function can be considered the product of two other functions. For example,

$$f(x) = (x^4 - 3x^2 + 2)(x^2 - x + 5)$$

can be thought of as

$$f(x) = u(x)v(x)$$

where

$$u(x) = x^4 - 3x^2 + 2 \quad \text{and} \quad v(x) = x^2 - x + 5$$

In general, suppose we have the product form

$$f(x) = u(x)v(x) \quad \text{where} \quad u'(x) \text{ and } v'(x) \text{ exist}$$

We will develop a derivative formula for $f'(x)$ in terms of $u'(x)$ and $v'(x)$. We proceed as follows:

$$f'(x) = \lim_{\Delta x \to 0} \frac{f(x + \Delta x) - f(x)}{\Delta x}$$

$$= \lim_{\Delta x \to 0} \frac{u(x + \Delta x)v(x + \Delta x) - u(x)v(x)}{\Delta x}$$

We now add zero in a special form to the numerator. That is, we subtract and add $u(x + \Delta x)v(x)$ in the middle of the numerator to obtain

$$f'(x) = \lim_{\Delta x \to 0} \frac{u(x + \Delta x)v(x + \Delta x) - u(x + \Delta x)v(x) + u(x + \Delta x)v(x) - u(x)v(x)}{\Delta x}$$

$$= \lim_{\Delta x \to 0} \frac{u(x + \Delta x)[v(x + \Delta x) - v(x)] + v(x)[u(x + \Delta x) - u(x)]}{\Delta x}$$

$$= \lim_{\Delta x \to 0} \left[u(x + \Delta x)\frac{v(x + \Delta x) - v(x)}{\Delta x} + v(x)\frac{u(x + \Delta x) - u(x)}{\Delta x} \right]$$

$$= \lim_{\Delta x \to 0} u(x + \Delta x)\frac{v(x + \Delta x) - v(x)}{\Delta x} + \lim_{\Delta x \to 0} v(x)\frac{u(x + \Delta x) - u(x)}{\Delta x}$$

$$= u(x)v'(x) + v(x)u'(x)$$

Thus, **the derivative of a product is the first times the derivative of the second plus the second times the derivative of the first.**

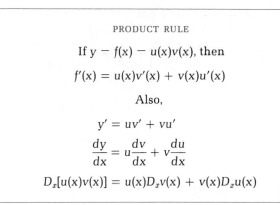

PRODUCT RULE

If $y = f(x) = u(x)v(x)$, then

$$f'(x) = u(x)v'(x) + v(x)u'(x)$$

Also,

$$y' = uv' + vu'$$

$$\frac{dy}{dx} = u\frac{dv}{dx} + v\frac{du}{dx}$$

$$D_x[u(x)v(x)] = u(x)D_xv(x) + v(x)D_xu(x)$$

Example 22 (A) Find $f'(x)$ for $f(x) = 2x^2(3x^4 - 2)$ two ways.

(B) Find $D_x[(x^2 - 2x + 1)(3x^3 + x - 5)]$.

(C) Find $D_x[(\sqrt[3]{x} + 2)(2\sqrt{x} - 1)]$.

Solutions (A) Method I.   Use the product rule:

$$
\begin{aligned}
f'(x) &= 2x^2(3x^4 - 2)' + (3x^4 - 2)(2x^2)' \qquad &\text{First times derivative}\\
&= 2x^2(12x^3) + (3x^4 - 2)(4x) &\text{of second plus second}\\
&= 24x^5 + 12x^5 - 8x &\text{times derivative of first}\\
&= 36x^5 - 8x
\end{aligned}
$$

Method II.   Multiply first; then take derivatives:

$$f(x) = 2x^2(3x^4 - 2) = 6x^6 - 4x^2$$
$$f'(x) = 36x^5 - 8x$$

(B) $D_x[(x^2 - 2x + 1)(3x^3 + x - 5)]$

$$
\begin{aligned}
&= (x^2 - 2x + 1)D_x(3x^3 + x - 5) + (3x^3 + x - 5)D_x(x^2 - 2x + 1)\\
&= (x^2 - 2x + 1)(9x^2 + 1) + (3x^3 + x - 5)(2x - 2)\\
&= 9x^4 - 18x^3 + 10x^2 - 2x + 1 + 6x^4 - 6x^3 + 2x^2 - 12x + 10\\
&= 15x^4 - 24x^3 + 12x^2 - 14x + 11
\end{aligned}
$$

How we write the final answer depends on what we want to do with it; we might have chosen to leave the answer in the unsimplified form two steps back for certain purposes

(C) $D_x[(\sqrt[3]{x} + 2)(2\sqrt{x} - 1)]$

$$
\begin{aligned}
&= D_x[(x^{1/3} + 2)(2x^{1/2} - 1)] \qquad &\text{Change radicals to fractional}\\
& &\text{exponent form}\\
&= (x^{1/3} + 2)D_x(2x^{1/2} - 1) + (2x^{1/2} - 1)D_x(x^{1/3} + 2)\\
&= (x^{1/3} + 2)x^{-1/2} + (2x^{1/2} - 1)\tfrac{1}{3}x^{-2/3}\\
&= \frac{\sqrt[3]{x} + 2}{\sqrt{x}} + \frac{2\sqrt{x} - 1}{3\sqrt[3]{x^2}}
\end{aligned}
$$

Problem 22   Find:

(A) $f'(x)$ for $f(x) = 3x^3(2x^2 - 3x + 1)$ two ways

(B) y' for $y = (2x^2 - 3x + 5)(x^2 + 3x + 1)$

(C) $D_x[(\sqrt{x} - 2)(\sqrt[3]{x^2} + 5)]$

Answers   (A) $30x^4 - 36x^3 + 9x^2$

(B) $(2x^2 - 3x + 5)(2x + 3) + (x^2 + 3x + 1)(4x - 3) = 8x^3 + 9x^2 - 4x + 12$

(C) $(x^{1/2} - 2)(\tfrac{2}{3}x^{-1/3}) + (x^{2/3} + 5)(\tfrac{1}{2}x^{-1/2})$ or $\dfrac{2(\sqrt{x} - 2)}{3\sqrt[3]{x}} + \dfrac{\sqrt[3]{x^2} + 5}{2\sqrt{x}}$

**DERIVATIVES
OF QUOTIENTS**

Suppose

$$f(x) = \frac{u(x)}{v(x)} \qquad \text{where} \quad u'(x) \text{ and } v'(x) \text{ exist}$$

Then, proceeding in a way similar to that used to obtain the product rule, we obtain

$$f'(x) = \frac{v(x)u'(x) - u(x)v'(x)}{[v(x)]^2}$$

Thus, **the derivative of a quotient is the denominator times the derivative of the numerator minus the numerator times the derivative of the denominator, all over the denominator squared.**

QUOTIENT RULE

If

$$y = f(x) = \frac{u(x)}{v(x)}$$

then

$$f'(x) = \frac{v(x)u'(x) - u(x)v'(x)}{[v(x)]^2}$$

Also,

$$y' = \frac{vu' - uv'}{v^2}$$

$$\frac{dy}{dx} = \frac{v(du/dx) - u(dv/dx)}{v^2}$$

$$D_x \frac{u(x)}{v(x)} = \frac{v(x)D_x u(x) - u(x)D_x v(x)}{[v(x)]^2}$$

Example 23   (A)   If

$$f(x) = \frac{x^2}{2x - 1}$$

find $f'(x)$.

(B)   Find

$$D_x \frac{x^2 - x}{x^3 + 1}$$

(C)   Find

$$D_x \frac{x^{1/2} - 3}{x^{1/2}}$$

by using the quotient rule and also by splitting the fraction into two fractions.

Solutions   (A) $f'(x) = \dfrac{(2x - 1)(x^2)' - x^2(2x - 1)'}{(2x - 1)^2}$

$$= \frac{(2x - 1)(2x) - x^2(2)}{(2x - 1)^2}$$

$$= \frac{4x^2 - 2x - 2x^2}{(2x - 1)^2}$$

$$= \frac{2x^2 - 2x}{(2x - 1)^2}$$

The bottom times the derivative of the top minus the top times the derivative of the bottom, all over the bottom squared

(B) $D_x \dfrac{x^2 - x}{x^3 + 1} = \dfrac{(x^3 + 1)D_x(x^2 - x) - (x^2 - x)D_x(x^3 + 1)}{(x^3 + 1)^2}$

$$= \frac{(x^3 + 1)(2x - 1) - (x^2 - x)(3x^2)}{(x^3 + 1)^2}$$

$$= \frac{2x^4 - x^3 + 2x - 1 - 3x^4 + 3x^3}{(x^3 + 1)^2}$$

$$= \frac{-x^4 + 2x^3 + 2x - 1}{(x^3 + 1)^2}$$

(C)   Method I.   Use the quotient rule:

$$D_x \frac{x^{1/2} - 3}{x^{1/2}} = \frac{x^{1/2}D_x(x^{1/2} - 3) - (x^{1/2} - 3)D_x x^{1/2}}{(x^{1/2})^2}$$

$$= \frac{x^{1/2}(\tfrac{1}{2}x^{-1/2}) - (x^{1/2} - 3)\tfrac{1}{2}x^{-1/2}}{x}$$

$$= \frac{\tfrac{1}{2} - \tfrac{1}{2} + \tfrac{3}{2}x^{-1/2}}{x}$$

$$= \frac{3}{2x(x^{1/2})} = \frac{3}{2x^{3/2}}$$

Method II.   Split into two fractions:

$$\frac{x^{1/2} - 3}{x^{1/2}} = \frac{x^{1/2}}{x^{1/2}} - \frac{3}{x^{1/2}} = 1 - 3x^{-1/2}$$

$$D_x(1 - 3x^{-1/2}) = 0 + \tfrac{3}{2}x^{-3/2} = \frac{3}{2x^{3/2}}$$

Comparing methods I and II, we see that it may sometimes pay to change an expression algebraically before blindly using a derivative formula.

Problem 23   Find:

(A) $f'(x)$ for $f(x) = \dfrac{2x}{x^2 + 3}$ (B) y' for $y = \dfrac{x^3 - 3x}{x^2 - 4}$

(C) $D_x\dfrac{2 + x^{1/3}}{x^{1/3}}$ two ways

Answers   (A) $\dfrac{(x^2 + 3)2 - (2x)(2x)}{(x^2 + 3)^2} = \dfrac{6 - 2x^2}{(x^2 + 3)^2}$

(B) $\dfrac{(x^2 - 4)(3x^2 - 3) - (x^3 - 3x)(2x)}{(x^2 - 4)^2} = \dfrac{3x^4 - 2x^3 - 9x^2 + 12}{(x^2 - 4)^2}$

(C) $\dfrac{-2}{3x^{4/3}}$

EXERCISE 2-6

A   For $f(x)$ as given, find $f'(x)$.

1. $f(x) = 2x^3(x^2 - 2)$ 2. $f(x) = 5x^2(x^3 + 2)$

3. $f(x) = (x - 3)(2x - 1)$ 4. $f(x) = (3x + 2)(4x - 5)$

5. $f(x) = \dfrac{x}{x - 3}$ 6. $f(x) = \dfrac{3x}{2x + 1}$

7. $f(x) = \dfrac{2x + 3}{x - 2}$ 8. $f(x) = \dfrac{3x - 4}{2x + 3}$

9. $f(x) = (x^2 + 1)(2x - 3)$ 10. $f(x) = (3x + 5)(x^2 - 3)$

11. $f(x) = \dfrac{x^2 + 1}{2x - 3}$ 12. $f(x) = \dfrac{3x + 5}{x^2 - 3}$

B   Find each of the following (Problems 21–31 do not have to be simplified):

13. $f'(x)$ for $f(x) = (2x + 1)(x^2 - 3x)$

14. y' for $y = (x^3 + 2x^2)(3x - 1)$

15. $\dfrac{dy}{dx}$ for $y = (2x - x^2)(5x + 2)$

16. $D_x[(3 - x^3)(x^2 - x)]$

17. y' for $y = \dfrac{5x - 3}{x^2 + 2x}$

18. $f'(x)$ for $f(x) = \dfrac{3x^2}{2x - 1}$

19. $D_x \dfrac{x^2 - 3x + 1}{x^2 - 1}$

20. $\dfrac{dy}{dx}$ for $y = \dfrac{x^4 - x^3}{3x - 1}$

21. $f'(x)$ for $f(x) = (2x^4 - 3x^3 + x)(x^2 - x + 5)$

22. $\dfrac{dy}{dx}$ for $y = (x^2 - 3x + 1)(x^3 + 2x^2 - x)$

23. $D_x \dfrac{3x^2 - 2x + 3}{4x^2 + 5x - 1}$

24. y' for $y = \dfrac{x^3 - 3x + 4}{2x^2 + 3x - 2}$

25. $\dfrac{dy}{dx}$ for $y = 9x^{1/3}(x^3 + 5)$

26. $D_x[(4x^{1/2} - 1)(3x^{1/3} + 2)]$

27. $f'(x)$ for $f(x) = \dfrac{6\sqrt[3]{x}}{x^2 - 3}$

28. y' for $y = \dfrac{2\sqrt{x}}{x^2 - 3x + 1}$

C

29. $D_x \dfrac{x^3 - 2x^2}{\sqrt[3]{x^2}}$

30. $\dfrac{dy}{dx}$ for $y = \dfrac{x^2 - 3x + 1}{\sqrt[4]{x}}$

31. $f'(x)$ for $f(x) = \dfrac{(2x^2 - 1)(x^2 + 3)}{x^2 + 1}$

32. y' for $\dfrac{2x - 1}{(x^3 + 2)(x^2 - 3)}$

APPLICATIONS

BUSINESS & ECONOMICS

33. *Price–demand function.* According to classical economic theory, the demand $d(x)$ for a commodity in a free market decreases as the price x increases. Suppose that the number $d(x)$ of transistor radios people are willing to buy per week in a given city at a price $\$x$ is given by

$$d(x) = \frac{50,000}{x^2 + 10x + 25} \qquad \$5 \le x \le \$15$$

(A) Find $d'(x)$, the rate of change of demand with respect to price change.

(B) Find $d'(5)$ and $d'(10)$.

LIFE SCIENCES

34. *Drug sensitivity.* One hour after x milligrams of a particular drug are given to a person, the change in body temperature $T(x)$ in degrees Fahrenheit is given approximately by

$$T(x) = x^2\left(1 - \frac{x}{9}\right) \qquad 0 \le x \le 6$$

The rate at which T changes with respect to the size of the dosage x, $T'(x)$, is called the *sensitivity* of the body to the dosage.

(A) Find $T'(x)$, using the product rule.

(B) Find $T'(1)$, $T'(3)$, and $T'(6)$.

SOCIAL SCIENCES   35. *Learning.*   In the early days of quantitative learning theory (around 1917), L. L. Thurstone found that a given person successfully accomplished $N(x)$ acts after x practice acts, as given by

$$N(x) = \frac{100x + 200}{x + 32}$$

(A) Find the rate of change of learning, $N'(x)$, with respect to the number of practice acts x.

(B) Find $N'(4)$ and $N'(68)$.

2-7   DERIVATIVES OF COMPOSITE FUNCTIONS; CHAIN RULE

We now come to perhaps the most important derivative rule of all—the *chain rule*. This rule will enable us to determine the derivatives of some fairly complicated functions in terms of derivation of more elementary functions.

Suppose you were asked to find the derivative of

$$y = h(x) = (x^2 - 2)^8$$

The derivative rules at our disposal will not be of great help. Of course, you could expand $(x^2 - 2)^8$, but what would you do if 8 were replaced with 100 or 1,000 or $\frac{2}{3}$?

We note that h is actually a **composite function** (see Section 1-3); that is,

$$y = h(x) = f[g(x)]$$

where

$$y = f(u) = u^8 \qquad \text{and} \qquad u = g(x) = x^2 - 2$$

Now, recalling that

$$\frac{dy}{dx} = \lim_{\Delta x \to 0} \frac{\Delta y}{\Delta x} \tag{1}$$

we might be tempted to write

$$\frac{\Delta y}{\Delta x} = \frac{\Delta y}{\Delta u} \frac{\Delta u}{\Delta x}$$

and substitute this into equation (1) to obtain

$$\frac{dy}{dx} = \lim_{\Delta x \to 0} \frac{\Delta y}{\Delta u} \frac{\Delta u}{\Delta x}$$

Note that $\Delta u \to 0$ as $\Delta x \to 0$ (assuming the functions involved are sufficiently well-behaved). We may then conclude that

$$\frac{dy}{dx} = \left(\lim_{\Delta u \to 0} \frac{\Delta y}{\Delta u} \right)\left(\lim_{\Delta x \to 0} \frac{\Delta u}{\Delta x} \right)$$

$$= \frac{dy}{du} \frac{du}{dx}$$

The result is correct under rather general conditions, and is called the *chain rule*, but our "derivation" is superficial, because it ignores a number of hidden problems. Since a formal proof of the **chain rule** is beyond the scope of this book, we simply state it as follows:

CHAIN RULE

If $y = f(u)$ and $u = g(x)$ define the composite function
$y = h(x) = f[g(x)]$, then

$$\frac{dy}{dx} = \frac{dy}{du} \frac{du}{dx} \qquad \text{provided } \frac{dy}{du} \text{ and } \frac{du}{dx} \text{ exist}$$

Example 24 Find dy/dx, given:
(A) $y = (x^2 - 2)^8$ (B) $y = \sqrt[3]{x^3 - 3x + 1}$

Solutions (A) Let $y = u^8$ and $u = x^2 - 2$:

$$\frac{dy}{dx} = \frac{dy}{du} \frac{du}{dx}$$

$$= 8u^7(2x)$$

$$= 8(x^2 - 2)^7(2x) \qquad \text{Since } u = x^2 - 2$$

$$= 16x(x^2 - 2)^7$$

Gradually, you will want to be able to do most of these steps in your head and simply write

$$D_x[(x^2 - 2)^8] = 8(x^2 - 2)^7(2x)$$
$$= 16x(x^2 - 2)^7$$

(B) $y = \sqrt[3]{x^3 - 3x + 1} = (x^3 - 3x + 1)^{1/3}$

Let $y = u^{1/3}$ and $u = x^3 - 3x + 1$:

$$\boxed{\begin{aligned}\frac{dy}{dx} &= \frac{dy}{du}\frac{du}{dx}\\ &= \tfrac{1}{3}u^{-2/3}(3x^2 - 3)\end{aligned}}$$

$$= \tfrac{1}{3}(x^3 - 3x + 1)^{-2/3}(3x^2 - 3) \qquad \text{Since } u = x^3 - 3x + 1$$

$$= (x^3 - 3x + 1)^{-2/3}(x^2 - 1)$$

Or, doing the boxed steps mentally,

$$D_x[(x^3 - 3x + 1)^{1/3}] = \tfrac{1}{3}(x^3 - 3x + 1)^{-2/3}(3x^2 - 3)$$
$$= (x^3 - 3x + 1)^{-2/3}(x^2 - 1)$$

Problem 24   Find dy/dx, given:

(A) $y = (x^2 - 2x + 1)^4$ (B) $y = \sqrt{x^2 + 2x}$

Answers   (A) $8(x^2 - 2x + 1)^3(x - 1)$ (B) $(x^2 + 2x)^{-1/2}(x + 1)$

Example 25   Find dy/dx, given:

(A) $y = 3x^2(x^2 - 1)^5$ (B) $y = \dfrac{(x^2 + 1)^4}{2x^3}$

Solutions   (A) $\dfrac{dy}{dx} = D_x[3x^2(x^2 - 1)^5]$

$$= 3x^2 D_x[(x^2 - 1)^5] + (x^2 - 1)^5 D_x(3x^2)$$
$$= 3x^2[5(x^2 - 1)^4(2x)] + (x^2 - 1)^5(6x)$$
$$= 30x^3(x^2 - 1)^4 + 6x(x^2 - 1)^5 \qquad \text{Note that } (x^2 - 1)^4 \text{ is a common factor}$$
$$= (x^2 - 1)^4[30x^3 + 6x(x^2 - 1)] \qquad \text{Simplify expression within brackets}$$
$$= (x^2 - 1)^4(36x^3 - 6x)$$

(B) $\dfrac{dy}{dx} = D_x \dfrac{(x^2 + 1)^4}{2x^3} = \dfrac{2x^3 D_x[(x^2 + 1)^4] - (x^2 + 1)^4 D_x(2x^3)}{(2x^3)^2}$

$$= \frac{2x^3[4(x^2 + 1)^3(2x)] - (x^2 + 1)^4(6x^2)}{4x^6}$$

$$= \frac{16x^4(x^2 + 1)^3 - 6x^2(x^2 + 1)^4}{4x^6}$$

$$= \frac{(x^2 + 1)^3[16x^4 - 6x^2(x^2 + 1)]}{4x^6} \qquad \text{Note that } (x^2 + 1)^3 \text{ is a common factor}$$

$$= \frac{(x^2 + 1)^3(10x^4 - 6x^2)}{4x^6} \qquad \text{Simplify expression within brackets}$$

$$= \frac{(x^2 + 1)^3(5x^2 - 3)}{2x^4}$$

Problem 25 Find dy/dx, given

(A) $y = 2x^2(x^2 + 2)^6$ (B) $y = \dfrac{3x^2}{(x^2 + 5)^4}$

Answers (A) $(x^2 + 2)^5(28x^3 + 8x)$ (B) $\dfrac{30x - 18x^3}{(x^2 + 5)^5}$

EXERCISE 2-7

A   Write each composite function in the form $y = f(u)$ and $u = g(x)$. For example, for $y = (x - 1)^4$, we would write $y = u^4$ and $u = x - 1$.

1. $y = (2x + 5)^3$ 2. $y = (3x - 7)^5$
3. $y = (x^3 - x^2)^8$ 4. $y = (2x^2 - 3x + 1)^4$
5. $y = (x^3 + 3x)^{1/3}$ 6. $y = (x^2 - 6)^{3/2}$

Find dy/dx using the chain rule, given:

7. $y = (2x + 5)^3$ 8. $y = (3x - 7)^5$
9. $y = (x^3 - x^2)^8$ 10. $y = (2x^2 - 3x + 1)^4$
11. $y = (x^3 + 3x)^{1/3}$ 12. $y = (x^2 - 6)^{3/2}$

B   Find dy/dx using the chain rule, given:

13. $y = 3(x^2 - 2)^4$ 14. $y = 2(x^3 + 6)^5$
15. $y = 2(x^2 + 3x)^{-3}$ 16. $y = 3(x^3 + x^2)^{-2}$
17. $y = \sqrt{x^2 + 6}$ 18. $y = \sqrt[3]{3x - 7}$
19. $y = \sqrt[3]{3x + 4}$ 20. $y = \sqrt{2x - 5}$
21. $y = (x^2 - 4x + 2)^{1/2}$ 22. $y = (2x^2 + 2x - 3)^{1/2}$

23. $y = \dfrac{1}{2x + 4}$ 24. $y = \dfrac{1}{(x^2 - 3)^8}$

$\left[\text{Hint:}\quad y = \dfrac{1}{2x + 4} = (2x + 4)^{-1} \right]$

25. $y = \dfrac{1}{4x^2 - 4x + 1}$ 26. $y = \dfrac{1}{2x^2 - 3x + 1}$

Find each of the following and simplify:

27. $D_x[3x(x^2 + 1)^3]$ 28. $D_x[2x^2(x^3 - 3)^4]$
29. $D_x\dfrac{(x^3 - 7)^4}{2x^3}$ 30. $D_x\dfrac{3x^2}{(x^2 + 5)^3}$

C   Find each of the following and simplify:

31. $D_x[(2x - 3)^2(2x^2 + 1)^3]$ 32. $D_x[(x^2 - 1)^3(x^2 - 2)^2]$

33. $D_x[4x^2\sqrt{x^2-1}]$ **34.** $D_x[3x\sqrt{2x^2+3}]$

35. $D_x \dfrac{2x}{\sqrt{x-3}}$ **36.** $D_x \dfrac{x^2}{\sqrt{x^2+1}}$

37. Find the equation of the tangent line to the graph of

$$y = \frac{4}{2x^2-3x+3} = 4(2x^2-3x+3)^{-1}$$

at (1, 2), using the chain rule to find the slope. Write the anwer in the form $y = mx + b$.

38. Find the equation of the tangent line to the graph of

$$y = \frac{6}{\sqrt{x^2-3x}} = 6(x^2-3x)^{-1/2}$$

at (4, 3), using the chain rule to find the slope. Write the answer in the form $Ax + By = C$, with A, B, and C integers and $A > 0$.

APPLICATIONS

BUSINESS & ECONOMICS

39. *Marginal average cost.* A manufacturer of skis finds that the average cost $\overline{C}(x)$ per pair of skis at an output level of x thousand skis is

$$\overline{C}(x) = (2x - 8)^2 + 25$$

(A) Find the marginal average cost $\overline{C}'(x)$ using the chain rule.
(B) Find $\overline{C}'(2)$, $\overline{C}'(4)$, and $\overline{C}'(6)$.

LIFE SCIENCES

40. *Pollution.* A small lake in a resort area became contaminated with a harmful bacteria because of excessive septic tank seepage. After treating the lake with a bactericide, the Department of Public Health esimated the bacteria concentration (number per cubic centimeter) after t days to be given by

$$C(t) = 500(8-t)^2 \qquad 0 \le t \le 7$$

(A) Find $C'(t)$ using the chain rule.
(B) Find $C'(1)$ and $C'(6)$, and interpret.

SOCIAL SCIENCES

41. *Learning.* In 1930, L. L. Thurstone developed the following formula to indicate how learning time T depends on the length of a list n:

$$T = f(n) = \frac{c}{k} n \sqrt{n-a}$$

where a, c, and k are empirical constants. Suppose for a particular person, time T in minutes for learning a list of length n is

$$T = f(n) = 2n\sqrt{n-2}$$

(A) Find dT/dn, the rate of change in time with respect to n.
(B) Find $f'(11)$ and $f'(27)$, and interpret.

2-8   MARGINAL ANALYSIS IN BUSINESS AND ECONOMICS

One important use of calculus in business and economics is in marginal analysis. We introduced the concept of marginal cost earlier. There is no reason to stop there. Economists also talk about **marginal revenue** and **marginal profit.** Recall that the word *marginal* refers to a rate of change, that is, a derivative. Thus, we define the following:

MARGINAL COST, REVENUE, AND PROFIT

If x is the number of units of product produced in some time interval, then

$$\text{Total cost} = C(x)$$
$$\text{Marginal cost} = C'(x)$$
$$\text{Total revenue} = R(x)$$
$$\text{Marginal revenue} = R'(x)$$
$$\text{Total profit} = P(x) = R(x) - C(x)$$
$$\text{Marginal profit} = P'(x) = R'(x) - C'(x)$$
$$= (\text{Marginal revenue}) - (\text{Marginal cost})$$

In words, the marginal cost is the change in cost per unit change in production at a given output level; the marginal revenue is the change in revenue per unit change in production at a given output level; and the marginal profit is the change in profit per unit change in production at a given output level. Or, stated more simply, **the marginal cost, revenue, and profit represent the change in cost, revenue, and profit, respectively, that result from a unit increase in production.**

We now present an example in market research to show how all these ideas are tied together.

Production strategy. The market research department of a company recommends that the company manufacture and market a new transistor radio. After extensive surveys, the research department presents the following **demand equation:**

$$x = 10,000 - 1,000p \qquad \text{x is demand at \$p per radio} \qquad (1)$$

or

$$p = 10 - \frac{x}{1,000} \qquad (2)$$

where x is the number of radios retailers are likely to buy per week at \$p per radio. Equation (2) is simply equation (1) solved for p in terms of x. Notice that as price goes up, demand goes down.

The financial department provides the following **cost equation:**

$$C(x) = 5{,}000 + 2x \qquad\qquad (3)$$

where \$5,000 is the estimated fixed costs (tooling and overhead) and \$2 is the estimated variable costs (cost per unit for materials, labor, marketing, transportation, storage, etc.).

The **marginal cost** is

$$C'(x) = 2$$

which means it costs an additional \$2 to produce one more radio at all production levels.

The **revenue equation** [the amount of money $R(x)$ received by the company for manufacturing and selling x units at \$$p$ per unit] is

$$R(x) = (\text{Number of units sold})(\text{Price per unit})$$

$$= xp$$

$$= x\left(10 - \frac{x}{1{,}000}\right) \qquad \text{Using equation (2) above} \qquad (4)$$

$$= 10x - \frac{x^2}{1{,}000}$$

The **marginal revenue** is

$$R'(x) = 10 - \frac{x}{500}$$

For production levels of $x = 2{,}000, 5{,}000$ and $7{,}000$, we have

$$R'(2{,}000) = 6 \qquad R'(5{,}000) = 0 \qquad R'(7{,}000) = -4$$

This means that at production levels of 2,000, 5,000, and 7,000, the respective change in revenue per unit change in production is \$6, \$0, and $-\$4$. That is, at the 2,000 output level revenue increases as production increases; at the 5,000 output level revenue does not change with a "small" change in production; and at the 7,000 output level revenue decreases with an increase in production. Figure 10 illustrates these results.

Finally, the **profit equation** is

$$P(x) = R(x) - C(x)$$

$$= \left(10x - \frac{x^2}{1{,}000}\right) - (5{,}000 + 2x)$$

$$= -\frac{x^2}{1{,}000} + 8x - 5{,}000$$

The **marginal profit** is

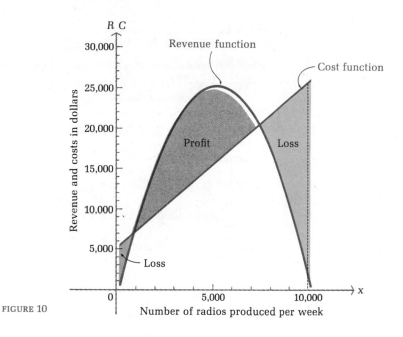

FIGURE 10

$$P'(x) = -\frac{x}{500} + 8$$

For production levels of 1,000, 4,000, and 6,000, we have

$$P'(1,000) = 6 \qquad P'(4,000) = 0 \qquad P'(6,000) = -4$$

This means that at production levels of 1,000, 4,000, and 6,000, the respective changes in profit per unit change in production are $6, $0, and −$4. That is, at the 1,000 output level profit will be increased if production is increased; at the 4,000 output level profit does not change for "small" changes in production; at the 6,000 output level profits will decrease if production is increased. It seems the best production level to produce a maximum profit is 4,000. **In the next chapter we will develop a systematic procedure for finding the production level [and, using the demand equation (2), the selling price] that will maximize profit.**

The above example warrants careful study, since a number of important ideas in economics and calculus are involved.

Sometimes it is desirable to carry out marginal analysis relative to **average cost (cost per unit), average revenue (revenue per unit), and average profit (profit per unit).** The relevant definitions are summarized in the following box:

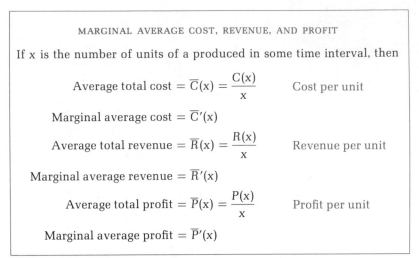

MARGINAL AVERAGE COST, REVENUE, AND PROFIT

If x is the number of units of a produced in some time interval, then

$$\text{Average total cost} = \overline{C}(x) = \frac{C(x)}{x} \qquad \text{Cost per unit}$$

$$\text{Marginal average cost} = \overline{C}'(x)$$

$$\text{Average total revenue} = \overline{R}(x) = \frac{R(x)}{x} \qquad \text{Revenue per unit}$$

$$\text{Marginal average revenue} = \overline{R}'(x)$$

$$\text{Average total profit} = \overline{P}(x) = \frac{P(x)}{x} \qquad \text{Profit per unit}$$

$$\text{Marginal average profit} = \overline{P}'(x)$$

In the above example,

$$\overline{C}(x) = \frac{C(x)}{x} = \frac{5{,}000 + 2x}{x} = \frac{5{,}000}{x} + 2 \qquad \begin{array}{l}\text{As production goes up,}\\ \text{cost per unit goes down}\end{array}$$

$$\overline{C}'(x) = -5{,}000x^{-2} = \frac{-5{,}000}{x^2}$$

$$\overline{C}'(100) = \frac{-5{,}000}{(100)^2} = -\$0.50 \qquad \begin{array}{l}\text{Cost is decreasing at a}\\ \text{rate of 50¢ per unit at a}\\ \text{production level of 100}\\ \text{units per week}\end{array}$$

$$\overline{C}'(1{,}000) = \frac{-5{,}000}{(1{,}000)^2} = \$0.005 \qquad \begin{array}{l}\text{Cost is decreasing at a}\\ \text{rate of 0.5¢ per unit at}\\ \text{a production level of}\\ \text{1,000 units per week}\end{array}$$

Similar interpretations are given to $\overline{R}(x)$ and $\overline{R}'(x)$, and to $\overline{P}(x)$ and $\overline{P}'(x)$.

EXERCISE 2-8   **APPLICATIONS**

BUSINESS & ECONOMICS

1. In the production strategy problem discussed in this section, suppose we have the demand equation

$$x = 6{,}000 - 30p \qquad \text{or} \qquad p = 200 - \frac{x}{30}$$

and the cost equation

$$C(x) = 72{,}000 + 60x$$

(A)   Find the marginal cost.
(B)   Find the revenue equation in terms of x.
(C)   Find the marginal revenue.
(D)   Find $R'(1,500)$ and $R'(4,500)$, and interpret.
(E)   Graph the cost function and the revenue function on the same coordinate system. Indicate regions of loss and profit. Use $0 \leq x \leq 6,000$.
(F)   Find the profit equation in terms of x.
(G)   Find the marginal profit.
(H)   Find $P'(1,500)$ and $P'(3,000)$, and interpret.

2. In the production strategy problem discussed in this section, suppose we have the demand equation

$$x = 9,000 - 30p \qquad \text{or} \qquad p = 300 - \frac{x}{30}$$

and the cost equation

$$C(x) = 90,000 + 30x$$

(A)   Find the marginal cost.
(B)   Find the revenue equation in terms of x.
(C)   Find the marginal revenue.
(D)   Find $R'(3,000)$ and $R'(6,000)$, and interpret.
(E)   Graph the cost function and the revenue function on the same coordinate system for $0 \leq x \leq 9,000$. Indicate regions of loss and profit.
(F)   Find the profit equation in terms of x.
(G)   Find the marginal profit.
(H)   Find $P'(1,500)$ and $P'(4,500)$, and interpret.

3. Referring to Problem 1, find:
(A) $\overline{C}(x)$, $\overline{R}(x)$, and $\overline{P}(x)$
(B) $\overline{C}'(x)$, $\overline{R}'(x)$, and $\overline{P}'(x)$
(C) $\overline{P}'(1,000)$ and $\overline{P}'(6,000)$, and interpret

4. Referring to Problem 2, find:
(A) $\overline{C}(x)$, $\overline{R}(x)$, and $\overline{P}(x)$
(B) $\overline{C}'(x)$, $\overline{R}'(x)$, and $\overline{P}'(x)$
(C) $\overline{P}'(1,000)$ and $\overline{P}'(2,000)$, and interpret

2-9   CHAPTER REVIEW

IMPORTANT TERMS AND SYMBOLS

2-2 Limits and continuity.   continuous at a point,   discontinuous at a point,   continuous over an interval,   limit, $\lim_{x \to c} f(x) = L$

2-3 Increments, tangent lines, and rates of change.   increment,   increment in x,   increment in y,   secant line,   tangent line,   secant line slope, tangent line slope, slope of the graph of $y = f(x)$ at $(x_1, f(x_1))$, equation of a tangent line to the graph of $y = f(x)$ at $(x_1, f(x_1))$, average velocity, instantaneous velocity, average rate of change, instantaneous rate of change, $\Delta x = x_2 - x_1$, $\Delta y = f(x_2) - f(x_1)$,

$$\frac{\Delta y}{\Delta x} = \frac{f(x_1 + \Delta x) - f(x_1)}{\Delta x}, \quad \lim_{\Delta x \to 0} \frac{\Delta y}{\Delta x} = \lim_{\Delta x \to 0} \frac{f(x_1 + \Delta x) - f(x_1)}{\Delta x}$$

2-4 Definition of the derivative. derivative of f at x, slope function, instantaneous velocity function, instantaneous rate function, marginal cost, $f'(x) = \lim_{\Delta x \to 0} \dfrac{f(x + \Delta x) - f(x)}{\Delta x}$, $m = f'(x)$, $v = f'(x)$, $C'(x)$

2-5 Derivatives of constants, power forms, sums, and differences. derivative of a constant, derivative of a power form, derivative of a constant times a function, derivative of a sum and difference, $f'(x)$,

y', $\dfrac{dy}{dx}$, $D_x f(x)$, $D_x C = 0$, $D_x x^n = nx^{n-1}$, $D_x kf(x) = kD_x f(x)$,

$D_x[u(x) \pm v(x)] = D_x u(x) \pm D_x v(x)$

2-6 Derivatives of products and quotients. derivative of a product, derivative of a quotient, $D_x[u(x)v(x)] = u(x)D_x v(x) + v(x)D_x u(x)$,

$$D_x \frac{u(x)}{v(x)} = \frac{v(x)D_x u(x) - u(x)D_x v(x)}{[v(x)]^2}$$

2-7 Derivatives of composite functions; chain rule. composite functions, chain rule, $\dfrac{dy}{dx} = \dfrac{dy}{du}\dfrac{du}{dx}$

2-8 Marginal analysis in business and economics. demand equation, cost equation, marginal cost, revenue equation, marginal revenue, profit equation, marginal profit, average cost, marginal average cost, average revenue, marginal average revenue, average profit, marginal average profit, $C'(x)$, $\overline{C}'(x)$, $R'(x)$, $\overline{R}'(x)$, $P'(x)$, $\overline{P}'(x)$

EXERCISE 2-9   CHAPTER REVIEW

Work through all the problems in this chapter review and check your answers in the back of the book. (Answers to all review problems are there.) Where weaknesses show up, review appropriate sections in the text. When you are satisfied that you know the material, take the practice test following this review.

A   In Problems 1– 10 find $f'(x)$ for $f(x)$ as given.

1. $f(x) = 3x^4 - 2x^2 + 1$ 2. $f(x) = 2x^{1/2} - 3x$

3. $f(x) = 5$ 4. $f(x) = \frac{2}{3}$

5. $f(x) = (2x - 1)(3x + 2)$ 6. $f(x) = (x^2 - 1)(x^3 - 3)$

7. $f(x) = \dfrac{2x}{x^2 + 2}$ 8. $f(x) = \dfrac{1}{3x + 2}$

9. $f(x) = (2x - 3)^3$ 10. $f(x) = (x^2 + 2)^{-2}$

B   In Problems 11–18 find the indicated derivatives.

11. $\dfrac{dy}{dx}$ for $y = 3x^4 - 2x^{-3} + 5$

12. y' for $y = (2x^2 - 3x + 2)(x^2 + 2x - 1)$

13. $f'(x)$ for $f(x) = \dfrac{2x - 3}{(x - 1)^2}$

14. y' for $y = 2\sqrt{x} + \dfrac{4}{\sqrt{x}}$

15. $D_x[(x^2 - 1)(2x + 1)^2]$

16. $D_x\sqrt[3]{x^3 - 5}$

17. $\dfrac{dy}{dx}$ for $y = \dfrac{1}{\sqrt[3]{3x^2 - 2}}$

18. $D_x\dfrac{(x^2 + 2)^4}{2x - 3}$

19. For $y = f(x) = x^2 + 4$, find:
    (A) The slope of the graph at $x = 1$
    (B) The equation of the tangent line at $x = 1$ in the form $y = mx + b$

20. For $y = f(x) = 10x - x^2$, find:
    (A) The slope function
    (B) The value(s) of x such that the slope is zero

21. If an object moves along the y axis (scale in feet) so that it is at $y = f(x) = 16x^2 - 4x$ at time x (in seconds), find:
    (A) The instantaneous velocity function
    (B) The velocity at time $x = 3$ seconds

22. An object moves along the y axis (scale in feet) so that at time x (in seconds) it is at $y = f(x) = 96x - 16x^2$. Find:
    (A) The instantaneous velocity function
    (B) The time(s) when the velocity is zero

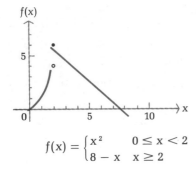

$$f(x) = \begin{cases} x^2 & 0 \le x < 2 \\ 8 - x & x \ge 2 \end{cases}$$

Problems 23 and 24 refer to the function f described at the left:

23.   (A) $f(5) = ?$ (B) $\lim_{x \to 5} f(x) = ?$
       (C)   Is f continuous at $x = 5$?

24.   (A) $f(2) = ?$ (B) $\lim_{x \to 2} f(x) = ?$
       (C)   Is f continuous at $x = 2$?

In Problems 25–28 find points of discontinuity, if any exist.

25. $f(x) = 2x^2 - 3x + 1$ 26. $f(x) = \dfrac{1}{x + 5}$

27. $f(x) = \dfrac{x + 7}{(x - 3)(x + 2)}$ 28. $\sqrt{x - 3} \quad x \ge 3$

In Problems 29–32 find each limit, if it exists.

29. $\lim_{x \to 3} \dfrac{2x - 3}{x + 5}$ 30. $\lim_{x \to 3} (2x^2 - x + 1)$

31. $\lim_{x \to 0} \dfrac{2x}{3x^2 - 2x}$ 32. $\lim_{\Delta x \to 0} \dfrac{f(2 + \Delta x) - f(2)}{\Delta x}$
                                   for $f(x) = x^2 + 4$

33.   Use the definition of a derivative to find $f'(x)$ for $f(x) = x^2 - 2$.

C   Problems 34 and 35 refer to the function specified by

$$f(x) = \frac{2x^2 - 3x - 2}{3x^2 - 4x - 4}$$

34.   For what values of x is f discontinuous?
35. $\lim_{x \to 2} f(x) = ?$

APPLICATIONS

BUSINESS & ECONOMICS

36.   *Marginal average cost.*   Suppose a firm manufactures items having an average cost per item in hundreds of dollars given by

$$\overline{C}(x) = x^2 - 10x + 30$$

where x is the number of items manufactured.
(A)   Find the marginal average cost $\overline{C}'(x)$.
(B)   Find the marginal average cost at $x = 3, 5,$ and 7, and interpret.

LIFE SCIENCES

37.   *Pollution.*   A sewage treatment plant disposes of its effluent through a pipeline that extends 1 mile toward the center of a large lake. The concentration of effluent $C(x)$, in parts per million, x meters from the end of the pipe is given approximately by

$$C(x) = 500(x + 1)^{-2}$$

What is the instantaneous rate of change of concentration at 9 meters? At 99 meters?

SOCIAL SCIENCES   **38.** *Learning.*   If a person learns N items in t hours, as given by

$$N(t) = 20\sqrt{t}$$

find the rate of learning after:
(A)   1 hour   (B)   4 hours

PRACTICE TEST: **CHAPTER 2**

In Problems 1–4 find $f'(x)$ for $f(x)$ as given.

1. $f(x) = 3x^2 - 2x^{1/2} - 3$ 2. $f(x) = (x^2 + 2)(2x - 3)$

3. $f(x) = \dfrac{3x^2 - 5}{x^3 + 1}$ 4. $f(x) = (2x^3 - 3x + 1)^3$

In Problems 5 and 6 find the indicated derivatives.

5. $D_x[(x^2 - 1)^3(2x + 1)]$ 6. $\dfrac{dy}{dx}$ for $y = \dfrac{1}{\sqrt[4]{2x^2 - 3}}$

7.   Given $y - f(x) - 8x - x^2$. Find:
   (A)   The slope function
   (B)   The slope at $x = 2$
   (C)   The equation of the tangent line at $x = 2$ in the form of $y = mx + b$
   (D)   The value(s) of x that produces a zero slope

8.   An object moves along the y axis (scale in feet) so that its position at time x (in seconds) is given by $y = f(x) = 20 + 80x - 10x^2$. Find:
   (A)   The instantaneous velocity function
   (B)   The velocity at $x = 3$ seconds
   (C)   The time(s) when the velocity is zero

9.   Find

$$\lim_{\Delta x \to 0} \frac{f(x + \Delta x) - f(x)}{\Delta x}$$

   for $f(x) = 3 - x$.

10.   Find $\lim_{x \to 2} f(x)$ if:

   (A) $f(x) = 3x^2 - 2x + 1$ (B) $f(x) = \dfrac{x - 3}{x + 3}$

11. Given

$$f(x) = \frac{4x}{x^2 - 2x}$$

(A) Where is f discontinuous?

(B) $\lim\limits_{x \to 0} f(x) = ?$

12. Find the marginal average profit at a production level of x, given the average profit equation

$$\overline{P}(x) = 400 - \frac{x}{20} - \frac{4,000}{x}$$

CHAPTER 3
ADDITIONAL TOPICS ON THE DERIVATIVE

CONTENTS

3 ADDITIONAL TOPICS ON THE DERIVATIVE

IMPLICIT DIFFERENTIATION

Consider the equation

$$3x^2 + y - 2 = 0 \qquad (1)$$

and the equation obtained by solving (1) for y in terms of x,

$$y = 2 - 3x^2 \qquad (2)$$

Both equations define the same function using x as the independent variable and y as the dependent variable. For (2) we can write

$$y = f(x)$$

where

$$f(x) = 2 - 3x^2 \qquad (3)$$

and we have an **explicit** (clearly stated) rule that enables us to determine y for each value of x. On the other hand, the y in equation (1) is the same y as in equation (2), and equation (1) **implicitly** gives (implies though not plainly expresses) y as a function of x. Thus, we say that equations (2) and (3) define the function f explicitly and equation (1) defines f implicitly.

There are cases where equations of the form

$$F(x, y) = 0 \qquad (4)$$

are either difficult or impossible to solve for y explicitly in terms of x (try it for $x^2y^5 - 3xy + 5 = 0$), yet it can be shown that under fairly general conditions on F, equation (4) will define one or more functions where y is a dependent variable and x is an independent variable. Suppose we have an equation of the form shown in (4). How can we find y' without solving for y in terms of x? The answer is: We **differentiate implicitly.** We illustrate the process with equation (1) first.

Starting with

$$3x^2 + y - 2 = 0$$

we think of y as a function of x, that is, $y = y(x)$, and write

$$3x^2 + y(x) - 2 = 0$$

and differentiate both sides with respect to x:

$$D_x[3x^2 + y(x) - 2] = D_x 0$$
$$D_x 3x^2 + D_x y(x) - D_x 2 = 0$$
$$6x + y' - 0 = 0$$

Since y is a function of x, but is not explicitly given, we simply write $D_x y(x) = y'$ to indicate its derivative

Now we solve for y':

$$y' = -6x$$

Note that we get the same result if we start with equation (2) and differentiate directly.

Before we consider a more involved example, let us observe that if $y = y(x)$, then, using the chain rule,

$$D_x[y(x)]^n = n[y(x)]^{n-1}y'(x)$$

Or, stated more concisely,

If $y = y(x)$, then

$$\frac{dy^n}{dx} = ny^{n-1}\frac{dy}{dx} \qquad \text{or} \qquad D_x y^n = ny^{n-1}y'$$

Example 1   Find y' for $y = y(x)$ defined implicitly by

$$y^3 - 2y^2 + x^4 - 16 = 0 \tag{5}$$

Solution   We differentiate both sides of (5) with respect to x and solve for y':

$$D_x(y^3 - 2y^2 + x^4 - 16) = D_x(0) \qquad \text{Think } y = y(x)$$

$$D_x y^3 - D_x 2y^2 + D_x x^4 - D_x 16 = 0$$

$$3y^2 y' - 4yy' + 4x^3 = 0 \qquad \text{Now solve for } y'$$

$$3y^2 y' - 4yy' = -4x^2 \qquad \text{by getting all terms}$$

$$(3y^2 - 4y)y' = -4x^3 \qquad \begin{array}{l} \text{involving } y' \text{ on} \\ \text{one side} \end{array}$$

$$y' = \frac{-4x^3}{3y^2 - 4y}$$

$$= \frac{4x^3}{4y - 3y^2} \qquad \begin{array}{l} \text{Leave answer in} \\ \text{terms of } x \text{ and } y \end{array}$$

We have found y' without solving (5) for y in terms of x. The fact that y' is given in terms of both x and y is no great disadvantage. We have only to make certain that when we want **to evaluate y' for a particular value of x and y, say (x_0, y_0), the ordered pair must satisfy the original equation.** For example, at the point $(2, 2)$, which satisfies equation (5),

$$y'|_{(2, 2)} = \frac{-4(2^3)}{3(2^2) - 4(2)} = \frac{-32}{4} = -8$$

The symbol

$$y'|_{(a, b)}$$

is used to indicate that we are evaluating y' at $x = a$ and $y = b$. The number -8 is the slope of the graph of equation (5) at $(2, 2)$.

Problem 1   Find y' for $y = y(x)$ defined implicitly by

$$y^4 + 3y^2 - x^3 + 4 = 0$$

and evaluate y' at $(2, 1)$.

Answer $y' = \dfrac{3x^2}{4y^3 + 6y} \qquad y'|_{(2, 1)} = \frac{6}{5}$

Example 2   Find y' for $y = y(x)$ defined implicitly by

$$y^2 + 3x^2y - 10 = 0$$

and evaluate y' at $(1, 2)$.

Solution   Differentiate both sides with respect to x, using the product rule on the middle term, $3x^2y$, since it is the product of two functions, $3x^2$ and $y(x)$:

$$D_x y^2 + D_x 3x^2 y - D_x 10 = D_x 0$$

$$2yy' + \underbrace{3x^2 y' + y6x}_{\text{Result of product rule}} = 0$$

$$(2y + 3x^2)y' = -6xy$$

$$y' = \frac{-6xy}{2y + 3x^2}$$

$$y'|_{(1,2)} = \frac{-6(1)(2)}{2(2) + 3(1)^2} = \frac{-12}{7}$$

Problem 2   Find y' for $y = y(x)$ defined implicitly by

$$2y + 2x^3y^2 - 24 = 0$$

and evaluate y' at $(1, 3)$.

Answer $y' = \dfrac{-6x^2y^2}{2 + 4x^3y}$ $y'|_{(1,3)} = -27/7$

Example 3   Find the equation(s) of the tangent line(s) to the graph of

$$y - xy^2 + x^2 + 1 = 0 \qquad (6)$$

at the point(s) where $x = 1$.

Solution   We first find y when $x = 1$:

$$y - xy^2 + x^2 + 1 = 0$$
$$y - (1)y^2 + (1)^2 + 1 = 0$$
$$y - y^2 + 2 = 0$$
$$y^2 - y - 2 = 0$$
$$(y - 2)(y + 1) = 0$$
$$y = -1, 2$$

Thus, there are two points on the graph of (6) where $x = 1$; namely,

$$(1, -1) \quad \text{and} \quad (1, 2)$$

We next find the slope of the graph at these two points by differentiating (6) implicitly:

$$y - xy^2 + x^2 + 1 = 0$$
$$D_x y - D_x xy^2 + D_x x^2 + D_x 1 = D_x 0$$
$$y' - (x2yy' + y^2) + 2x = 0$$
$$y' - 2xyy' - y^2 + 2x = 0 \qquad \text{Solve for } y'$$
$$y' - 2xyy' = y^2 - 2x$$
$$(1 - 2xy)y' = y^2 - 2x$$
$$y' = \frac{y^2 - 2x}{1 - 2xy}$$

Now find the slope at each point:

$$y'|_{(1,-1)} = \frac{(-1)^2 - 2(1)}{1 - 2(1)(-1)} = \frac{1-2}{1+2} = \frac{-1}{3} = -\frac{1}{3}$$

$$y'|_{(1,2)} = \frac{2^2 - 2(1)}{1 - 2(1)(2)} = \frac{4-2}{1-4} = \frac{2}{-3} = -\frac{2}{3}$$

Equation of the tangent line at $(1, -1)$:

$$y - y_1 = m(x - x_1)$$
$$y + 1 = -\tfrac{1}{3}(x - 1)$$
$$y + 1 = -\tfrac{1}{3}x + \tfrac{1}{3}$$
$$y = -\tfrac{1}{3}x - \tfrac{2}{3}$$

Equation of the tangent line at $(1, 2)$:

$$y - y_1 = m(x - x_1)$$
$$y - 2 = -\tfrac{2}{3}(x - 1)$$
$$y - 2 = -\tfrac{2}{3}x + \tfrac{2}{3}$$
$$y = -\tfrac{2}{3}x + \tfrac{8}{3}$$

Problem 3 Repeat Example 3 for $x^2 + y^2 - xy - 7 = 0$ at $x = 1$.

Answer $y = \tfrac{4}{5}x - \tfrac{14}{5}, \quad y = \tfrac{1}{5}x + \tfrac{14}{5}$

EXERCISE 3-1

In Problems 1–14 find y' without solving for y in terms of x (use implicit differentiation). Evaluate y' at the indicated point.

A
1. $y - 3x^2 + 5 = 0, \quad (1, -2)$
2. $3x^4 + y - 2 = 0, \quad (1, -1)$
3. $y^2 - 3x^2 + 8 = 0, \quad (2, 2)$
4. $3y^2 + 2x^3 - 14 = 0, \quad (1, 2)$
5. $y^2 + y - x = 0, \quad (2, 1)$
6. $2y^3 + y^2 - x = 0, \quad (-1, 1)$

B
7. $xy - 6 = 0, \quad (2, 3)$
8. $3xy - 2x - 2 = 0, \quad (2, 1)$
9. $2xy + y + 2 = 0, \quad (-1, 2)$
10. $2y + xy - 1 = 0, \quad (-1, 1)$
11. $x^2y - 3x^2 - 4 = 0, \quad (2, 4)$
12. $2x^3y - x^3 + 5 = 0, \quad (-1, 3)$
13. $y^2 - x^2y + x^3 + 11 = 0, \quad (-2, 1)$
14. $y^3 - xy^2 - 4 = 0, \quad (-3, -2)$

C
Find the equation(s) of the tangent line(s) to the graphs of the indicated equations at the point(s) with abscissas as indicated.

15. $xy - x - 4 = 0, \quad x = 2$
16. $3x + xy + 1 = 0, \quad x = -1$
17. $y^2 - xy - 6 = 0, \quad x = 1$
18. $xy^2 - y - 2 = 0, \quad x = 1$

19. The graph of $x^2 + y^2 = 25$ is a circle of radius 5 with center at the origin.
   (A) Find y' using implicit differentiation.
   (B) Find the equations of the tangent lines at $(3, 4)$ and $(3, -4)$. Write the equations in the form $Ax + By = C$.
   (C) Sketch the tangent lines on the circle.

20. The graph of $xy - y = 1$ looks like this:

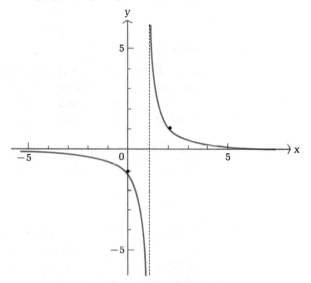

   (A) Find y' using implicit differentiation.
   (B) Find the equations of the tangent lines at $(0, -1)$ and $(2, 1)$. Write the equations in the form $y = mx + b$.
   (C) Sketch the tangent lines on the graph.

Find y' and the slope of the tangent line to the graph of each equation at the indicated point.

21. $(1 + y)^3 + y = x + 7$, $(2, 1)$
22. $(y - 3)^4 - x = y$, $(-3, 4)$
23. $(x - 2y)^3 = 2y^2 - 3$, $(1, 1)$
24. $(2x - y)^4 - y^3 = 8$, $(-1, -2)$

APPLICATIONS

BUSINESS & ECONOMICS   *For the demand equations in Problems 25–28, find the rate of change of p with respect to x by differentiating implicitly (x is the number of items that can be sold at a price of $\$p$).*

25. $x = p^2 - 2p + 1,000$ 26. $x = p^3 - 3p^2 + 200$
27. $x = \sqrt{10,000 - p^2}$ 28. $x = \sqrt[3]{1,500 - p^3}$

LIFE SCIENCES

29. *Biophysics.* In biophysics, the equation

$$(L + m)(V + n) = k$$

is called the *fundamental equation of muscle contraction*, where m, n, and k are constants, and V is the velocity of the shortening of muscle fibers for a muscle subjected to a load of L. Find dL/dV using implicit differentiation.

3-2   **RELATED RATES**

In applications we often encounter two (or more) variables that are differentiable functions of time, say $x = x(t)$ and $y = y(t)$, but $x = x(t)$ and $y = y(t)$ may not be explicitly given. In addition, x and y may be related by some equation such as

$$x^2 + y^2 = 25 \tag{1}$$

Differentiating both sides of (1) with respect to t, we obtain

$$2x \frac{dx}{dt} + 2y \frac{dy}{dt} = 0 \tag{2}$$

The derivatives dx/dt and dy/dt are related by equation (2); hence, they are referred to as **related rates.** If one of the rates and the value of one variable are both known, we can use equation (1) to find the value of the other variable and then we can use equation (2) to find the other rate. The following examples will illustrate how related rates can be used to solve certain types of practical problems.

Example 4   Suppose two motor boats leave from the same point at the same time. If one travels north at 15 miles per hour and the other travels east at 20 miles per hour, how fast will the distance between them be changing after 2 hours?

Solution   First, draw a picture.

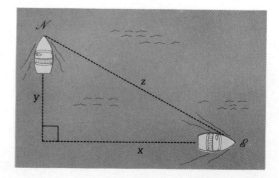

All variables, x, y, and z, are changing with time. Hence, they can be thought of as functions of time; $x = x(t)$, $y = y(t)$, and $z = z(t)$, given implicitly. It now makes sense to take derivatives of each variable with respect to time. From the Pythagorean theorem,

$$z^2 = x^2 + y^2 \tag{3}$$

We also know that

$$\frac{dx}{dt} = 20 \text{ miles per hour} \quad \text{and} \quad \frac{dy}{dt} = 15 \text{ miles per hour}$$

We would like to find dz/dt at the end of 2 hours; that is, when $x = 40$ miles and $y = 30$ miles. To do this we differentiate both sides of (3) with respect to t and solve for dz/dt:

$$2z\frac{dz}{dt} = 2x\frac{dx}{dt} + 2y\frac{dy}{dt} \tag{4}$$

We have everything we need except z. When $x = 40$ and $y = 30$, we find z from (3) to be 50. Substituting the known quantities into (4), we obtain

$$2(50)\frac{dz}{dt} = 2(40)(20) + 2(30)(15)$$

$$\frac{dz}{dt} = 25 \text{ miles per hour}$$

Problem 4   Repeat Example 4 for the situation at the end of 3 hours.

Answer $dz/dt = 25$ miles per hour

Example 5   Suppose that for a company manufacturing transistor radios, the cost, revenue, and profit equations are given by

$$C = 5{,}000 + 2x \qquad \text{Cost equation}$$

$$R = 10x - \frac{x^2}{1{,}000} \qquad \text{Revenue equation}$$

$$P = R - C \qquad \text{Profit equation}$$

where the production output in 1 week is x radios. If production is increasing at the rate of 500 radios per week when production is 2,000 radios, find the rate of increase in:
(A) Cost   (B) Revenue   (C) Profit

Solutions   If production x is a function of time (it must be since it is changing with respect to time), then C, R, and P must also be functions of time. They are implicitly (rather than explicitly) given. Letting t represent time in weeks, we differentiate both sides of each of the three equations above with respect to t, and then substitute $x = 2,000$ and $dx/dt = 500$ to find the desired rates.

(A) $C = 5,000 + 2x$ Think $C = C(t)$ and $x = x(t)$

$$\frac{dC}{dt} = \frac{d}{dt}(5,000) + \frac{d}{dt}(2x)$$ Differentiate both sides with respect to t

$$\frac{dC}{dt} = 0 + 2\frac{dx}{dt} = 2\frac{dx}{dt}$$

Since $dx/dt = 500$ when $x = 2,000$,

$$\frac{dC}{dt} = 2(500) = \$1,000 \text{ per week}$$

Cost is increasing at a rate of $1,000 per week.

(B) $R = 10x - \dfrac{x^2}{1,000}$

$$\frac{dR}{dt} = \frac{d}{dt}(10x) - \frac{d}{dt}\frac{x^2}{1,000}$$

$$\frac{dR}{dt} = 10\frac{dx}{dt} - \frac{x}{500}\frac{dx}{dt}$$

$$\frac{dR}{dt} = \left(10 - \frac{x}{500}\right)\frac{dx}{dt}$$

Since $dx/dt = 500$ when $x = 2,000$,

$$\frac{dR}{dt} = \left(10 - \frac{2,000}{500}\right)(500) = \$3,000 \text{ per week}$$

Revenue is increasing at a rate of $3,000 per week.

(C) $P = R - C$

$$\frac{dP}{dt} = \frac{dR}{dt} - \frac{dC}{dt}$$

$$= \$3,000 - \$1,000 \qquad \text{Results from parts A and B}$$

$$= \$2,000 \text{ per week}$$

Profit is increasing at a rate of $2,000 per week.

Problem 5   Repeat Example 5 for a production level of 6,000 radios per week.

Answers   (A) $dC/dt = \$1,000$ per week   (B) $dR/dt = -\$1,000$ per week
             (C) $dP/dt = -\$2,000$ per week

EXERCISE 3-2

A   *In Problems 1–6 assume $x = x(t)$ and $y = y(t)$. Find the indicated rate, given the other information.*

1. $y = 2x^2 - 1$, $dy/dt = ?$, $dx/dt = 2$ when $x = 30$
2. $y = 2x^{1/2} + 3$, $dy/dt = ?$, $dx/dt = 8$ when $x = 4$
3. $x^2 + y^2 = 25$, $dy/dt = ?$, $dx/dt = -3$ when $x = 3$ and $y = 4$
4. $y^2 + x = 3$, $dx/dt = ?$, $dy/dt = -2$ when $x = 2$ and $y = 3$
5. $x^2 + xy - 2 = 0$, $dy/dt = ?$, $dx/dt = -1$ when $x = 2$ and $y = -3$
6. $y^2 + xy - 3x = 5$, $dx/dt = ?$, $dy/dt = -2$ when $x = 1$ and $y = 0$

B 7. A rock is thrown into a still pond and causes a circular ripple. If the radius of the ripple is increasing at 2 feet per second, how fast is the area changing when the radius is 10 feet? (Use $A = \pi R^2$, $\pi \approx 3.14$.)

8. In Problem 7 how fast is the circumference of a circular ripple changing when the radius is 10 feet? (Use $C = 2\pi R$, $\pi \approx 3.14$.)

9. The radius of a spherical balloon is increasing at the rate of 3 centimeters per minute. How fast is the volume changing when the radius is 10 centimeters? (Use $V = \frac{4}{3}\pi R^3$, $\pi \approx 3.14$.)

10. In Problem 9 how fast is the surface area of the sphere increasing? (Use $S = 4\pi R^2$, $\pi \approx 3.14$.)

11. A 10 foot ladder is placed against a vertical wall. Suppose the bottom slides away from the wall at a constant rate of 3 feet per second. How fast is the top sliding down the wall (negative rate) when the bottom is 6 feet from the wall? [*Hint:* Use the Pythagorean theorem: $a^2 + b^2 = c^2$, where c is the length of the hypotenuse of a right triangle and a and b are the lengths of the two shorter sides.]

12. A weather balloon is rising vertically at the rate of 5 meters per second. An observer is standing on the ground 300 meters from the point where the balloon was released. At what rate is the distance between the observer and the balloon changing when the balloon is 400 meters high?

C 13. A streetlight is on top of a 20 foot pole. A 5 foot tall person walks away from the pole at the rate of 5 feet per second. At what rate is the tip of the person's shadow moving away from the pole when he is 20 feet from the pole?

14. In Problem 13 at what rate is the person's shadow growing when he is 20 feet from the pole?

APPLICATIONS

BUSINESS & ECONOMICS
15. *Cost, revenue, and profit rates.*   Suppose that for a company manufacturing hand calculators, the cost, revenue, and profit equations are given by

$$C = 90,000 + 30x$$

$$R = 300x - \frac{x^2}{30}$$

$$P = R - C$$

where the production output in 1 week is x calculators. If production is increasing at a rate of 500 calculators per week when production output is 6,000 calculators, find the rate of increase (decrease) in:

(A) Cost   (B)   Revenue   (C)   Profit

16. *Cost, revenue, and profit rates.*   Repeat Problem 15 for

$$C = 72{,}000 + 60x$$

$$R = 200x - \frac{x^2}{30}$$

$$P = R - C$$

where production is increasing at a rate of 500 calculators per week at a production level of 1,500 calculators.

LIFE SCIENCES

17. *Pollution.*   An oil tanker aground on a reef is leaking oil that forms a circular oil slick about 0.1 foot thick. To estimate the rate (in cubic feet per minute, dV/dt) at which the oil is leaking from the tanker, it was found that the radius of the slick was increasing at 0.32 foot per minute ($dR/dt = 0.32$) when the radius was 500 feet ($R = 500$). Find dV/dt, using $\pi \approx 3.14$.

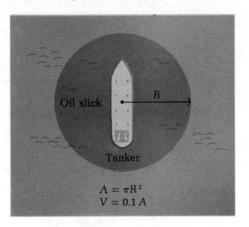

Oil slick   R
Tanker

$$A = \pi R^2$$
$$V = 0.1\,A$$

SOCIAL SCIENCES

18. *Learning.*   A person who is new on an assembly line performs an operation in T minutes after x performances of the operations, as given by

$$T = 6\left(1 + \frac{1}{\sqrt{x}}\right)$$

If

$$\frac{dx}{dt} = 6 \text{ operations per hour}$$

where t is time in hours, find dT/dt after thirty-six performances of the operation.

3-3   HIGHER-ORDER DERIVATIVES

HIGHER-ORDER DERIVATIVES FOR EXPLICITLY DEFINED FUNCTIONS

If we start with the function f defined by

$$f(x) = 3x^3 - 4x^2 - x + 1$$

and take the derivative, we obtain a new function f' defined by

$$f'(x) = 9x^2 - 8x - 1$$

Now if we take another derivative, called the **second derivative,** we obtain a new function f'' defined by

$$f''(x) = 18x - 8$$

And taking still another derivative produces the **third derivative** f''' defined by

$$f'''(x) = 18$$

and so on.

In general, the successive derivatives for a function f are denoted by

$$f', f'', f''', f^{(4)}, \ldots, f^{(n)}$$

Example 6   Find f', f'', and f''' for $f(x) = 3x^{-1} + x^2$.

Solution $f'(x) = -3x^{-2} + 2x$ $f''(x) = 6x^{-3} + 2$ $f'''(x) = -18x^{-4}$

Problem 6   Find f', f'', and f''' for $f(x) = 2 - 3x^2 + x^{-2}$.

Answer $f'(x) = -6x - 2x^{-3}$ $f''(x) = -6 + 6x^{-4}$ $f'''(x) = -24x^{-5}$

Along with the various other symbols for the first derivative that we considered earlier, we have corresponding symbols for higher-order derivatives. For example, if

$$y = f(x)$$

then

$$\frac{dy}{dx} = f'(x)$$

and the second derivative is given by

$$\frac{d}{dx}\left(\frac{dy}{dx}\right) = f''(x)$$

or, in short,

$$\frac{d^2y}{dx^2} = f''(x)$$ Note how the 2's are placed

Similarly,

$$\frac{d^3y}{dx^3} = f'''(x)$$

and so on. We summarize some of the more commonly used higher-derivative forms in the box.

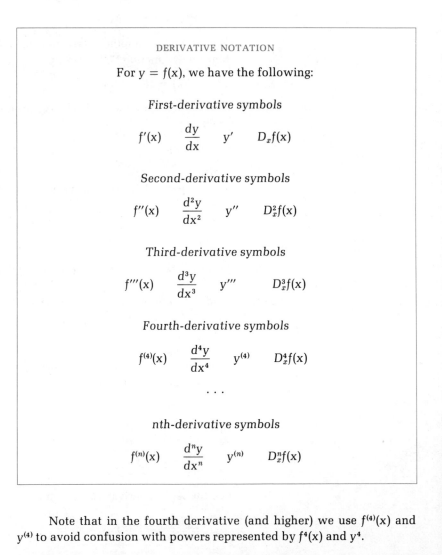

DERIVATIVE NOTATION

For $y = f(x)$, we have the following:

First-derivative symbols

$$f'(x) \qquad \frac{dy}{dx} \qquad y' \qquad D_x f(x)$$

Second-derivative symbols

$$f''(x) \qquad \frac{d^2y}{dx^2} \qquad y'' \qquad D_x^2 f(x)$$

Third-derivative symbols

$$f'''(x) \qquad \frac{d^3y}{dx^3} \qquad y''' \qquad D_x^3 f(x)$$

Fourth-derivative symbols

$$f^{(4)}(x) \qquad \frac{d^4y}{dx^4} \qquad y^{(4)} \qquad D_x^4 f(x)$$

$$\cdots$$

nth-derivative symbols

$$f^{(n)}(x) \qquad \frac{d^ny}{dx^n} \qquad y^{(n)} \qquad D_x^n f(x)$$

Note that in the fourth derivative (and higher) we use $f^{(4)}(x)$ and $y^{(4)}$ to avoid confusion with powers represented by $f^4(x)$ and y^4.

Example 7 If $y = 4x^{1/2}$, then

$$y' = 2x^{-1/2} \qquad \frac{dy}{dx} = 2x^{-1/2} \qquad D_x 4x^{1/2} = 2x^{-1/2}$$

$$y'' = -x^{-3/2} \qquad \frac{d^2y}{dx^2} = -x^{-3/2} \qquad D_x^2 4x^{1/2} = -x^{-3/2}$$

$$y''' = \tfrac{3}{2}x^{-5/2} \qquad \frac{d^3y}{dx^3} = \tfrac{3}{2}x^{-5/2} \qquad D_x^3 4x^{1/2} = \tfrac{3}{2}x^{-5/2}$$

Problem 7 If $y = 27x^{4/3}$, find:

(A) d^2y/dx^2 (B) $D_x^3 27x^{4/3}$ (C) $y^{(4)}$

Answers (A) $12x^{-2/3}$ (B) $-8x^{-5/3}$ (C) $\tfrac{40}{3}x^{-8/3}$

SECOND-ORDER DERIVATIVES FOR IMPLICITLY DEFINED FUNCTIONS

Suppose we have a function $y = y(x)$ defined implicitly by an equation of the form $F(x, y) = 0$. How can we find y'' without solving for y in terms of x? We will illustrate the process with an example.

Example 8 Find y'' for $y = y(x)$ defined implicitly by $x^2 + y^2 = 4$.

$$x^2 + y^2 = 4 \tag{1}$$

Solution Differentiate both sides with respect to x and solve for y'.

$$2x + 2y\, y' = 0$$
$$2y\, y' = -2x$$
$$y' = \frac{-x}{y} \tag{2}$$

Now differentiate both sides again with respect to x, thinking of $y = y(x)$, to obtain

$$y'' = \frac{y(-1) - (-x)\, y'}{y^2} \tag{3}$$

We are almost there! Substituting (2) into (3) we obtain

$$y'' = \frac{-y + x(-x/y)}{y^2}$$

$$= \frac{-y^2 - x^2}{y^3}$$

$$= \frac{-(x^2 + y^2)}{y^3}$$

Since $x^2 + y^2 = 4$ from our original equation, we obtain a further simplification:

$$y'' = \frac{-4}{y^3}$$

Problem 8   Find y'' for $y = y(x)$ defined implicitly by $3x^2 - y^2 = 9$.

Answer $y'' = -27/y^3$

In Sections 3-5 and 3-6 we will see how second derivatives provide a very useful tool in sketching graphs of equations and solving maxima–minima problems.

EXERCISE 3-3

Find the indicated derivative for each function.

A   1. $f''(x)$ for $f(x) = x^3 - 2x^2 - 1$
2. $g''(x)$ for $g(x) = x^4 - 3x^2 + 5$
3. $f'''(x)$ for $f(x) = 3x - 16x^2$
4. $g'''(x)$ for $g(x) = 1 - x - 2x^4$
5. d^2y/dx^2 for $y = 2x^5 - 3$
6. d^2y/dx^2 for $y = 3x^4 - 7x$
7. d^3y/dx^3 for $y = 120 - 30x^2$
8. d^3y/dx^3 for $y = 1 + 2x^2 - 4x^4$

9. $D_x^3(x^{-1})$ 10. $D_x^3(x^{-2})$
11. $D_x^2(1 - 2x + x^3)$ 12. $D_x^4(3x^2 - x^3)$

B   13. $D_x^2(3x^{-1} + 2x^{-2} + 5)$ 14. d^2y/dx^2 for $y = x^2 - \sqrt[3]{x}$
15. $y^{(4)}$ for $y = \sqrt{2x - 1}$ 16. $f^{(4)}(x)$ for $f(x) = 27\sqrt[3]{x^2}$
17. $D_x^2(1 - 2x)^3$ 18. $D_x^3(3 - x)^4$
19. y'' for $y = (x^2 - 1)^3$ 20. y'' for $y = (x^2 + 4)^4$

Use implicit differentiation to find y'' for each of the following:

21. $4x^2 - y^2 = 3$ 22. $2x^3 - 3y^2 = 4$
23. $y^3 + x^2 = 7$ 24. $3xy - x^2 = 2$

C   25. Find: $D_x^3 \dfrac{x}{2x - 1}$

26. Find y''' for $y = (2x - 1)(x^2 + 1)$.
27. Find y''' for $x^2 + y^2 = 4$.
28. Find y''' for $4x^2 - y^2 = 3$.

3-4 **DERIVATIVES AND GRAPHS**

Since the derivative is associated with the slope of the graph of a function at a point, we might expect that it is also associated with other properties of a graph. As we will see in this and the next section, the first and second derivatives can tell us a great deal about the shape of the graph of a function. In addition, this investigation will lead to methods for finding absolute maximum and minimum values for functions that do not require graphing. Companies can use these methods to find production levels that will minimize cost or maximize profit. Pharmacologists can use them to find levels of drug dosages that will produce maximum sensitivity to a drug. And so on.

INTERVAL NOTATION

Where appropriate in the discussions that follow we will use **interval notation.** Table 1 summarizes this notation.

One form of interval notation may confuse you a little at first. If, for example, we write $(-2, 3)$, then you will have to determine from the context in which it appears whether we mean the interval $-2 < x < 3$ or the coordinates of a point in a rectangular system. Generally, only one of these meanings will make sense in context.

INCREASING AND DECREASING FUNCTIONS

Graphs of functions generally have rising or falling sections as we move from left to right. It would be an aid to graphing if we could figure out where these sections occur. Suppose the graph of a function f is as indicated in Figure 1. As we move from left to right, we see that on the interval (a, b) the graph of f is rising, $f(x)$ is increasing, and the slope of the graph is

TABLE 1

INTERVAL NOTATION	INEQUALITY NOTATION	LINE GRAPH
$[a, b]$	$a \leq x \leq b$	x a b
$[a, b)$	$a \leq x < b$	x a b
$(a, b]$	$a < x \leq b$	x a b
(a, b)	$a < x < b$	x a b
$(-\infty, a]$	$x \leq a$	x a
$(-\infty, a)$	$x < a$	x a
$[b, \infty)$	$x \geq b$	x b
(b, ∞)	$x > b$	x b

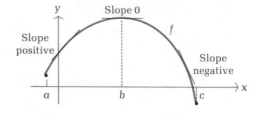

FIGURE 1

positive [$f'(x) > 0$]. On the other hand, on the interval (b, c) the graph of f is falling, $f(x)$ is decreasing, and the slope of the graph is negative [$f'(x) < 0$].

In general, we can prove that if $f'(x) > 0$ (is positive) on the interval (a, b), then $f(x)$ increases (↗) and the graph of f rises as we move from left to right over the interval; if $f'(x) < 0$ (is negative) on an interval (a, b), then $f(x)$ decreases (↘) and the graph of f falls as we move from left to right over the interval. Of course, if $f'(b) = 0$, then there is a horizontal tangent line at $(b, f(b))$, as shown at $x = b$ in Figure 1. We summarize these important results in the box.

INCREASING AND DECREASING FUNCTIONS

For the interval (a, b):

$f'(x)$	$f(x)$	Graph of f	Examples
$+$	Increases ↗	Rises ↗	
$-$	Decreases ↘	Falls ↘	

Example 9

Given $f(x) = 8x - x^2$:

(A) For which values of x is $f(x)$ increasing? Decreasing?

(B) Which values of x correspond to horizontal tangent lines?

(C) Sketch a graph of $f(x) = 8x - x^2$. Add horizontal tangent lines.

Solutions

(A) Take the derivative of f and determine which values of x make $f'(x) > 0$ and which values make $f'(x) < 0$.

$$
\begin{array}{cc}
f'(x) > 0 & f'(x) < 0 \\
8 - 2x > 0 & 8 - 2x < 0 \\
-2x > -8 & -2x < -8 \\
x < 4 & x > 4
\end{array}
$$

Thus,

x	$f'(x)$	$f(x)$	Graph of f
$(-\infty, 4)$	$+$	Increasing	Rising
$(4, \infty)$	$-$	Decreasing	Falling

These results can also be conveniently summarized as follows:

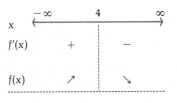

	$-\infty$	4	∞
x			Real number line (x axis)
$f'(x)$	$+$	$-$	Intervals over which $f'(x)$ is positive or negative
$f(x)$	↗	↘	Intervals over which the graph of f rises or falls

(B) A horizontal tangent line occurs where the slope of the curve is zero. So we must find x such that $f'(x) = 0$.

$$f'(x) = 8 - 2x = 0$$
$$-2x = -8$$
$$x = 4$$

Thus, a horizontal tangent line exists at $x = 4$ only.

(C)

x	$f(x)$
0	0
2	12
4	16
6	12
8	0

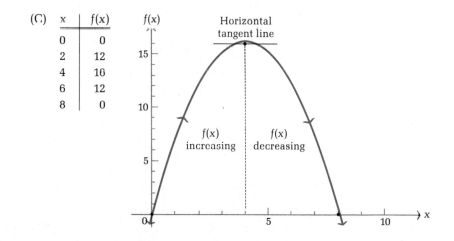

Problem 9   Repeat Example 9 for $f(x) = x^2 - 6x + 10$.

Answers   (A) $f(x)$ is decreasing for $(-\infty, 3)$; $f(x)$ is increasing for $(3, \infty)$
(B) Horizontal tangent line at $x = 3$

(C) $f(x)$

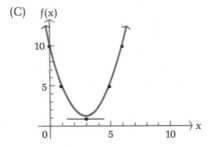

**CONCAVITY AND
INFLECTION POINTS**

A curve that lies below its tangent line at each point on the interval (a, b) is said to be **concave downward** over (a, b). If it lies above the tangent line at each point on (a, b), then it is said to be **concave upward** over (a, b). A point on a graph that separates a concave downward portion of a curve from a concave upward portion is called an **inflection point.** Figure 2 illustrates each of these forms.

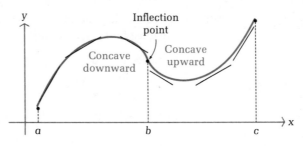

FIGURE 2

The second derivative can be useful in determining regions over which the graph of $y = f(x)$ is concave upward or concave downward. Suppose we find that $f''(x) < 0$ over (a, b). Then we can conclude that $f'(x)$ is decreasing over (a, b). That is, the slopes of the tangents to the graph of $y = f(x)$ decrease as x increases from a to b. This is precisely what happens in a concave downward portion of a curve (see Fig. 2). Similarly, if $f''(x) > 0$ over (a, b), then $f'(x)$ increases over (a, b), and the slopes of the tangent lines increase as x increases from a to b. Figure 3 illustrates several typical cases.

In general, the following can be proved:

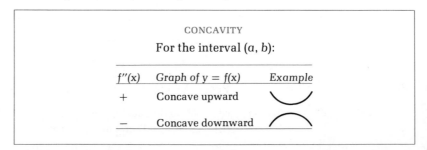

$f''(x) > 0$ over (a, b)
CONCAVE UPWARD

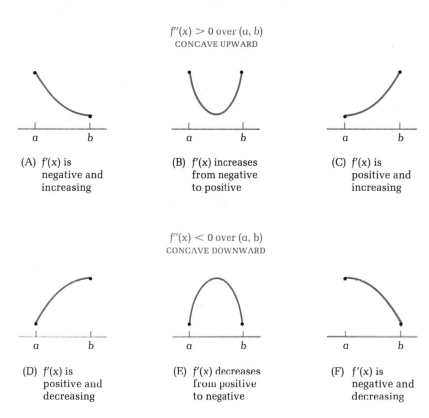

(A) $f'(x)$ is
negative and
increasing

(B) $f'(x)$ increases
from negative
to positive

(C) $f'(x)$ is
positive and
increasing

$f''(x) < 0$ over (a, b)
CONCAVE DOWNWARD

(D) $f'(x)$ is
positive and
decreasing

(E) $f'(x)$ decreases
from positive
to negative

(F) $f'(x)$ is
negative and
decreasing

FIGURE 3   Concavity.

Example 10   We are given the graph of $y = f(x)$ shown below.

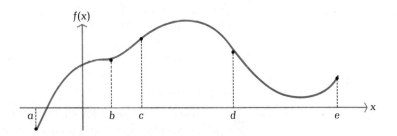

(A)   Identify intervals over which the graph of f is concave downward.
Concave upward.

(B)   Identify inflection points.

Solutions   (A)   Concave downward for intervals (a, b) and (c, d); concave upward
for intervals (b, c) and (d, e).

(B)   Inflection points are at $b, c,$ and d.

Problem 10   Repeat Example 10 for the graph below.

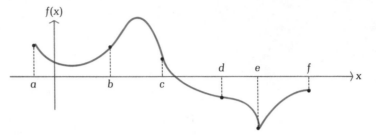

Answers   (A)   Concave upward for (a, b) and (c, d); concave downward for (b, c), (d, e), and (e, f)

(B) b, c, and d

Example 11   Given $f(x) = x^2 - 4x + 5$, $x \geq 0$, find intervals over which the graph of f is rising, falling, concave upward, and concave downward, and locate horizontal tangents. Sketch a graph.

Solutions   Take first and second derivatives:

$$f(x) = x^2 - 4x + 5$$
$$f'(x) = 2x - 4 = 2(x - 2)$$
$$f''(x) = 2$$

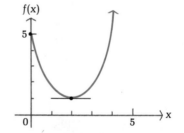

Thus, we see that since $f'(x)$ is negative for x on the interval $(0, 2)$ and positive for x on the interval $(2, \infty)$, $f(x)$ decreases for x in $(0, 2)$ and increases for x in $(2, \infty)$. Since $f'(2) = 0$, there is a horizontal tangent line at $x = 2$. And since $f''(x) = 2 > 0$, f is concave upward everywhere.

Problem 11   Repeat Example 11 for $f(x) = 4x - x^2$, $0 \leq x \leq 4$.

Answer   The function f increases for x in $(0, 2)$, decreases for x in $(2, 4)$, has a horizontal tangent at $x = 2$, and is concave downward for all x.

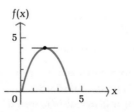

EXERCISE 3-4

A   Problems 1–6 refer to the following graph of $y = f(x)$:

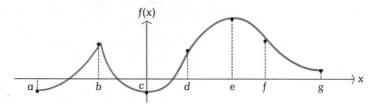

1. Identify intervals over which $f(x)$ is increasing.
2. Identify intervals over which $f(x)$ is decreasing.
3. Identify intervals over which the graph of f is concave upward.
4. Identify intervals over which the graph of f is concave downward.
5. Identify inflection points.
6. Locate horizontal tangents.

Find intervals over which:
(A)   The first derivative of each function is positive; negative
(B)   The second derivative of each function is positive; negative

7. $g(x) = x^2$
8. $F(x) = 1 - x^2$
9. $G(x) = x^2 - 4x + 5$
10. $h(x) = 2 + 6x - x^2$

B   For each function, find intervals over which the graph of f is rising, falling, concave upward, and concave downward, and locate horizontal tangents. Sketch a graph. Identify inflection points, if any.

11. $f(x) = x^2 - 2$
12. $f(x) = 4 - x^2$
13. $f(x) = 4 + 8x - x^2$
14. $f(x) = 2x^2 - 8x + 9$
15. $f(x) = x^3$
16. $f(x) = (x - 2)^3$
17. $f(x) = \sqrt[3]{x}$
18. $f(x) = \sqrt[3]{x^2}$

C   19. $f(x) = x^3 - 3x + 1$
20. $f(x) = x^3 - 12x + 2$
21. $f(x) = 2x^3 - 3x^2 - 12x - 5$
22. $f(x) = 2x^3 + 3x^2 - 12x - 1$

APPLICATIONS

BUSINESS & ECONOMICS

23. *Average cost functions.* The average cost per record album (in dollars) is given by

$$\overline{C}(x) = x^2 - 6x + 12 \qquad 0 \le x \le 6$$

where x is production in thousands.
(A)   For what values of x is $\overline{C}$ increasing? Decreasing?
(B)   Sketch a graph of $\overline{C}$ using any aids discussed in this section.

24. *Profit function.* If the profit $P(x)$ in dollars for an output of x units is given by

$$P(x) = -\frac{x^2}{30} + 140x - 72{,}000 \qquad x \geq 0$$

find production levels for which P is increasing and levels for which P is decreasing.

LIFE SCIENCES 25. *Pulse rate.* A person x inches tall has a pulse rate of y beats per minute given approximately by

$$y = 590x^{-1/2} \qquad 30 \leq x \leq 75$$

Is the rate increasing or decreasing in the indicated interval?

26. *Drug sensitivity.* One hour after x milligrams of a particular drug are given to a person, the change in body temperature $T(x)$ in degrees Fahrenheit is given by

$$T(x) = x^2\left(1 - \frac{x}{9}\right) \qquad 0 \leq x \leq 6$$

The rate at which T changes with respect to the size of the dosage x, $T'(x)$, is called the *sensitivity* of the body to the dosage. For what values of x is $T'(x)$ increasing? Decreasing? [*Hint:* Use $T''(x)$.]

SOCIAL SCIENCES 27. *Learning.* If a learning curve is given by

$$N(t) = 20t - t^2 \qquad 0 \leq t \leq 9$$

where $N(t)$ is the number of basic phrases learned in t hours, determine where N increases and decreases for $(0, 9)$.

3-5 **LOCAL MAXIMA AND MINIMA**

Being able to locate high and low points on a graph is also of help in graphing functions. In Figure 4, high points occur at c_3 and c_6, and low points occur at c_2 and c_4. In general, we call a point $(c, f(c))$ a **local maximum** if there exists an interval (m, n) containing c such that

$$f(x) \leq f(c)$$

for all x in (m, n). A point $(c, f(c))$ is called a **local minimum** if there exists an interval (m, n) containing c such that

$$f(x) \geq f(c)$$

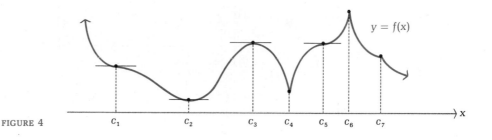

$y = f(x)$

FIGURE 4

c_1 c_2 c_3 c_4 c_5 c_6 c_7 x

for all x in (m, n). Thus, in Figure 4 we see that local maxima occur at c_3 and c_6 and local minima occur at c_2 and c_4.

How can we locate local maxima and minima if we are given the equation for the function and not its graph? Figure 4 suggests an approach. It appears that local maxima and minima occur among those values of x such that $f'(x) = 0$ or $f'(x)$ does not exist, that is, among the values $c_1, c_2, c_3, c_4, c_5, c_6$, and c_7. [Recall from Section 2-4 that $f'(x)$ is not defined* at sharp points or corners on a graph.] It is possible to prove the following:

EXISTENCE OF LOCAL EXTREMA

If f is a continuous function over the interval (a, b), then local maxima or minima, if they exist, must occur at values of x, called **critical values,** such that $f'(x) = 0$ or $f'(x)$ does not exist (is not defined).

Our strategy is now clear. We find all critical values for f and test each one to see if it is a local maximum, a local minimum, or neither. There are two derivative tests that can be used for this purpose. The first test we will discuss is called the *first-derivative test*, and it will work in all cases. The second test, called the *second-derivative test*, is often easier to use, but it does not work in all cases.

FIRST-DERIVATIVE TEST FOR LOCAL MAXIMA OR MINIMA

Since the sign of $f'(x)$ tells whether $f(x)$ is increasing or decreasing, we look at the sign of $f'(x)$ on either side of a critical value c. If the sign of $f'(x)$ changes as we go from one side of c to the other, then $f(c)$ is a local maximum or minimum; if the sign of $f'(x)$ does not change, then $f(c)$ is neither. For the former possibility we have two cases to consider:

*"Not defined" and "does not exist" are used interchangeably. If $f'(x) = x/(x - 1)$, for example, then $f'(1)$ does not exist (is not defined), but $f'(0) = 0$ does exist (is defined).

Case 1 $f'(c) = 0$

x	m	c	n	Number line (x axis)
$f'(x)$	$-$	0	$+$	
$f(x)$	$\searrow$		$\nearrow$	$f(c)$ is a local minimum $\smile$
$f'(x)$	$+$	0	$-$	
$f(x)$	$\nearrow$		$\searrow$	$f(c)$ is a local maximum $\frown$

Case 2 $f'(c)$ is not defined [but $f(c)$ is defined]

x	m	c	n	Number line (x axis)
$f'(x)$	$-$	ND	$+$	
$f(x)$	$\searrow$		$\nearrow$	$f(c)$ is a local minimum $\vee$
$f'(x)$	$+$	ND	$-$	
$f(x)$	$\nearrow$		$\searrow$	$f(c)$ is a local maximum $\wedge$

ND = Not defined

Example 12   Find local maxima and minima for $f(x) = x^2 - 4x + 5$ and use any other aids we have discussed (that are convenient) to graph f for $0 \le x \le 4$.

Solution   Find critical values:

$$f'(x) = 2x - 4 = 2(x - 2)$$
$$f'(2) = 0$$

Thus, $x = 2$ is the only critical value. Now check the sign of $f'(x)$ on either side of $x = 2$:

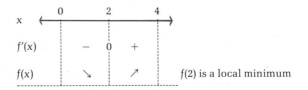

x	0	2	4	
$f'(x)$	$-$	0	$+$	
$f(x)$	$\searrow$		$\nearrow$	$f(2)$ is a local minimum

We also note that

$$f''(x) = 2$$

for all x; hence, the graph of f is concave upward everywhere. Using all the above information, we can sketch the graph of f rather easily:

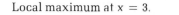
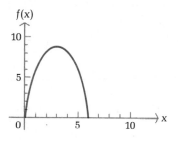

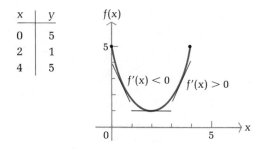

x	y
0	5
2	1
4	5

Problem 12   Repeat Example 12 for $f(x) = 6x - x^2, 0 \leq x \leq 6$.

Answer   Local maximum at $x = 3$.

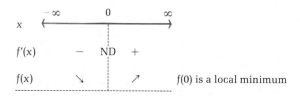

Example 13   Find local maxima and minima for $f(x) = x^{2/3}$ and use any other aids we have discussed to graph f.

Solution   Find critical values:

$$f'(x) = \tfrac{2}{3}x^{-1/3} = \frac{2}{3\sqrt[3]{x}}$$

The first derivative $f'(x)$ is not zero for any x, but when $x = 0$, $f'(0)$ is not defined. Hence, $x = 0$ is the only critical value. Now check the sign of $f'(x)$ on either side of $x = 0$:

	$-\infty$	0	∞	
x	$\longleftarrow$		$\longrightarrow$	
$f'(x)$	$-$	ND	$+$	
$f(x)$	$\searrow$		$\nearrow$	$f(0)$ is a local minimum

Now find $f''(x)$:

$$f''(x) = -\tfrac{2}{9}x^{-4/3} = -\frac{2}{9x^{4/3}}$$

Thus, $f''(x) < 0$ for all x except $x = 0$, and we conclude that the graph of f is concave downward for $(-\infty, 0)$ and $(0, \infty)$. Putting all the above information together, we obtain the graph of f, as shown on the next page.

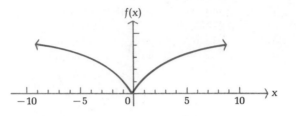

Graph for Example 13.

Problem 13   Repeat Example 13 for $f(x) = x^{1/3}$.

Answer   No local maxima or minima.

Example 14   Use the first-derivative test to find local maxima and minima for $f(x) = x^3 - 6x^2 + 9x + 1$.

Solution
$$f(x) = x^3 - 6x^2 + 9x + 1$$
$$f'(x) = 3x^2 - 12x + 9 = 3(x^2 - 4x + 3) = 3(x - 1)(x - 3)$$

Critical values: $x = 1, 3$

x	$-\infty$		1		3		∞
$(x - 1)$	$-$		0	$+$		$+$	
$(x - 3)$	$-$			$-$	0	$+$	
$f'(x)$	$+$		0	$-$	0	$+$	
$f(x)$	$\nearrow$			$\searrow$		$\nearrow$	

Thus,
$$f(1) = 5 \text{ is a local maximum}$$
$$f(3) = 1 \text{ is a local minimum}$$

Problem 14   Use the first-derivative test to find local maxima and minima for $f(x) = x^3 - 9x^2 + 24x$.

Answer $f(2) = 20$ is a local maximum; $f(4) = 16$ is a local minimum

SECOND-DERIVATIVE TEST   You probably observed in Figure 4 and Example 12 that if a curve has a horizontal tangent line at $x = c$ and is concave upward (or downward) in an interval containing the point of tangency, then $f(c)$ is a local minimum (or maximum). This observation leads to a simple second-derivative test that is effective in some cases. Before stating the test we note that if $f''(c) > 0$, then it follows from the properties of limits that $f''(x) > 0$ in some interval (m, n) containing c. This means that the graph of f is concave upward in an interval about c if $f''(c) > 0$. A similar statement can be made about $f''(c) < 0$.

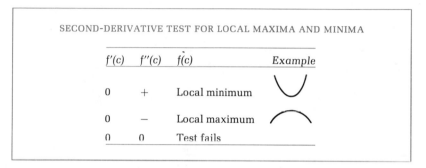

SECOND-DERIVATIVE TEST FOR LOCAL MAXIMA AND MINIMA

$f'(c)$	$f''(c)$	$f(c)$	Example
0	+	Local minimum	
0	−	Local maximum	
0	0	Test fails	

Example 15

Find local maxima and minima for $f(x) - x^3 - 6x^2 + 9x + 1$ and use any other aids that are convenient to graph f.

Solution

Take first and second derivatives and find critical values:

$$f(x) = x^3 - 6x^2 + 9x + 1$$
$$f'(x) = 3x^2 - 12x + 9 = 3(x - 1)(x - 3)$$
$$f''(x) = 6x - 12 = 6(x - 2)$$

Critical values are $x - 1, 3$.

$$f''(1) = -6 \qquad \text{Therefore, } f(1) \text{ is a local maximum}$$
$$f''(3) = 6 \qquad \text{Therefore, } f(3) \text{ is a local minimum}$$

Note that

$$f''(x) < 0 \quad \text{for } (-\infty, 2) \qquad \text{Concave downward}$$
$$f''(x) > 0 \quad \text{for } (2, \infty) \qquad \text{Concave upward}$$

Thus, there is an inflection point at $x = 2$. Locate the critical values and the inflection point, and sketch the graph.

x	y
0	1
1	5
2	3
3	1
4	5

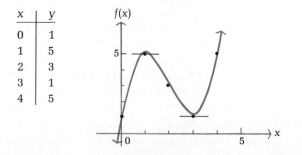

Problem 15   Repeat Example 15 for $f(x) = x^3 - 9x^2 + 24x$.

Answer   Local maximum at $x = 2$; local minimum at $x = 4$.

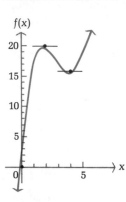

SUMMARY   It is important to remember that the second-derivative test can only be used for critical values c such that $f'(c) = 0$. If $f'(c)$ does not exist [and $f(c)$ does], then we must use the first-derivative test to find out whether $f(c)$ is a local maximum or a local minimum. Also, if we use the second-derivative test and find that $f''(c) = 0$ for a critical value c, then the test fails (gives no information) and we must return to the first-derivative test.

EXERCISE 3-5

A   Use the figure below for Problems 1–4.

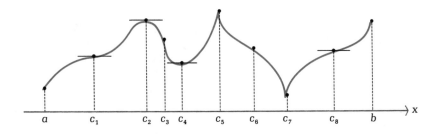

1. Find local maxima.
2. Find local minima.
3. Which points are critical values because of $f'(c) = 0$?
4. Which points are critical values because $f'(c)$ is not defined?

In Problems 5–10 replace question marks in the tables with "Local maximum," "Local minimum," or "Neither," as appropriate. Assume f is continuous over (m, n) unless otherwise stated. (Sketching pictures may help you decide.)

5.

	$f'(c)$	$f'(x)$ (m, c)	$f'(x)$ (c, n)	$f(c)$
(A)	0	−	+	?
(B)	0	−	−	?

6.

	$f'(c)$	$f'(x)$ (m, c)	$f'(x)$ (c, n)	$f(c)$
(A)	0	+	−	?
(B)	0	+	+	?

7.

	$f'(c)$	$f(c)$	$f'(x)$ (m, c)	$f'(x)$ (c, n)	$f(c)$
(A)	Not defined	Defined	+	−	?
(B)	Not defined	Defined	+	+	?
(C)	Not defined	Not defined	−	+	?

8.

	$f'(c)$	$f(c)$	$f'(x)$ (m, c)	$f'(x)$ (c, n)	$f(c)$
(A)	Not defined	Defined	−	+	?
(B)	Not defined	Defined	−	−	?
(C)	Not defined	Not defined	+	−	?

9.

	$f'(c)$	$f''(c)$	$f(c)$
(A)	0	+	?
(B)	1	−	?

10.

	$f'(c)$	$f''(c)$	$f(c)$
(A)	0	−	?
(B)	−1	+	?

B Find all local maxima and minima using either the first-derivative test or the second-derivative test, whichever is applicable or convenient (do not graph).

11. $f(x) = 2x^2 - x^4$

12. $f(x) = 2x - x^3$

13. $f(x) = x^3 - 3x^2 - 24x + 7$

14. $f(x) = x^3 + 3x^2 - 9x + 5$

15. $f(x) = (x - 1)^4 + 2$ 16. $f(x) = 2 - (x + 1)^6$

17. $f(x) = x + \dfrac{4}{x}$ 18. $f(x) = \dfrac{9}{x} + x$

19. $f(x) = 1 + \dfrac{1}{x} + \dfrac{1}{x^2}$ 20. $f(x) = 3 - \dfrac{4}{x} - \dfrac{2}{x^2}$

21. $f(x) = (1 - x)^{2/3}$ 22. $f(x) = (1 - x)^{1/3}$

23. $f(x) = \dfrac{x^2}{x - 2}$ 24. $f(x) = \dfrac{x^2}{x + 1}$

Find local maxima and minima using either the first-derivative test or the second-derivative test, whichever is most convenient. Sketch a graph of each function using any other aids we have discussed.

25. $f(x) = x^2 - 6x + 10$ 26. $f(x) = 4 + 8x - x^2$

27. $f(x) = 4x - x^2, \quad 0 \le x \le 4$ 28. $f(x) = x^2 - 8x, \quad 0 \le x \le 8$

29. $f(x) = (x - 2)^{2/3}$ 30. $f(x) = (x + 3)^{2/3}$

31. $f(x) = x^3 - 27x + 4$ 32. $f(x) = x^3 - 48x + 5 \cdot$

33. $f(x) = x^4 + 1$ 34. $f(x) = 4 - x^4$

35. $f(x) = 2x^3 - 3x^2 - 36x + 6$ 36. $f(x) = 2x^3 + 3x^2 - 72x + 10$

C   37. $f(x) = x + \dfrac{1}{x}$ 38. $f(x) = x^2 + \dfrac{1}{x^2}$

39. $f(x) = 3x^{2/3} + x$ 40. $f(x) = 2\sqrt{x} - x, \quad 0 \le x \le 9$

3-6   ABSOLUTE MAXIMUM AND MINIMUM—APPLICATIONS

We are now ready to consider one of the most important applications of the derivative, that is, the use of derivative to find the *absolute maximum* or *minimum* value of a function. As we mentioned earlier, an economist may be interested in the price or production level of a commodity that will bring a maximum profit; a doctor may be interested in the time it takes for a drug to reach its maximum concentration in the bloodstream after an injection; and a city planner might be interested in the location of heavy industry in a city to produce minimum pollution in residential and business areas. Before we launch an attack on problems of this type, we have to say a few words about the procedures needed to find absolute maximum and absolute minimum values of functions. We have most of the tools we need from the previous sections.

ABSOLUTE MAXIMUM
AND MINIMUM

First, what do we mean by *absolute maximum* and *absolute minimum*? We say that $f(c)$ is an **absolute maximum** of f if

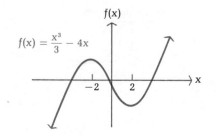

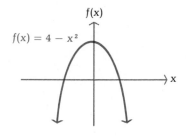

(A) No absolute maximum or minimum
One local maximum at $x = -2$
One local minimum at $x = 2$

(B) Absolute maximum at $x = 0$
No absolute minimum

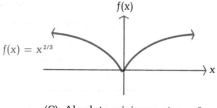

(C) Absolute minimum at $x = 0$
No absolute maximum

FIGURE 5

$$f(c) \geq f(x)$$

for all x in the domain of f. Similarly, $f(c)$ is called an **absolute minimum** of f if

$$f(c) \leq f(x)$$

for all x in the domain of f. Figure 5 illustrates several typical examples.

In many practical problems, the domain of a function is restricted because of practical or physical considerations. If the domain is restricted to some closed interval, as is often the case, then Theorem 1 can be proved.

Theorem 1 A function f continuous on a closed interval $[a, b]$ assumes both an absolute maximum and an absolute minimum on that interval.

To get an idea of where an absolute maximum or minimum may occur on a closed interval, look at Figure 6 on the next page. Points a and b are end points and c_1, c_2, c_3, and c_4 are critical values because $f'(c_1) = f'(c_2) = f'(c_3) = 0$ and $f'(c_4)$ is not defined. We see that between any of these special points, the curve is either rising or falling. The absolute maximum and the absolute minimum each occurs at a critical value or an end point. Thus, to find the absolute maximum or minimum value of a continuous function on a closed interval, we simply identify the end points and the critical values, evaluate each, and then choose the largest and smallest values out of this group.

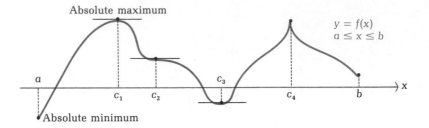

FIGURE 6

STEPS IN FINDING ABSOLUTE MAXIMUM
AND MINIMUM VALUES OF CONTINUOUS FUNCTIONS

1. Check to make certain that f is continuous over $[a, b]$.
2. List end points and critical values: a, b, c_1, c_2, . . . , c_n.
3. Evaluate $f(a)$, $f(b)$, $f(c_1)$, $f(c_2)$, . . . , $f(c_n)$.
4. The absolute maximum $f(x)$ on $[a, b]$ is the largest of the values found in step 3.
5. The absolute minimum $f(x)$ on $[a, b]$ is the smallest of the values found in step 3.

It often happens in practical situations that there is only one critical value c for f. If this is the case and $f''(c)$ exists, then we have the following second-derivative test:

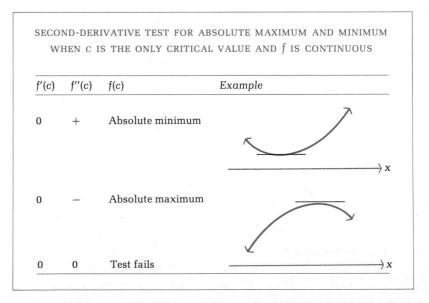

SECOND-DERIVATIVE TEST FOR ABSOLUTE MAXIMUM AND MINIMUM
WHEN c IS THE ONLY CRITICAL VALUE AND f IS CONTINUOUS

$f'(c)$	$f''(c)$	$f(c)$	Example
0	+	Absolute minimum	
0	−	Absolute maximum	
0	0	Test fails	

Example 16   Find the absolute maximum and minimum for:
(A) $f(x) = x^2 - 6x$
(B) $f(x) = x^3 + 3x^2 - 9x - 7$, $-6 \le x \le 2$

Solutions   (A)   The function $f(x) = x^2 - 6x$ is continuous for all x, so we find

$$f'(x) = 2x - 6 = 2(x - 3)$$

There is only one critical value, which is at $x = 3$ $[f'(3) = 0]$. We can now use the second-derivative test. Computing the second derivative, we find

$$f''(x) = 2 > 0 \qquad \text{for all } x$$

Since $x = 3$ is the only critical value, and since $f'(3) = 0$ and $f''(3) > 0$, according to the second-derivative test, $f(3) = -9$ is the absolute minimum for $f(x)$. There is no absolute maximum.

(B)   The function $f(x) = x^3 + 3x^2 - 9x - 7$ is continuous for all x on $[-6, 2]$.

$$\begin{aligned} f'(x) &= 3x^2 + 6x - 9 \\ &= 3(x - 1)(x + 3) \end{aligned}$$

Critical values are at $x = 1$ and -3. Evaluate f at the end points and critical values ($-6, -3, 1,$ and 2), and choose the maximum and minimum from these:

$$\begin{aligned} f(-6) &= -61 & \text{Absolute minimum} \\ f(-3) &= 20 & \text{Absolute maximum} \\ f(1) &= -12 \\ f(2) &= -5 \end{aligned}$$

Problem 16   Repeat Example 16 for:
(A) $f(x) = 8x - x^2$
(B) $f(x) = x^3 - 12x$, $-3 \le x \le 4$

Answers   (A)   Absolute maximum $= f(4) = 16$; no absolute minimum
(B)   Absolute maximum $= f(4) = 16$ and $f(-2) = 16$;
absolute minimum $= f(2) = -16$

APPLICATIONS   We are now ready to consider several significant applications. Many more are given in the exercises.

Example 17   A company manufactures and sells x transistor radios per week. If the weekly cost and revenue equations are

$$C(x) = 5,000 + 2x$$
$$R(x) = 10x - \frac{x^2}{1,000} \qquad 0 \le x \le 8,000$$

find for each week:

(A) The minimum possible cost
(B) The maximum possible revenue
(C) The maximum possible profit

Solutions

(A) $C(x) = 5,000 + 2x$
$C'(x) = 2$

There are no critical values; therefore, we look at the end points. Minimum C [denoted Min $C(x)$] is obviously at $x = 0$. Thus,

Min $C(x) = C(0) = \$5,000$ per week

(B) $R(x) = 10x - \dfrac{x^2}{1,000}$

$R'(x) = 10 - \dfrac{x}{500}$

$10 - \dfrac{x}{500} = 0$

$x = 5,000$ Only critical value

Use the second-derivative test:

$R''(x) = -\dfrac{1}{500} < 0$ for all x

Hence,

Max $R(x) = R(5,000) = \$25,000$ per week

(C) Profit $=$ Revenue $-$ Cost

$P(x) = R(x) - C(x)$

$= 10x - \dfrac{x^2}{1,000} - 5,000 - 2x$

$= 8x - \dfrac{x^2}{1,000} - 5,000$

$P'(x) = 8 - \dfrac{x}{500}$

$8 - \dfrac{x}{500} = 0$

$x = 4,000$

$P''(x) = -\dfrac{1}{500} < 0$ for all x

Hence,

Max $P(x) = P(4,000) = \$11,000$ per week

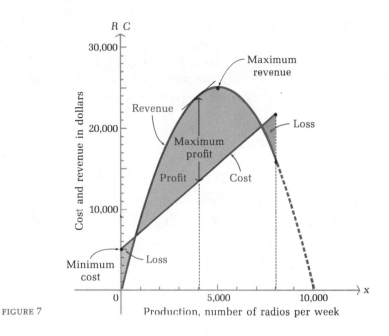

FIGURE 7

Thus, a maximum profit of $11,000 per week is realized at a production level of 4,000 radios per week.

All the results in this example are illustrated in Figure 7. We also note that profit is maximum when

$$P'(x) = R'(x) - C'(x) = 0$$

that is, when the marginal revenue is equal to the marginal cost (the rate of increase in revenue is the same as the rate of increase in cost at the 4,000 output level—notice that the slopes of the two curves are the same at this point).

Problem 17   Repeat Example 17 for

$$C(x) = 90,000 + 30x$$

$$R(x) = 300x - \frac{x^2}{30} \qquad 0 \le x \le 9,000$$

Answers   (A)   Min $C(x) = C(0) = \$90,000$
(B)   Max $R(x) = R(4,500) = \$675,000$
(C)   Max $P(x) = P(4,050) = \$456,750$

Example 18   A walnut grower estimates from past records that if twenty trees are planted per acre, each tree will average 60 pounds of nuts per year. If for each additional tree planted per acre (up to fifteen) the average yield per tree drops 2 pounds, how many trees should be planted to maximize the yield per acre? What is the maximum yield?

Solution   Let x be the number of additional trees planted per acre. Then

$$20 + x = \text{Total number of trees per acre}$$
$$60 - 2x = \text{Yield per tree}$$
$$\text{Yield per acre} = (\text{Total number of trees per acre}) (\text{Yield per tree})$$
$$Y(x) = (20 + x)(60 - 2x)$$
$$= 1{,}200 + 20x - 2x^2 \qquad 0 \le x \le 15$$
$$Y'(x) = 20 - 4x$$
$$20 - 4x = 0$$
$$x = 5$$
$$Y''(x) = -4 \qquad \text{for all x}$$

Hence,

$$\text{Max } Y(x) = Y(5) = 1{,}250 \text{ pounds per acre}$$

Thus, a maximum yield of 1,250 pounds of nuts per acre is realized if twenty-five trees are planted per acre.

Problem 18   Repeat Example 18 starting with thirty trees per acre and a reduction of 1 pound per tree for each additional tree planted.

Answer   Max $Y(x) = Y(15) = 2{,}025$ pounds per acre

Example 19   A Wyoming rancher has 20 miles of fencing to fence in a rectangular piece of grazing land along a river. If no fence is required along the river, what is the dimension of the rectangle that will give the maximum area? What is the maximum area?

Solution

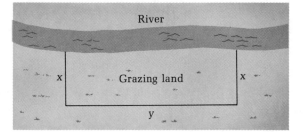

$$\text{Area} = xy$$
$$2x + y = 20 \text{ miles of fencing}$$
$$y = 20 - 2x \qquad 0 \le x \le 10$$

Thus,

$$A(x) = x(20 - 2x)$$
$$= 20x - 2x^2 \qquad 0 \le x \le 10$$
$$A'(x) = 20 - 4x$$
$$20 - 4x = 0$$
$$x = 5 \qquad \qquad \text{Only one critical value}$$
$$A''(x) = -4 < 0$$

Hence, a maximum area of 50 square miles is given by a 5 mile by 10 mile rectangle.

A graph of $A(x) = x(20 - 2x)$ gives a dramatic confirmation of this result. Note how the area of the grazing land moves away from the maximum as x moves away from 5:

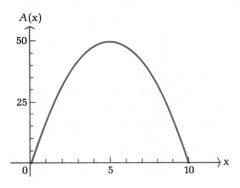

Problem 19	Repeat Example 19 using 32 miles of fencing.
Answer	8 by 16 miles; 128 square miles

EXERCISE 3-6

Find the absolute maximum and absolute minimum, if either exists, for each function.

A   1. $f(x) = x^2 - 4x + 5$ 2. $f(x) = x^2 + 6x + 7$
        3. $f(x) = 10 + 8x - x^2$ 4. $f(x) = 6 - 8x - x^2$

B   5. $f(x) = x^2 - 4x + 6, \quad 0 \le x \le 5$
        6. $f(x) = 8 + 8x - x^2, \quad 0 \le x \le 6$
        7. $f(x) = x^3 - 6x^2 + 9x + 6, \quad 0 \le x \le 4$
        8. $f(x) = 2x^3 - 3x^2 - 12x + 24, \quad -2 \le x \le 3$

C   9. $f(x) = (x - 1)(x - 5)^3 + 1, \quad 0 \le x \le 6$
      10. $f(x) = 1 + x^{2/3}, \quad -1 \le x \le 8$
      11. $f(x) = x^{1/3} - 1, \quad -1 \le x \le 1$
      12. $f(x) = 2\sqrt{x} - x, \quad 0 \le x \le 9$

PRELIMINARY WORD
PROBLEMS

13. How would you divide a 10 inch line so that the product of the two lengths is maximum?

14. If the same quantity is added to 5 and subtracted from 5, how much should it be to produce the maximum product of the results?

15. Find two numbers whose difference is 30 and whose product is minimum.

16. Find two positive numbers whose sum is 60 and whose product is maximum.

17. Find the dimensions of a rectangle with perimeter 100 centimeters that has maximum area. Find the maximum area.

18. Find the dimensions of a rectangle of area 225 square centimeters that has the least perimeter. What is the perimeter?

APPLICATIONS

Many of the following applications are patterned after actual situations analyzed in professional journals and books associated with the given subjects. The material has been simplified to make it accessible to readers of this book. The more difficult problems are marked with two stars (**), the moderately difficult problems with one star (*), and the easier problems are not marked.

BUSINESS & ECONOMICS

19. *Average costs.* If the average manufacturing cost (in dollars) per pair of sunglasses is given by

$$\overline{C}(x) = x^2 - 6x + 12 \qquad 0 \le x \le 6$$

where x is the number of pairs manufactured in thousands, how many pairs of glasses should be manufactured to minimize the average cost per pair? What is the minimum average cost per pair?

20. *Maximum revenue and profit.* A company manufactures and sells x televisions per month. If the cost and revenue equations are

$$C(x) = 72{,}000 + 60x$$
$$R(x) = 200x - \frac{x^2}{30} \qquad 0 \le x \le 6{,}000$$

find:

(A) The minimum possible cost   (B) The maximum revenue

(C) The maximum profit

*21. *Car rental.* A car rental agency rents 100 cars per day at a rate of $10 per day. For each $1 increase in rate, five less cars are rented. At what rate should the cars be rented to produce the maximum income? What is the maximum income?

*22. *Rental income.* A ninety room hotel in Las Vegas is filled to capacity every night at $25 a room. For each $1 increase in rent, three fewer rooms are rented. If each rented room costs $3 to service per day, how much should the management charge for each room to maximize gross profit? What is the maximum gross profit?

*23. *Agriculture.* A commercial cherry grower estimates from past records that if thirty trees are planted per acre, each tree will yield an average of 50 pounds of cherries per season. If for each additional tree planted per acre (up to twenty) the average yield per tree is reduced by 1 pound, how many trees should be planted per acre to obtain the maximum yield per acre? What is the maximum yield?

*24. *Agriculture.*   A commercial pear grower must decide on the optimum time to have fruit picked and sold. If the pears are picked now, they will bring 30¢ per pound, with each tree yielding an average of 60 pounds of salable pears. If the average yield per tree increases 6 pounds per tree per week for the next 4 weeks, but the price drops 2¢ per pound per week, when should the pears be picked to realize the maximum return per tree? What is the maximum return?

*25. *Manufacturing.*   A candy box is to be made out of a piece of cardboard that measures 8 by 12 inches. Equal sized squares will be cut out of each corner, and then the ends and sides will be folded up to form a rectangular box. What size square should be cut from each corner to obtain a maximum volume?

*26. *Packaging.*   A parcel delivery service will only deliver packages with length plus girth (distance around) not exceeding 108 inches.
(A)   Find the dimensions of a rectangular box with square ends of maximum volume. What is the maximum volume?
(B)   Find the dimensions (radius and height) of a cylindrical container of maximum volume. What is the maximum volume?

27. *Construction costs.*   A fence is to be built to enclose a rectangular area of 800 square feet. The fence along three sides is to be made of material that costs $2 a foot. The material for the fourth side costs $6 a foot. Find the dimensions of the rectangle for the most economical fence.

**28. *Operational costs.*   The cost per hour for fuel for running a train is $\frac{1}{4}v^2$ dollars, where v is the speed in miles per hour. (Note that the cost goes up as the square of the speed.) Other costs, including labor, are $300 per hour. How fast should the train travel on a 360 mile trip to minimize total cost for the trip?

**29. *Construction costs.*   A freshwater pipeline is to be run from a source on the edge of a lake to a small resort community on an island 5 miles off-shore, as indicated in the figure on the next page. If it costs 1.4 times as much to lay the pipe in the lake as it does on land, what should x be in miles to minimize the total cost of the project? (Compare with Problem 34 below.)

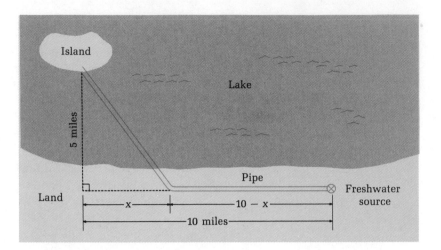

*30. *Manufacturing costs.* A manufacturer wishes to produce cans that will hold 12 ounces (approximately 22 cubic inches) in the form of a right circular cylinder. Find the dimensions (radius of an end and height) of the can that will use the smallest amount of material. Assume the circular ends are cut out of squares, with the corner portions wasted, and the sides are made from rectangles, with no waste.

LIFE SCIENCES

31. *Bacteria control.* A recreational swimming lake is treated periodically to control harmful bacteria growth. Suppose t days after a treatment, the concentration of bacteria per cubic centimeter is given by

$$C(t) = 30t^2 - 240t + 500 \qquad 0 \le t \le 8$$

How many days after a treatment will the concentration be minimal? What is the minimum concentration?

*32. *Drug concentration.* The concentration $C(t)$ in milligrams per cubic centimeter of a particular drug in a patient's bloodstream is given by

$$C(t) = \frac{0.16t}{t^2 + 4t + 4}$$

where t is the number of hours after the drug is taken. How many hours after the drug is given will the concentration be maximum? What is the maximum?

33. *Drug sensitivity.* One hour after x milligrams of a particular drug are given to a person, the change in body temperature in degrees

Fahrenheit is given by

$$T(x) = x^2\left(1 - \frac{x}{6}\right) \qquad 0 \le x \le 6$$

The rate at which T changes with respect to the size of the dosage, $T'(x)$, is the sensitivity of the body to the dosage. Find the dosage size that will produce the maximum sensitivity. [*Hint:* Maximize $T'(x)$.]

**34. *Bird flights.* Some birds tend to avoid flights over large bodies of water during daylight hours. It is speculated that more energy is required to fly over water than land because air generally rises over land and falls over water during the day. Suppose an adult bird with these tendencies is taken from its nesting area on the edge of a large lake to an island 5 miles off-shore, and is then released. If it takes 1.4 times as much energy to fly over water as over land, how far up the shore (x, in miles) should the bird head in order to minimize the total energy expended in returning to the nesting area? (Compare with Problem 29 above.)

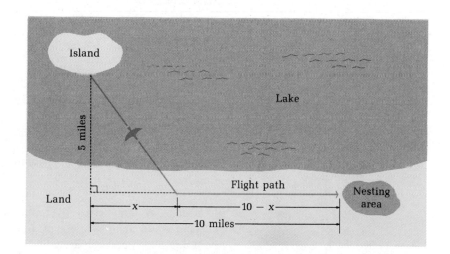

35. *Botany.* If it is found from carefully conducted experiments that the height in feet of a given plant after t months is given approximately by

$$H(t) = 4t^{1/2} - 2t \qquad 0 \le t \le 2$$

how long, on the average, will it take a plant to reach its maximum height? What is the maximum height?

*36. *Pollution.* Two heavy industrial areas are located 10 miles apart, as indicated in the figure. If the concentration of particulate matter in parts per million decreases as the reciprocal of the square of the distance from the source, and area A_1 emits eight times the particulate matter as A_2, then the concentration of particulate matter at any point between the two areas is given by

$$C(x) = \frac{8k}{x^2} + \frac{k}{(10 - x)^2} \qquad 0.5 \le x \le 9.5, \quad k > 0$$

How far from A_1 will the concentration of particulate matter be at a minimum?

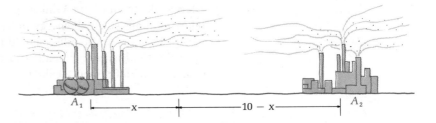

SOCIAL SCIENCES

37. *Politics.* In a newly incorporated city it is estimated that the voting population (in thousands) will increase according to

$$N(t) = 30 + 12t^2 - t^3 \qquad 0 \le t \le 8$$

where t is time in years. When will the rate of increase be most rapid?

38. *Learning.* A large grocery chain found that on the average a checker could memorize $P\%$ of a given price list in x continuous hours, as given approximately by

$$P(x) = 96x - 24x^2 \qquad 0 \le x \le 3$$

How long should a checker plan to take to memorize the maximum percentage? What is the maximum?

3-7   THE DIFFERENTIAL

In Chapter 2 we introduced the concept of increment. Recall that for a function defined by

$$y = f(x)$$

we said that Δx represents a change in the independent variable x; that is,

$$\Delta x = x_2 - x_1 \quad \text{or} \quad x_2 = x_1 + \Delta x$$

And Δy represents the corresponding change in the dependent variable y; that is,

$$\Delta y = f(x_1 + \Delta x) - f(x_1)$$

We then defined the derivative of f at x_1 to be

$$\frac{dy}{dx} = \lim_{\Delta x \to 0} \frac{\Delta y}{\Delta x}$$

If the limit exists, then it follows that

$$\frac{\Delta y}{\Delta x} \approx \frac{dy}{dx} \quad \text{for small } \Delta x$$

or

$$\Delta y \approx \frac{dy}{dx} \Delta x \tag{1}$$

We used dy/dx as an alternate symbol for $f'(x)$. We will now give dy and dx special meaning, and we will show how dy can be used to approximate Δy. This turns out to be quite useful, since a number of practical problems require the computation of Δy, and we will be able to use the more readily computed dy. The symbols dy and dx are called **differentials** and are defined below.

If $y = f(x)$ defines a differentiable function, then:

1. The **differential dx** of the independent variable x is an arbitrary real number.
2. The **differential dy** of the dependent variable y is defined as the product of $f'(x)$ and dx; that is,

$$dy = f'(x) \, dx \tag{2}$$

The differential dy is actually a function involving two independent variables, x and dx—a change in either one or both will affect dy.

Example 20   Find dy for $f(x) = x^2 + 3x$. Evaluate dy for x = 2 and $dx = 0.1$, for x = 3 and $dx = 0.1$, and for x = 1 and $dx = 0.02$.

Solution $dy = f'(x)\,dx$
$\quad\quad\quad = (2x + 3)\,dx$

When $x = 2$ and $dx = 0.1$,

$\quad dy = [2(2) + 3]0.1 = 0.7$

When $x = 3$ and $dx = 0.1$,

$\quad dy = [2(3) + 3]0.1 = 0.9$

When $x = 1$ and $dx = 0.02$,

$\quad dy = [2(1) + 3]0.02 = 0.1$

Problem 20 Find dy for $f(x) = \sqrt{x} + 3$. Evaluate dy for $x = 4$ and $dx = 0.1$, for $x = 9$ and $dx = 0.12$, and for $x = 1$ and $dx = 0.01$.

Answer $dy = \dfrac{1}{2\sqrt{x}}\,dx$

0.025, 0.02, 0.005

If you compare the right-hand sides of (1) and (2) you will see what motivated the definition of dy. The differential concept has a very clear geometric interpretation, as is indicated in Figure 8 (study it carefully).

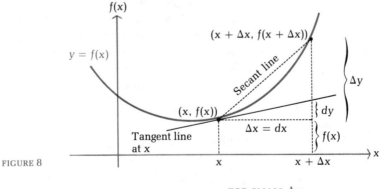

FIGURE 8

FOR SMALL Δx

Slope of secant line $\approx$ Slope of tangent line

$$\frac{\Delta y}{\Delta x} \approx \frac{dy}{dx}$$

$$\Delta y \approx dy = f'(x)\,dx$$

Example 21   Find Δy and dy for $f(x) = 6x - x^2$ when $x = 2$ and $\Delta x = dx = 0.1$.

Solution
$$\Delta y = f(x + \Delta x) - f(x)$$
$$= f(2.1) - f(2)$$
$$= [6(2.1) - (2.1)^2] - [6(2) - 2^2]$$
$$= 8.19 - 8$$
$$= 0.19$$
$$dy = f'(x)\,dx$$
$$= (6 - 2x)\,dx$$
$$= [6 - 2(2)](0.1)$$
$$= 0.2$$

Notice that dy and Δy differ by only 0.01 in this case.

Problem 21   Repeat Example 21 for $x = 4$ and $\Delta x = dx = 0.2$.

Answer
$$\Delta y = -0.44 \qquad dy = -0.4$$

DIFFERENTIAL APPROXIMATION

If $f'(x)$ exists, then for small Δx

$$\Delta y \approx dy$$

and

$$f(x + \Delta x) = f(x) + \Delta y$$
$$\approx f(x) + dy$$
$$= f(x) + f'(x)\,dx$$

We will use these relationships in the examples that follow. (Before proceeding, however, it should be mentioned that even though differentials provide a convenient and fast way of approximating certain quantities, the error can be substantial in certain cases.)

Example 22   A formula relating the approximate weight, W, in pounds of an average person and their height, h, in inches is

$$W = 0.0005h^3 \qquad 30 \le h \le 74$$

What is the approximate change in weight for a height increase from 40 to 42 inches?

Solution   We are actually interested in finding ΔW, the change in weight brought about by the change in height from 40 to 42 inches ($\Delta h = 2$). We will use the differential dW to approximate ΔW, since Δh is small. The problem is now to find dW for $h = 40$ and $dh = \Delta h = 2$.

$$W(h) = 0.0005h^3$$
$$dW = W'(h)\, dh$$
$$= 0.0015h^2\, dh$$
$$= 0.0015(40)^2(2)$$
$$= 4.8 \text{ pounds}$$

Thus, a child growing from 40 inches to 42 inches would expect to increase in weight by approximately 4.8 pounds. Notice that using the differential is somewhat easier than finding $\Delta W = W(42) - W(40)$.

Problem 22   Approximate the change in weight resulting from a height increase from 70 to 72 inches.

Answer   14.7 pounds

Example 23   A company manufactures and sells x transistor radios per week. If the weekly cost and revenue equations are

$$C(x) = 5,000 + 2x$$

$$R(x) = 10x - \frac{x^2}{1,000} \qquad 0 \le x \le 8,000$$

find the approximate change in revenue and profit if production is increased from 2,000 to 2,010 units per week.

Solution   We will approximate ΔR and ΔP with dR and dP, respectively, using $x = 2,000$ and $dx = \Delta x = 2,010 - 2,000 = 10$.

$$R(x) = 10x - \frac{x^2}{1,000}$$

$$dR = R'(x)\, dx$$

$$= \left(10 - \frac{x}{500}\right) dx$$

$$= \left(10 - \frac{2,000}{500}\right) 10$$

$$= \$60 \text{ per week}$$

$$P(x) = R(x) - C(x)$$

$$= 10x - \frac{x^2}{1,000} - 5,000 - 2x$$

$$= 8x - \frac{x^2}{1,000} - 5,000$$

$$dP = P'(x)\, dx$$

$$= \left(8 - \frac{x}{500}\right) dx$$

$$= \left(8 - \frac{2,000}{500}\right) 10$$

$$= \$40 \text{ per week}$$

Problem 23   Repeat Example 23 with production increasing from 6,000 to 6,010.

Answer $dR = -\$20 \text{ per week} \qquad dP = -\$40 \text{ per week}$

Comparing the results in Example 23 and Problem 23, we see that an increase in production results in a revenue and profit increase at the 2,000 production level, but a revenue and profit loss at the 6,000 production level.

Now we will consider a slightly different type of problem involving differential approximations.

Example 24   Use differentials to approximate $\sqrt[3]{27.54}$.

Solution   Even though the problem is trivial using a hand calculator, its solution using differentials will help increase the understanding of this concept. Form the function

$$y = f(x) = \sqrt[3]{x} = x^{1/3}$$

and note that we can compute $f(27)$ and $f'(27)$ exactly. Thus, if we let $x = 27$ and $dx = \Delta x = 0.54$ and use

$$f(x + \Delta x) = f(x) + \Delta y$$
$$\approx f(x) + dy$$
$$= f(x) + f'(x)\,dx$$

we will obtain an approximation for $f(27.54) = \sqrt[3]{27.54}$ that is easy to compute.

$$f(x + \Delta x) \approx f(x) + f'(x)\,dx$$

$$(x + \Delta x)^{1/3} \approx x^{1/3} + \frac{1}{3x^{2/3}}\,dx$$

$$(27 + 0.54)^{1/3} \approx 27^{1/3} + \frac{1}{3(27)^{2/3}}(0.54)$$

Thus,

$$\sqrt[3]{27.54} \approx 3 + \frac{0.54}{27} = 3.02 \qquad \text{(Calculator value} = 3.0199)$$

Problem 24 .   Use differentials to approximate $\sqrt{36.72}$.

Answer   6.06

We close this section by listing a number of differential rules that will be of use to us in the next chapter. These rules follow directly from the definition of the differential and the derivative rules discussed earlier.

DIFFERENTIAL RULES

If u and v are differentiable functions and c is a constant, then:

1. $dc = 0$
2. $du^n = nu^{n-1}\,du$
3. $d(u \pm v) = du \pm dv$
4. $d(uv) = u\,dv + v\,du$
5. $d\left(\dfrac{u}{v}\right) = \dfrac{v\,du - u\,dv}{v^2}$

To illustrate how these rules are established, we derive rule 4 as an example:

$$y = f(x) = u(x)v(x)$$
$$dy = f'(x)\,dx$$
$$= [u(x)v'(x) + v(x)u'(x)]\,dx$$
$$= u(x)v'(x)\,dx + v(x)u'(x)\,dx$$
$$= u\,dv + v\,du$$

EXERCISE 3-7

A   Find dy for each function.

1. $y = 30 + 12x^2 - x^3$

2. $y = 200x - \dfrac{x^2}{30}$

3. $y = x^2\left(1 - \dfrac{x}{9}\right)$

4. $y = x^3(60 - x)$

5. $y = f(x) = \dfrac{590}{\sqrt{x}}$

6. $y = 52\sqrt{x}$

7. $y = 75\left(1 - \dfrac{2}{x}\right)$

8. $y = 100\left(x - \dfrac{4}{x^2}\right)$

B   Evaluate dy and Δy for each function at the indicated values.

9. $y = f(x) = x^2 - 3x + 2, \quad x = 5, \quad \Delta x = dx = 0.2$

10. $y = f(x) = 30 + 12x^2 - x^3, \quad x = 2, \quad \Delta x = dx = 0.1$

11. $y = f(x) = 75\left(1 - \dfrac{2}{x}\right), \quad x = 5, \quad dx = \Delta x = 0.5$

12. $y = f(x) = 100\left(x - \dfrac{4}{x^2}\right), \quad x = 2, \quad \Delta x = dx = 0.1$

Use differentials to approximate the indicated roots.

13. $\sqrt[4]{17}$ 14. $\sqrt{83}$ 15. $\sqrt[3]{28}$ 16. $\sqrt[5]{34}$

17. A cube with sides 10 inches long is covered with a 0.2 inch thick coat of fiberglass. Use differentials to estimate the volume of the fiberglass shell.

18. A sphere with a radius of 5 centimeters is coated with ice 0.1 centimeter thick. Use differentials to estimate the volume of the ice $(V = \tfrac{4}{3}\pi r^3, \pi \approx 3.14)$.

C   19. Find dy if $y = \sqrt[3]{3x^2 - 2x + 1}$.

20. Find dy if $y = (2x^2 - 4)\sqrt{x + 2}$.

21. Find dy and Δy for $y = 52\sqrt{x}$, $x = 4$, and $\Delta x = dx = 0.3$.

22. Find dy and Δy for $y = 590/\sqrt{x}$, $x = 64$, and $\Delta x = dx = 1$.

APPLICATIONS

Use differential approximations in the following problems.

BUSINESS & ECONOMICS

23. *Advertising.* Using past records, it is estimated that a company will sell N units of a product after spending x thousand dollars in advertising, as given by

$$N = 60x - x^2 \qquad 5 \le x \le 30$$

Estimate the increase in sales that will result in increasing the advertising budget from $10,000 to $11,000. From $20,000 to $21,000.

24. *Price–demand.* Suppose in a grocery chain the daily demand in pounds for chocolate candy at $x per pound is given by

$$D = 1,000 - 40x^2 \qquad 1 \le x \le 5$$

If the price is increased from $3.00 per pound to $3.20 per pound, what is the approximate change in demand?

25. *Average cost.* For a company that manufactures tennis rackets, the average cost per racket, $\overline{C}$, is found to be

$$\overline{C} = x^2 - 20x + 110 \qquad 6 \le x \le 14$$

where x is the number of rackets produced per hour. Approximate the change in cost per racket if production is increased from seven per hour to eight per hour. From twelve per hour to thirteen per hour.

26. *Revenue and profit.* A company manufactures and sells x televisions per month. If the cost and revenue equations are

$$C(x) = 72,000 + 60x$$
$$R(x) = 200x - \frac{x^2}{30} \qquad 0 \le x \le 6,000$$

find the approximate change in revenue and profit if production is increased from 1,500 to 1,501. From 4,500 to 4,501.

LIFE SCIENCES

27. *Pulse rate.* The average pulse rate y in beats per minute of a healthy person x inches tall is given approximately by

$$y = \frac{590}{\sqrt{x}} \qquad 30 \le x \le 75$$

Approximate the change in pulse rate for a height change from 36 to 37 inches. From 64 to 65 inches.

28. *Measurement.* An egg of a particular bird is very nearly spherical. If the radius to the inside of the shell is 5 millimeters and the radius

to the outside of the shell is 5.3 millimeters, approximately what is the volume of the shell? (Remember that $V = \frac{4}{3}\pi r^3$ and use $\pi \approx 3.14$.)

29. *Medicine.*   A drug is given to a patient to dilate her arteries. If the radius of an artery is increased from 2 to 2.1 millimeters, approximately how much is a cross-sectional area increased? (Assume the cross-section of the artery is circular; $A = \pi r^2$ and $\pi \approx 3.14$.)

30. *Drug sensitivity.*   One hour after x milligrams of a particular drug are given to a person, the change in body temperature T in degrees Fahrenheit is given by

$$T = x^2\left(1 - \frac{x}{9}\right) \qquad 0 \le x \le 6$$

Approximate the changes in body temperature produced by the following changes in drug dosages:
(A)   From 2 to 2.1 milligrams
(B)   From 3 to 3.1 milligrams
(C)   From 4 to 4.1 milligrams

SOCIAL SCIENCES

31. *Learning.*   A particular person learning to type has an achievement record given approximately by

$$N = 75\left(1 - \frac{2}{t}\right) \qquad 3 \le t \le 20$$

where N is the number of words per minute typed after t weeks of practice. What is the approximate improvement from 5 to 5.5 weeks of practice?

32. *Learning.*   If a person learns y items in x hours, as given approximately by

$$y = 52\sqrt{x} \qquad 0 \le x \le 9$$

what is the approximate increase in the number of items learned when x changes from 1 to 1.1 hours? From 4 to 4.1 hours?

33. *Politics.*   In a newly incorporated city it is estimated that the voting population (in thousands) will increase according to

$$N(t) = 30 + 12t^2 - t^3 \qquad 0 \le t \le 8$$

where t is time in years. Find the approximate change in votes for the following time changes:
(A)   From 1 to 1.1 years
(B)   From 4 to 4.1 years
(C)   From 7 to 7.1 years

3-8   CHAPTER REVIEW

EXERCISE 3-8   CHAPTER REVIEW

Work through all the problems in this chapter review and check your answers in the back of the book. (Answers to all review problems are there.) Where weaknesses show up, review appropriate sections in the text. When you are satisfied that you know the material, take the practice test following this review.

A
1. Differentiate implicitly to find dy/dx if $2y^2 - 3x^3 - 5 = 0$. Evaluate dy/dx at $(1, 2)$.
2. For $y = 3x^2 - 5$, find dy/dt if $dx/dt = 3$ when $x = 12$.
3. Find $f''(x)$ if $f(x) = 2 - 3x^2 - 3x^5$.

Problems 4–11 refer to the following graph of $y = f(x)$:

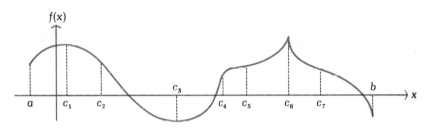

Identify the points on the x axis that produce an indicated behavior.

4. f is rising
5. $f'(x) < 0$
6. Graph of f is concave downward
7. Relative minima
8. Absolute maximum
9. $f'(x)$ appears to be zero
10. $f'(x)$ does not exist
11. Inflection points

Problems 12–17 refer to the function $y = f(x) = x^2 - 6x + 10, 0 \le x \le 5$.

12. Find intervals over which f is increasing. Decreasing.
13. Find intervals over which the graph of f is concave upward. Downward.
14. Identify critical points.
15. Find relative maxima and minima.
16. Find the absolute maximum and minimum.
17. Graph f.
18. Find dy if $y = 10 - x + 3x^2$.

B
19. Find dy/dx if $x^2y - 2x^3 + 4 = 0$, and evaluate at $(2, 3)$. Differentiate implicitly.
20. For $x^2 + y^2 = 25$, find dy/dt if $dx/dt = -2$ when $x = -3$ and $y = 4$.
21. Find d^2y/dx^2 if $y = x^3 - \sqrt{x}$.
22. Find d^2y/dx^2 if $2x^2 - y^2 = 3$, using implicit differentiation.

Problems 23–28 refer to the function $y = f(x) = x^{2/3} - 1$.

23. Find intervals over which f is decreasing. Increasing.
24. Find intervals over which f is concave downward. Upward.
25. Identify critical points.
26. Find relative maxima and minima.
27. Find the absolute maximum and absolute minimum.
28. Graph f.
29. Find relative maxima and minima for $y = f(x) = x^2(x - 3)$, and graph f for $-2 \le x \le 4$ using any other derivative aids to graphing you wish.
30. Find the absolute maximum and minimum for $y = f(x) = x^3 - 12x + 12, -3 \le x \le 5$.
31. Find dy and Δy for $y = f(x) = x^3 - 2x + 1, x = 5$, and $\Delta x = dx = 0.1$.
32. Approximate $\sqrt{17}$ using differentials.

C   33. Find the equation of the tangent line to the circle $x^2 + y^2 = 25$ at $(-4, 3)$. Use implicit differentiation.
34. Find d^3y/dx^3 if $x^2 + y^2 = 6$ using implicit differentiation.
35. Find the absolute maximum and minimum for $y = f(x) = 3\sqrt[3]{x} - x$, $0 \le x \le 8$. Graph f.
36. Find dy and Δy for $y = (2/\sqrt{x}) + 8, x = 16, \Delta x = dx = 0.2$.

APPLICATIONS

BUSINESS & ECONOMICS

37. *Revenue.* If the revenue equation for the weekly production of x units is $R = 300x - (x^2/30)$, and if $dx/dt = 30$ when $x = 1,500$, find dR/dt at this production level (t is time in weeks).

38. *Profit.* The profit for a company manufacturing and selling x units per month is given by

$$P(x) = 150x - \frac{x^2}{40} - 50,000 \qquad 0 \le x \le 5,000$$

What production level will produce the maximum profit? What is the maximum?

39. *Rental income.* A 100 room hotel in San Francisco is filled to capacity every night at a rate of $20 per room. For each $1 increase in the nightly rate, two fewer rooms are rented. If each rented room costs $4 a day to service, how much should the management charge per room in order to maximize gross income?

LIFE SCIENCES

40. *Ecology.* A coal-burning electrical generating plant emits sulfur dioxide into the surrounding air. The concentration C in parts per million is given approximately by

$$C = 0.1x^{-2}$$

where x is the distance from the plant in miles. Approximate the change in concentration when the distance is increased from 1 to 1.05 miles, using differentials.

41. *Bacteria control.* If t days after a treatment the bacteria count per cubic centimeter in a body of water is given by

$$C(t) = 20t^2 - 120t + 800 \qquad 0 \leq t \leq 9$$

in how many days will the count be a minimum?

SOCIAL SCIENCES

42. *Learning.* If a person learns y items in x hours, as given approximately by

$$y = 24\sqrt{x} \qquad 0 \leq x \leq 4$$

use differentials to approximate the increase in items learned when x changes from x = 1 to x = 1.2 hours.

43. *Politics.* In a new suburb it is estimated that the number of registered voters will grow according to

$$N = 10 + 6t^2 - t^3 \qquad 0 \leq t \leq 5$$

where t is time in years and N is in thousands. When will the rate of increase be maximum?

PRACTICE TEST: **CHAPTER 3**

1. Find y' if $x^2 - 3xy + 4y^2 = 22$ using implicit differentiation.
2. Find $D_x^3 \sqrt{1 - 2x}$.

Problems 3–5 refer to the function $y = f(x) = 2x^3 - 9x^2 + 7$.

3. Find all local maxima and minima.
4. Find intervals over which the graph of f is concave upward. Concave downward. Indicate the x coordinate(s) of any inflection point(s).
5. Sketch a graph of f.
6. Find all local maxima and minima for $f(x) = (x - 4)^{2/3} - 1$.
7. Find the absolute maximum and minimum for $f(x)$ in Problem 6 over the interval $[-4, 5]$.

8. A point is moving around the circle given by $x^2 + y^2 = 25$. Find dx/dt if $dy/dt = 2$ feet per second when $x = 4$ feet and $y = 3$ feet.

9. Find two positive numbers whose sum is 12 and whose product is maximum. What is the maximum product?

10. Find dy and Δy for $y = f(x) = x^2 + 1$, $x = 3$, and $\Delta x = dx = 0.1$.

11. A wooden cube 2 inches on a side is coated with an ivory shell 0.1 inch thick. Use differentials to approximate the volume of·the ivory coating.

12. A cable television company has 3,600 subscribers in a city, each paying $10 per month for the service. A survey indicates that for each 50¢ reduction in rate, 300 more people will subscribe (and none of the original subscribers will be lost). What rate will maximize revenue? What is the maximum revenue and how many subscribers will produce this revenue?

CHAPTER 4
INTEGRATION

CONTENTS

4 INTEGRATION

The last two chapters dealt with differential calculus. We now begin the development of the second main part of calculus, called *integral calculus*. Two types of integrals will be introduced, the *indefinite integral* and the *definite integral*; each is quite different from the other. But through the remarkable *fundamental theorem of calculus*, we will show that not only are the two integral forms intimately related, but both are intimately related to differentiation.

4-1 ANTIDERIVATIVES AND INDEFINITE INTEGRALS

Many operations in mathematics have reverses—compare addition and subtraction, multiplication and division, and powers and roots. The function $f(x) = \frac{1}{3}x^3$ has the derivative $f'(x) = x^2$. Reversing this process is referred to as *antidifferentiation*. Thus,

$$\frac{x^3}{3}$$

is an antiderivative of

$$x^2$$

since

$$D_x\left(\frac{x^3}{3}\right) = x^2$$

In general, we say that $F(x)$ is an **antiderivative** of $f(x)$ over an open interval (a, b) if

$$F'(x) = f(x)$$

over (a, b).

Note that

$$D_x\left(\frac{x^3}{3} + 2\right) = x^2 \qquad D_x\left(\frac{x^3}{3} - \pi\right) = x^2 \qquad D_x\left(\frac{x^3}{3} + \sqrt{5}\right) = x^2$$

Hence,

$$\frac{x^3}{3} + 2 \qquad \frac{x^3}{3} - \pi \qquad \frac{x^3}{3} + \sqrt{5}$$

are also antiderivatives of x^2, since each has x^2 as a derivative. In fact, it appears that

$$\frac{x^3}{3} + C$$

for any real number C, is an antiderivative of x^2, since

$$D_x\left(\frac{x^3}{3} + C\right) = x^2$$

Thus, antidifferentiation of a given function does not, in general, lead to a unique function, but to a whole set of functions.

Does the expression

$$\frac{x^3}{3} + C$$

with C any real number, include all antiderivatives of x^2? Theorem 1 (which we state without proof) indicates that the answer is yes.

Theorem 1   If F and G are differentiable functions on the interval (a, b) and $F'(x) = G'(x)$, then $F(x) = G(x) + k$ for some constant k.

In words, the theorem states that **if the derivatives of two functions are equal, then the functions differ by at most a constant.** We use the symbol

$$\int f(x) \, dx$$

called the **indefinite integral,** to represent all antiderivatives of $f(x)$, and we write

$$\int f(x) \, dx = F(x) + C$$

where $F'(x) = f(x)$; that is, if $F(x)$ is any antiderivative of $f(x)$. The symbol $\int$ is called an **integral sign,** $f(x)$ is called the **integrand,** and dx indicates the variable with respect to which the integration is being performed (we will have more to say about the symbol dx later). The arbitrary constant C is called the **constant of integration.**

Just as with differentiation, we can develop formulas and special properties that will enable us to find indefinite integrals of many frequently encountered functions. To start, we list some formulas that can be established using the definitions of antiderivative and indefinite integral, and the many properties of derivatives considered in Chapters 2 and 3.

INDEFINITE INTEGRAL PROPERTIES

For f and g continuous over (a, b), and k and C constants:

1. $\displaystyle \int k\,dx = kx + C$

2. $\displaystyle \int x^n\,dx = \frac{x^{n+1}}{n+1} + C \qquad n \neq -1$

3. $\displaystyle \int kf(x)\,dx = k\int f(x)\,dx$

4. $\displaystyle \int [f(x) \pm g(x)]\,dx = \int f(x)\,dx \pm \int g(x)\,dx$

We will establish properties 2 and 3 here (the others may be shown to be true in a similar manner). To establish property 2, we simply differentiate the right side to obtain the integrand on the left side. Thus,

$$D_x\left(\frac{x^{n+1}}{n+1} + C\right) = \frac{(n+1)x^n}{(n+1)} + 0 = x^n \qquad n \neq -1$$

(The case when $n = -1$ will be considered later.) To establish property 3, let F be a function such that $F'(x) = f(x)$. Then

$$k\int f(x)\,dx = k\int F'(x)\,dx = k[F(x) + C_1] = kF(x) + kC_1$$

and since $(kF(x))' = kF'(x) = kf(x)$,

$$\int kf(x)\,dx = \int kF'(x)\,dx = kF(x) + C_2$$

But $kF(x) + kC_1$ and $kF(x) + C_2$ describe the same set of functions, since C_1 and C_2 are arbitrary real numbers. It is important to remember that property 3 states that a constant factor can be moved across an integral sign; a variable factor cannot be moved across an integral sign.

Now let us put the four properties to use.

Example 1   (A) $\int 5\,dx = 5x + C$

(B) $\int x^4\,dx = \dfrac{x^{4+1}}{4+1} + C = \dfrac{x^5}{5} + C$

(C) $\int 5x^7\,dx = 5\int x^7\,dx = 5\dfrac{x^8}{8} + C = \tfrac{5}{8}x^8 + C$

(D) $\int (4x^3 + 2x - 1)\,dx = \int 4x^3\,dx + \int 2x\,dx - \int dx$

$= 4\int x^3\,dx + 2\int x\,dx - \int dx$

$= \dfrac{4x^4}{4} + \dfrac{2x^2}{2} - x + C$

$= x^4 + x^2 - x + C$

(E) $\int \dfrac{3\,dx}{x^2} = \int 3x^{-2}\,dx = \dfrac{3x^{-2+1}}{-2+1} + C = -3x^{-1} + C$

(F) $\int 5\sqrt[3]{x^2}\,dx = 5\int x^{2/3}\,dx = 5\dfrac{x^{(2/3)+1}}{(2/3)+1} + C$

$= 5\dfrac{x^{5/3}}{5/3} + C = 3x^{5/3} + C$

To check any of these, we differentiate the final result to obtain the integrand in the original indefinite integral.

Problem 1   Find each of the following:

(A) $\int dx$ (B) $\int 3x^4\,dx$ (C) $\int (2x^5 - 3x^2 + 1)\,dx$

(D) $\int 4\sqrt[5]{x^3}\,dx$ (E) $\int \left(2x^{2/3} - \dfrac{3}{x^4}\right)dx$

Answers   (A) $x + C$ (B) $\tfrac{3}{5}x^5 + C$ (C) $x^6/3 - x^3 + x + C$
(D) $\tfrac{5}{2}x^{8/5} + C$ (E) $\tfrac{6}{5}x^{5/3} + x^{-3} + C$

We will consider additional techniques of integration in the next section. Let us now consider some applications of the indefinite integral to see why we are interested in finding antiderivatives of functions.

APPLICATIONS

Example 2   Find the equation of the curve that passes through (2, 5) if its slope is given by $dy/dx = 2x$ at any point x.

Solution   We are interested in finding a function $y = f(x)$ such that

$$\frac{dy}{dx} = 2x \tag{1}$$

and

$$y = 5 \quad \text{when } x = 2 \tag{2}$$

If

$$\frac{dy}{dx} = 2x$$

then

$$y = \int 2x \, dx$$

$$= x^2 + C \tag{3}$$

Since y must be 5 when $x = 2$, we determine the *particular* value of C so that

$$5 = 2^2 + C$$

Thus,

$$C = 1$$

and

$$y = x^2 + 1$$

is the particular antiderivative out of all those possible from (3) that satisfies both (1) and (2). See Figure 1.

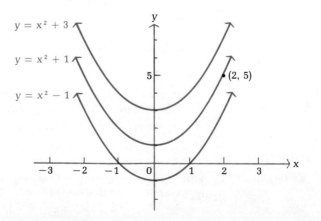

FIGURE 1 $y = x^2 + C$

Problem 2   Find the equation of the curve that passes through (2, 6) if the slope of the curve at any point x is given by $dy/dx = 3x^2$.

Answer $y = x^3 - 2$

In certain situations it is easier to determine the rate at which something happens than how much of it has happened in a given length of time (e.g., population growth rates, business growth rates, rate of healing of a wound, rates of learning or forgetting). If a rate function (derivative) is given and we know the value of the dependent variable for a given value of the independent variable, then—if the rate function is not too complicated—we can often find the original function by integration.

Example 3   If the marginal cost of producing x units is given by

$$C'(x) = 3x^2 - 2x$$

and the fixed cost is $2,000, find the cost function $C(x)$ and the cost of producing twenty units.

Solution   Recall that marginal cost is the derivative of the cost function and that fixed cost is cost at a zero production level. Thus, the mathematical problem is to find $C(x)$ given

$$C'(x) = 3x^2 - 2x \qquad C(0) = 2,000$$

We now find the indefinite integral of $3x^2 - 2x$ and determine the arbitrary integration constant using $C(0) = 2,000$.

$$C'(x) = 3x^2 - 2x$$

$$C(x) = \int (3x^2 - 2x)\, dx$$

$$= x^3 - x^2 + k$$

But

$$C(0) = 0^3 - 0^2 + k = 2,000$$

Thus,

$$k = 2,000$$

and the particular cost function is

$$C(x) = x^3 - x^2 + 2,000$$

We now find C(20), the cost of producing twenty units:

$$C(20) = 20^3 - 20^2 + 2,000$$
$$= \$9,600$$

Problem 3   Find the revenue function $R(x)$ when the marginal revenue is

$$R'(x) = 400 - 0.4x$$

and no revenue results at a zero production level. What is the revenue at a production level of 1,000 units?

Answer $R(x) = 400x - 0.2x^2$ $R(1,000) = \$200,000$

Example 4   An oil tanker aground on a reef is losing oil and producing an oil slick that is radiating out at a rate given approximately by

$$\frac{dR}{dt} = \frac{25}{\sqrt{t}} \qquad t \geq 1$$

where R is the radius of the circular slick in feet after t minutes. Find the radius of the slick after 49 minutes if the radius is 50 feet after 1 minute.

Solution   First find the radius function $R(t)$; that is, solve

$$\frac{dR}{dt} = \frac{25}{\sqrt{t}} \qquad R(1) = 50$$

for R. Integrating, that is, finding the indefinite integral, we obtain

$$R = \int 25t^{-1/2}\, dt = \frac{25t^{-(1/2)+1}}{-(\frac{1}{2}) + 1} + C$$

Thus,

$$R(t) = 50t^{1/2} + C$$

But, since $R(1) = 50$, we find C as follows:

$$R(1) = 50(1^{1/2}) + C = 50$$

and

$$C = 0$$

Hence,

$$R(t) = 50\sqrt{t}$$

and the radius of the slick after 49 minutes is

$$R(49) = 50\sqrt{49} = 350 \text{ feet}$$

Problem 4   An influenza epidemic has hit a city and it is estimated that the rate of change of people without influenza with respect to time is given by

$$\frac{dW}{dt} = 400t - 12,000$$

Find $W(t)$, the number of people without influenza in t days, if $W(0) = 500,000$. Then find the number of people without influenza 30 days after the start of the epidemic.

Answer $W(t) = 200t^2 - 12,000t + 500,000$ $W(30) = 320,000$

EXERCISE 4-1

A   *Find each indefinite integral. (Check by differentiating.)*

1. $\int 7\,dx$ 2. $\int \pi\,dx$

3. $\int x^6\,dx$ 4. $\int x^3\,dx$

5. $\int 8t^3\,dt$ 6. $\int 10t^4\,dt$

7. $\int (2u + 1)\,du$ 8. $\int (1 - 2u)\,du$

9. $\int (3x^2 + 2x - 5)\,dx$ 10. $\int (2 + 4x - 6x^2)\,dx$

11. $\int (s^4 - 8s^5)\,ds$ 12. $\int (t^5 + 6t^3)\,dt$

Find all antiderivatives for each derivative.

13. $dy/dx - 200x^4$ 14. $dx/dt = 42t^5$
15. $dP/dx = 24 - 6x$ 16. $dy/dx = 3x^2 - 4x^3$
17. $dy/du = 2u^5 - 3u^2 - 1$ 18. $dA/dt = 3 - 12t^3 - 9t^5$

B   *Find each indefinite integral. (Check by differentiation.)*

19. $\int 6x^{1/2}\,dx$ 20. $\int 8t^{1/3}\,dt$

21. $\int 8x^{-3}\,dx$ 22. $\int 12u^{-4}\,du$

23. $\int \dfrac{du}{\sqrt{u}}$ 24. $\int \dfrac{dt}{\sqrt[3]{t}}$

25. $\int \dfrac{dx}{4x^3}$ 26. $\int \dfrac{6\,dm}{m^2}$

27. $\int \dfrac{du}{2u^5}$ 28. $\int \dfrac{dy}{3y^4}$

29. $\int (3x^{1/2} - x^{-1/2})\,dx$ 30. $\int (4x^{1/3} + 2x^{-1/3})\,dx$

31. $\int (10x^{2/3} - 8x^{1/3} - 2)\,dx$ 32. $\int (6x^{-4} - 2x^{-3} + 1)\,dx$

33. $\int \left(3\sqrt{x} + \dfrac{2}{\sqrt{x}}\right)dx$ 34. $\int \left(\dfrac{2}{\sqrt[3]{x}} - \sqrt[3]{x^2}\right)dx$

35. $\int \left(\sqrt[3]{x^2} - \dfrac{4}{x^3}\right)dx$ 36. $\int \left(\dfrac{12}{x^5} - \dfrac{1}{\sqrt[3]{x^2}}\right)dx$

Find the particular antiderivative of each derivative that satisfies the second condition.

37. $dy/dx = 2x - 3, \quad y(0) = 5$
38. $dy/dx = 5 - 4x, \quad y(0) = 20$
39. $C'(x) = 6x^2 - 4x, \quad C(0) = 3{,}000$
40. $R'(x) = 600 - 0.6x, \quad R(0) = 0$
41. $dx/dt = 20/\sqrt{t}, \quad x(1) = 40$
42. $dR/dt = 100/t^2, \quad R(1) = 400$
43. Find the equation of the curve that passes through $(2, 3)$ if its slope is given by

$$\frac{dy}{dx} = 4x - 3$$

for each x.
44. Find the equation of the curve that passes through $(1, 3)$ if its slope is given by

$$\frac{dy}{dx} = 12x^2 - 12x$$

for each x.

C *Find each indefinite integral.*

45. $\displaystyle\int \frac{2x^4 - x}{x^3}\, dx$

46. $\displaystyle\int \frac{x^{-1} - x^4}{x^2}\, dx$

47. $\displaystyle\int \frac{x^5 - 2x}{x^4}\, dx$

48. $\displaystyle\int \frac{1 - 3x^4}{x^2}\, dx$

Find the antiderivative of each of the derivatives that satisfies the second condition.

49. $\dfrac{dM}{dt} = \dfrac{\sqrt{t} - 1}{\sqrt{t}}, \quad M(4) = 5$

50. $\dfrac{dR}{dx} = \dfrac{1 - \sqrt[3]{x^2}}{\sqrt[3]{x}}, \quad R(8) = 4$

51. $\dfrac{dy}{dx} = \dfrac{5x + 2}{\sqrt[3]{x}}, \quad y(1) = 0$

52. $\dfrac{dx}{dt} = \dfrac{\sqrt{t^3} - t}{\sqrt{t^3}}, \quad x(9) = 4$

APPLICATIONS

BUSINESS & ECONOMICS

53. *Profit function.* If the marginal profit for producing x units is given by

$$P'(x) = 50 - 0.04x \qquad P(0) = 0$$

where $P(x)$ is the profit in dollars, find the profit function P and the profit on 100 units of production.

54. *Natural resources.* The world demand for wood is increasing. In 1975 the demand was 12.6 billion cubic feet, and the rate of increase in demand is given approximately by

$$d'(t) = 0.009t$$

where t is time in years after 1975. Noting that $d(0) = 12.6$, find $d(t)$. Also find $d(25)$, the demand in the year 2000.

55. *Resale value.* A company purchased an airplane for $300,000. The resale value is expected to decrease over a 16 year period at a rate that changes with time and is estimated to be

$$v'(t) = \frac{-25}{\sqrt{t}}$$

where $v(t)$ is the resale value of the plane in thousands of dollars after t years. Noting that $v(0) = \$300,000$, find $v(t)$. Also find $v(15)$, the resale value in thousands of dollars after 15 years.

LIFE SCIENCES

56. *Pollution.* A lake, contaminated because of septic tank leakage, is treated with a bactericide. It is found that the rate of decrease in harmful bacteria t days after the treatment is given by

$$\frac{dN}{dt} = -2,000 + 200t \qquad 0 \le t \le 9$$

where $N(t)$ is the number of bacteria per milliliter of water. Find $N(t)$ and then find the bacteria count after 9 days, assuming the initial count is 10,000 per milliliter.

57. *Weight–height.* The rate of change of an average person's weight with respect to their height h (in inches) is given approximately by

$$\frac{dW}{dh} = 0.0015h^2$$

Find $W(h)$ if $W(60) = 108$ pounds. Also find the average weight for a person that is 5 feet 10 inches tall.

58. *Wound healing.* If the area of a healing wound changes at a rate given approximately by

$$\frac{dA}{dt} = -4t^{-3} \qquad 1 \le t \le 10$$

where t is in days and $A(1) = 2$ square centimeters, what will the area of the wound be in 10 days?

SOCIAL SCIENCES

59. *Urban growth.* A suburban area of Chicago incorporated into a city. The growth rate t years after incorporation is estimated to be

$$\frac{dN}{dt} = 400 + 600\sqrt{t} \qquad 0 \le t \le 9$$

If the current population is 5,000, what will the population be 9 years from now?

60. *Learning.* In an experiment on memorizing vocabulary from a foreign language, it is found that the rate of learning during a study session increases and then decreases because of saturation. A typical rate might be given by

$$v'(t) = 0.04t - 0.0003t^2$$

where $v(t)$ is the amount of vocabulary learned after t minutes of study. Find $v(t)$ if $v(0) = 0$. Then find how many words are learned after 60 minutes of study.

61. *Retention.* In an experiment on retention of information by witnesses, a group of people were shown a film of an accident. Then, without any written records, they were tested orally every few days over a 36 day period to see if they correctly recalled twenty relevant facts. We expect that the people will forget at a faster rate at the beginning of the period than at the end. In this case, the rate of retention (the negative of the rate of forgetting) was found to be given by

$$N'(t) = -\frac{1}{\sqrt{t}} \qquad t \ge 1$$

where $N(t)$ is the number of facts remembered after t days. Find $N(t)$ if $N(1) = 18$. Then find the number of facts retained after 36 days.

4-2   INTEGRATION BY SUBSTITUTION

The properties of the indefinite integral discussed in the last section enable us to find the indefinite integrals of a fairly large set of functions, but there are still many functions with indefinite integrals that cannot be found by the application of these properties. In this section we will substantially increase our capability of finding indefinite integrals by a method called *substitution*.

In the notation

$$\frac{dy}{dx} = f'(x)$$

dy and dx can be defined as separate quantities, allowing us to write

$$dy = f'(x)\, dx$$

where x and dx are independent variables. For example, if

$$y = f(x) = x^7$$

then

$$\frac{dy}{dx} = f'(x) = 7x^6 \qquad \text{or} \qquad dy = f'(x)\, dx = 7x^6\, dx$$

And if

$$u = g(x) = x^{1/2}$$

then

$$du = g'(x)\, dx = \tfrac{1}{2}x^{-1/2}\, dx$$

In the last section, we said that the symbol du in

$$\int f(u)\, du$$

is used to indicate the variable for which the integration is to take place. Now we will see a more significant use of the symbol du in the very powerful **substitution rule.** The rule can be proved by showing (using the chain rule) that the derivative of the right side is the integrand on the left side.

SUBSTITUTION RULE

If we let $u = u(x)$ and $du = u'(x)\, dx$, then:

1. General rule

$$\int \underbrace{f(u(x))}_{u}\underbrace{u'(x)\, dx}_{du} = \int f(u)\, du = F(u) + C = F(u(x)) + C$$
$$F'(u) = f(u)$$

2. Power form of the general rule

$$\int \underbrace{[u(x)]^n}_{u}\underbrace{u'(x)\, dx}_{du} = \int u^n\, du = \frac{u^{n+1}}{n+1} + C = \frac{[u(x)]^{n+1}}{n+1} + C \quad n \neq -1$$

We will use the power form of the substitution rule in this chapter and the general form in the next chapter.

Let us prove the power form of the substitution rule. Differentiating the right-hand member with respect to x, using the chain rule, we obtain

$$D_x\left(\frac{u^{n+1}}{n+1} + C\right) = (n+1)\frac{u^n}{n+1}u'(x) = [u(x)]^n u'(x)$$

which is the integrand of the left-hand member.

Example 5   Integrate $\int (x^2 - 1)^{10} 2x\ dx$.

Solution   This integral is a special case of

$$\int [u(x)]^n u'(x)\ dx = \int u^n\ du$$

where

$$u(x) = x^2 - 1 \qquad \text{and} \qquad du = 2x\ dx$$

Thus, using the substitution rule, we obtain

$$\int \underbrace{(x^2 - 1)^{10}}_{u}\ \underbrace{2x\ dx}_{du} = \int u^{10}\ du$$

$$= \frac{u^{11}}{11} + C$$

$$= \frac{(x^2 - 1)^{11}}{11} + C \qquad \text{Since } u = x^2 - 1$$

Check: $D_x\left[\dfrac{(x^2-1)^{11}}{11} + C\right] = \dfrac{11(x^2-1)^{10}}{11}D_x(x^2 - 1) = (x^2 - 1)^{10} 2x$

Problem 5   Integrate $\int (x^3 + 5)^6 3x^2\ dx$ and check by differentiating.

Answer $\dfrac{(x^3 + 5)^7}{7} + C$

Example 6   Integrate $\int x^2\sqrt{x^3 - 10}\ dx$.

Solution   Rewrite the integrand in the power form

$$\int (x^3 - 10)^{1/2} x^2\ dx$$

Now, if we let

$$u = x^3 - 10$$

then

$$du = 3x^2\ dx$$

The original integral needs a factor of 3 in the integrand to qualify as a $\int u^n\, du$ form. A missing *constant* factor can always be taken care of as follows:

$$\int (x^3 - 10)^{1/2} x^2\, dx = \int (x^3 - 10)^{1/2} \frac{3}{3} x^2\, dx$$

$$= \frac{1}{3} \int \underbrace{(x^3 - 10)^{1/2}}_{u}\, \underbrace{3x^2\, dx}_{du} = \frac{1}{3} \int u^{1/2}\, du$$

$$= \frac{1}{3} \frac{u^{3/2}}{\frac{3}{2}} + C = \frac{2}{9}(x^3 - 10)^{3/2} + C$$

Check: $\quad D_x[\frac{2}{9}(x^3 - 10)^{3/2} + C] = (\frac{2}{9})(\frac{3}{2})(x^3 - 10)^{1/2}D_x(x^3 - 10)$

$$= \frac{1}{3}(x^3 - 10)^{1/2}3x^2$$

$$= (x^3 - 10)^{1/2}x^2$$

Problem 6 Integrate $\int x \sqrt{x^2 + 5}\, dx$ and check.

Answer $\frac{1}{3}(x^2 + 5)^{3/2} + C$

Example 7 Integrate: $\quad \displaystyle\int \frac{x - 1}{(x^2 - 2x + 3)^3}\, dx$

Solution $\quad \displaystyle\int \frac{x - 1}{(x^2 - 2x + 3)^3}\, dx = \int (x^2 - 2x + 3)^{-3}(x - 1)\, dx$

If we let $u = x^2 - 2x + 3$, then $du = (2x - 2)\, dx = 2(x - 1)\, dx$. We are missing a factor of 2 in the integrand. We introduce the factor as in Example 6:

$$\int (x^2 - 2x + 3)^{-3}(x - 1)\, dx = \int (x^2 - 2x + 3)^{-3}\frac{2}{2}(x - 1)\, dx$$

$$= \frac{1}{2} \int \underbrace{(x^2 - 2x + 3)^{-3}}_{u}\, \underbrace{2(x - 1)\, dx}_{du}$$

$$= \frac{1}{2} \int u^{-3}\, du$$

$$= \frac{1}{2} \frac{u^{-2}}{-2} + C$$

$$= -\frac{1}{4}u^{-2} + C$$

$$= -\frac{1}{4}(x^2 - 2x + 3)^{-2} + C$$

Problem 7   Integrate: $\displaystyle\int \frac{x^2 - x}{\sqrt{2x^3 - 3x^2 + 5}}\,dx$

Answer $\frac{1}{3}\sqrt{2x^3 - 3x^2 + 5} + C$

A CONSTANT FACTOR CAN BE MOVED BACK AND FORTH
ACROSS AN INTEGRAL SIGN, BUT A VARIABLE FACTOR CANNOT

YES   **NO**

$$\int kf(x)\,dx = k \int f(x)\,dx \qquad \int xf(x)\,dx = x \int f(x)\,dx$$

(*k* a constant)   (*x* a variable)

If we try to use the substitution rule

$$\int \sqrt{x^2 + 1}\,dx$$

by letting

$$u = x^2 + 1 \qquad \text{and} \qquad du = 2x\,dx$$

we find we are missing a factor of $2x$ in the integrand of the original integral. We cannot introduce the $2x$ in the integrand as we did the 3 in Example 6, since $1/2x$ cannot be taken outside the integral. This integral requires methods we will not consider in this book.

EXERCISE 4-2

Find each indefinite integral. (Check by differentiating the result.)

A   1. $\displaystyle\int (x^2 - 4)^5 2x\,dx$ 2. $\displaystyle\int (x^3 + 1)^4 3x^2\,dx$

3. $\displaystyle\int \sqrt{2x^2 - 1}\,4x\,dx$ 4. $\displaystyle\int \sqrt[3]{2x^3 + 5}\,6x^2\,dx$

5. $\displaystyle\int (3x - 2)^7\,dx$ 6. $\displaystyle\int (5x + 3)^9\,dx$

B   7. $\displaystyle\int (x^2 + 3)^7 x\,dx$ 8. $\displaystyle\int (x^3 - 5)^4 x^2\,dx$

9. $\int x\sqrt{3x^2 + 7}\,dx$

10. $\int x^2\sqrt{2x^3 + 1}\,dx$

11. $\int \dfrac{x^3}{\sqrt{2x^4 + 3}}\,dx$

12. $\int \dfrac{x^2}{\sqrt{4x^3 - 1}}\,dx$

13. $\int (x - 1)\sqrt{x^2 - 2x - 3}\,dx$

14. $\int (x^3 - x)\sqrt{x^4 - 2x^2 + 7}\,dx$

15. $\int \dfrac{t}{(3t^2 + 1)^4}\,dt$

16. $\int \dfrac{t^2}{(t^3 - 2)^5}\,dt$

C 17. $\int \dfrac{x^2}{\sqrt{4 - x^3}}\,dx$

18. $\int \dfrac{x}{(5 - 2x^2)^5}\,dx$

19. $\int \dfrac{x^3 + x}{(x^4 + 2x^2 + 1)^4}\,dx$

20. $\int \dfrac{x^2 - 1}{\sqrt[3]{x^3 - 3x + 7}}\,dx$

APPLICATIONS

BUSINESS & ECONOMICS

21. *Revenue function.* If the marginal revenue in thousands of dollars of producing x units is given by

$$R'(x) = x(x^2 + 9)^{-1/2}$$

and no revenue results from a zero production level, find the revenue function $R(x)$. Find the revenue at a production level of four units.

LIFE SCIENCES

22. *Pollution.* An oil tanker aground on a reef is losing oil and producing an oil slick that is radiating outward at a rate given approximately by

$$\dfrac{dR}{dt} = \dfrac{60}{\sqrt{t + 9}} \qquad t \geq 0$$

where R is the radius in feet of the circular slick after t minutes. Find the radius of the slick after 16 minutes if the radius is zero when $t = 0$.

SOCIAL SCIENCES

23. *College enrollment.* The projected rate of increase in enrollment in a new college is estimated by

$$\dfrac{dE}{dt} = 5{,}000(t + 1)^{-3/2} \qquad t \geq 0$$

where $E(t)$ is the projected enrollment in t years. If enrollment when $t = 0$ is 2,000, find the projected enrollment 15 years from now.

4-3   DEFINITE INTEGRALS

We start this discussion with a simple example, out of which will evolve a definition of a definite integral.

Suppose a manufacturing company's marginal cost equation for a given product is given by

$$C'(x) = 2 - 0.2x \qquad 0 \le x \le 8$$

where the marginal cost is in thousands of dollars and production is x units per day. What is the total change in cost per day going from a production level of 2 units per day to 6 units per day? If $C = C(x)$ is the cost function, then

$$\begin{pmatrix} \text{Total net change in cost} \\ \text{between } x = 2 \text{ and } x = 6 \end{pmatrix} = C(6) - C(2) = C(x)|_2^6 \tag{1}$$

The special symbol $C(x)|_2^6$ is a convenient way of representing the center expression that will prove useful to us later.

To evaluate (1), we need to find the antiderivative of $C'(x)$, that is,

$$C(x) = \int (2 - 0.2x)\, dx = 2x - 0.1x^2 + K \tag{2}$$

Thus, we are within a constant of knowing the original marginal cost function. However, we do not need to know the constant K to solve the original problem (1). We compute $C(6) - C(2)$ for $C(x)$ found in (2):

$$C(6) - C(2) = [2(6) - 0.1(6)^2 + K] - [2(2) - 0.1(2)^2 + K]$$

$$= 12 - 3.6 + \cancel{K} - 4 + 0.4 - \cancel{K}$$

$$= \$4.8 \text{ thousand per day increase in costs for a}$$
$$\text{production increase from 2 to 6 units per day}$$

The unknown constant K canceled out! Thus, we conclude that any antiderivative of $C'(x) = 2 - 0.2x$ will do, since antiderivatives of a given function can differ by at most a constant (see Section 4-1). Thus, we really do not have to find the original cost function to solve the problem.

Since $C(x)$ is an antiderivative of $C'(x)$, the above discussion suggests the following notation:

$$C(6) - C(2) = C(x)|_2^6 = \int_2^6 C'(x)\, dx \tag{3}$$

The integral form on the right in (3) is called a *definite integral*—it represents the number found by evaluating an antiderivative of the integrand at 6 and 2 and taking the difference as indicated.

Since definite integrals are used in many different fields, we formally define the concept as follows:

DEFINITE INTEGRAL

The **definite integral** of a continuous function f over an interval from $x = a$ to $x = b$ is the net change of an antiderivative of f over the interval. Symbolically, if $F(x)$ is an antiderivative of $f(x)$, then

$$\int_a^b f(x)\,dx = F(x)\big|_a^b = F(b) - F(a) \qquad \text{where} \quad F'(x) = f(x)$$

Integrand: $f(x)$ **Upper limit:** b **Lower limit:** a

Example 8   Evaluate $\int_{-1}^{2} (3x^2 - 2x)\,dx$.

Solution   We choose the simplest antiderivative of $(3x^2 - 2x)$, namely $(x^3 - x^2)$, since any antiderivative will do.

$$\int_{-1}^{2} (3x^2 - 2x)\,dx = (x^3 - x^2)\big|_{-1}^{2}$$

$$= [2^3 - 2^2] - [(-1)^3 - (-1)^2]$$

$$= 4 - (-2) = 6$$

Problem 8   Evaluate $\int_{-2}^{2} (2x - 1)\,dx$.

Answer -4

REMARK

Do not confuse a definite integral with an indefinite integral. The definite integral $\int_a^b f(x)\,dx$ is a real number; the indefinite integral $\int f(x)\,dx$ is a whole set of functions—all the antiderivatives of $f(x)$.

Example 9   A steel ball is dropped from a tower. Its velocity t seconds later is $v(t) = 32t$ feet per second. How far will the ball fall from the end of 2 seconds to the end of 4 seconds?

Solution   The antiderivative of a velocity function is a position function $s = s(t)$, and we are looking for $s(4) - s(2)$:

$$s(4) - s(2) = \int_{2}^{4} 32t\,dt = 16t^2\big|_2^4 = 256 - 64 = 192 \text{ feet}$$

Problem 9   Repeat Example 9 with $v(t) = 32t - 10$.

Answer   172 feet

Some of the properties of the definite integral are similar to those for the indefinite integral listed in Section 4-1, as given in the box.

DEFINITE INTEGRAL PROPERTIES

1. $\displaystyle\int_a^a f(x)\,dx = 0$

2. $\displaystyle\int_a^b f(x)\,dx = -\int_b^a f(x)\,dx$

3. $\displaystyle\int_a^b Kf(x)\,dx = K\int_a^b f(x)\,dx$ K a constant

4. $\displaystyle\int_a^b [f(x) \pm g(x)]\,dx = \int_a^b f(x)\,dx \pm \int_a^b g(x)\,dx$

5. $\displaystyle\int_a^b f(x)\,dx = \int_a^c f(x)\,dx + \int_c^b f(x)\,dx$

These properties are justified as follows: If $F'(x) = f(x)$ and $G'(x) = g(x)$, then

1. $\displaystyle\int_a^a f(x)\,dx = F(x)\big|_a^a = F(a) - F(a) = 0$

2. $\displaystyle\int_a^b f(x)\,dx = F(x)\big|_a^b = F(b) - F(a) = -\,[F(a) - F(b)] = -\int_b^a f(x)\,dx$

3. $\displaystyle\int_a^b Kf(x)\,dx = KF(x)\big|_a^b = KF(b) - KF(a) = K[F(b) - F(a)] = K\int_a^b f(x)\,dx$

and so on.

Example 10   Evaluate $\int_0^1 [(2x - 1)^3 + 2x]\,dx$.

Solution

$$\int_0^1 [(2x - 1)^3 + 2x]\,dx = \int_0^1 (2x - 1)^3\,dx + \int_0^1 2x\,dx$$

$$= \frac{1}{2}\int_0^1 \underbrace{(2x - 1)^3}_{u}\underbrace{2\,dx}_{du} + 2\int_0^1 x\,dx$$

$$= \frac{1}{2}\cdot\frac{(2x-1)^4}{4}\bigg|_0^1 + 2\cdot\frac{x^2}{2}\bigg|_0^1$$

$$= \frac{(2\cdot 1 - 1)^4}{8} - \frac{(2\cdot 0 - 1)^4}{8} + (1^2 - 0^2)$$

$$= \frac{1^4}{8} - \frac{(-1)^4}{8} + 1 = 1$$

Problem 10   Evaluate $\int_1^2 [3x^2 - x\sqrt{x^2 - 1}]\, dx$.

Answer $7 - \sqrt{3}$

Example 11   A large factory on the Mississippi River dischages pollutants into the river at a rate that is estimated by a water quality control agency to be

$$P'(t) = R(t) = t\sqrt{t^2 + 1} \qquad 0 \le t \le 5$$

where $P(t)$ is the total number of tons of pollutants discharged into the river after t years of operation. What quantity of pollutants will be discharged into the river during the first 3 years of operation?

Solution
$$P(3) - P(0) = \int_0^3 t\sqrt{t^2 + 1}\, dt$$

$$= \int_0^3 (t^2 + 1)^{1/2} t\, dt$$

$$= \frac{1}{2} \int_0^3 (t^2 + 1)^{1/2} 2t\, dt$$

$$= \frac{1}{2} \cdot \frac{(t^2 + 1)^{3/2}}{3/2} \Big|_0^3$$

$$= \frac{1}{3}(t^2 + 1)^{3/2} \Big|_0^3$$

$$= \frac{1}{3}(3^2 + 1)^{3/2} - \frac{1}{3}(0^2 + 1)^{3/2}$$

$$= \frac{1}{3}(10^{3/2} - 1) \approx 10.2 \text{ tons}$$

Problem 11   Repeat Example 11 for the time interval from 3 to 5 years.

Answer $\frac{1}{3}(26^{3/2} - 10^{3/2}) \approx 33.7$ tons

EXERCISE 4-3

Evaluate.

A   1. $\int_2^3 2x\, dx$

2. $\int_1^2 3x^2\, dx$

3. $\int_3^4 5\, dx$

4. $\int_{12}^{20} dx$

5. $\int_1^3 (2x - 3)\, dx$

6. $\int_1^3 (6x + 5)\, dx$

7. $\int_0^4 (3x^2 - 4)\, dx$

8. $\int_0^2 (6x^2 - 2x)\, dx$

9. $\displaystyle\int_{-3}^{4} (4 - x^2)\, dx$ 10. $\displaystyle\int_{-1}^{2} (x^2 - 4x)\, dx$

11. $\displaystyle\int_{0}^{1} 24x^{11}\, dx$ 12. $\displaystyle\int_{0}^{2} 30x^5\, dx$

B   13. $\displaystyle\int_{1}^{2} (2x^{-2} - 3)\, dx$ 14. $\displaystyle\int_{1}^{2} (5 - 16x^{-3})\, dx$

15. $\displaystyle\int_{1}^{4} 3\sqrt{x}\, dx$ 16. $\displaystyle\int_{4}^{25} \frac{2}{\sqrt{x}}\, dx$

17. $\displaystyle\int_{2}^{3} 12(x^2 - 4)^5 x\, dx$ 18. $\displaystyle\int_{0}^{1} 32(x^2 + 1)^7 x\, dx$

19. $\displaystyle\int_{1}^{9} \sqrt[3]{x - 1}\, dx$ 20. $\displaystyle\int_{-1}^{0} \sqrt[5]{x + 1}\, dx$

C   21. $\displaystyle\int_{2}^{3} x\sqrt{2x^2 - 3}\, dx$ 22. $\displaystyle\int_{0}^{1} x\sqrt{3x^2 + 2}\, dx$

23. $\displaystyle\int_{0}^{1} \frac{x - 1}{\sqrt[3]{x^2 - 2x + 3}}\, dx$ 24. $\displaystyle\int_{1}^{2} \frac{x + 1}{\sqrt{2x^2 + 4x - 2}}\, dx$

APPLICATIONS

BUSINESS & ECONOMICS

25. *Marginal analysis.* A company's marginal cost, revenue, and profit equations (in thousands of dollars per day) are:

$$\left.\begin{array}{l} C'(x) = 1 \\ R'(x) = 10 - 2x \\ P'(x) = R'(x) - C'(x) \end{array}\right\} 0 \le x \le 10$$

where x is the number of units produced per day. Find the change in
(A) Cost   (B) Revenue   (C) Profit
in going from a production level of 2 units per day to 4 units per day.

26. *Marginal analysis.* Repeat Problem 25 with $C'(x) = 2$ and $R'(x) = 12 - 2x$.

27. *Salvage value.* A new piece of industrial equipment will depreciate in value rapidly at first, then less rapidly as time goes on. Suppose the rate (in dollars per year) at which the book value of a new milling machine changes is given approximately by

$$V'(t) = f(t) = 500(t - 12)\qquad 0 \le t \le 10$$

where $V(t)$ is the value of the machine after t years. Find the total loss in value of the machine in the first 5 years. In the second 5 years. Set up appropriate integrals and solve.

28. *Maintenance costs.* Maintenance costs for an apartment house generally increase as the building gets older. From past records, a managerial service determines that the rate of increase in maintenance costs (in dollars per year) for a particular apartment complex is given approximately by

$$M'(x) = f(x) = 90x^2 + 5,000$$

where x is the age of the apartment in years and $M(x)$ is the total (accumulated) cost of maintenance for x years. Write a definite integral that will give the total maintenance costs from 2 to 7 years after the apartment house was built, and evaluate it.

LIFE SCIENCES

29. *Pulse rate versus height.* The rate of change of an average person's pulse rate with respect to height is given approximately by

$$P'(x) = f(x) = -295x^{-3/2} \qquad 30 \le x \le 75$$

where x is height in inches. Find the total change in pulse rate for a child growing from 49 to 64 inches. Set up an appropriate definite integral and solve.

30. *Drug sensitivity.* One hour after x milligrams of a particular drug are given to a person, the rate of change of temperature in degrees Fahrenheit, $T'(x)$, with respect to dosage x (called *sensitivity*) is given approximately by

$$T'(x) = 2x - \frac{x^2}{3} \qquad 0 \le x \le 6$$

What total change in temperature results from a dosage change from 0 to 2 milligrams? From 2 to 3 milligrams? Set up definite integrals and evaluate.

31. *Natural resource depletion.* The instantaneous rate of change of demand for wood in the United States since 1970 ($t = 0$) in billions of cubic feet per year is estimated to be given by

$$Q'(t) = 12 + 0.006t^2 \qquad 0 \le t \le 50$$

where $Q(t)$ is the total amount of wood consumed in billions of cubic feet t years after 1970. How many billions of cubic feet of wood will be consumed from 1980 to 1990?

32. *Natural resource depletion.* Repeat Problem 31 for the time interval from 1990 to 2000.

SOCIAL SCIENCES

33. *Learning.* A person learns N items at a rate given approximately by

$$N'(t) = f(t) = \frac{25}{\sqrt{t}} \qquad 1 \le t \le 9$$

where t is the number of hours of continuous study. Use a definite integral to determine the total number of items learned from $t = 1$ to $t = 9$ hours of study.

4-4 AREA AND THE DEFINITE INTEGRAL

AREA UNDER A CURVE

Consider the graph of $f(x) = x$ from $x = 0$ to $x = 4$ (Fig. 2). We can easily compute the area of the triangle bounded by $f(x) = x$, the x axis ($y = 0$), and the line $x = 4$, using the formula for the area of a triangle:

$$A = \frac{bh}{2} = \frac{4 \cdot 4}{2} = 8$$

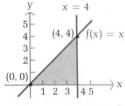

FIGURE 2

Let us integrate $f(x)$ from $x = 0$ to $x = 4$:

$$\int_0^4 x \, dx = \frac{x^2}{2} \Big|_0^4 = \frac{4^2}{2} - \frac{0^2}{2} = 8$$

We get the same result! It turns out that this is not a coincidence. In general, we can prove the following:

AREA UNDER A CURVE

If f is continuous and $f(x) \ge 0$ over the interval $[a, b]$, then the area bounded by $y = f(x)$, the x axis ($y = 0$), and the vertical lines $x = a$ and $x = b$ is given exactly by

$$A = \int_a^b f(x) \, dx$$

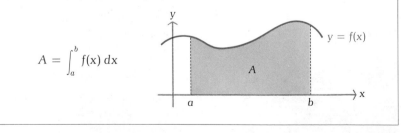

Let us see why the definite integral gives us the area exactly. Let $A(x)$ be the area under the graph of $y = f(x)$ from a to x, as indicated in Figure 3.

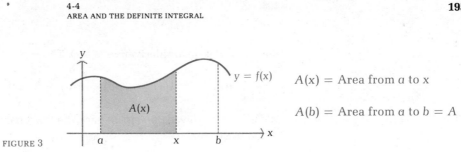

FIGURE 3

$A(x)$ = Area from a to x

$A(b)$ = Area from a to b = A

If we can show that $A(x)$ is an antiderivative of $f(x)$, then we can write

$$\int_a^b f(x)\,dx = A(x)\big|_a^b = A(b) - A(a)$$

$$= \left(\begin{array}{c}\text{Area from}\\ x = a \text{ to } x = b\end{array}\right) - \left(\begin{array}{c}\text{Area from}\\ x = a \text{ to } x = a\end{array}\right)$$

$$= A - 0 = A$$

To show that $A(x)$ is an antiderivative of $f(x)$—that is, $A'(x) = f(x)$—we use the definition of a derivative (Section 2-4) and write

$$A'(x) = \lim_{\Delta x \to 0} \frac{A(x + \Delta x) - A(x)}{\Delta x}$$

Geometrically, $A(x + \Delta x) - A(x)$ is the area from x to $x + \Delta x$ (see Fig. 4). This area is given approximately by the area of the rectangle $\Delta x \cdot f(x)$, and the smaller Δx is, the better the approximation. Using

$$A(x + \Delta x) - A(x) \approx \Delta x \cdot f(x)$$

and dividing both sides by Δx, we obtain

$$\frac{A(x + \Delta x) - A(x)}{\Delta x} \approx f(x)$$

Now, if we let $\Delta x \to 0$, then the left side has $A'(x)$ as a limit, which is equal to the right side. Hence,

$$A'(x) = f(x)$$

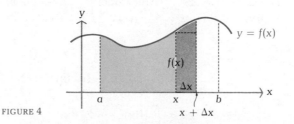

FIGURE 4

that is, $A(x)$ is an antiderivative of $f(x)$. Thus,

$$\int_a^b f(x)\,dx = A(x)\big|_a^b = A(b) - A(a) = A - 0 = A$$

We have now solved, at least in part, the third basic problem of calculus stated in Section 2-1.

Example 12 Find the area bounded by $f(x) = 6x - x^2$ and $y = 0$ for $1 \le x \le 4$.

Solution We sketch a graph of the region first:

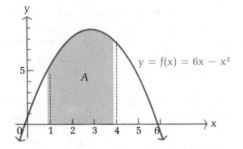

$$A = \int_1^4 (6x - x^2)\,dx = \left(3x^2 - \frac{x^3}{3}\right)\bigg|_1^4$$

$$= \left[3(4)^2 - \frac{4^3}{3}\right] - \left[3(1)^2 - \frac{1^3}{3}\right]$$

$$= 48 - \frac{64}{3} - 3 + \frac{1}{3}$$

$$= 48 - 21 - 3 = 24$$

Problem 12 Find the area bounded by $f(x) = x^2 + 1$ and $y = 0$ for $-1 \le x \le 3$.

Answer $13\frac{1}{3}$

AREA BETWEEN TWO CURVES Consider the area bounded by $y = f(x)$ and $y = g(x)$ for $a \le x \le b$, as indicated in Figure 5.

$$\begin{pmatrix} \text{Area } A \text{ between} \\ f(x) \text{ and } g(x) \end{pmatrix} = \begin{pmatrix} \text{Area under} \\ f(x) \end{pmatrix} - \begin{pmatrix} \text{Area under} \\ g(x) \end{pmatrix}$$

Areas are from $x = a$ to $x = b$ above the x axis

$$= \int_a^b f(x)\,dx - \int_a^b g(x)\,dx$$

From definite integral property 4 (Section 4-3)

$$= \int_a^b [f(x) - g(x)]\,dx$$

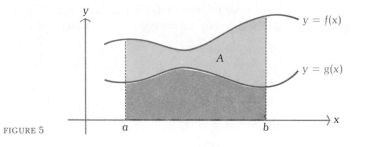

FIGURE 5

It can be shown that the above result does not require $f(x)$ or $g(x)$ to remain positive over the interval $[a, b]$. A more general result is stated in the box:

AREA BETWEEN TWO CURVES

If f and g are continuous and $f(x) \geq g(x)$ over the interval $[a, b]$, then the area bounded by $y = f(x)$ and $y = g(x)$ for $a \leq x \leq b$ is given exactly by

$$A = \int_a^b [f(x) - g(x)]\, dx$$

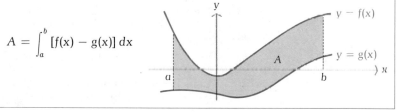

Example 13   Find the area bounded by $f(x) = 5 - x^2$ and $g(x) = x^2 - 3$.

Solution

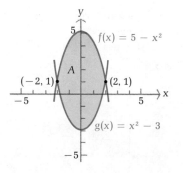

The graphs are two parabolas, one opening up and the other down, as shown in the figure. To find the points of intersection (hence, the upper and lower limits of integration), we solve $y = 5 - x^2$ and $y = x^2 - 3$ simultaneously by setting $5 - x^2$ equal to $x^2 - 3$ (substitution method):

$$5 - x^2 = x^2 - 3$$
$$2x^2 - 8 = 0$$
$$x^2 - 4 = 0$$
$$x = \pm 2$$

Thus,

$$A = \int_{-2}^{2} [(5 - x^2) - (x^2 - 3)]\, dx$$

$$= \int_{-2}^{2} (8 - 2x^2)\, dx$$

$$= \left(8x - \frac{2x^3}{3}\right)\Big|_{-2}^{2}$$

$$= \left[8(2) - \frac{2(2)^3}{3}\right] - \left[8(-2) - \frac{2(-2)^3}{3}\right]$$

$$= 16 - \frac{16}{3} + 16 - \frac{16}{3} = \frac{64}{3}$$

Problem 13 Find the area bounded by $f(x) = 6 - x^2$ and $g(x) = x$.

Answer $20\frac{5}{6}$

SIGNED AREAS Consider the area bounded by $f(x) = x$, the x axis ($y = 0$), $x = -2$, and $x = 2$, as indicated in Figure 6. Integrating $f(x) = x$ from $x = -2$ to $x = 2$, we obtain

$$\int_{-2}^{2} x\, dx = \frac{x^2}{2}\Big|_{-2}^{2} = \frac{(2)^2}{2} - \frac{(-2)^2}{2} = 2 - 2 = 0$$

which is not the actual area indicated in Figure 6. But now consider the following two integrals:

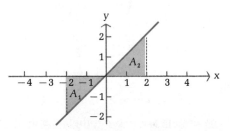

FIGURE 6

$$\int_{-2}^{0} x \, dx = \frac{x^2}{2}\Big|_{-2}^{0} = \frac{0^2}{2} - \frac{(-2)^2}{2} = -2$$

$$\int_{0}^{2} x \, dx = \frac{x^2}{2}\Big|_{0}^{2} = \frac{2^2}{2} - \frac{0^2}{2} = 2$$

We interpret the results as signed areas: area A_2 above the x axis is positive and area A_1 below the x axis is negative. The actual area can then be obtained by adding the absolute value of the negative area to the positive area:

$$\text{Total area} = \left|\int_{-2}^{0} x \, dx\right| + \int_{0}^{2} x \, dx = |-2| + 2 = 4$$

Note that the integral from -2 to 2 is the algebraic sum of the signed areas:

$$\int_{-2}^{2} x \, dx = \int_{-2}^{0} x \, dx + \int_{0}^{2} x \, dx = -2 + 2 = 0$$

From definite integral property 5 (Section 4-3)

We summarize these observations as follows:

SIGNED AREAS AND THE DEFINITE INTEGRAL

The **area** bounded by $y = f(x)$, the x axis ($y = 0$), $x = a$, and $x = b$ is **positive** where it is above the x axis and **negative** where it is below the x axis. The definite integral of $f(x)$ from $x = a$ to $x = b$ can always be interpreted as the algebraic sum of these signed areas:

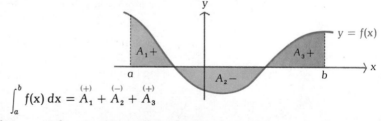

$$\int_{a}^{b} f(x) \, dx = \overset{(+)}{A_1} + \overset{(-)}{A_2} + \overset{(+)}{A_3}$$

If we want the **actual bounded area**, then we add the absolute value of each negative area to the sum of the positive areas.

Example 14   (A)   Find the finite area bounded by $f(x) = 1 - x^2$ and $y = 0, 0 \leq x \leq 2$.
(B)   Find the definite integral of $f(x)$ from $x = 0$ to $x = 2$.

Solutions   (A)   We need to sketch a graph first to see if negative areas are involved.

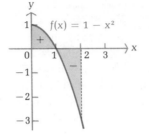

$$\text{Actual area} = \int_0^1 (1 - x^2)\, dx + \left| \int_1^2 (1 - x^2)\, dx \right|$$

$$= \frac{2}{3} + \left| -\frac{4}{3} \right| = \frac{2}{3} + \frac{4}{3} = 2$$

(B) $\displaystyle \int_0^2 (1 - x^2)\, dx = \left(x - \frac{x^3}{3} \right)\Big|_0^2 = 2 - \frac{8}{3} = -\frac{2}{3}$ This is the algebraic sum of the signed areas $\frac{2}{3}$ and $-\frac{4}{3}$

Problem 14

(A) Find the area bounded by $f(x) = x^2 - 1$, $y = 0$, $x = -1$, and $x = 2$.

(B) Evaluate the definite integral of $f(x)$ from $x = -1$ to $x = 2$.

Answers

(A) $\frac{8}{3}$ (B) 0

CONSUMERS' AND PRODUCERS' SURPLUS

If we graph the supply and demand functions $p = S(x)$ and $p = D(x)$ and locate the equilibrium point (a, b) (the point at which supply is equal to demand), then the area between $p = b$ and $p = D(x)$ from $x = 0$ to $x = a$ is called **consumers' surplus.** The area between $p = S(x)$ and $p = b$ is called **producers' surplus** (see Fig. 7):

$$\text{Consumers' surplus} = \int_0^a [D(x) - b]\, dx$$

$$\text{Produers' surplus} = \int_0^a [b - S(x)]\, dx$$

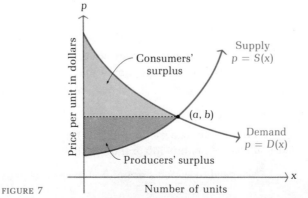

FIGURE 7

In other words, if the price stabilizes at b per unit, then there is still a demand by some people at higher prices, and people who are willing to pay a higher price benefit by only having to pay the equilibrium price. The total of these benefits over $[0, a]$ is the consumers' surplus. On the other hand, there are still some producers who are willing to supply at a lower price, and these people benefit by receiving the equilibrium price. The total of these benefits for the produers over the interval $[0, a]$ is the producers' surplus.

Example 15   Find the consumers' surplus for

$$p = D(x) = -\frac{x}{2} + 11 \quad \text{and} \quad p = S(x) = x + 2$$

Solution   Sketch a graph:

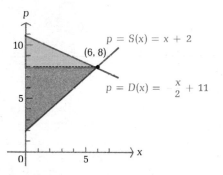

To find the equilibrium point, set $(x + 2)$ equal to $[(-x/2) + 11]$.

$$x + 2 = -\frac{x}{2} + 11$$

$$2x + 4 = -x + 22$$

$$3x = 18$$

$$x = 6$$

$$p = x + 2$$

$$= 6 + 2 = 8$$

Therefore, the equilibrium point is $(6, 8)$, as shown in the figure. Now,

$$\text{Consumers' surplus} = \int_0^a [D(x) - b]\, dx$$

$$= \int_0^6 \left(-\frac{x}{2} + 11 - 8 \right) dx$$

$$= \int_0^6 \left(-\frac{x}{2} + 3 \right) dx$$

$$= \left(-\frac{x^2}{4} + 3x \right) \Bigg|_0^6 = 9$$

Problem 15   Find the producers' surplus for Example 15.

Answer   18

EXERCISE 4-4

Find the area bounded by the graphs of the indicated equations.

A   1. $y = 2x + 4$, $y = 0$, $1 \leq x \leq 3$
     2. $y = -2x + 6$, $y = 0$, $0 \leq x \leq 2$
     3. $y = 3x^2$, $y = 0$, $1 \leq x \leq 2$
     4. $y = 4x^3$, $y = 0$, $1 \leq x \leq 2$
     5. $y = x^2 + 2$, $y = 0$, $-1 \leq x \leq 0$
     6. $y = 3x^2 + 1$, $y = 0$, $-2 \leq x \leq 0$
     7. $y = 4 - x^2$, $y = 0$, $-1 \leq x \leq 2$
     8. $y = 12 - 3x^2$, $y = 0$, $-2 \leq x \leq 1$

B   9. $y = 12$, $y = -2x + 8$, $-1 \leq x \leq 2$
     10. $y = 3$, $y = 2x + 6$, $-1 \leq x \leq 2$

     11. $y = 3x^2$, $y = 12$ 12. $y = x^2$, $y = 9$
     13. $y = 4 - x^2$, $y = -5$ 14. $y = x^2 - 1$, $y = 3$

     15. $y = x^2 + 1$, $y = 2x - 2$, $-1 \leq x \leq 2$
     16. $y = x^2 - 1$, $y = x - 2$, $-2 \leq x \leq 1$
     17. $y = -x$, $y = 0$, $-2 \leq x \leq 1$
     18. $y = -x + 1$, $y = 0$, $-1 \leq x \leq 2$

C   19. $y = x^2 - 4$, $y = 0$, $0 \leq x \leq 3$
     20. $y = 4\sqrt[3]{x}$, $y = 0$, $-1 \leq x \leq 8$

     21. $y = x^2 + 2x + 3$, $y = 2x + 4$ 22. $y = 8 + 4x - x^2$, $y = x^2 - 2x$

APPLICATIONS

BUSINESS & ECONOMICS   23. *Consumers' and producers' surplus.*   Find the consumers' surplus and the producers' surplus for

$$p = D(x) = -\frac{x}{2} + 2$$

$$p = S(x) = \frac{x^2}{4}$$

24. *Consumers' and producers' surplus.*   Find the consumers' surplus and the producers' surplus for

$$p = D(x) = 50 - x^2$$
$$p = S(x) = x^2 + 2x + 10$$

25. *Marginal analysis.* A company has a vending machine with the following marginal cost and revenue equations (in thousands of dollars per year):

$$C'(t) = 2 \atop R'(t) = 12 - 2t \qquad 0 \le t \le 10$$

where $C(t)$ and $R(t)$ represent total accumulated costs and revenues, respectively, t years after the machine is put into use. The area between the graphs of the marginal equations for the time period such that $R'(t) \ge C'(t)$ represents the total accumulated profit for the useful life of the machine. What is the useful life of the machine and what is the total profit?

26. *Marginal analysis.* Repeat Problem 25 for $C'(t) = 0.5t + 2$ and $R'(t) = 10 - 0.5t, 0 \le t \le 20$.

4-5 DEFINITE INTEGRAL AS A LIMIT OF A SUM

Up to this point, in order to evaluate a definite integral

$$\int_a^b f(x) \, dx$$

we need to find an antiderivative of the function f so that we can write

$$\int_a^b f(x) \, dx \; F(x) \Big|_a^b = F(b) - F(a) \qquad F'(x) = f(x)$$

But suppose we cannot find an antiderivative of f (it may not even exist in a convenient or closed form). For example, how would you evaluate the following?

$$\int_2^8 \sqrt{x^3 + 1} \, dx \qquad \text{or} \qquad \int_1^5 \left(\frac{x}{x+1}\right)^3 dx$$

We now introduce the *rectangle rule* for approximating definite integrals, and out of this discussion will evolve a new way of looking at definite integrals.

RECTANGLE RULE
FOR APPROXIMATING
DEFINITE INTEGRALS

In the last section we saw that any definite integral of a continuous function f over an interval $[a, b]$ can always be interpreted as the algebraic sum of the signed areas bounded by $y = f(x), y = 0, x = a$, and $x = b$ (see Fig. 8). What we need is a way of approximating such areas, given $y = f(x)$ and an interval $[a, b]$.

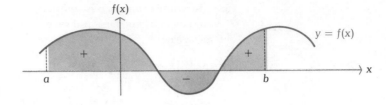

FIGURE 8

Let us start with a concrete example and generalize from the experience. We will start with a simple definite integral we can evaluate exactly:

$$\int_1^5 (x^2 + 3)\, dx = \left(\frac{x^3}{3} + 3x\right)\Big|_1^5$$

$$= \left[\frac{5^3}{3} + 3(5)\right] - \left[\frac{1^3}{3} + 3(1)\right]$$

$$= \left(\frac{125}{3} + 15\right) - \left(\frac{1}{3} + 3\right)$$

$$= \frac{160}{3} = 53\tfrac{1}{3}$$

This integral represents the area bounded by $y = x^2 + 3$, $y = 0$, $x = 1$, and $x = 5$, as indicated in Figure 9.

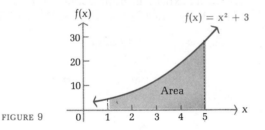

FIGURE 9

Since areas of rectangles are easy to compute, we cover the area in Figure 9 with rectangles so that the top of each rectangle has a point in common with the graph of $y = f(x)$. As our first approximation, we divide the interval $[1, 5]$ into two equal subintervals, each with length $(b - a)/2 = (5 - 1)/2 = 2$, and use the midpoint of each subinterval to compute the altitude of the rectangle sitting on top of that subinterval (see Fig. 10):

$$\int_1^5 (x^2 + 3)\, dx \approx f(2)\cdot 2 + f(4)\cdot 2$$
$$= 2[f(2) + f(4)]$$
$$= 2[7 + 19] = 52$$

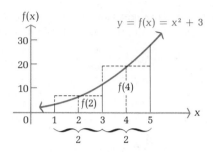

FIGURE 10

This approximation is less than 3% off of the exact area we found above (53⅓).

Now let us divide the interval [1, 5] into four equal subintervals, each of length $(b - a)/4 = (5 - 1)/4 = 1$, and use the midpoint* of each subinterval to compute the altitude of the rectangle corresponding to that subinterval (see Fig. 11):

$$\int_1^5 (x^2 + 3)\, dx \approx f(1.5) \cdot 1 + f(2.5) \cdot 1 + f(3.5) \cdot 1 + f(4.5) \cdot 1$$
$$= f(1.5) + f(2.5) + f(3.5) + f(4.5)$$
$$= 5.25 + 9.25 + 15.25 + 23.25$$
$$= 53$$

Now we are less than 1% off of the exact area (53⅓).

We would expect the approximations to continue to improve as we use more and more rectangles with smaller and smaller bases. We now

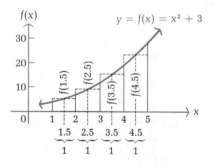

FIGURE 11

*We actually do not need to choose the midpoint of each subinterval; any point from each subinterval will do. The midpoint is often a convenient point to choose, because then the rectangle tops are usually above part of the graph and below part of the graph. This tends to cancel some of the error that occurs.

state the rectangle rule for approximating definite integrals of a continuous function f over the interval from $x = a$ to $x = b$.

RECTANGLE RULE

Divide the interval from $x = a$ to $x = b$ into n equal subintervals of length $\Delta x = (b - a)/n$. Let x_k be any point on the kth subinterval. Then

$$\int_a^b f(x)\, dx \approx f(x_1)\Delta x + f(x_2)\Delta x + \cdots + f(x_n)\Delta x$$

$$= \Delta x[f(x_1) + f(x_2) + \cdots + f(x_n)]$$

The *fundamental theorem of calculus* states that the approximations in the rectangle rule (for f continuous over the interval from $x = a$ to $x = b$) approach the definite integral as a limit as $n \to \infty$; that is, as $\Delta x \to 0$. We state the fundamental theorem without proof.

FUNDAMENTAL THEOREM OF CALCULUS

For f continuous over the interval from $x = a$ to $x = b$, and x_k and Δx determined as in the rectangle rule, then*

$$\lim_{n \to \infty} [f(x_1)\Delta x + f(x_2)\Delta x + \cdots + f(x_n)\Delta x] = \int_a^b f(x)\, dx$$

$$= F(b) - F(a) \quad \text{where} \quad F'(x) = f(x)$$

This remarkable theorem states that the limit of the sum of rectangles over the interval from $x = a$ to $x = b$ can be found exactly by evaluating an antiderivative only at the end points of the interval and then taking the difference of these values. We will have more to say about the fundamental theorem shortly. Let us first work another example using the rectangle rule in order to standardize the process.

Example 16   Use the rectangle rule to approximate

$$\int_2^{10} \frac{x}{x + 1}\, dx$$

---

*Limit at infinity (informal definition): We write $\lim_{x \to \infty} f(x) = L$ if we can make $f(x)$ as close to L as we please by increasing x sufficiently. Similarly, we write $\lim_{x \to -\infty} f(x) = L$ if we can make $f(x)$ as close to L as we please by decreasing x sufficiently.

to three significant figures using $n = 4$ and x_k the midpoint of each sub-interval.

Solution   **Step 1.** *Find Δx, the length of each subinterval.*

$$\Delta x = \frac{b - a}{n} = \frac{10 - 2}{4} = \frac{8}{4} = 2$$

Step 2. *Select x_k, the midpoint of each subinterval.*

Subintervals: [2, 4],   [4, 6],   [6, 8],   [8, 10]
Midpoints: $x_1 = 3$, $x_2 = 5$, $x_3 = 7$, $x_4 = 9$

Step 3. *Use the rectangle rule with $n = 4$.*

$$\int_a^b f(x)\,dx \approx f(x_1)\Delta x + f(x_2)\Delta x + f(x_3)\Delta x + f(x_4)\Delta x$$

$$= \Delta x[f(x_1) + f(x_2) + f(x_3) + f(x_4)]$$
$$= 2[f(3) + f(5) + f(7) + f(9)]$$
$$= 2[0.750 + 0.833 + 0.875 + 0.900]$$
$$= 2[3.358] = 6.72 \qquad \text{To three significant figures}$$

Problem 16   Approximate

$$\int_2^{14} \frac{x}{x - 1}\,dx$$

to three significant figures with $n = 4$ and x_k the midpoint of each sub-interval.

Answer   14.4

RECOGNIZING A DEFINITE INTEGRAL   Because of the fundamental theorem, we now have another way of look-ing at definite integrals. If the form

$$f(x_1)\Delta x + f(x_2)\Delta x + \cdots + f(x_n)\Delta x \tag{1}$$

should arise in a problem, and if Δx and the function f meet the condi-tions stated in the rectangle rule, then we know (because of the funda-mental theorem) that

$$f(x_1)\Delta x + f(x_2)\Delta x + \cdots + f(x_n)\Delta x \approx \int_a^b f(x)\,dx$$

and in the limit they are exactly equal. Let us see how form (1) comes up in a natural way in two specific situations.

AVERAGE VALUE OF A CONTINUOUS FUNCTION   Suppose the temperature T (in degrees Fahrenheit) in the middle of a small shallow lake from 8 AM ($t = 0$) to 6 PM ($t = 10$) during the month of May is given approximately as shown in Figure 12. How can we compute

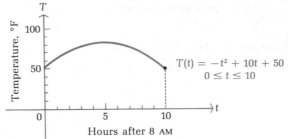

FIGURE 12

the average temperature from 8 AM to 6 PM? We know that the average of a finite number of values

$$a_1, a_2, \ldots, a_n$$

is given by

$$\text{Average} = \frac{a_1 + a_2 + \cdots + a_n}{n}$$

But how can we handle a continuous function with infinitely many values? It would seem reasonable to divide the time interval [0, 10] into n equal subintervals, compute the temperature at a point in each subinterval, and then use the average of these values as an approximation of the average value of the continuous function $T = T(t)$ over [0, 10]. We would expect the approximations to improve as n increases. In fact, we would be inclined to define the limit of the average for n values as $n \to \infty$ as *the average value of* T over [0, 10], if the limit exists. This is exactly what we will do:

$$\binom{\text{Average temperature}}{\text{for } n \text{ values}} = \frac{1}{n}[T(t_1) + T(t_2) + \cdots + T(t_n)] \qquad (2)$$

where t_k is a point on the kth subinterval. We will call the limit of (2) as $n \to \infty$ *the average temperature over the time interval* [0, 10].

Form (2) looks sort of like form (1), but we are missing the $\Delta t = (b - a)/n$. We take care of this by multiplying (2) by $(b - a)/(b - a)$, which will change the form of (2) without changing its value.

$$\frac{b - a}{b - a} \cdot \frac{1}{n}[T(t_1) + T(t_2) + \cdots + T(t_n)]$$

$$= \frac{1}{b - a} \cdot \frac{b - a}{n}[T(t_1) + T(t_2) + \cdots + T(t_n)]$$

$$= \frac{1}{b - a} \cdot \left[T(t_1)\frac{b - a}{n} + T(t_2)\frac{b - a}{n} + \cdots + T(t_n)\frac{b - a}{n} \right]$$

$$= \frac{1}{b - a}[T(t_1)\Delta t + T(t_2)\Delta t + \cdots + T(t_n)\Delta t]$$

Now the part in brackets is of form (1). If we take its limit as $n \to \infty$, then by the fundamental theorem, we obtain

$$\begin{pmatrix} \text{Average temperature} \\ \text{over } [a, b] = [0, 10] \end{pmatrix} = \frac{1}{b - a} \int_a^b T(t)\, dt$$

$$= \frac{1}{10 - 0} \int_0^{10} (-t^2 + 10t + 50)\, dt$$

$$= \frac{1}{10} \left(-\frac{t^3}{3} + 5t^2 + 50t \right) \Big|_0^{10}$$

$$= \frac{200}{3} \approx 67°F$$

In general, proceeding as above for an arbitrary continuous function f over an interval $[a, b]$, we obtain:

AVERAGE VALUE OF A CONTINUOUS FUNCTION f OVER $[a, b]$

$$\frac{1}{b - a} \int_a^b f(x)\, dx$$

Example 17   Find the average value of $f(x) = x - 3x^2$ over the interval $[-1, 2]$.

Solution
$$\frac{1}{b - a} \int_a^b f(x)\, dx = \frac{1}{2 - (-1)} \int_{-1}^{2} (x - 3x^2)\, dx$$

$$= \frac{1}{3} \left(\frac{x^2}{2} - x^3 \right) \Big|_{-1}^{2} = -\frac{5}{2}$$

Problem 17   Find the average value of $g(t) = 6t^2 - 2t$ over the interval $[-2, 3]$.

Answer   13

     In developing the formula for the average value of a continuous function, we observe the limit of a sum of the form of expression (1). Then, from the fundamental theorem, we know we are dealing with a definite integral. We can now evaluate the integral using an antiderivative of f if it can be found, or we can approximate it using the rectangle rule for a sufficiently large n.

VOLUME OF REVOLUTION   Let us consider another application in which expression (1) occurs naturally. Suppose we start out with the region bounded by $y = f(x) = \sqrt{x}$, $y = 0$, and $x = 9$ (see Fig. 13). This is the upper half of a parabola opening to the right.

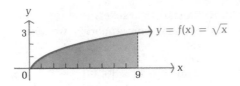

FIGURE 13

If we rotate the shaded area in Figure 13 around the x axis, we obtain a solid called a **volume of revolution.** Figure 14 shows the result, which is called a **paraboloid.** What is its volume?

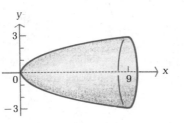

FIGURE 14

Let us cover the region in Figure 13 with rectangles (as we did earlier in the section using the rectangle rule) and rotate the rectangles around the x axis (see Fig. 15). We can then use the stacked cylinders to give an approximation of the volume—the more rectangles we use, the better the approximation.

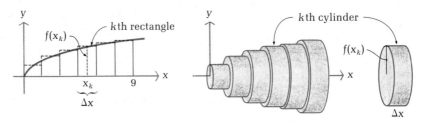

FIGURE 15

Volumes of cylinders are easy to compute:

$$V = (\text{Area of circular base})(\text{Height}) = \pi R^2 h$$

In terms of the kth cylinder in Figure 15, we have:

$$V_k = \pi [f(x_k)]^2 \Delta x$$

The volume of n cylinders is

$$\pi[f(x_1)]^2\Delta x + \pi[f(x_2)]^2\Delta x + \cdots + \pi[f(x_n)]^2\Delta x =$$
$$\pi\{[f(x_1)]^2\Delta x + [f(x_2)]^2\Delta x + \cdots + [f(x_n)]^2\Delta x\}$$

Again we recognize form (1) within the braces. And, from the fundamental theorem, in the limit we have a definite integral. Thus, the exact volume is given by

$$V = \pi \int_a^b [f(x)]^2 \, dx$$

$$= \pi \int_0^9 (\sqrt{x})^2 \, dx = \pi \int_0^9 x \, dx = \pi \frac{x^2}{2}\Big|_0^9 = \frac{81\pi}{2} \approx 127$$

In general, proceeding as above, we have:

VOLUME OF REVOLUTION

The **volume of revolution** obtained by revolving the region bounded by $y = f(x)$, $y = 0$, $x = a$, and $x = b$ about the x axis is given by

$$V = \pi \int_a^b [f(x)]^2 \, dx$$

Example 18   Find the volume of revolution formed by rotating the region bounded by $y = x^2$, $y = 0$, and $x = 2$ about the x axis.

Solution   Sketch a graph of the region first:

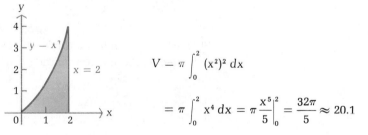

$$V = \pi \int_0^2 (x^2)^2 \, dx$$

$$= \pi \int_0^2 x^4 \, dx = \pi \frac{x^5}{5}\Big|_0^2 = \frac{32\pi}{5} \approx 20.1$$

Problem 18   Find the volume of revolution formed by rotating the region bounded by $y = x^2$, $y = 0$, $x = 1$, and $x = 2$ about the x axis.

Answer $31\pi/5 \approx 19.5$

EXERCISE 4-5

For Problems 1–12:

(A) Use the rectangle rule to approximate each definite integral (to three significant figures) for the indicated number of subintervals n. Choose x_k as the midpoint of each subinterval.

(B)   Evaluate each integral exactly using an antiderivative. If an antiderivative cannot be found by methods we have considered, say so.

A   1. $\int_1^5 3x^2\,dx, \quad n = 2$ 2. $\int_2^6 x^2\,dx, \quad n = 2$

     3. $\int_1^5 3x^2\,dx, \quad n = 4$ 4. $\int_2^5 x^2\,dx, \quad n = 4$

B   5. $\int_0^4 (4 - x^2)\,dx, \quad n = 2$ 6. $\int_0^4 (3x^2 - 12)\,dx, \quad n = 2$

     7. $\int_0^4 (4 - x^2)\,dx, \quad n = 4$ 8. $\int_0^4 (3x^2 - 12)\,dx, \quad n = 4$

     9. $\int_0^4 \left(\dfrac{x}{x + 1}\right)^2 dx, \quad n = 2$ 10. $\int_1^7 \dfrac{1}{x}\,dx, \quad n = 3$

     11. $\int_0^4 \left(\dfrac{x}{x + 1}\right)^2 dx, \quad n = 4$ 12. $\int_1^7 \dfrac{1}{x}\,dx, \quad n = 6$

Find the average value of each function over the indicated interval.

13. $f(x) = 500 - 50x, \quad [0, 10]$ 14. $g(x) = 2x + 7, \quad [0, 5]$
15. $f(t) = 3t^2 - 2t, \quad [-1, 2]$ 16. $g(t) = 4t - 3t^2, \quad [-2, 2]$
17. $f(x) = \sqrt[3]{x}, \quad [1, 8]$ 18. $g(x) = \sqrt{x + 1}, \quad [3, 8]$

Find the volume of revolution of each region bounded by the graphs of the
indicated equations. Rotate the region about the x axis and express the
answer in terms of π.

19. $y = \sqrt{3}\,x, \quad y = 0, \quad x = 1, \quad x = 3$
20. $y = x + 1, \quad y = 0, \quad x = 1, \quad x = 2$

21. $y = \sqrt{2x}, \quad y = 0, \quad x = 8$ 22. $y = \sqrt{5}\,x^2, \quad y = 0, \quad x = 2$
23. $y = \sqrt{4 - x^2}, \quad y = 0$ 24. $y = \sqrt{9 - x^2}, \quad y = 0$

C   Use the rectangle rule to estimate each quantity to three significant
figures. Use $n = 4$ and x_k the midpoint of each subinterval.

25. The average value of $f(x) = 1/x$ for $[1, 9]$
26. The average value of $f(x) = x/(x + 1)$ for $[0, 4]$
27. The volume of revolution of the region bounded by $y = 1/\sqrt{x}, y = 0,$
     $x = 1,$ and $x = 9$; rotate around the x axis and use $\pi \approx 3.14$
28. The volume of revolution of the region bounded by $y = x/(x + 1)$,
     $y = 0,$ and $x = 8$; rotate around the x axis and use $\pi \approx 3.14$

APPLICATIONS

BUSINESS & ECONOMICS

29. *Inventory.* A store orders 600 units of a product every 3 months. If the product is steadily depleted to zero by the end of each 3 months, the inventory on hand, I, at any time t during the year is illustrated as follows:

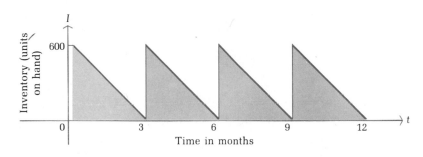

(A) Write an inventory function (assume it is continuous) for the first 3 months. [The graph is a straight line joining (0, 600) and (3, 0).]

(B) What is the average number of units on hand for a 3 month period?

30. Repeat Problem 29 with an order of 1,200 units every 4 months.

31. *Cash reserves.* Suppose cash reserves (in thousands of dollars) are approximated by

$$C(x) = 1 + 12x - x^2 \qquad 0 \le x \le 12$$

where x is the number of months after the first of the year. What is the average cash reserve for the first quarter?

32. Repeat Problem 31 for the second quarter.

LIFE SCIENCES

33. *Temperature.* If the temperature $C(t)$ in an artificial habitat was made to change according to

$$C(t) = t^3 - 2t + 10 \qquad 0 \le t \le 2$$

(in degrees Celsius) over a 2 hour period, what is the average temperature over this period?

SOCIAL SCIENCES

34. *Population composition.* Because of various factors (such as birth rate expansion, then contraction; family flights from urban areas; etc.), the number of children in a large city was found to increase and then decrease rather drastically. If the number of children over a 6 year period was found to be given approximately by

$$N(t) = -\tfrac{1}{4}t^2 + t + 4 \qquad 0 \le t \le 6$$

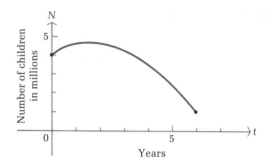

what was the average number of children in the city over the 6 year time period? [Assume $N = N(t)$ is continuous.]

4-6   CHAPTER REVIEW

IMPORTANT TERMS AND SYMBOLS

4-1 Antiderivatives and indefinite integrals.   antiderivative,   indefinite integral,   integral sign,   integrand,   constant of integration, indefinite integral properties, $\int f(x)\, dx = F(x) + C$ where $F'(x) = f(x)$

4-2 Integration by substitution.   substitution rule,

$$\int f(\underbrace{u(x)})\underbrace{u'(x)\, dx} = \int f(u)\, du = F(u) + C = F(u(x)) + C$$
$$\quad\;\; u \qquad\;\; du \qquad\qquad\qquad F'(u) = f(u)$$

4-3 Definite integrals.   definite integral,   integrand,   upper limit, lower limit,   definite integral properties,

$$\int_a^b f(x)\, dx = F(x)\Big|_a^b = F(b) - F(a) \quad \text{where } F'(x) = f(x)$$

4-4 Area and the definite integral.   area under a curve,   area between two curves,   signed areas,   definite integral interpreted as the algebraic sum of signed areas,   consumers' surplus,   producers' surplus

4-5 Definite integral as a limit of a sum.   rectangle rule for approximating a definite integral,   fundamental theorem of calculus,   average value of a continuous function,   volume of revolution,

$$\lim_{n \to \infty} [f(x_1)\Delta x + f(x_2)\Delta x + \cdots + f(x_n)\Delta x] = \int_a^b f(x)\, dx = F(b) - F(a)$$
$$\text{where } F'(x) = f(x)$$

EXERCISE 4-6   CHAPTER REVIEW

Work through all the problems in this chapter review and check your answers in the back of the book. (Answers to all review problems are there.) Where weaknesses show up, review appropriate sections in the text. When you are satisfied that you know the material, take the practice test following this review.

A   *Find each of the following:*

1. $\int (3t^2 - 2t)\, dt$

2. $\int_2^5 (2x - 3)\, dx$

3. $\int (3t^{-2} - 3)\, dt$

4. $\int_1^4 x\, dx$

5.   Find a function $y = f(x)$ that satisfies both conditions:

$$\frac{dy}{dx} = 3x^2 - 2 \qquad f(0) = 4$$

6.   Find the area bounded by the graphs of $y = 3x^2 + 1$, $y = 0$, $x = -1$, and $x = 2$.

7.   Approximate $\int_1^5 (x^2 + 1)\, dx$ using the rectangle rule with $n = 2$ and x_k the midpoint of the kth subinterval.

B   *Find each of the following:*

8. $\int \sqrt[3]{6x - 5}\, dx$

9. $\int_0^1 10(2x - 1)^4\, dx$

10. $\int \left(\frac{2}{x^2} - \sqrt[3]{x^2}\right) dx$

11. $\int_0^4 \sqrt{x^2 + 4}\, x\, dx$

12.   Find a function $y = f(x)$ that satisfies both conditions:

$$\frac{dy}{dx} = 3\sqrt{x} - x^{-2} \qquad f(1) = 5$$

13.   Find the equation of the curve that passes through $(2, 10)$ if its slope is given by

$$\frac{dy}{dx} = 6x + 1$$

for each x.

C   14.   Approximate $\int_{-2}^4 (x^2 - 4)\, dx$ using the rectangle rule with $n = 3$ and x_k the midpoint of the kth subinterval.

15.   Find the average value of $f(x) = 3x^{1/2}$ over the interval $[1, 9]$.

16. Find the volume of revolution of the region bounded by the graphs of $y = 1/x$, $y = 0$, $x = 1$, and $x = 2$ rotated about the x axis. State the answer in terms of π.

17. Find the actual area bounded by $y = x^2 - 4$, $y = 0$, and $x = 4$.

Find each of the following:

18. $\displaystyle\int_0^5 \sqrt[3]{x^2 - 2x}(x - 1)\, dx$

19. $\displaystyle\int \frac{\sqrt{x} - 2x^{-2}}{x}\, dx$

20. Find a function $y = f(x)$ that satisfies both conditions:

$$\frac{dy}{dx} = x^2\sqrt{x^3 + 4} \qquad f(0) = 2$$

21. Find the area bounded by the graphs of $y = 6 - x^2$, $y = x^2 - 2$, $x = 0$, and $x = 3$. Be careful!

22. Approximate the average value of $f(x) = 1/(x + 1)$ over the interval $[0, 4]$ using the rectangle rule with $n = 4$ and x_k the midpoint of the kth subinterval.

23. (A) Approximate the volume of revolution obtained by rotating the region bounded by the graphs of $y = \sqrt{16 - x^2}$, $y = 0$, and $x = 0$ about the x axis. Use the rectangle rule with $n = 4$ and x_k the midpoint of the kth subinterval. (Use $\pi \approx 3.14$.)

    (B) Find the volume of revolution exactly.

APPLICATIONS

BUSINESS & ECONOMICS

24. *Profit function.* If the marginal profit for producing x units per day is given by

$$P'(x) = 100 - 0.02x \qquad P(0) = 0$$

where $P(x)$ is the profit in dollars, find the profit function P and the profit on ten units of production per day.

25. *Resource depletion.* An oil well starts out producing oil at the rate of 60,000 barrels of oil per year, but the production rate is expected to decrease by 4,000 barrels per year. Thus, if $P(t)$ is the total production in t years, then

$$P'(t) = f(t) = 60 - 4t \qquad 0 \le t \le 15$$

Write a definite integral that will give the total production after 15 years of operation. Evaluate it.

26. *Profit and production.* The weekly marginal profit for an output of x units is given approximately by

$$P'(x) = 150 - \frac{x}{10} \qquad 0 \le x \le 40$$

What is the total change in profit for a production change from ten units per week to forty units? Set up a definite integral and evaluate it.

27. *Inventory.* Suppose the inventory of a certain item t months after the first of the year is given approximately by

$$I(t) = 10 + 36t - 3t^2 \qquad 0 \le t \le 12$$

What is the average inventory for the second quarter of the year?

LIFE SCIENCES

28. *Wound healing.* The area of a small, healing surface wound changes at a rate given approximately by

$$\frac{dA}{dt} = -5t^{-2} \qquad 1 \le t \le 5$$

where t is in days and $A(1) = 5$ square centimeters. What will the area of the wound be in 5 days?

29. *Height–weight relationship.* For an average person, the rate of change of weight $W'(h)$ (in pounds) per unit change in height h (in inches) is given approximately by

$$W'(h) = 0.0015h^2$$

What is the expected total change in weight in a child growing from 50 to 60 inches? Set up an appropriate definite integral and evaluate.

SOCIAL SCIENCES

30. *School enrollment.* The student enrollment in a new high school is expected to grow at a rate that is estimated to be

$$\frac{dN}{dt} = 200 + 300t \qquad 0 \le t \le 4$$

where $N(t)$ is the number of students t years after opening. If the initial enrollment ($t = 0$) is 2,000, what will be the enrollment 4 years from now?

31. *Politics.* In a newly incorporated city, it is estimated that the rate of change of the voting population, $N'(t)$, with respect to time t in years is given by

$$N'(t) = 12t - 3t^2 \qquad 0 \le t \le 4$$

where $N(t)$ is in thousands. What is the total increase in the voting population during the first 4 years? Set up an appropriate definite integral and evaluate.

PRACTICE TEST: CHAPTER 4

Find each of the following:

1. $\displaystyle\int_1^2 (5t^{-3} - t)\, dt$

2. $\displaystyle\int x^2 \sqrt{x^3 + 9}\, dx$

3. $\displaystyle\int \frac{4 + x^4}{x^3}\, dx$

4. $\displaystyle\int_1^5 \sqrt{2x - 1}\, dx$

5. Find the equation of a function whose graph passes through the point $(3, 10)$ and whose slope is given by

   $$f'(x) = 6 - 2x$$

6. Find the area bounded by the graphs of $y = x^2$ and $y = \sqrt{x}$.

7. Find the finite area bounded by $y = 1 - x^2$ and $y = 0$, $0 \le x \le 2$.

8. Approximate (to three significant figures) $\displaystyle\int_0^6 (x^2 - 4)\, dx$ using the rectangle rule with $n = 3$ and x_k the midpoint of the kth subinterval. Also, evaluate the integral exactly.

9. Find the volume of revolution of the region bounded by $y = \sqrt{x}$, $y = 0$, $x = 2$, and $x = 4$. Rotate the region about the x axis and write the answer in terms of π.

10. *Inventory.* Suppose the inventory of a certain item t months after the first of the year is given approximately by

    $$I(t) = -2t + 36 \qquad 0 \le t \le 12$$

    What is the average inventory for the second quarter of the year?

11. *Resource depletion.* The instantaneous rate of change of production for a gold mine, in thousands of ounces of gold per year, is estimated to be given by

    $$Q'(t) = 40 - 4t \qquad 0 \le t \le 10$$

    where $Q(t)$ is the total quantity (in thousands of ounces) of gold produced after t years of operation. How much gold is produced during the first 2 years of operation? During the next 2 years?

CHAPTER 5
LOGARITHMIC AND EXPONENTIAL FUNCTIONS AND RELATED TOPICS

CONTENTS

5 LOGARITHMIC AND EXPONENTIAL FUNCTIONS AND RELATED TOPICS

5-1   INTRODUCTION

We now know how to differentiate the large class of functions called *algebraic functions*; that is, functions that can be defined using the algebraic operations of addition, subtraction, multiplication, division, powers, and roots. Furthermore, we can find the antiderivatives of many of these functions—but not all! It may surprise you to learn that the antiderivative of an algebraic function is not necessarily an algebraic function. For example, the one case not covered by the integration formula

$$\int x^n \, dx = \frac{x^{n+1}}{n+1} + C \qquad n \neq -1$$

that is, the case where $n = -1$,

$$\int x^{-1} \, dx = \int \frac{1}{x} \, dx$$

involves an antiderivative (as we will see) that is a *logarithmic function*. Also, the indefinite integrals

$$\int \frac{dx}{x^2 + 4} \qquad \int \sqrt{x^2 + 9} \, dx \qquad \int \frac{dx}{x \sqrt{x^2 - 1}}$$

all have algebraic integrands but involve antiderivatives that are not algebraic functions. We will not consider integrals of the latter type in this book.

In this chapter we will demonstrate how to differentiate and integrate forms that involve the exponential and logarithmic functions discussed in Sections 1-5 and 1-6. (You might wish to review some of the properties of these functions before proceeding further.) Then we will turn our attention to some of the vast number of applications that can be solved using these newly acquired tools. First, a word or two about the irrational number e.

220

5-2 THE IRRATIONAL NUMBER *e*

In Chapter 1 we introduced the special irrational number *e* as a particularly suitable base for both exponential and logarithmic functions. In this chapter we will see why this is so. We said earlier that *e* can be approximated as closely as we like by $[1 + (1/n)]^n$ by taking *n* sufficiently large. Now that we have the limit concept, we formally define *e* by either of the following two limits:

THE NUMBER *e*

1. $e = \lim_{n \to \infty}\left(1 + \dfrac{1}{n}\right)^n$

or, alternately

2. $e = \lim_{s \to 0}(1 + s)^{1/s}$

$e = 2.7182818\ldots$

We will use both these forms. [*Note:* If $s = 1/n$, then as $n \to \infty$, $s \to 0$.]

The proof that the indicated limits exist and represent an irrational number between 2 and 3 is not easy and is omitted here. Many people reason (incorrectly) that the limits are 1, since "$(1 + s)$ approaches 1 as $s \to 0$, and 1 to any power is 1." A little experimentation with a pocket calculator can convince you otherwise. Consider the table of values for *s* and $f(s) = (1 + s)^{1/s}$ and the graph shown in Figure 1 for *s* close to zero.

s	$f(s) = (1 + s)^{1/s}$
−0.5	4.0000
−0.2	3.0518
−0.1	2.8680
−0.01	2.7320
0.01	2.7048
0.1	2.5937
0.2	2.4883
0.5	2.2500

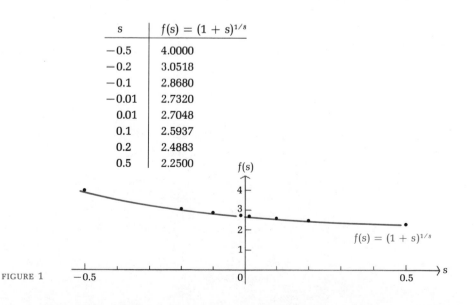

FIGURE 1

Compute some of the table values with a calculator yourself and also try several values of s even closer to zero. Note that the function is discontinuous at $s = 0$.

Exactly who discovered e is still being debated. It is named after the great mathematician Leonhard Euler (1707–1783), who computed e to twenty-three decimal places using $[1 + (1/n)]^n$.

5-3   LOGARITHMIC FUNCTIONS

We are now ready to investigate derivative and integral forms that involve logarithmic functions. We start by deriving a derivative formula for

$$f(x) = \log_b x \qquad b > 0, \quad b \neq 1, \quad x > 0$$

using the definition of the derivative,

$$f'(x) = \lim_{\Delta x \to 0} \frac{f(x + \Delta x) - f(x)}{\Delta x}$$

We simplify the difference quotient first, and then compute the limit:

$$\frac{f(x + \Delta x) - f(x)}{\Delta x} = \frac{\log_b(x + \Delta x) - \log_b x}{\Delta x}$$

$$= \frac{1}{\Delta x}[\log_b(x + \Delta x) - \log_b x]$$

$$= \frac{1}{\Delta x}\log_b \frac{x + \Delta x}{x} \qquad \text{Property of logs}$$

$$= \frac{1}{x}\left(\frac{x}{\Delta x}\right)\log_b\left(1 + \frac{\Delta x}{x}\right) \qquad \text{Multiply by } \frac{x}{x} = 1$$

$$= \frac{1}{x}\log_b\left(1 + \frac{\Delta x}{x}\right)^{x/\Delta x} \qquad \text{Property of logs}$$

Now, if we let $s = \Delta x/x$, then for x fixed, if $\Delta x \to 0$, so must $s \to 0$. Thus,

$$D_x \log_b x = \lim_{\Delta x \to 0} \frac{f(x + \Delta x) - f(x)}{\Delta x}$$

$$= \lim_{\Delta x \to 0} \frac{1}{x}\log_b\left(1 + \frac{\Delta x}{x}\right)^{x/\Delta x}$$

$$= \lim_{s \to 0} \frac{1}{x}\log_b(1 + s)^{1/s}$$

$$= \frac{1}{x}\log_b[\lim_{s \to 0}(1 + s)^{1/s}] \qquad \text{Properties of limits and continuity of log functions}$$

$$= \frac{1}{x}\log_b e \qquad \text{Definition of e}$$

This derivative formula takes on a particularly simple form for one particular base. What base? Since $\log_b b = 1$ for any permissible base b, then

$$\log_e e = 1$$

Thus, for the natural logarithmic function

$$\ln x = \log_e x$$

we have

$$D_x \ln x = D_x \log_e x = \frac{1}{x} \log_e e = \frac{1}{x} \cdot 1 = \frac{1}{x}$$

Now you see why we might want the complicated irrational number e as a base—it, of all possible bases, provides the simplest derivative formula for logarithmic functions.

Once we have a derivative formula, then we automatically have a corresponding integral formula. Thus,

$$\int \frac{1}{x} dx = \ln x + C \qquad x > 0$$

or if $x < 0$, we can show that

$$\int \frac{1}{x} dx = \ln |x| + C \qquad x < 0$$

Both cases can be combined into the one form

$$\int \frac{1}{x} dx = \ln |x| + C \qquad x \neq 0$$

If we are dealing with the composite function

$$y = \ln u \qquad \text{where } u = u(x)$$

then, using the chain rule, we obtain

$$D_x \ln u = \frac{1}{u} \frac{du}{dx}$$

and the corresponding integral formula

$$\int \frac{du}{u} = \ln |u| + C$$

In most of the cases we will be dealing with, u will be positive; hence, $\int du/u = \ln u + C$, since $|u| = u$ for $u > 0$.

What about the indefinite integral of $\ln x$?

$$\int \ln x \, dx$$

We postpone a discussion of this integral until Section 5-6, where we will be able to find it using a technique called *integration by parts*.

Let us summarize the above results and consider several examples.

LOGARITHMIC FUNCTIONS

1. $\ln u = \log_e u$

2. $D_x \ln u = \dfrac{1}{u} D_x u$

3. $\displaystyle\int \dfrac{du}{u} = \ln |u| + C$

Example 1 Differentiate.

(A) $D_x \ln (x^2 + 1)$ (B) $D_x(\ln x)^4$ (C) $D_x \ln x^4$

Solutions (A) $\ln(x^2 + 1)$ is a composite function of the form

$$y = \ln u \qquad u = u(x) = x^2 + 1$$

Formula 2 applies; thus,

$$D_x \ln(x^2 + 1) = \frac{1}{x^2 + 1} D_x(x^2 + 1)$$

$$= \frac{2x}{x^2 + 1}$$

(B) $(\ln x)^4$ is a composite function of the form

$$y = u^p \qquad u = u(x) = \ln x$$

Hence, $D_x u^p = p u^{p-1} D_x u$, and

$$D_x(\ln x)^4 = 4(\ln x)^3 D_x \ln x \qquad \text{Power rule}$$

$$= 4(\ln x)^3 \left(\frac{1}{x}\right) \qquad \text{Formula 2}$$

$$= \frac{4(\ln x)^3}{x}$$

(C) We work this problem two ways. The second method takes particular advantage of logarithmic properties.

Method I. $D_x \ln x^4 = \dfrac{1}{x^4} D_x x^4 = \dfrac{4x^3}{x^4} = \dfrac{4}{x}$

Method II. $D_x \ln x^4 = D_x(4 \ln x) = 4 D_x \ln x = \dfrac{4}{x}$

Problem 1   Differentiate.

(A) $D_x \ln(x^3 + 5)$ (B) $D_x(\ln x)^{-3}$ (C) $D_x \ln x^{-3}$

Answers   (A) $\dfrac{3x^2}{x^3 + 5}$ (B) $\dfrac{-3(\ln x)^{-4}}{x}$ (C) $\dfrac{-3}{x}$

Example 2   Find:

$$D_x \ln \frac{x^5}{\sqrt{x + 1}}$$

Solution   Using the chain rule directly results in a messy operation. (Try it.) Instead, we first take advantage of logarithmic properties to write

$$\ln \frac{x^5}{(x + 1)^{1/2}} = \ln x^5 - \ln(x + 1)^{1/2} = 5 \ln x - \tfrac{1}{2} \ln(x + 1)$$

Then,

$$D_x \ln \frac{x^5}{(x + 1)^{1/2}} = 5D_x \ln x - \tfrac{1}{2}D_x \ln(x + 1)$$

$$= \frac{5}{x} - \frac{1}{2(x + 1)}$$

Problem 2   Find $D_x \ln[(x - 1)^2 \sqrt{x + 2}]$.

Answer $\dfrac{2}{x - 1} + \dfrac{1}{2(x + 2)}$

Example 3   Find:

(A) $\displaystyle\int \frac{2x}{x^2 + 1} dx$ (B) $\displaystyle\int \left(2x + \frac{1}{x}\right) dx$ (C) $\displaystyle\int \frac{dx}{2x - 1}$

Solutions   (A) Mentally note that if $u = x^2 + 1$, then $du = 2x\, dx$; thus,

$$\int \frac{2x}{x^2 + 1} dx$$

is of the form

$$\int \frac{du}{u}$$

and formula 3 applies. Hence,

$$\int \frac{2x}{x^2 + 1} dx = \ln |x^2 + 1| + C$$

$$= \ln(x^2 + 1) + C \qquad \text{Since } x^2 + 1 > 0 \text{ for all } x$$

(B)   Using the general property of the indefinite integral,

$$\int [f(x) \pm g(x)]\, dx = \int f(x)\, dx \pm \int g(x)\, dx$$

we obtain

$$\int \left(2x + \frac{1}{x}\right) dx = \int 2x\, dx + \int \frac{dx}{x}$$

$$= x^2 + \ln |x| + C$$

(C)   This integral is almost of the form $\int du/u$ since if $u = 2x - 1$, then $du = 2\,dx$ (the constant factor 2 is missing in the integrand). We transform $\int dx/(2x - 1)$ into the form $\int du/u$ as follows:

$$\int \frac{dx}{2x - 1} = \int \frac{2}{2} \frac{dx}{2x - 1}$$

$$= \frac{1}{2} \int \frac{2\, dx}{2x - 1} \qquad \text{Form } \frac{1}{2} \int \frac{du}{u}$$

$$= \frac{1}{2} \ln |2x - 1| + C$$

Problem 3   Find:

(A) $\displaystyle \int \frac{3x^2}{x^3 + 5}\, dx$ (B) $\displaystyle \int \left(\frac{1}{x + 1} - 3x^2\right) dx$ (C) $\displaystyle \int \frac{x\, dx}{x^2 + 4}$

Answers   (A) $\ln |x^3 + 5| + C$ (B) $\ln |x + 1| - x^3 + C$
(C) $\frac{1}{2}\ln(x^2 + 4) + C$

  Of course, when we can determine indefinite integrals, we are in a position to evaluate some definite integrals. Consider Example 4 and Problem 4.

Example 4   Evaluate.

(A) $\displaystyle \int_1^e \frac{dx}{x}$ (B) $\displaystyle \int_1^4 \frac{dx}{x + 1}$

Solutions   (A) $\displaystyle \int_1^e \frac{dx}{x} = \ln x \Big|_1^e = \ln e - \ln 1 = 1 - 0 = 1$

  Recall that for any permissible base b, $\log_b 1 = 0$ and $\log_b b = 1$. (Convert to equivalent exponential form to see why.)

(B) $\displaystyle \int_1^4 \frac{dx}{x + 1} = \ln(x + 1) \Big|_1^4 = \ln 5 - \ln 2 \approx 0.9163$

  Note that we did not need the absolute value sign, since x is restricted to the interval $[1, 4]$ and $(x + 1) > 0$ over this interval.

Problem 4 Evaluate.

(A) $\displaystyle\int_{e^2}^{1} \frac{dx}{x}$ (B) $\displaystyle\int_{2}^{7} \frac{dx}{x - 1}$

Answers (A) -2 (B) $\ln 6 - \ln 1 \approx 1.7918$

EXERCISE 5-3

A *Find each derivative.*

1. $D_t \ln t$
2. $D_z \ln z$
3. $D_x \ln(x - 3)$
4. $D_w \ln(w + 100)$
5. $D_x 3 \ln(x - 1)$
6. $D_x 5 \ln x$
7. $D_z (z^2 + 3 \ln z)$
8. $D_t (2t^3 - 5 \ln t)$

Find each indefinite integral.

9. $\displaystyle\int \frac{dt}{t}$
10. $\displaystyle\int \frac{dz}{z}$

11. $\displaystyle\int \frac{dx}{x - 3}$
12. $\displaystyle\int \frac{dw}{w + 100}$

13. $\displaystyle\int \frac{3}{x} dx$
14. $\displaystyle\int \frac{2}{x + 1} dx$

15. $\displaystyle\int \left(1 + \frac{1}{x}\right) dx$
16. $\displaystyle\int \left(2x - \frac{1}{x}\right) dx$

Evaluate each definite integral.

17. $\displaystyle\int_{1}^{e} \frac{dt}{t}$
18. $\displaystyle\int_{e^2}^{1} \frac{dz}{z}$

19. $\displaystyle\int_{4}^{7} \frac{dx}{x - 3}$
20. $\displaystyle\int_{0}^{10} \frac{dw}{w + 100}$

B *Find each derivative.*

21. $D_x \ln x^7$
22. $D_x \ln x^{-3}$
23. $D_x \ln \sqrt{x}$
24. $D_x \ln \sqrt[3]{x}$
25. $D_x(\sqrt{\ln x} + \ln \sqrt{x})$
26. $D_x[(\ln x)^5 - \ln x^5]$
27. $D_x \ln(x + 1)^4$
28. $D_x \ln(x + 1)^{-3}$
29. $D_t \ln(t^2 + 3t)$
30. $D_x \ln(x^3 - 3x^2)$
31. $D_x[2x^3 + \ln(x^2 + 1)]$
32. $D_x[\ln(x^2 - 5) + 4x^3]$
33. $D_x \dfrac{\ln x}{x^2}$
34. $D_x \dfrac{\ln 3x}{x^3}$

35. $D_x(x \ln x - x)$

36. $D_x(x^2 \ln x)$

37. $D_x[(x^2 + x) \ln (x^2 + x)]$

38. $D_x[(x^3 + x^2) \ln x]$

39. $D_x \dfrac{\ln x^2}{\ln x^4}$

40. $D_x \dfrac{\ln \sqrt{x}}{\ln x^3}$

Find each indefinite integral.

41. $\displaystyle\int \dfrac{dt}{5t - 3}$

42. $\displaystyle\int \dfrac{dw}{7w + 2}$

43. $\displaystyle\int \dfrac{x \, dx}{2x^2 + 1}$

44. $\displaystyle\int \dfrac{x^2 \, dx}{4x^3 - 5}$

Evaluate each definite integral.

45. $\displaystyle\int_1^e \left(3t^2 + \dfrac{1}{t}\right) dt$

46. $\displaystyle\int_1^e \left(2w + \dfrac{1}{w}\right) dw$

47. $\displaystyle\int_2^{11} \dfrac{dx}{x - 1}$

48. $\displaystyle\int_{-1}^3 \dfrac{dt}{t + 2}$

49. $\displaystyle\int_{-2}^{-1} \dfrac{dx}{x}$

50. $\displaystyle\int_0^2 \dfrac{dt}{t - 3}$

C *Find each derivative.*

51. $D_x \ln(x^2 + 1)^{1/2}$

52. $D_x \ln(x^2 + 5)^4$

53. $D_x \log_{10}(3x^2 - 2x)$

54. $D_x \log_{10}(x^3 - 1)$

55. $D_x \ln \dfrac{(x - 1)^2}{(x + 1)^3}$

56. $D_x \ln \dfrac{\sqrt{x}}{(x + 1)^2}$

57. $D_x \ln[(x - 1)^2 \sqrt{x}]$

58. $D_x \ln(x^4 \sqrt{x - 1})$

Find each indefinite integral.

59. $\displaystyle\int \dfrac{x - 2}{x^2 - 4x + 3} \, dx$

60. $\displaystyle\int \dfrac{2x^2 - x}{4x^3 - 3x^2} \, dx$

61. $\displaystyle\int \dfrac{2x}{2x^2 + 5} \, dx$

62. $\displaystyle\int \dfrac{x^2 - 2x}{x^3 - 3x^2} \, dx$

Find each definite integral.

63. $\displaystyle\int_{-1}^0 \dfrac{x - 2}{x^2 - 4x + 1} \, dx$

64. $\displaystyle\int_1^2 \dfrac{2x^2 - x}{4x^3 - 3x^2} \, dx$

65. $\displaystyle\int_1^2 \dfrac{2x}{2x^2 + 5} \, dx$

66. $\displaystyle\int_1^2 \dfrac{x^2 - 2x}{x^3 - 3x^2} \, dx$

67. Find the area between the curve $y = 1/t$ and the t axis from $t = 1$ to $t = 2$; from $t = 1$ to $t = 3$; from $t = 1$ to $t = x$, $x \geq 1$. (From this we conclude that the area under the curve $y = 1/t$ from $t = 1$ to $t = x$ is ln x, as is indicated in the figure.)

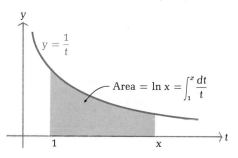

5-4   EXPONENTIAL FUNCTIONS

Recall from Chapter 1 that an exponential function is a function of the form

$$y = b^x \qquad b > 0, \quad b \neq 1 \tag{1}$$

To derive a derivative formula for exponential functions, instead of starting with the basic definition of a derivative, as we did in the last section, we can take advantage of the formulas derived in that section and use implicit differentiation (see Section 3-1).

We start by taking the natural logarithm of both sides of the equation in (1) to obtain

$$\begin{aligned} \ln y &= \ln b^x \\ &= x \ln b \end{aligned} \tag{2}$$

Now, thinking of y as a function of x,

$$y = y(x)$$

we differentiate both sides of (2) with respect to x:

$$D_x \ln y = D_x(x \ln b)$$

Using formula 2 from Section 5-3 and implicit differentiation, we arrive at

$$\frac{1}{y}\frac{dy}{dx} = \ln b$$

and we solve for dy/dx:

$$\frac{dy}{dx} = y \ln b$$

Recall that $y = b^x$ from equation (1), and we have

$$D_x b^x = b^x \ln b \tag{3}$$

We ask, as before, for what number b will the derivative formula (3) be the simplest? If $b = e$, then (3) becomes

$$D_x e^x = e^x \ln e = e^x \cdot 1 = e^x$$

and we find that the derivative of the exponential function with base e is the function itself. Thus, all higher-order derivatives of e^x are e^x; that is,

$$D_x^n e^x = e^x \tag{4}$$

for all natural numbers n.

If we have a composite function

$$y = e^u \qquad u = u(x)$$

or

$$y = b^u \qquad u = u(x)$$

then, using the chain rule, we obtain

$$D_x e^u = e^u \frac{du}{dx} \tag{5}$$

and

$$D_x b^u = b^u \ln b \frac{du}{dx} \tag{6}$$

From (5) and (6) we can immediately write the corresponding integral formulas

$$\int e^u \, du = e^u + C$$

and

$$\int b^u \, du = \frac{b^u}{\ln b} + C$$

We summarize the above results for convenient reference. Then, we will consider several examples. Formulas 1 and 2 in the box are used far more frequently than formulas 3 and 4; hence, they are given more attention in the examples and exercises that follow.

EXPONENTIAL FUNCTIONS

For $b > 0$ and $b \neq 1$:

1. $D_x e^u = e^u D_x u$

2. $\displaystyle\int e^u \, du = e^u + C$

3. $D_x b^u = b^u \ln b \, D_x u$

4. $\displaystyle\int b^u \, du = \frac{b^u}{\ln b} + C$

Example 5   Differentiate.

(A) $D_x e^{2x-1}$ (B) $D_x e^{-x^2}$ (C) $D_x 3^{2x}$

Solutions   (A) e^{2x-1} is a composite function of the form

$$y = e^u \qquad u = u(x) = 2x - 1$$

and formula 1 applies. Thus,

$$\begin{aligned}
D_x e^{2x-1} &= e^{2x-1} D_x(2x - 1) \\
&= e^{2x-1}(2) \\
&= 2e^{2x-1}
\end{aligned}$$

(B) e^{-x^2} is also a composite function of the form

$$y = e^u \qquad u = u(x) = -x^2$$

and, using formula 1, we obtain

$$\begin{aligned}
D_x e^{-x^2} &= e^{-x^2} D_x(-x^2) \\
&= e^{-x^2}(-2x) \\
&= -2x e^{-x^2}
\end{aligned}$$

(C) 3^{2x} is a composite function of the form b^u, $u = u(x) = 2x$; hence, formula 3 is used to obtain

$$\begin{aligned}
D_x 3^{2x} &= 3^{2x}(\ln 3) D_x(2x) \\
&= 3^{2x}(\ln 3)(2) \\
&= 2(3^{2x})(\ln 3) = 3^{2x} \ln 3^2 = 3^{2x} \ln 9
\end{aligned}$$

[Note: $2(3^{2x}) \neq 6^{2x}$ (Why?) and $D_x 3^{2x} \neq (2x)3^{2x-1}$ (Why?)]

Problem 5   Differentiate.

(A) $D_x e^{3x}$ (B) $D_x e^{x^2-x}$ (C) $D_x 10^{5x+2}$

Answers   (A) $3e^{3x}$ (B) $(2x - 1)e^{x^2-x}$ (C) $5(10^{5x+2})(\ln 10)$

Example 6 Find:

(A) $\int 3e^{3x}\,dx$ (B) $\int xe^{x^2}\,dx$ (C) $\int 5^x\,dx$

Solutions (A) Mentally note that if $u = 3x$, then $du = 3\,dx$. Thus,

$$\int e^{3x}3\,dx$$

is of the form

$$\int e^u\,du \qquad \text{Where } u = 3x \text{ and } du = 3\,dx$$

and formula 2 applies. Hence,

$$\int e^{3x}3\,dx = e^{3x} + C$$

(B) Again, mentally note that if $u = x^2$, then $du = 2x\,dx$, and the integral differs from the form $\int e^u\,du$ only by a factor of 2 in the integrand. We introduce this factor as in the past, and then use formula 2:

$$\int xe^{x^2}\,dx = \int e^{x^2}(\tfrac{2}{2})x\,dx$$

$$= \tfrac{1}{2}\int e^{x^2}2x\,dx \qquad \tfrac{1}{2}\int e^u\,du \text{ form}$$

$$= \tfrac{1}{2}e^{x^2} + C$$

(C) First, note that the functions 5^x and x^5 are very different. While it is true that

$$\int x^5\,dx = \frac{x^{5+1}}{5+1} + C = \frac{x^6}{6} + C$$

it is not true that

$$\int 5^x\,dx = \frac{5^{x+1}}{x+1} + C$$

The integral $\int 5^x\,dx$ is really a special case of formula 4; thus,

$$\int 5^x\,dx = \frac{5^x}{\ln 5} + C$$

(Use formula 3 to check this result by differentiating the right side to obtain the integrand on the left side.)

Problem 6 Find:

(A) $\int 5e^{5x}\,dx$ (B) $\int xe^{3x^2}\,dx$ (C) $\int 10^x\,dx$

Answers   (A) $e^{5x} + C$ (B) $\frac{1}{6}e^{3x^2} + C$ (C) $\frac{10^x}{\ln 10} + C$

Example 7   Find:

(A) $D_x(x - e^x \ln x)$ (B) $D_x\frac{1 - e^{-x}}{x^2 + 1}$

Solutions   (A) $D_x(x - e^x \ln x) = D_x x - D_x e^x \ln x$

$$= 1 - (e^x D_x \ln x + \ln x\, D_x e^x)$$

$$= 1 - \frac{e^x}{x} - e^x \ln x$$

(B) $D_x\frac{1 - e^{-x}}{x^2 + 1} = \frac{(x^2 + 1)D_x(1 - e^{-x}) - (1 - e^{-x})D_x(x^2 + 1)}{(x^2 + 1)^2}$

$$= \frac{e^{-x}(x^2 + 1) - 2x(1 - e^{-x})}{(x^2 + 1)^2}$$

Problem 7   Find:

(A) $D_x(x^2 e^x + \ln x)$ (B) $D_x\frac{e^{2x}}{x + 1}$

Answers   (A) $x^2 e^x + 2xe^x + \frac{1}{x}$ (B) $\frac{e^{2x}(2x + 1)}{(x + 1)^2}$

Example 0   Evaluate.

(A) $\int_0^2 e^{3x-5}\, dx$ (B) $\int_{-1}^2 (e^{x+1} + 5)\, dx$

Solutions   (A) $\int_0^2 e^{3x-5}\, dx = \frac{1}{3}\int_0^2 e^{3x-5}\, 3\, dx$

$$= \frac{1}{3}e^{3x-5}\Big|_0^2$$

$$= \frac{1}{3}(e^1 - e^{-5}) \approx 0.9038$$

(B) $\int_{-1}^2 (e^{x+1} + 5)\, dx = (e^{x+1} + 5x)\Big|_{-1}^2$

$$= (e^3 + 10) - (e^0 - 5)$$

$$= e^3 + 10 - 1 + 5$$

$$= e^3 + 14 \approx 34.0855$$

Problem 8   Evaluate.

(A) $\int_0^1 e^{2x+1}\, dx$ (B) $\int_{-2}^0 (2x + e^{x+2})\, dx$

Answers   (A) $\frac{1}{2}(e^3 - e) \approx 8.6836$ (B) $e^2 - 5 \approx 2.3891$

EXERCISE 5-4

A   Find each derivative.

1. $D_t e^t$ 2. $D_z e^z$ 3. $D_x e^{8x}$ 4. $D_x e^{5x-1}$

5. $D_x 3e^{2x}$ 6. $D_y 2e^{3y}$ 7. $D_t 2e^{-4t}$ 8. $D_r 6e^{-3r}$

Find each indefinite integral.

9. $\int e^t \, dt$

10. $\int e^z \, dz$

11. $\int 8e^{8x} \, dx$

12. $\int 5e^{5x-1} \, dx$

13. $\int e^{3x} \, dx$

14. $\int e^{4t} \, dt$

15. $\int e^{-2t} \, dt$

16. $\int e^{-4t} \, dt$

Find each definite integral.

17. $\int_0^1 e^t \, dt$

18. $\int_0^2 e^z \, dz$

19. $\int_0^{1/4} 8e^{8x} \, dx$

20. $\int_0^1 5e^{5x-1} \, dx$

B   Find each derivative.

21. $D_x e^{3x^2-2x}$

22. $D_x e^{x^3-3x^2+1}$

23. $D_x \dfrac{e^x - e^{-x}}{2}$

24. $D_x \dfrac{e^x + e^{-x}}{2}$

25. $D_x(e^{2x} - 3x^2 + 5)$

26. $D_x(2e^{3x} - 2e^{2x} + 5x)$

27. $D_x(xe^x)$

28. $D_x(x - 1)e^x$

29. $D_x 100e^{-0.03x}$

30. $D_t 1{,}000e^{0.06t}$

31. $D_x(e^{2x} - 1)^4$

32. $D_x(e^{x^2} + 3)^5$

33. $D_x 7^x$

34. $D_x 2^x$

35. $D_x[(e^x)^4 + e^{x^4}]$

36. $D_x(\sqrt{e^x} + e^{\sqrt{x}})$

37. $D_x \dfrac{e^{2x}}{x^2 + 1}$

38. $D_x \dfrac{e^{x+1}}{x + 1}$

39. $D_x(x^2 + 1)e^{-x}$

40. $D_x(1 - x)e^{2x}$

41. $D_x e^{-x} \ln x$

42. $D_x \dfrac{\ln x}{e^x + 1}$

Find each indefinite integral.

43. $\displaystyle \int xe^{x^2+1}\,dx$

44. $\displaystyle \int xe^{-x^2}\,dx$

45. $\displaystyle \int \frac{e^x - e^{-x}}{2}\,dx$

46. $\displaystyle \int \frac{e^x + e^{-x}}{2}\,dx$

47. $\displaystyle \int 4^x\,dx$

48. $\displaystyle \int 12^x\,dx$

Evaluate each definite integral.

49. $\displaystyle \int_0^1 te^{-t^2}\,dt$

50. $\displaystyle \int_{-1}^1 we^{w^2+1}\,dw$

C   *Find each derivative.*

51. $D_x xe^x \ln x$

52. $D_x \dfrac{e^x - e^{-x}}{e^x + e^{-x}}$

53. $D_x 10^{x^2+x}$

54. $D_x 8^{1-2x^2}$

Find each indefinite integral.

55. $\displaystyle \int \frac{x}{e^{x^2-1}}\,dx$

56. $\displaystyle \int (e^{2x} - 1)^3 e^{2x}\,dx$

57. $\displaystyle \int x6^{x^2}\,dx$

58. $\displaystyle \int x3^{-x^2}\,dx$

59. $\displaystyle \int \frac{e^x - e^{-x}}{e^x + e^{-x}}\,dx$

60. $\displaystyle \int \frac{e^{2x} + e^{-2x}}{e^{2x} - e^{-2x}}\,dx$

Compute.

61. $D_x \displaystyle \int_1^x e^t\,dt$

62. $D_x \displaystyle \int_1^x e^{t-1}\,dt$

63. $D_x \displaystyle \int_1^x \frac{t}{e^{t^2}}\,dt$

64. $D_x \displaystyle \int_1^x te^{t^2}\,dt$

5-5   APPLICATIONS

In this section we will consider a number of applications that involve the use of logarithmic and exponential functions. We will go through several types of problems in detail and will include many more in the exercises.

**CONTINUOUS
COMPOUND INTEREST**

Let us start with simple interest, move on to compound interest, and then to continuous compound interest. If a principal P is borrowed at an annual rate r, then after t years at simple interest the borrower will owe the lender an amount A given by

$$A = P + Prt = P(1 + rt) \qquad \text{Simple interest} \qquad (1)$$

On the other hand, if interest is compounded n times a year, the borrower will owe the lender an amount A given by

$$A = P\left(1 + \frac{r}{n}\right)^{nt} \qquad \text{Compound interest} \qquad (2)$$

Suppose P, r, and t in (2) are held fixed and n is increased. Will the amount A increase without bound or will it tend to some limiting value?

Let us perform a calculator experiment before we attack the general limit problem. If $P = \$100$, $r = 0.06$, and $t = 2$ years, then

$$A = 100\left(1 + \frac{0.06}{n}\right)^{2n}$$

We compute A for several values of n in Table 1. The biggest gain appears in the first step; then the gains slow down as n increases. In fact, it appears that A might be tending to something close to \$112.75 as n gets larger and larger.

TABLE 1

COMPOUNDING FREQUENCY	n	$A = 100\left(1 + \dfrac{0.06}{n}\right)^{2n}$
Annually	1	\$112.3600
Semiannually	2	112.5509
Quarterly	4	112.6493
Weekly	52	112.7419
Daily	365	112.7486
Hourly	8,760	112.7491

Now we turn back to the general problem for a moment. Keeping P, r, and t fixed, we compute the following limit and observe an interesting and useful result:

$$\lim_{n \to \infty} P\left(1 + \frac{r}{n}\right)^{nt} = P \lim_{n \to \infty} \left(1 + \frac{r}{n}\right)^{(n/r)rt} \qquad \text{Insert } r/r \text{ in the exponent}$$

$$= P[\lim_{s \to 0}(1 + s)^{1/s}]^{rt} \qquad \text{Let } s = r/n$$

$$= Pe^{rt} \qquad \text{Definition of } e \\ \text{(see Section 5-2)}$$

Again, we encounter the special constant e. The resulting formula is called the **continuous compound interest formula.**

CONTINUOUS COMPOUND INTEREST

$A = Pe^{rt}$ P = Principal

r = Annual rate compounded continuously

t = Time in years

A = Amount at time t

Example 9   If \$100 is invested at 6% compounded continuously, what amount will be in the account after 2 years?

Solution
$$A = Pe^{rt}$$
$$= 100e^{(0.06)(2)}$$
$$\approx \$112.7497$$

(Compare this result with the values calculated in Table 1.)

Problem 9   What amount (to the nearest cent) will an account have after 5 years if \$100 is invested at 8% compounded annually? Semiannually? Continuously?

Answer   \$146.93; \$148.02; \$149.18

Let us approach the continuous compound interest problem from a point of view that will enable us to generalize the concept and apply it to other problems.

Suppose we say that the amount of money A in an account grows at a rate proportional to the amount present, and the amount in the account at the start is P. Mathematically,

$$\frac{dA}{dt} = rA \qquad A(0) = P$$

where r is an appropriate constant. We would like to find a function $A = A(t)$ that satisfies both conditions. Treating dA and dt as separate quantities and multiplying both sides of the first equation by dt/A, we obtain

$$\frac{dA}{A} = r \, dt$$

Now we integrate both sides,

$$\int \frac{dA}{A} = \int r \, dt$$
$$\ln A = rt + c$$

and convert this last equation into the equivalent exponential form

$$A = e^{rt+c} \qquad \text{From Section 1-6, } y = \ln x \text{ if and only if } x = e^y$$
$$= e^c e^{rt} \qquad \text{Property of exponents: } b^m b^n = b^{m+n}$$

Since $A(0) = P$, we evaluate $A(t) = e^c e^{rt}$ at $t = 0$ and set it equal to P:

$$A(0) = e^c e^0 = e^c = P$$

Hence, $e^c = P$, and we can rewrite $A = e^c e^{rt}$ in the form

$$A = Pe^{rt}$$

This is the same continuous compound interest formula obtained earlier, where the principal P is invested at an annual rate of r compounded continuously for t years.

EXPONENTIAL GROWTH LAW

In general, if a quantity Q changes at a rate proportional to the amount present and $Q(0) = Q_0$, then proceeding in exactly the same way as above, we obtain the following:

EXPONENTIAL GROWTH LAW

If $\dfrac{dQ}{A} = rQ$ and $Q(0) = Q_0$, then $Q = Q_0 e^{rt}$.

$Q_0 = $ Amount at $t = 0$
$r = $ Annual rate compounded continuously
$t = $ Time
$Q = $ Quantity at time t

This law not only applies to money invested at interest compounded continuously, but to many other types of problems—population growth, radioactive decay, natural resource depletion, and so on.

POPULATION GROWTH

The world population is growing at an ever-increasing rate, as illustrated in Figure 2. Population growth over certain periods of time can often be approximated by the exponential growth law described above.

Example 10

India had a population of 500 million people in 1966 ($t = 0$) and a growth rate of 3% per year (which we will assume is compounded continuously). If P is the population in millions t years after 1966, and the same growth rate continues, then

$$\frac{dP}{dt} = 0.03P \qquad P(0) = 500$$

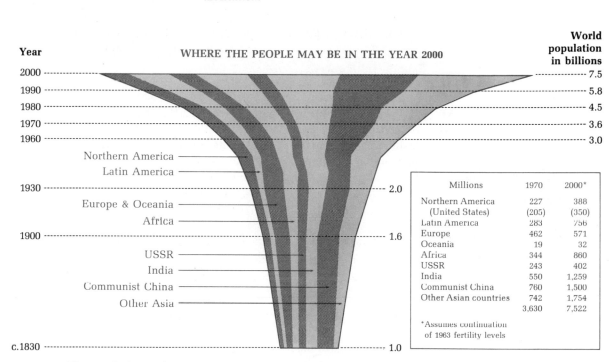

FIGURE 2 The population explosion. *Source:* United States State Department

Thus, using the exponential growth law,

$$P = 500e^{0.03t}$$

With this result, we can estimate the population of India in 1986 ($t = 20$) to be

$$P(20) = 500e^{0.03(20)}$$
$$\approx 911 \text{ million people}$$

Problem 10 Assuming the same rate of growth, what will India's population be in the year 2001?

Answer 1,429 million people

Example 11 If the USSR has a growth rate of 1.1% per year (assumed compounded continuously), how long will it take the population to double?

Solution Given $P = P_0e^{rt}$ with $r = 0.011$, find t so that $P = 2P_0$.

$$2P_0 = P_0e^{0.011t}$$
$$2 = e^{0.011t}$$
$$\ln 2 = \ln e^{0.011t}$$
$$\ln 2 = 0.011t$$
$$t = \frac{\ln 2}{0.011} \approx 63 \text{ years}$$

Problem 11   Find the doubling time for the population in Mexico if it continues growing at 3.2% per year (assumed compounded continuously).

Answer   Approximately 22 years

RADIOACTIVE DECAY AND CARBON-14 DATING

In 1946, Willard Libby, who later received a Nobel Prize in Chemistry, found that as long as a plant or animal is alive, radioactive carbon-14 is maintained at a constant level in its tissues. Once dead, however, the radioactive carbon-14 diminishes by radioactive decay at a rate proportional to the amount present. Thus,

$$\frac{dQ}{dt} = rQ \qquad Q(0) = Q_0$$

and we have another example of the exponential growth law. The rate of decay for radioactive carbon-14 is found to be 0.0001238; thus, $r = -0.0001238$, since decay is negative growth.

Example 12   A piece of human bone was found at an archaeological site in Africa. If 10% of the original amount of radioactive carbon-14 was present, estimate the age of the bone.

Solution   Using the exponential growth law for

$$\frac{dQ}{dt} = -0.0001238Q \qquad Q(0) = Q_0$$

we find that

$$Q = Q_0 e^{-0.0001238t}$$

and our problem is to find t so that $Q = 0.1Q_0$ (the amount of carbon-14 present now is 10% of the amount present, Q_0, at the death of the person). Thus,

$$0.1Q_0 = Q_0 e^{-0.0001238t}$$
$$0.1 = e^{-0.0001238t}$$
$$\ln 0.1 = \ln e^{-0.0001238t}$$
$$t = \frac{\ln 0.1}{-0.0001238} \approx 18{,}600 \text{ years}$$

Problem 12   Estimate the age of the bone in Example 12 if 50% of the original amount of carbon-14 was present.

Answer   Approximately 5,600 years

LEARNING   In learning certain skills such as typing and swimming, a mathematical model often used is one that assumes there is a maximum skill attainable, say M, and the rate of improving is proportional to the difference between that achieved, y, and that attainable, M. Mathematically,

$$\frac{dy}{dt} = k(M - y) \qquad y(0) = 0$$

We solve this using the same technique that was used to obtain the exponential growth law. First, multiply both sides of the first equation by $dt/(M - y)$ to obtain

$$\frac{dy}{M - y} = k \, dt$$

and then integrate both sides:

$$\int \frac{dy}{M - y} = \int k \, dt$$
$$-\ln(M - y) = kt + c$$

Change this last equation to equivalent exponential form:

$$M - y = e^{-kt-c}$$
$$M - y = e^{-c}e^{-kt}$$
$$y = M - e^{-c}e^{-kt}$$

Now $y(0) = 0$; hence,

$$y(0) = M - e^{-c}e^0 = 0$$

Solving for e^{-c}, we obtain

$$e^{-c} = M$$

and our final solution is

$$y = M - Me^{-kt} = M(1 - e^{-kt})$$

Example 13   For a particular person who is learning to swim, it is found that the distance y (in feet) the person is able to swim in 1 minute after t hours of practice is given approximately by

$$y = 50(1 - e^{-0.04t})$$

What is the rate of improvement after 10 hours of practice?

Solution
$$y = 50 - 50e^{-0.04t}$$
$$y'(t) = 2e^{-0.04t}$$
$$y'(10) = 2e^{-0.04(10)}$$
$$\approx 1.34 \text{ feet per hour of practice}$$

Problem 13   In Example 13, what is the rate of improvement after 50 hours of practice?

Answer   Approximately 0.27 foot per hour

**A COMPARISON OF
EXPONENTIAL GROWTH
PHENOMENA**

The following graphs and equations compare several widely used growth models. These are divided basically into two groups: unlimited growth and limited growth. Following each graph and equation is a short, incomplete list of areas in which the models are used. This only touches on a subject that has been extensively developed and which you are likely to encounter in greater depth in the future.

UNLIMITED GROWTH

$$\frac{dy}{dx} = ay \qquad a > 0, \quad x > 0$$

$$y(0) = c$$

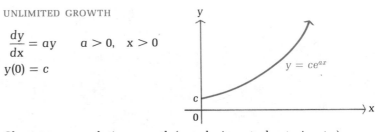

Short-term population growth (people, insects, bacteria, etc.)
Growth of money at continuous compound interest
Short-term consumption of some natural resources
FIGURE 3 — Price–supply curves

DECAY

$$\frac{dy}{dx} = -ay \qquad a > 0, \quad x > 0$$

$$y(0) = c$$

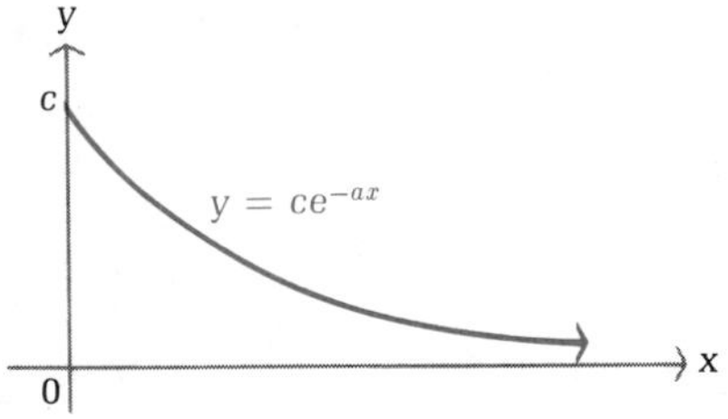

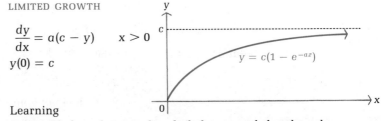

Radioactive decay
Light absorption in water
Atmospheric pressure (x is altitude)
FIGURE 4 — Price–demand curves

LIMITED GROWTH

$$\frac{dy}{dx} = a(c - y) \qquad x > 0$$

$$y(0) = c$$

Learning
Sales of fad products such as hula hoops and skateboards
FIGURE 5 — Depreciation of vehicles and equipment

LIMITED GROWTH
(LOGISTIC CURVE)

$$\frac{dy}{dx} = ky(M - y) \qquad x > 0$$

Learning
Population growth—long-term
Epidemics
FIGURE 6   Sales of new products

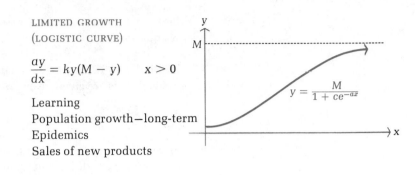

$$y = \frac{M}{1 + ce^{-ax}}$$

EXERCISE 5-5   APPLICATIONS

BUSINESS & ECONOMICS

1. *Interest.* If $1,000 is invested at 8% compounded continuously, how much will be in the account at the end of 20 years?

2. *Interest.* The earth weighs approximately 2.11×10^{26} ounces. If an ounce of gold is worth $1,000, what would be the value of an account in the year 2000 in terms of solid gold earths if $1 had been invested at 4% interest compounded continuously at the birth of Christ? What would be the value in dollars at simple interest?

3. *Present value.* A note will pay $20,000 at maturity 10 years from now. How much should you be willing to pay for the note now if money is worth 7% compounded continuously?

4. *Present value.* A note will pay $50,000 at maturity 5 years from now. How much should you be willing to pay for the note now if money is worth 8% compounded continuously?

5. *Marginal analysis.* Suppose the price–demand equation for x units of a commodity is

   $$p(x) = 100e^{-0.05x}$$

   Then the revenue equation is

   $$R(x) = xp(x) = x100e^{-0.05x}$$

   Find the marginal revenue.

6. *Price–supply.* Suppose the price–supply equation for x units of a commodity is

   $$p(x) = 10e^{0.0005x}$$

   where p is dollars and x is in thousands of units. What price will result if 100,000 units are available?

7. *Marginal analysis.* Find the marginal supply, given the price–supply equation in Problem 6.

8. *Marginal analysis.* If a company's daily marginal revenue and marginal costs are given, respectively, by

$$R'(t) = 250e^{0.05t}$$
$$C'(t) = 100 - 8t$$

where t is time in days, what is the total profit for the first month's operation ($t = 30$ days)? Set up an appropriate definite integral and evaluate.

9. *Salvage value.* The salvage value of a company airplane after t years is estimated to be given by

$$S(t) = 300,000e^{-0.1t}$$

What is the rate of depreciation in dollars per year after 1 year? 5 years? 10 years?

10. *Advertising.* A company is trying to expose a new product to as many people as possible through television advertising. Suppose the rate of exposure to new people is proportional to those who have not seen the product out of L possible viewers. If no one is aware of the product at the start of the campaign and after 10 days 40% of L are aware of the product, solve

$$\frac{dN}{dt} = k(L - N) \qquad N(0) = 0, \quad N(10) = 0.4L$$

for $N = N(t)$, the number of people who are aware of the product after t days of advertising.

LIFE SCIENCES

11. *Population growth.* How long will it take the earth's population to double at a growth rate of 2% per year, assuming this rate continues?

12. *Population growth.* The world population of 4 billion people is now estimated to be growing at 2% per year. If it continues to grow at this rate, how long will it be before there is only 1 square yard of land per person? The estimated land area of the earth is 1.7×10^{14} square yards.

13. *Ecology.* For relatively clear bodies of water, light intensity is reduced according to

$$\frac{dI}{dx} = -kI \qquad I(0) = I_0$$

where I is the intensity of light at x feet below the surface. For the Sargasso Sea off the West Indies, $k = 0.00942$. Find I in terms of x and find the depth at which the light is reduced to half of that at the surface.

14. *Blood pressure.* It can be shown under certain assumptions that blood pressure P in the largest artery in the human body (the aorta) changes between beats with respect to time t according to

$$\frac{dP}{dt} = -aP \qquad P(0) = P_0$$

Without referring to the text, find $P = P(t)$ that satisfies both conditions.

15. *Drug concentrations.* A single injection of a drug is administered to a patient. The amount Q in the body then decreases at a rate proportional to the amount present, and for this particular drug the rate is 4% per hour. Thus,

$$\frac{dQ}{dt} = -0.04Q \qquad Q(0) = Q_0$$

where t is time in hours. If the initial injection is 3 milliliters $[Q(0) = 3]$, find $Q = Q(t)$ that satisfies both conditions. How many milliliters of the drug are still in the body after 10 hours?

16. *Simple epidemic.* A community of 1,000 individuals is assumed to be homogeneously mixed. One individual who has just returned from another community has influenza. Assume the home community has not had influenza shots and all are susceptible. One mathematical model for an influenza epidemic assumes that influenza tends to spread at a rate in direct proportion to those who have it, N, and to those who have not contracted it, in this case, $1,000 - N$. Mathematically,

$$\frac{dN}{dt} = kN(1,000 - N) \qquad N(0) = 1$$

where N is the number of people who contract influenza after t days. For $k = 0.0004$, it can be shown that $N(t)$ is given by

$$N(t) = \frac{1,000}{1 + 999e^{-0.4t}}$$

See Figure 6 for the characteristic graph.

(A) How many people have contracted influenza after 10 days? After 20 days?

(B)   How many days will it take until half the community has contracted influenza?

(C)   Find $\lim_{t \to \infty} N(t)$.

SOCIAL SCIENCES

17.   *Archaeology.*   A skull from an ancient tomb was discovered and was found to have 5% of the original amount of radioactive carbon-14 present. Estimate the age of the skull. (See Example 12.)

18.   *Learning.*   For a particular person learning to type, it was found that the number of words per minute, N, the person was able to type after t hours of practice was given approximately by

$$N = 100(1 - e^{-0.02t})$$

See Figure 5 for a characteristic graph. What is the rate of improvement after 10 hours of practice? After 40 hours of practice?

19.   *Small group analysis.*   In a stuy on small group dynamics, sociologists Stephan and Mischler found that, when the members of a discussion group of ten were ranked according to the numer of times each participated, the number of times $N(k)$ the kth-ranked person participated was given approximately by

$$N(k) = N_1 e^{-0.11(k-1)} \qquad 1 \le k \le 10$$

where N_1 is the number of times the first-ranked person participated in the discussion. If, in a particular discussion group of ten people, $N_1 = 180$, estimate how many times the sixth-ranked person participated. The tenth-ranked person.

20.   *Perception.*   One of the oldest laws in mathematical psychology is the Weber–Fechner law (discovered in the middle of the nineteenth century). It concerns a person's sensed perception of various strengths of stimulation involving weights, sound, light, shock, taste, and so on. One form of the law states that the rate of change of sensed sensation S with respect to stimulus R is inversely proportional to the strength of the stimulus R. Thus,

$$\frac{dS}{dR} = \frac{k}{R}$$

If we let R_0 be the threshold level at which the stimulus R can be detected (the least amount of sound, light, weight, and so on that can be detected), then it is appropriate to write

$$S(R_0) = 0$$

Find a function S in terms of R that satisfies the above conditions.

21. *Rumor spread.* A group of 400 parents, relatives, and friends are waiting anxiously at Kennedy Airport for a student charter to return after a year in Europe. It is stormy and the plane is late. A particular parent thought he had heard that the plane's radio had gone out and related this news to some friends, who in turn passed it on to others, and so on. Sociologists have studied rumor propagation and have found that a rumor tends to spread at a rate in direct proportion to those who have heard it, x, and to those who have not, $P - x$, where P is the total population. Mathematically, for our case, $P = 400$ and

$$\frac{dx}{dt} = 0.001x(400 - x) \qquad x(0) = 1$$

where t is time in minutes. From this, it can be shown that

$$x(t) = \frac{400}{1 + 399e^{-0.4t}}$$

See Figure 6 for a characteristic graph.

(A) How many people have heard the rumor after 5 minutes? 20 minutes?

(B) Find $\lim_{t \to \infty} x(t)$.

5-6   INTEGRATION BY PARTS

In Section 5-3 we said that we would return to the indefinite integral

$$\int \ln x \, dx$$

later, since none of the integration techniques considered up to that time could be used to find an antiderivative for ln x. We will now develop a very useful technique, called *integration by parts*, that will not only enable us to find the above integral, but also many others, including integrals such as

$$\int x \ln x \, dx \qquad \text{and} \qquad \int xe^x \, dx$$

The integration by parts technique is also used to derive many integration formulas that are tabulated in mathematical handbooks.

The method of integration by parts is based on the product formula for derivatives. If f and g are differentiable functions, then

$$D_x[f(x)g(x)] = f(x)g'(x) + g(x)f'(x)$$

which can be written in the equivalent form

$$f(x)g'(x) = D_x[f(x)g(x)] - g(x)f'(x)$$

Integrating both sides, we obtain

$$\int f(x)g'(x)\,dx = \int D_x[f(x)g(x)]\,dx - \int g(x)f'(x)\,dx$$

The first integral to the right of the equal sign is $f(x)g(x) + C$. (Why?) We will leave out the constant of integration for now, since we can add it after integrating the second integral to the right of the integral sign. So we have

$$\int f(x)g'(x)\,dx = f(x)g(x) - \int g(x)f'(x)\,dx$$

This last form can be transformed into a more convenient form by letting $u = f(x)$ and $v = g(x)$; then $du = f'(x)\,dx$ and $dv = g'(x)\,dx$. Making these substitutions, we obtain the **integration by parts formula:**

INTEGRATION BY PARTS FORMULA

$$\int u\,dv = uv - \int v\,du$$

This formula may be very useful when the integral on the left is difficult to integrate using standard formulas. If u and dv are chosen with care, then the integral on the right side may be easier to integrate than the one on the left. A couple of examples will demonstrate the use of the formula.

Example 14   Find $\int x \ln x\,dx$, $x > 0$, using integration by parts.

Solution   First, write the integration by parts formula

$$\int u\,dv = uv - \int v\,du$$

Then try to identify u and dv in $\int x \ln x\,dx$ (this is the key step) so that when $\int u\,dv$ is written in the form $uv - \int v\,du$, the new integral will be easier to integrate.

Suppose we choose

$$u = x \quad \text{and} \quad dv = \ln x\,dx$$

Then

$$du = dx \quad v = ?$$

We do not know an antiderivative of $\ln x$ yet, so we change our choice for u and dv to

$$u = \ln x \quad dv = x\,dx$$

Then

$$du = \frac{1}{x}dx \qquad v = \frac{x^2}{2}$$

Any constant may be added to v (we choose zero for simplicity). There are cases where it is convenient to add a constant other than zero, but in most cases zero will do. The general arbitrary constant of integration will be added at the end of the process.

Using the chosen u, du, dv, and v in the integration by parts formula, we obtain

$$\overset{u}{\int} (\ln x)\, x\, dx = \overset{u}{(\ln x)}\overset{v}{\left(\frac{x^2}{2}\right)} - \int \overset{v}{\left(\frac{x^2}{2}\right)}\overset{du}{\frac{1}{x}dx}$$

$$= \frac{x^2}{2}\ln x - \int \frac{x}{2}dx \qquad \text{This new integral is easy to integrate}$$

$$= \frac{x^2}{2}\ln x - \frac{x^2}{4} + C$$

To check this result, show that

$$D_x\left(\frac{x^2}{2}\ln x - \frac{x^2}{4} + C\right) = x\ln x$$

which is the integrand in the original integral.

Problem 14   Find $\int x \ln 2x\, dx$.

Answer $\dfrac{x^2}{2}\ln 2x - \dfrac{x^2}{4} + C$

Example 15   Find $\int_1^e \ln x\, dx$.

Solution   First, find $\int \ln x\, dx$; then return to the definite integral. We start by writing the integration by parts formula

$$\int u\, dv = uv - \int v\, du$$

and make the choice for u and dv:

$$u = \ln x \qquad dv = dx$$

Then,

$$du = \frac{1}{x}dx \qquad v = x$$

Hence,

$$\int \ln x \, dx = (\ln x)(x) - \int (x)\frac{1}{x} dx$$
$$= x \ln x - x + C$$

Thus,

$$\int_1^e \ln x \, dx = (x \ln x - x)\Big|_1^e$$
$$= (e \ln e - e) - (1 \ln 1 - 1)$$
$$= (e - e) - (0 - 1)$$
$$= 1$$

Problem 15   Find $\int_1^2 \ln 3x \, dx$.

Answer $2 \ln 6 - \ln 3 - 1 \approx 1.4849$

EXERCISE 5-6

A   *Integrate using integration by parts. Assume x > 0 whenever the natural log function is involved.*

1. $\int xe^x \, dx$ 2. $\int xe^{4x} \, dx$

3. $\int x^2 \ln x \, dx$ 4. $\int x^3 \ln x \, dx$

B   *Problems 5–18 are mixed—some require integration by parts and others can be solved using techniques we have considered earlier. Integrate as indicated, assuming x > 0 whenever the natural log function is involved.*

5. $\int xe^{-x} \, dx$ 6. $\int (x - 1)e^{-x} \, dx$

7. $\int xe^{x^2} \, dx$ 8. $\int xe^{-x^2} \, dx$

9. $\int_0^1 (x - 3)e^x \, dx$ 10. $\int_0^2 (x + 5)e^x \, dx$

11. $\int_1^3 \ln 2x \, dx$ 12. $\int_2^3 \ln 7x \, dx$

13. $\int \frac{2x}{x^2 + 1} dx$ 14. $\int \frac{x^2}{x^3 + 5} dx$

15. $\int \dfrac{\ln x}{x} \, dx$

16. $\int \dfrac{e^x}{e^x + 1} \, dx$

17. $\int x \sqrt{x + 1} \, dx$

18. $\int x \sqrt{2x - 1} \, dx$

C *Some of these problems may require using the integration by parts formula more than once. Assume $x > 0$ whenever the natural log function is involved.*

19. $\int x^2 e^x \, dx$

20. $\int x^3 e^x \, dx$

21. $\int x e^{ax} \, dx, \quad a \neq 0$

22. $\int \ln(ax) \, dx, \quad a > 0$

23. $\int_1^e \dfrac{\ln x}{x^2} \, dx$

24. $\int_1^2 x^3 e^{x^2} \, dx$

25. $\int (\ln x)^2 \, dx$

26. $\int x(\ln x)^2 \, dx$

APPLICATIONS

BUSINESS & ECONOMICS

27. *Marginal profit.*   If the marginal profit per year in millions of dollars is given by

$$P'(t) = 2t - te^{-t}$$

where t is time in years and the profit at time zero is zero, find $P = P(t)$.

28. *Production.*   An oil field is estimated to produce $R(t)$ thousand barrels of oil per month t months from now, as given by

$$R(t) = 10te^{-0.1t}$$

Estimate the total production in the first year of operation by use of an appropriate definite integral.

5-7   **IMPROPER INTEGRALS**

We are now going to consider an integral form that has wide application in probability studies as well as other areas. Earlier, when we introduced the idea of a definite integral,

$$\int_a^b f(x)\,dx \tag{1}$$

We required f to be continuous over a closed interval $[a, b]$. Now we are going to extend the meaning of (1) so that the interval $[a, b]$ may become infinite in length.

Let us investigate a particular example that will motivate several general definitions. What would be a reasonable interpretation for the following expression?

$$\int_1^\infty \frac{dx}{x^2}$$

Sketching a graph of $f(x) = 1/x^2$, $x \geq 1$ (Fig. 7), we note that for any fixed $b > 1$, $\int_1^b f(x)\,dx$ is the area between the curve $y = 1/x^2$, the x axis, $x = 1$, and $x = b$.

FIGURE 7

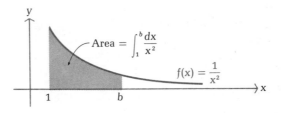

Let us see what happens when we let $b \to \infty$; that is, when we compute the following limit:

$$\lim_{b \to \infty} \int_1^b \frac{dx}{x^2} = \lim_{b \to \infty} \left[(-x^{-1}) \Big|_1^b \right]$$

$$= \lim_{b \to \infty} \left(-\frac{1}{b} + 1 \right) = 1$$

Did you expect this result? No matter how large b is taken, the area under the curve from $x = 1$ to $x = b$ never exceeds 1, and in the limit it is 1. This suggests that we write

$$\int_1^\infty \frac{dx}{x^2} = \lim_{b \to \infty} \int_1^b \frac{dx}{x^2} = 1$$

This integral is an example of an *improper integral*. In general, the forms

$$\int_{-\infty}^b f(x)\,dx \qquad \int_a^\infty f(x)\,dx \qquad \int_{-\infty}^\infty f(x)\,dx$$

where f is continuous over the indicated interval, are called *improper integrals*. (There are also other types of improper integrals that will not be considered here. These involve certain types of points of discontinuity within the interval of integration.) Each type of **improper integral** above is formally defined in the box:

IMPROPER INTEGRALS

If f is continuous over the indicated interval and the limit exists, then:

1. $\displaystyle\int_a^\infty f(x)\,dx = \lim_{b \to \infty} \int_a^b f(x)\,dx$

2. $\displaystyle\int_{-\infty}^b f(x)\,dx = \lim_{a \to -\infty} \int_a^b f(x)\,dx$

3. $\displaystyle\int_{-\infty}^\infty f(x)\,dx = \int_{-\infty}^c f(x)\,dx + \int_c^\infty f(x)\,dx$

where c is any point on $(-\infty, \infty)$

If the indicated limit exists, then the improper integral is said to exist or **converge;** if the limit does not exist, then the improper integral is said not to exist or **diverge** (and no value is assigned to it).

Example 16　　Evaluate $\int_2^\infty dx/x$ if it converges.

Solution

$$\int_2^\infty \frac{dx}{x} = \lim_{b \to \infty} \int_2^b \frac{dx}{x}$$

$$= \lim_{b \to \infty}(\ln x)\Big|_2^b$$

$$= \lim_{b \to \infty}(\ln b - \ln 2)$$

Since $\ln b \to \infty$ as $b \to \infty$ (a property of the natural log function that we will not prove), the limit does not exist. Hence, the improper integral diverges.

Problem 16　　Evaluate $\int_3^\infty dx/(x - 1)^2$ if it converges.

Answer　　$\frac{1}{2}$

Example 17　　Evaluate $\int_{-\infty}^2 e^x\,dx$ if it converges.

Solution

$$\int_{-\infty}^2 e^x\,dx = \lim_{a \to -\infty} \int_a^2 e^x\,dx$$

$$= \lim_{a \to -\infty}(e^x\big|_a^2)$$

$$= \lim_{a \to -\infty}(e^2 - e^a) = e^2 - 0 = e^2 \qquad \text{The integral converges}$$

Problem 17   Evaluate $\int_{-\infty}^{-1} x^{-2}\, dx$ if it converges.

Answer   1

Example 18   Evaluate

$$\int_{-\infty}^{\infty} \frac{2x}{(1 + x^2)^2}\, dx$$

if it converges.

Solution $\displaystyle\int_{-\infty}^{\infty} \frac{2x}{(1 + x^2)^2}\, dx = \int_{-\infty}^{0} (1 + x^2)^{-2} 2x\, dx + \int_{0}^{\infty} (1 + x^2)^{-2} 2x\, dx$

$$= \lim_{a \to -\infty} \int_{a}^{0} (1 + x^2)^{-2} 2x\, dx + \lim_{b \to \infty} \int_{0}^{b} (1 + x^2)^{-2} 2x\, dx$$

$$= \lim_{a \to -\infty} \left[\frac{(1 + x^2)^{-1}}{-1} \Big|_{a}^{0} \right] + \lim_{b \to \infty} \left[\frac{(1 + x^2)^{-1}}{-1} \Big|_{0}^{b} \right]$$

$$= \lim_{a \to -\infty} \left[-1 + \frac{1}{1 + a^2} \right] + \lim_{b \to \infty} \left[-\frac{1}{1 + b^2} + 1 \right]$$

$$= -1 + 1 = 0 \qquad \text{The integral converges}$$

Problem 18   Evaluate $\int_{-\infty}^{\infty} dx/e^x$ if it converges.

Answer   Diverges

PROBABILITY DENSITY FUNCTIONS

We will now take a brief look at the use of improper integrals relative to probability density functions. The approach will be intuitive and informal. Hopefully, when you next encounter these concepts in a more formal setting, you will have a better idea how calculus enters into the subject.

Suppose an experiment is designed in such a way that any real number x on the interval $[a, b]$ is a possible outcome. For example, x may represent an IQ score, the height of a person in inches, or the life of a light bulb in hours.

In certain situations it is possible to find a function f with x as an independent variable that can be used to determine the probability that x will assume a value on a given subinterval of $(-\infty, \infty)$. Such a function, called a **probability density function,** must satisfy the following three conditions (see Fig. 8):

1. $f(x) \geq 0$ for all $x \in (-\infty, \infty)$
2. $\int_{-\infty}^{\infty} f(x)\, dx = 1$
3. If $[c, d]$ is a subinterval of $(-\infty, \infty)$, then

$$\text{Probability}(c \leq x \leq d) = \int_{c}^{d} f(x)\, dx$$

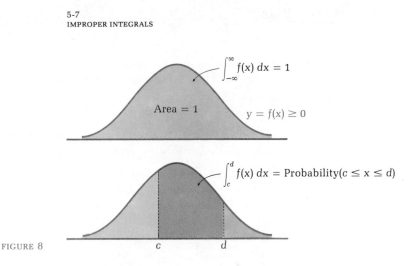

FIGURE 8

Example 19 A sailing club has a race over the same course twice a month. The races always start at 12 noon on Sunday, and the boats finish according to the probability density function (where x is hours after noon):

$$f(x) = \begin{cases} -\dfrac{x}{2} + 2 & 2 \le x \le 4 \\ 0 & \text{otherwise} \end{cases}$$

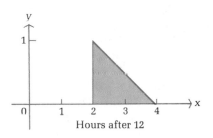

Hours after 12

Note that

$$f(x) \ge 0$$

and

$$\int_{-\infty}^{\infty} f(x)\, dx = \int_{2}^{4} \left(-\frac{x}{2} + 2\right) dx = \left(-\frac{x^2}{4} + 2x\right)\Big|_{2}^{4} = 1$$

The probability that a boat selected at random from the sailing fleet will finish between 2 and 3 hours after the start is given by

$$\text{Probability}(2 \le x \le 3) = \int_{2}^{3} \left(-\frac{x}{2} + 2\right) dx$$

$$= \left(-\frac{x^2}{4} + 2x\right)\Big|_{2}^{3} = .75$$

which is the area under the curve from x = 2 to x = 3.

Problem 19   In Example 19, find the probability that a boat selected at random from the fleet will finish between 2:30 and 3:30 PM.

Answer   .5

Example 20   Suppose the length of telephone calls (in minutes) in a public telephone booth has the probability density function

$$f(t) = \begin{cases} \frac{1}{4}e^{-t/4} & t \geq 0 \\ 0 & \text{otherwise} \end{cases}$$

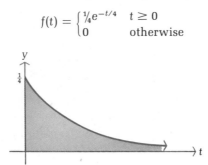

(A) Compute $\int_{-\infty}^{\infty} f(t)\, dt$.
(B) Determine the probability that a call selected at random will last between 2 and 3 minutes.

Solutions   (A) $\displaystyle\int_{-\infty}^{\infty} f(t)\, dt = \int_{-\infty}^{0} f(t)\, dt + \int_{0}^{\infty} f(t)\, dt$

$$= 0 + \int_{0}^{\infty} \frac{1}{4}e^{-t/4}\, dt$$

$$= \lim_{b \to \infty} \int_{0}^{b} \frac{1}{4}e^{-t/4}\, dt$$

$$= \lim_{b \to \infty} \left(-e^{-t/4} \Big|_{0}^{b} \right)$$

$$= \lim_{b \to \infty} \left(-e^{-b/4} + e^{0} \right)$$

$$= \lim_{b \to \infty} \left(-\frac{1}{e^{b/4}} + 1 \right)$$

$$= 0 + 1 = 1$$

(B)   Probability$(2 \leq t \leq 3) = \displaystyle\int_{2}^{3} \frac{1}{4}e^{-t/4}\, dt$

$$= \left(-e^{-t/4} \right) \Big|_{2}^{3}$$

$$= -e^{-3/4} + e^{-1/2} \approx .13$$

Problem 20   In Example 20, find the probability that a call selected at random will last longer than 4 minutes.

Answer $e^{-1} \approx .37$

The most important probability density function is the **normal probability density function** defined below and graphed in Figure 9.

$$f(x) = \frac{1}{\sigma\sqrt{2\pi}} e^{-(x-\mu)^2/2\sigma^2}$$ μ is the mean
σ is the standard deviation

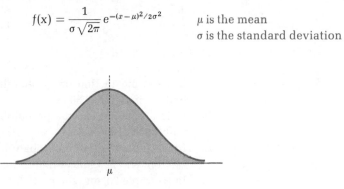

FIGURE 9.   Normal curve.

It can be shown, but not easily, that

$$\frac{1}{\sigma\sqrt{2\pi}} \int_{-\infty}^{\infty} e^{-(x-\mu)^2/2\sigma^2} \, dx = 1$$

Since $\int e^{-x^2} \, dx$ is nonintegrable in terms of elementary functions, probabilities such as

$$\text{Probability}(c \le x \le d) = \frac{1}{\sigma\sqrt{2\pi}} \int_{c}^{d} e^{-(x-\mu)^2/2\sigma^2} \, dx$$

are generally determined by making an appropriate substitution in the integrand and then using a table of areas under the standard normal curve. Such tables are readily available in most mathematical handbooks. A table can be constructed by using the rectangle rule discussed in Section 4-5; however, digital computers that use refined techniques are generally used for this purpose. Some hand calculators have the capability of computing normal curve areas directly.

EXERCISE 5-7

Find the value of each improper integral that converges.

A   1. $\int_{1}^{\infty} \frac{dx}{x^4}$ 2. $\int_{1}^{\infty} \frac{dx}{x^3}$

3. $\int_{0}^{\infty} e^{-x/2} \, dx$ 4. $\int_{0}^{\infty} e^{-x} \, dx$

B   5. $\int_{1}^{\infty} \frac{dx}{\sqrt{x}}$ 6. $\int_{1}^{\infty} \frac{dx}{\sqrt[3]{x}}$

7. $\int_{0}^{\infty} \frac{dx}{(x+1)^2}$ 8. $\int_{0}^{\infty} \frac{dx}{(x+1)^3}$

9. $\displaystyle\int_0^\infty \frac{dx}{(x+1)^{2/3}}$

10. $\displaystyle\int_0^\infty \frac{dx}{\sqrt{x+1}}$

11. $\displaystyle\int_1^\infty \frac{dx}{x^{0.99}}$

12. $\displaystyle\int_1^\infty \frac{dx}{x^{1.01}}$

13. $\displaystyle 0.3 \int_0^\infty e^{-0.3x}\, dx$

14. $\displaystyle 0.01 \int_0^\infty e^{-0.1x}\, dx$

15. In Example 19, find the probability that a randomly selected boat will finish before 3:30 PM.

16. In Example 19, find the probability that a randomly selected boat will finish after 2:30 PM.

17. In Example 20, find the probability that a telephone call selected at random will last longer than 1 minute.

18. In Example 20, find the probability that a telephone call selected at random will last less than 3 minutes.

C *Find the value of each improper integral that converges. Note that* $\lim_{x\to\infty} x^n e^{-x} = 0$ *for all integers n ≥ 0.*

19. $\displaystyle\int_0^\infty \frac{1}{k} e^{-x/k}\, dx, \quad k > 0$

20. $\displaystyle\int_0^\infty x e^{-x}\, dx$

21. $\displaystyle\int_{-\infty}^0 \frac{dx}{\sqrt{1-x}}$

22. $\displaystyle\int_{-\infty}^\infty x e^{-x^2}\, dx$

APPLICATIONS

BUSINESS & ECONOMICS

23. *Consumption.* The daily per capita use of water (in hundreds of gallons) for domestic purposes has a probability density function of the form

$$g(x) = \begin{cases} .05e^{-.05x} & x \geq 0 \\ 0 & \text{otherwise} \end{cases}$$

Find the probability that a person chosen at random will use at least 300 gallons of water per day.

24. *Warranty.* A manufacturer guarantees a product for 1 year. The time for failure of a new product after it is sold is given by the probability density function

$$f(t) = \begin{cases} .01e^{-.01t} & t \geq 0 \\ 0 & \text{otherwise} \end{cases}$$

where *t* is time in months. What is the probability that a buyer chosen at random will have a product failure during the warranty period?

LIFE SCIENCES

25. *Medicine.* If the length of stay for people in a hospital has a probability density function

$$g(t) = \begin{cases} .2e^{-.2t} & t \geq 0 \\ 0 & \text{otherwise} \end{cases}$$

where t is time in days, find the probability that a patient chosen at random will stay in the hospital less than 5 days.

26. *Medicine.* For a particular kind of disease, the length of time in days for recovery has a probability density function of the form

$$R(t) = \begin{cases} .03e^{-.03t} & t \geq 0 \\ 0 & \text{otherwise} \end{cases}$$

For a randomly selected person contracting this disease, what is the probability of that person taking at least 7 days to recover?

SOCIAL SCIENCES

27. *Politics.* In a particular election, the length of time each voter spent on campaigning for a candidate or issue was found to have a probability density function

$$F(x) = \begin{cases} \dfrac{1}{(x + 1)^2} & x \geq 0 \\ 0 & \text{otherwise} \end{cases}$$

where x is time in minutes. For a voter chosen at random, what is the probability of his or her spending at least 9 minutes on the campaign?

28. *Psychology.* In an experiment on conditioning, pigeons were required to recognize on a light display one pattern of dots out of five possible patterns to receive a food pellet. After the ninth successful trial, it was found that the probability density function for the length of time in seconds until success on the tenth trial is given by

$$f(t) = \begin{cases} e^{-t} & t \geq 0 \\ 0 & \text{otherwise} \end{cases}$$

What is the probability that a pigeon selected at random from those having successfully completed nine trials will take two or more seconds to complete the tenth trial successfully?

5-8 CHAPTER REVIEW

IMPORTANT TERMS
AND SYMBOLS

5-2 The irrational number e. $\quad e = 2.7182818\ldots, \quad e = \lim\limits_{n \to \infty}\left(1 + \dfrac{1}{n}\right)^n,$
$e = \lim\limits_{s \to 0}(1 + s)^{1/s}$

5-3 Logarithmic functions. $\ln u = \log_e u,\quad D_x \ln u = \dfrac{1}{u} D_x u,$

$$\int \frac{du}{u} = \ln|u| + C$$

5-4 Exponential functions. $D_x e^u = e^u D_x u,\quad \int e^u \, du = e^u + C,$

$D_x b^u = b^u \ln b \, D_x u,\quad \displaystyle\int b^u \, du = \dfrac{b^u}{\ln b} + C$

5-5 Applications. continuous compound interest,   exponential growth
law, $A = Pe^{rt},\quad Q = Q_0 e^{rt}$

5-6 Integration by parts. $\int u \, dv = uv - \int v \, du$

5-7 Improper integrals. improper integral,   converge,   diverge,
probability density function,   normal probability density function,
$\int_a^\infty f(x) \, dx = \lim\limits_{b \to \infty} \int_a^b f(x) \, dx,\quad \int_{-\infty}^b f(x) \, dx = \lim\limits_{a \to -\infty} \int_a^b f(x) \, dx,$
$\int_{-\infty}^\infty f(x) \, dx = \int_{-\infty}^c f(x) \, dx + \int_c^\infty f(x) \, dx$

EXERCISE 5-8   CHAPTER REVIEW

*Work through all the problems in this chapter review and check your
answers in the back of the book. (Answers to all review problems are
there.) Where weaknesses show up, review appropriate sections in the
text. When you are satisfied that you know the material, take the practice test following this review.*

Perform the indicated operations—if possible.

A 1. $D_x e^{2x-3}$ 2. $D_x \ln(x-2)$ 3. $\displaystyle\int \frac{dx}{x+1}$

  4. $\displaystyle\int e^{x-1} \, dx$ 5. $\displaystyle\int_0^1 e^{4x} \, dx$ 6. $\displaystyle\int_0^\infty e^{-2x} \, dx$

B 7. $\displaystyle\int e^{x^2} x \, dx$ 8. $D_x(e^{-2x} \ln 5x)$

  9. $\displaystyle\int_0^1 xe^{-x} \, dx$ 10. $\displaystyle\int x \ln x \, dx$

  11. $\displaystyle\int \frac{e^{-x}}{e^{-x}+3} \, dx$ 12. $\displaystyle\int \frac{e^x}{(e^x+2)^2} \, dx$

  13. $D_x[2\sqrt{x} + \ln(x^3 + 1)]$ 14. $\displaystyle\int_0^\infty \frac{dx}{(x+3)^2}$

  15. $\displaystyle\int_{-\infty}^0 e^x \, dx$

C   **16.** $\int \dfrac{(\ln x)^2}{x}\,dx$ **17.** $\int_0^\infty (x+1)e^{-x}\,dx$

       18. $D_x \log_5(x^2 - x)$ **19.** $D_x 5^{x^2-1}$

       20. $\int 10^{2x}\,dx$

APPLICATIONS

BUSINESS & ECONOMICS   **21.** *Doubling time.* If money is invested at 5% compounded continuously, how long will it take to double?

       22. *Marginal analysis.* If the price–demand equation for x units of a commodity is

$$p(x) = 1{,}000e^{-0.02x}$$

then the revenue equation is

$$R(x) = xp(x) = 1{,}000xe^{-0.02x}$$

Find the marginal revenue equation.

       23. *Marginal analysis.* Find the marginal demand equation for the demand equation in Problem 22.

       24. *Parts testing.* If in testing printed circuits for hand calculators, failures occur relative to time in hours according to the probability density function

$$F(t) = \begin{cases} .02e^{-.02t} & t \geq 0 \\ 0 & \text{otherwise} \end{cases}$$

what is the probability that a circuit chosen at random will fail in the first hour of testing?

LIFE SCIENCES   **25.** *Population growth.* If a bacteria culture is growing at a rate given by

$$N'(t) = 2{,}000e^{0.2t} \qquad N(0) = 10{,}000$$

where t is time in hours, find $N(t)$ and the number of bacteria after 10 hours.

       26. *Medicine.* For a particular doctor, the length of time in hours spent with a patient per office visit has the probability density function

$$f(t) = \begin{cases} \dfrac{4/_3}{(t+1)^2} & 0 \leq t \leq 3 \\ 0 & \text{otherwise} \end{cases}$$

For a patient selected at random, what is the probability that the doctor will spend more than 1 hour per visit?

SOCIAL SCIENCES

27. *Population growth.* Costa Rica is estimated to have a population doubling time of 20 years. What is its growth rate, assuming exponential growth?

28. *Psychology.* Rats were trained to go through a maze by rewarding them with a food pellet upon successful completion. After the seventh successful run, it was found that the probability density function for length of time in minutes until success on the eighth trial is given by

$$f(t) = \begin{cases} .5e^{-.5t} & t \geq 0 \\ 0 & \text{otherwise} \end{cases}$$

What is the probability that a rat selected at random after seven successful runs will take 2 or more minutes to complete the eighth run successfully?

PRACTICE TEST: CHAPTER 5

Perform the indicated operations in Problems 1–10.

1. $D_x e^{x^2 + 9x}$

2. $D_x \ln \sqrt{x + 1}$

3. $\displaystyle\int xe^{x^2}\, dx$

4. $\displaystyle\int_1^e \frac{x + 1}{x}\, dx$

5. $D_x[e^{-x} \ln(x + 1)]$

6. $\displaystyle\int (x + 3)e^x\, dx$

7. $\displaystyle\int \frac{x}{(x^2 + 3)^4}\, dx$

8. $\displaystyle\int_2^\infty e^{-8x}\, dx$

9. $\displaystyle\int \frac{(\ln x)^7}{x}\, dx$

10. $D_x[(\ln x)^3 + \ln x^3]$

11. If money is invested at 8% compounded continuously, how long will it take to double?

12. If the price–demand equation for x units of a commodity is

$$p(x) = 500e^{-0.003x}$$

where p is in dollars and x is in thousands of units, what is the marginal demand equation?

13. In Problem 12 what price will support sales of 10,000 units? 50,000 units?

CHAPTER 6
MULTIVARIABLE CALCULUS

CONTENTS

6 MULTIVARIABLE CALCULUS

6-1 FUNCTIONS OF SEVERAL VARIABLES

FUNCTIONS OF TWO OR MORE INDEPENDENT VARIABLES

In Section 1-3 we introduced the concept of a function with one independent variable. Now we will broaden the concept to include functions with more than one independent variable. We start with an example.

A small manufacturing company produces a standard type of surfboard and no other products. If fixed costs are $500 per week and variable costs are $70 per board produced, then the weekly cost function is given by

$$C(x) = 500 + 70x \tag{1}$$

where x is the number of boards produced per week. The cost function is a function of a single independent variable x. For each value of x from the domain of C there exists exactly one value of C(x) in the range of C.

Now, suppose the company decides to add a high-performance competition board to its line. If the fixed costs for the competition board are $200 per week and the variable costs are $100 per board, then the cost function (1) must be modified to

$$C(x, y) = 700 + 70x + 100y \tag{2}$$

where C(x, y) is the cost for weekly output of x standard boards and y competition boards. Equation (2) is an example of a function with two independent variables, x and y. Of course, as the company expands its product line even further, its weekly cost function must be modified to include more and more independent variables, one for each new product produced.

In general, an equation of the form

$$z = f(x, y)$$

will describe a **function of two independent variables** if for each ordered pair (x, y) from the domain of f there is one and only one value of z determined by $f(x, y)$ in the range of f. Unless otherwise stated, we will assume the domain of a function specified by an equation of the form $z = f(x, y)$ is the set of all ordered pairs of real numbers (x, y) such that $f(x, y)$ is also a real number. It should be noted, however, that certain conditions in practical problems often lead to futher restrictions of the domain of a function.

We can similarly define functions of three independent variables, $w = f(x, y, z)$; of four independent variables, $u = f(w, x, y, z)$; and so on. In this chapter, we will primarily concern ourselves with functions with two independent variables.

Example 1 For $C(x, y) = 700 + 70x + 100y$, find $C(10, 5)$.

Solution
$$C(10, 5) = 700 + 70(10) + 100(5)$$
$$= \$1{,}900$$

Problem 1 Find $C(20, 10)$ for the cost function in Example 1.

Answer $3,100

Example 2 For $f(x, y, z) = 2x^2 - 3xy + 3z + 1$, find $f(3, 0, -1)$.

Solution
$$f(3, 0, -1) = 2(3)^2 - 3(3)(0) + 3(-1) + 1$$
$$= 18 - 0 - 3 + 1$$
$$= 16$$

Problem 2 Find $f(-2, 2, 3)$ for f in Example 2.

Answer 30

FURTHER EXAMPLES A number of concepts we have already considered can be thought of in terms of functions of two or more variables. We list a few of these below, as well as others.

Area of a rectangle $A(x, y) = xy$

Volume of a box $V(x, y, z) = xyz$

Volume of a right $V(r, h) = \pi r^2 h$
circular cylinder

Simple interest	$A(P, r, t) = P(1 + rt)$	A = Amount
		P = Principal
		r = Annual rate
		t = Time in years

Compound interest	$A(P, r, t, n) = P\left(1 + \dfrac{r}{n}\right)^{nt}$	A = Amount
		P = Principal
		r = Annual rate
		t = Time in years
		n = Compound periods per year

IQ	$Q(M, C) = \dfrac{M}{C}(100)$	Q = IQ = Intelligence quotient
		M = MA = Mental age
		C = CA = Chronological age

Resistance for blood flow in a vessel	$R(L, r) = k\dfrac{L}{r^4}$	R = Resistance
		L = Length of vessel
		r = Radius of vessel
		k = Constant

GEOMETRIC INTERPRETATION

We now take a brief look at some graphs of functions of two independent variables. Since functions of the form $z = f(x, y)$ involve two independent variables, x and y, and one dependent variable, z, we need a three-dimensional coordinate system for their graphs. We take three mutually perpendicular number lines intersecting at their origins to form a rectangular coordinate system in three-dimensional space (see Fig. 1). In such a system, every ordered triplet of numbers (x, y, z) can be associated with a unique point, and conversely.

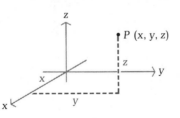

FIGURE 1 Rectangular coordinate system.

Example 3   Locate $(-3, 5, 2)$ in a rectangular coordinate system.

Solution

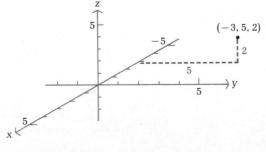

Problem 3   Find the coordinates of the corners A, C, G, and D of the rectangular box in the figure.

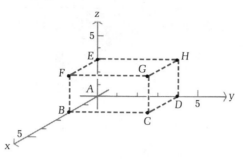

Answer $A(0, 0, 0)$, $C(2, 4, 0)$, $G(2, 4, 3)$, $D(0, 4, 0)$

What does the graph of $z = x^2 + y^2$ look like? If we let $x = 0$ and graph $z = 0^2 + y^2 = y^2$ in the yz plane, we obtain a parabola; if we let $y = 0$ and graph $z = x^2 + 0^2 = x^2$ in the xz plane, we obtain another parabola. It can be shown that the graph of $z = x^2 + y^2$ is just one of these parabolas rotated around the z axis (see Fig. 2). This cup-shaped figure is a **surface** and is called a **paraboloid.**

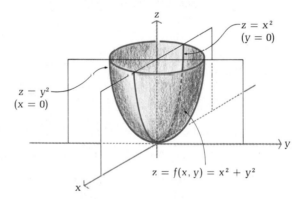

FIGURE 2   Paraboloid.

In general, the graph of any function of the form $z = f(x, y)$ is called a **surface.** The graph of such a function is the graph of all ordered triplets of numbers (x, y, z) that satisfy the equation. Graphing functions of two independent variables is often a very difficult task, and the general process will not be dealt with in this book. We present only a few simple graphs to suggest extensions of earlier geometric interpretations of the derivative and local maxima and minima to functions of two variables. Note that $z = f(x, y) = x^2 + y^2$ appears (Fig. 2) to have a local minimum at $(x, y) = (0, 0)$. Figure 3 shows a local maximum at $(x, y) = (0, 0)$, and Figure 4 shows a point at $(x, y) = (0, 0)$, called a **saddle point,** which is neither a local minimum nor a local maximum. More will be said about local maxima and minima in Section 6-4.

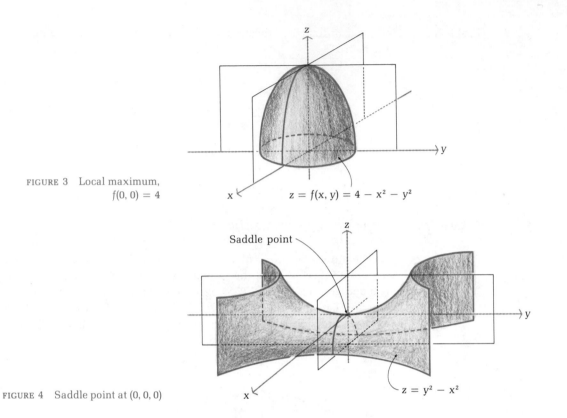

FIGURE 3 Local maximum, $f(0, 0) = 4$

$z = f(x, y) = 4 - x^2 - y^2$

Saddle point

$z = y^2 - x^2$

FIGURE 4 Saddle point at $(0, 0, 0)$

EXERCISE 6-1

A *For the functions*

$$f(x, y) = 10 + 2x - 3y \qquad g(x, y) = x^2 - 3y^2$$

find each of the following:

1. $f(0, 0)$ 2. $f(2, 1)$ 3. $f(-3, 1)$ 4. $f(2, -7)$
5. $g(0, 0)$ 6. $g(0, -1)$ 7. $g(2, -1)$ 8. $g(-1, 2)$

B *Find each of the following:*

9. $A(2, 3)$ for $A(x, y) = xy$

10. $V(2, 4, 3)$ for $V(x, y, z) = xyz$

11. $Q(12, 8)$ for $Q(M, C) = \dfrac{M}{C}(100)$

12. $T(50, 17)$ for $T(V, x) = \dfrac{33V}{x + 33}$

13. $V(2, 4)$ for $V(r, h) = \pi r^2 h$

14. $S(4, 2)$ for $S(x, y) = 5x^2 y^3$

15. $R(1, 2)$ for $R(x, y) = -5x^2 + 6xy - 4y^2 + 200x + 300y$

16. $P(2, 2)$ for $P(x, y) = -x^2 + 2xy - 2y^2 - 4x + 12y + 5$

17. $R(6, 0.5)$ for $R(L, r) = 0.002\dfrac{L}{r^4}$

18. $L(2,000, 50)$ for $L(w, v) = (1.25 \times 10^{-5})wv^2$

19. $A(100, 0.06, 3)$ for $A(P, r, t) = P + Prt$

20. $A(10, 0.04, 3, 2)$ for $A(P, r, t, n) = P\left(1 + \dfrac{r}{n}\right)^{tn}$

21. $A(100, 0.08, 10)$ for $A(P, r, t) = Pe^{rt}$

22. $A(1,000, 0.06, 8)$ for $A(P, r, t) = Pe^{rt}$

C   23. For the function $f(x, y) = x^2 + 2y^2$, find:

$$\frac{f(x + \Delta x, y) - f(x, y)}{\Delta x}$$

24. For the function $f(x, y) = x^2 + 2y^2$, find:

$$\frac{f(x, y + \Delta y) - f(x, y)}{\Delta y}$$

25. For the function $f(x, y) = 2xy^2$, find:

$$\frac{f(x + \Delta x, y) - f(x, y)}{\Delta x}$$

26. For the function $f(x, y) = 2xy^2$, find:

$$\frac{f(x, y + \Delta y) - f(x, y)}{\Delta y}$$

27. Find the coordinates of E and F in the figure for Problem 3 in the text.

28. Find the coordinates of B and H in the figure for Problem 3 in the text.

APPLICATIONS

BUSINESS & ECONOMICS

29. *Cost function.*   A small manufacturing company produces two models of a surfboard: a standard model and a competition model. If the standard model is produced at a variable cost of $70 each, the competition model at a variable cost of $100 each, and the total fixed costs per month are $2,000, then the monthly cost function is given by

$$C(x, y) = 2,000 + 70x + 100y$$

where x and y are the number of standard and competition models produced per month, respectively. Find $C(20, 10)$, $C(50, 5)$, and $C(30, 30)$.

30. *Advertising and sales.*   A company spends x thousand dollars per week on newspaper advertising and y thousand dollars per week on television advertising. Its weekly sales were found to be given by

$S(x, y) = 5x^2y^3$

Find $S(3, 2)$ and $S(2, 3)$.

31. *Revenue, cost, and profit functions.* A firm produces two types of calculators, x thousand of type A and y thousand of type B per year. The revenue and cost functions for the year are (in millions of dollars)

$R(x, y) = 2x + 3y$
$C(x, y) = x^2 - 2xy + 2y^2 + 6x - 9y + 5$

Find:
(A) $R(3, 2)$ (B) $C(2, 1)$ (C) $P(2, 1)$, where $P = R - C$

32. *Revenue function.* A supermarket sells two brands of coffee: brand A at $\$x$ per pound and brand B at $\$y$ per pound. The daily demand equations for brands A and B are, respectively,

$u = 200 - 5x + 4y$
$v = 300 - 4y + 2x$

(both in pounds). Thus, the daily revenue equation is

$R(x, y) = xu + yv$
$\qquad = x(200 - 5x + 4y) + y(300 - 4y + 2x)$
$\qquad = -5x^2 + 6xy - 4y^2 + 200x + 300y$

Find $R(2, 3)$ and $R(3, 2)$.

LIFE SCIENCES

33. *Marine biology.* In using scuba diving gear, a marine biologist estimates the time of a dive according to the equation

$$T(V, x) = \frac{33V}{x + 33}$$

where

$T = $ Time of dive in minutes
$V = $ Volume of air, at sea level pressure, compressed into tanks
$x = $ Depth of dive in feet

Find $T(70, 47)$ and $T(60, 27)$.

34. *Blood flow.* Poiseuille's law states that the resistance, R, for blood flowing in a blood vessel varies directly as the length of the vessel, L, and inversely as the fourth power of its radius, r. Stated as an equation,

$$R(L, r) = k\frac{L}{r^4} \qquad k \text{ a constant}$$

Find $R(8, 1)$ and $R(4, 0.2)$.

35. *Physical anthropology.* Anthropologists, in their study of race and human genetic groupings, often use an index called the *cephalic index*. The cephalic index, C, varies directly as the width, W, of the head, and inversely as the length, L, of the head (both viewed from the top). In terms of an equation,

$$C(W, L) = 100\frac{W}{L}$$

where

W = Width in inches
L = Length in inches

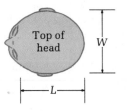

Find $C(6, 8)$ and $C(8.1, 9)$.

SOCIAL SCIENCES

36. *Safety research.* Under ideal conditions, if a person driving a car slams on the brakes and skids to a stop, the length of the skid marks is given by the formula

$$L(w, v) = kwv^2$$

where

k = Constant
w = Weight of car in pounds
v — Speed of car in miles per hour

For $k = 0.0000133$, find $L(2,000, 40)$ and $L(3,000, 60)$.

37. *Psychology.* Intelligence quotient (IQ) is defined to be the ratio of the mental age (MA), as determined by certain tests, and the chronological age (CA) multiplied by 100. Stated as an equation,

$$Q(M, C) = \frac{M}{C}100$$

where

Q = IQ
M = MA
C = CA

Find $Q(12, 10)$ and $Q(10, 12)$.

6-2   PARTIAL DERIVATIVES

We know how to differentiate many kinds of functions of one independent variable and how to interpret the results. What about functions with two or more independent variables? Let us return to the surfboard example considered at the beginning of the chapter.

For the company producing only the standard board, the cost function was

$$C(x) = 500 + 70x$$

Differentiating with respect to x, we obtain

$$C'(x) = 70$$

which is the marginal cost at any level of output. That is, $70 is the change in cost for one unit increase in production at any output level.

For the company producing two boards, a standard model and a competition model, the cost equation was

$$C(x, y) = 700 + 70x + 100y$$

Now suppose we differentiate with respect to x, holding y fixed, and denote this by $C_x(x, y)$; or we differentiate with respect to y, holding x fixed, and denote this by $C_y(x, y)$. Differentiating in this way, we obtain

$$C_x(x, y) = 70 \qquad C_y(x, y) = 100$$

Both these are called **partial derivatives** and, in this example, both represent marginal costs. The first is the change in cost due to one unit increase in production of the standard board with the production of the competition model held fixed. The second is the change in cost due to one unit increase in production of the competition board with the production of the standard board held fixed.

In general, if $z = f(x, y)$, then the **partial derivative of f with respect to x,** denoted by $\partial z / \partial x$, f_x, or $f_x(x, y)$, is defined by

$$\frac{\partial z}{\partial x} = \lim_{\Delta x \to 0} \frac{f(x + \Delta x, y) - f(x, y)}{\Delta x}$$

This is the ordinary derivative of f with respect to x, holding y constant. Thus, we are able to continue to use all the derivative rules and properties discussed in Chapters 2 and 3 for partials.

Similarly, the **partial derivative of f with respect to y,** denoted by $\partial z / \partial y$, f_y, or $f_y(x, y)$ is defined by

$$\frac{\partial z}{\partial y} = \lim_{\Delta y \to 0} \frac{f(x, y + \Delta y) - f(x, y)}{\Delta y}$$

which is the ordinary derivative with respect to y, holding x constant.

Parallel definitions and interpretations hold for functions with three or more independent variables.

Example 4   For $z = f(x, y) - 2x^2 - 3x^2y + 5y + 1$, find:

(A) $\dfrac{\partial z}{\partial x}$ (B) $f_x(2, 3)$

Solutions   (A) $z = 2x^2 - 3x^2y + 5y + 1$

Differentiating with respect to x, holding y constant (that is, treating y as a constant), we obtain

$$\frac{\partial z}{\partial x} = 4x - 6xy$$

(B) $f(x, y) = 2x^2 - 3x^2y + 5y + 1$
First differentiate with respect to x (part A) to obtain

$$f_x(x, y) = 4x - 6xy$$

Then evaluate at (2, 3). Thus,

$$f_x(2, 3) = 4(2) - 6(2)(3) = -28$$

Problem 4   For f in Example 4, find:

(A) $\dfrac{\partial z}{\partial y}$ (B) $f_y(2, 3)$

Answers   (A) $\dfrac{\partial z}{\partial y} = -3x^2 + 5$ (B) $f_y(2, 3) = -7$

Example 5   For $z = f(x, y) = e^{x^2 + y^2}$, find:

(A) $\dfrac{\partial z}{\partial x}$ (B) $f_y(2, 1)$

Solutions   (A) Using the chain rule [thinking of $z = e^u$, $u = u(x)$; y is held constant], we obtain

$$\frac{\partial z}{\partial x} = e^{x^2 + y^2}\frac{\partial (x^2 + y^2)}{\partial x}$$
$$= 2xe^{x^2 + y^2}$$

(B) $f_y(x, y) = 2ye^{x^2 + y^2}$
$f_y(2, 1) = 2(1)e^{2^2 + 1^2}$
$= 2e^5$

Problem 5   For $z = f(x, y) = (x^2 + 2xy)^5$, find:

(A) $\dfrac{\partial z}{\partial y}$ (B) $f_x(1, 0)$

Answers   (A) $10x(x^2 + 2xy)^4$ (B) 10

Partials have simple geometric interpretations, as indicated in Figure 5. If we hold x fixed, say $x = a$, then $f_y(a, y)$ is the slope of the curve obtained by intersecting the plane $x = a$ with the surface $z = f(x, y)$. A similar interpretation is given to $f_x(x, b)$.

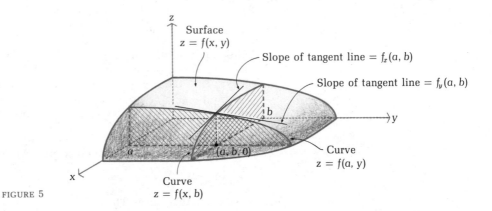

FIGURE 5

HIGHER-ORDER PARTIALS

Just as there are higher-order ordinary derivatives, there are higher-order partials, and we will be using some of these in the next section when we discuss local maxima and minima. The following second-order partials will be useful:

SECOND-ORDER PARTIALS

If $z = f(x, y)$, then

$$\frac{\partial^2 z}{\partial x^2} = \frac{\partial}{\partial x}\left(\frac{\partial z}{\partial x}\right) = f_{xx}(x, y) = f_{xx}$$

$$\frac{\partial^2 z}{\partial x\,\partial y} = \frac{\partial}{\partial x}\left(\frac{\partial z}{\partial y}\right) = f_{yx}(x, y) = f_{yx}$$

$$\frac{\partial^2 z}{\partial y\,\partial x} = \frac{\partial}{\partial y}\left(\frac{\partial z}{\partial x}\right) = f_{xy}(x, y) = f_{xy}$$

$$\frac{\partial^2 z}{\partial y^2} = \frac{\partial}{\partial y}\left(\frac{\partial z}{\partial y}\right) = f_{yy}(x, y) = f_{yy}$$

In the mixed partial $\partial^2 z/\partial x\,\partial y = f_{yx}$, we start with $z = f(x, y)$ and first differentiate with respect to y (holding x constant). Then we differentiate with respect to x (holding y constant). What is the order of differentiation for $\partial^2 x/\partial y\,\partial x = f_{xy}$? It can be shown that for functions we will consider, $f_{xy}(x, y) = f_{yx}(x, y)$.

Example 6 For $z = f(x, y) = 3x^2 - 2xy^3 + 1$, find:

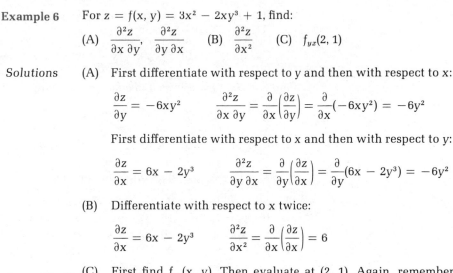

(A) $\dfrac{\partial^2 z}{\partial x\, \partial y}$, $\dfrac{\partial^2 z}{\partial y\, \partial x}$ (B) $\dfrac{\partial^2 z}{\partial x^2}$ (C) $f_{yx}(2, 1)$

Solutions (A) First differentiate with respect to y and then with respect to x:

$$\frac{\partial z}{\partial y} = -6xy^2 \qquad \frac{\partial^2 z}{\partial x\, \partial y} = \frac{\partial}{\partial x}\left(\frac{\partial z}{\partial y}\right) = \frac{\partial}{\partial x}(-6xy^2) = -6y^2$$

First differentiate with respect to x and then with respect to y:

$$\frac{\partial z}{\partial x} = 6x - 2y^3 \qquad \frac{\partial^2 z}{\partial y\, \partial x} = \frac{\partial}{\partial y}\left(\frac{\partial z}{\partial x}\right) = \frac{\partial}{\partial y}(6x - 2y^3) = -6y^2$$

(B) Differentiate with respect to x twice:

$$\frac{\partial z}{\partial x} = 6x - 2y^3 \qquad \frac{\partial^2 z}{\partial x^2} = \frac{\partial}{\partial x}\left(\frac{\partial z}{\partial x}\right) = 6$$

(C) First find $f_{yx}(x, y)$. Then evaluate at (2, 1). Again, remember that f_{yx} means to differentiate with respect to y first and then with respect to x. Thus,

$$f_y(x, y) = -6xy^2$$
$$f_{yx}(x, y) = -6y^2$$

and

$$f_{yx}(2, 1) = -6(1)^2 = -6$$

Problem 6 For the function in Example 6, find:

(A) $\dfrac{\partial^2 z}{\partial y\, \partial x}$ (B) $\dfrac{\partial^2 z}{\partial y^2}$ (C) $f_{xy}(2, 3)$ (D) $f_{yx}(2, 3)$

Answers (A) $-6y^2$ (B) $-12xy$ (C) -54 (D) -54

EXERCISES 6-2

A For $z = f(x, y) = 10 + 3x + 2y$, find each of the following:

1. $\dfrac{\partial z}{\partial x}$ 2. $\dfrac{\partial z}{\partial y}$ 3. $f_y(1, 2)$ 4. $f_x(1, 2)$

For $z = f(x, y) = 3x^2 - 2xy^2 + 1$, find each of the following:

5. $\dfrac{\partial z}{\partial y}$ 6. $\dfrac{\partial z}{\partial x}$ 7. $f_x(2, 3)$ 8. $f_y(2, 3)$

For $S(x, y) = 5x^2y^3$, find each of the following:

9. $S_x(x, y)$ 10. $S_y(x, y)$ 11. $S_y(2, 1)$ 12. $S_x(2, 1)$

B   For $C(x, y) = x^2 - 2xy + 2y^2 + 6x - 9y + 5$, find each of the following:

13. $C_x(x, y)$ 14. $C_y(x, y)$ 15. $C_x(2, 2)$ 16. $C_y(2, 2)$
17. $C_{xy}(x, y)$ 18. $C_{yx}(x, y)$ 19. $C_{xx}(x, y)$ 20. $C_{yy}(x, y)$

Find $z = f(x, y) = e^{2x+3y}$, find each of the following:

21. $\dfrac{\partial z}{\partial x}$ 22. $\dfrac{\partial z}{\partial y}$ 23. $\dfrac{\partial^2 z}{\partial x\, \partial y}$ 24. $\dfrac{\partial^2 z}{\partial y\, \partial x}$

25. $f_{xy}(1, 0)$ 26. $f_{yx}(0, 1)$ 27. $f_{xx}(0, 1)$ 28. $f_{yy}(1, 0)$

Find $f_x(x, y)$ and $f_y(x, y)$ for each function f given by:

29. $f(x, y) = (x^2 - y^3)^3$ 30. $f(x, y) = \sqrt{2x - y^2}$
31. $f(x, y) = (3x^2y - 1)^4$ 32. $f(x, y) = (3 + 2xy^2)^3$
33. $f(x, y) = \ln(x^2 + y^2)$ 34. $f(x, y) = \ln(2x - 3y)$
35. $f(x, y) = y^2 e^{xy^2}$ 36. $f(x, y) = x^3 e^{x^2y}$

37. $f(x, y) = \dfrac{x^2 - y^2}{x^2 + y^2}$ 38. $f(x, y) = \dfrac{2x^2y}{x^2 + y^2}$

C   39.   For

$$P(x, y) = -x^2 + 2xy - 2y^2 - 4x + 12y - 5$$

find values of x and y such that

$$P_x(x, y) = 0 \quad \text{and} \quad P_y(x, y) = 0$$

simultaneously.

40.   For

$$C(x, y) = 2x^2 + 2xy + 3y^2 - 16x - 18y + 54$$

find values of x and y such that

$$C_x(x, y) = 0 \quad \text{and} \quad C_y(x, y) = 0$$

simultaneously.

41.   For $f(x, y) = x^2 + 2y^2$, find:

(A) $\displaystyle\lim_{\Delta x \to 0} \dfrac{f(x + \Delta x, y) - f(x, y)}{\Delta x}$ (B) $\displaystyle\lim_{\Delta y \to 0} \dfrac{f(x, y + \Delta y) - f(x, y)}{\Delta y}$

42.   For $f(x, y) = 2xy^2$, find:

(A) $\displaystyle\lim_{\Delta x \to 0} \dfrac{f(x + \Delta x, y) - f(x, y)}{\Delta x}$ (B) $\displaystyle\lim_{\Delta y \to 0} \dfrac{f(x, y + \Delta y) - f(x, y)}{\Delta y}$

APPLICATIONS

BUSINESS & ECONOMICS   43.   *Cost function.*   For the company in Problem 29 in Exercise 6-1, find $C_x(20, 10)$ and $C_y(20, 10)$, and interpret.

44. *Advertising and sales.* For the company in Problem 30 in Exercise 6-1, find $S_x(3, 2)$ and $S_y(3, 2)$, and interpret.

45. *Profit function.* For the firm in Problem 31 in Exercise 6-1, find $P_x(1, 1)$ and $P_y(1, 1)$, and interpret.

46. *Marginal productivity.* A company has determined that its productivity (units per employee per week) is given approximately by

$$z(x, y) = 50xy - x^2 - 3y^2$$

where x is the size of the labor force in thousands and y is the amount of capital investment in millions of dollars.
   (A) Determine the marginal productivity of labor when $x = 5$ and $y = 4$. Interpret.
   (B) Determine the marginal productivity of capital when $x = 5$ and $y = 4$. Interpret.

LIFE SCIENCES

47. *Marine biology.* Using the equation for timing a dive with scuba gear (Problem 33 in Exercise 6-1), find $T_V(70, 47)$ and $T_x(70, 47)$, and interpret.

48. *Blood flow.* Using Poiseuille's law (Problem 34 in Exercise 6-1), find $R_L(4, 0.2)$ and $R_r(4, 0.2)$, and interpret.

49. *Physical anthropology.* For Problem 35 in Exercise 6-1, find $C_W(6, 8)$ and $C_L(6, 8)$ and interpret.

SOCIAL SCIENCES

50. *Safety research.* For Problem 36 in Exercise 6-1, find $L_w(2,500, 60)$ and $L_v(2,500, 60)$, and interpret.

51. *Psychology.* For Problem 37 in Exercise 6-1, find $Q_M(12, 10)$ and $Q_C(12, 10)$, and interpret.

6-3 TOTAL DIFFERENTIALS AND THEIR APPLICATIONS

Recall (Section 3-7) that for a function defined by

$$y = f(x)$$

the differential dx of the *independent variable* x is another independent variable, which can be viewed as Δx, the change in x. The differential dy of the *dependent variable* y is given by $dy = f'(x)\ dx$. Thus, the differential of a function with one independent variable is a function with *two* independent variables, x and dx. How can the differential concept be extended to functions with two or more independent variables?

Suppose $z = f(x, y)$ is a function with the independent variables x and y. We define the **total differential** of the dependent variable z to be

$$dz = f_x(x, y)\, dx + f_y(x, y)\, dy$$

Notice that dz is a function of *four* variables: the independent variables x and y, and their differentials dx and dy.

Example 7 Find dz for $f(x, y) = x^2y^3$. Evaluate dz for:
(A) $x = 2, y = -1, dx = 0.1$, and $dy = 0.2$
(B) $x = 1, y = 2, dx = -0.1$, and $dy = 0.05$
(C) $x = -2, y = 1, dx = 0.3$, and $dy = -0.1$

Solutions Since $f_x(x, y) = 2xy^3$ and $f_y(x, y) = 3x^2y^2$,

$$\begin{aligned} dz &= f_x(x, y)\, dx + f_y(x, y)\, dy \\ &= 2xy^3\, dx + 3x^2y^2\, dy \end{aligned}$$

(A) When $x = 2, y = -1, dx = 0.1$, and $dy = 0.2$,

$$dz = 2(2)(-1)^3(0.1) + 3(2)^2(-1)^2(0.2) = 2$$

(B) When $x = 1, y = 2, dx = -0.1$, and $dy = 0.05$,

$$dz = 2(1)(2)^3(-0.1) + 3(1)^2(2)^2(0.05) = -1$$

(C) When $x = -2, y = 1, dx = 0.3$, and $dy = -0.1$,

$$dz = 2(-2)(1)^3(0.3) + 3(-2)^2(1)^2(-0.1) = -2.4$$

Problem 7 Find dz for $f(x, y) = xy^2 + x^2$. Evaluate dz for:
(A) $x = 3, y = 1, dx = 0.05$, and $dy = -0.1$
(B) $x = -2, y = 2, dx = 0.2$, and $dy = 0.1$
(C) $x = 1, y = -2, dx = 0.1$, and $dy = -0.04$

Answers $dz = (y^2 + 2x)\, dx + 2xy\, dy$
(A) -0.25 (B) -0.8 (C) 0.76

If $w = f(x, y, z)$, then the **total differential** is

$$dw = f_x(x, y, z)\, dx + f_y(x, y, z)\, dy + f_z(x, y, z)\, dz$$

This time, dw is a function of *six* independent variables: the original independent variables x, y, and z, and their differentials dx, dy, and dz. Generalizations to functions with more than three independent variables follow the same pattern.

Example 8 Find dw for $f(x, y, z) = xyz^2$. Evaluate dw for:
(A) $x = 2, y = 3, z = -1, dx = 0.1, dy = -0.2$, and $dz = 0.05$
(B) $x = 1, y = -2, z = 0, dx = -0.1, dy = 0.1$, and $dz = 0$
(C) $x = -1, y = 1, z = 2, dx = 0.2, dy = 0.3$, and $dz = -0.4$

Solutions Since $f_x(x, y, z) = yz^2$, $f_y(x, y, z) = xz^2$, and $f_z(x, y, z) = 2xyz$,

$$dw = yz^2\,dx + xz^2\,dy + 2xyz\,dz$$

(A) When $x = 2$, $y = 3$, $z = -1$, $dx = 0.1$, $dy = -0.2$, and $dz = 0.05$,
$$dw = (3)(-1)^2(0.1) + (2)(-1)^2(-0.2) + 2(2)(3)(-1)(0.05) = -0.7$$

(B) When $x = 1$, $y = -2$, $z = 0$, $dx = -0.1$, $dy = 0.1$, and $dz = 0$,
$$dw = (-2)(0)^2(-0.1) + (1)(0)^2(0.1) + 2(1)(-2)(0)(0) = 0$$

(C) When $x = -1$, $y = 1$, $z = 2$, $dx = 0.2$, $dy = 0.3$, and $dz = -0.4$,
$$dw = (1)(2)^2(0.2) + (-1)(2)^2(0.3) + 2(-1)(1)(2)(-0.4) = 1.2$$

Problem 8 Find dw for $f(x, y, z) = xy + yz + zx$. Evaluate dw for:
(A) $x = 1$, $y = 1$, $z = 1$, $dx = 0.1$, $dy = 0.1$, and $dz = 0.1$
(B) $x = 2$, $y = 2$, $z = -2$, $dx = 0.5$, $dy = 0.5$, and $dz = 0$
(c) $x = 4$, $y = -3$, $z = 1$, $dx = 0.1$, $dy = -0.2$, and $dz = 0.4$

Answers $$dw = (y + z)\,dx + (x + z)\,dy + (y + x)\,dz$$

(A) 0.6   (B) 0   (C) −0.8

APPROXIMATIONS USING DIFFERENTIALS If $z = f(x, y)$ and Δx and Δy represent the changes in the independent variables x and y, then the corresponding change in the dependent variable z is given exactly by

$$\Delta z = f(x + \Delta x, y + \Delta y) - f(x, y)$$

For small values of Δx and Δy, the differential dz can be used to approximate the change Δz.

Example 9 Find Δz and dz for $f(x, y) = x^2 + y^2$ when $x = 3$, $y = 4$, $\Delta x = dx = 0.01$, and $\Delta y = dy = -0.02$.

Solution
$$\Delta z = f(x + \Delta x, y + \Delta y) - f(x, y)$$
$$= f(3.01, 3.98) - f(3, 4)$$
$$= [(3.01)^2 + (3.98)^2] - [3^2 + 4^2]$$
$$= 24.9005 - 25$$
$$= -0.0995 \longleftarrow$$

Note that dz is a good approximation for Δz, the exact change in z, and dz was easier to calculate

$$dz = f_x(x, y)\,dz + f_y(x, y)\,dy$$
$$= 2x\,dx + 2y\,dy$$
$$= 2(3)(0.01) + 2(4)(-0.02)$$
$$= -0.1 \longleftarrow$$

Problem 9 Repeat Example 9 for $x = 2$, $y = 5$, $\Delta x = dx = -0.01$, and $\Delta y = dy = 0.05$.

Answer $\Delta z = 0.4626$, $dz = 0.46$

In addition to approximating Δz, the differential can also be used to approximate $f(x + \Delta x, y + \Delta y)$. These approximations are summarized in the box and illustrated in the examples that follow.

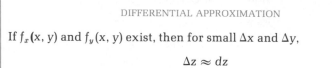

DIFFERENTIAL APPROXIMATION

If $f_x(x, y)$ and $f_y(x, y)$ exist, then for small Δx and Δy,

$$\Delta z \approx dz$$

and

$$
\begin{aligned}
f(x + \Delta x, y + \Delta y) &= f(x, y) + \Delta z \\
&\approx f(x, y) + dz \\
&= f(x, y) + f_x(x, y)\, dx + f_y(x, y)\, dy
\end{aligned}
$$

Example 10　Suppose the cost equation for a company producing standard and competition surfboards is

$$C(x, y) = 700 + 70x^{3/2} + 100y^{3/2} - 20x^{1/2}y^{1/2}$$

where x is the number of standard boards produced and y is the number of competition boards produced.

(A)　What is the cost of producing 100 boards of each type?

(B)　What is the approximate change in the cost if one fewer standard and two more competition boards are produced? Approximate the change using differentials.

Solutions　(A)　$C(100, 100) = 700 + 70(100)^{3/2} + 100(100)^{3/2} - 20(100)^{1/2}(100)^{1/2}$
$$= 700 + 70{,}000 + 100{,}000 - 2{,}000$$
$$= \$168{,}700$$

(B)　We will use dC to approximate ΔC as x changes from 100 to 99 and y changes from 100 to 102. We must evaluate dC for $x = 100$, $y = 100$, $dx = \Delta x = -1$, and $dy = \Delta y = 2$:

$$
\begin{aligned}
\Delta C &\approx dC \\
&= C_x(x, y)\, dx + C_y(x, y)\, dy \\
&= [105x^{1/2} - 10x^{-1/2}y^{1/2}]\, dx + [150y^{1/2} - 10x^{1/2}y^{-1/2}]\, dy \\
&= [105(100)^{1/2} - 10(100)^{-1/2}(100)^{1/2}](-1) \\
&\quad + [150(100)^{1/2} - 10(100)^{1/2}(100)^{-1/2}](2) \\
&= -1{,}040 + 2{,}980 \\
&= \$1{,}940
\end{aligned}
$$

Thus, decreasing the production of standard boards by one and increasing the production of competition boards by two will increase the cost by approximately \$1,940.

Problem 10　For the cost function in Example 10:

(A)　What is the cost of producing 25 standard boards and 100 competition boards?

(B)   What is the approximate change in the cost if three more standard boards and five fewer competition boards are produced?

Answers   (A) $108,450   (B)   −$5,960

Example 11   Approximate the hypotenuse of a right triangle with legs of length 6.02 and 7.97 inches.

Solution   If x and y are the lengths of the legs of a right triangle, then from the Pythagorean theorem we find the hypotenuse z to be

$$z = f(x, y) = \sqrt{x^2 + y^2}$$

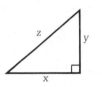

We could use a calculator to compute the value of $f(6.02, 7.97)$ directly, however, our purpose here is to illustrate the use of the differential to approximate the value of a function. Thus, we will proceed as though a calculator is not available. This means that we must select values of x and y that satisfy two conditions: First, they must be near 6.02 and 7.97; and second, we must be able to evaluate $\sqrt{x^2 + y^2}$ without using a calculator. Since

$$\sqrt{6^2 + 8^2} = \sqrt{36 + 64} = \sqrt{100} = 10$$

$x = 6$ and $y = 8$ satisfy both of these conditions. So, we let $x = 6$, $y = 8$, $dx = \Delta x = 0.02$, and $dy = \Delta y = -0.03$, and then we use

$$f(x + \Delta x, y + \Delta y) = f(x, y) + \Delta z$$
$$\approx f(x, y) + dz$$
$$= f(x, y) + f_x(x, y)\, dx + f_y(x, y)\, dy$$

Now we can obtain an approximation to $f(6.02, 7.97)$ that we can evaluate by hand:

$$f(x + \Delta x, y + \Delta y) \approx f(x, y) + f_x(x, y)\, dx + f_y(x, y)\, dy$$

$$\sqrt{(x + \Delta x)^2 + (y + \Delta y)^2} \approx \sqrt{x^2 + y^2} + \frac{x}{\sqrt{x^2 + y^2}}\, dx$$

$$+ \frac{y}{\sqrt{x^2 + y^2}}\, dy$$

$$\sqrt{(6 + 0.02)^2 + [8 + (-0.03)]^2} \approx \sqrt{6^2 + 8^2} + \frac{6}{\sqrt{6^2 + 8^2}}\, (0.02)$$

$$+ \frac{8}{\sqrt{6^2 + 8^2}}\, (-0.03)$$

$$\sqrt{(6.02)^2 + (7.97)^2} \approx 10 + 0.012 - 0.024 = 9.988$$

Problem 11 Approximate the hypotenuse of a right triangle with legs of length 2.95 and 4.02.

Answer 4.986

EXERCISE 6-3

A *Find dz for each function.*

1. $z = x^2 + y^2$

2. $z = 2x + xy + 3y$

3. $z = x^4 y^3$

4. $z = \sqrt{2x + 6y}$

5. $z = \sqrt{x} + \dfrac{5}{\sqrt{y}}$

6. $z = x\sqrt{1 + y}$

Find dw for each function.

7. $w = x^3 + y^3 + z^3$

8. $w = xy^2 z^3$

9. $w = xy + 2xz + 3yz$

10. $w = \sqrt{2x + 3y - z}$

B *Evaluate dz and Δz for each function at the indicated values.*

11. $z = f(x, y) = x^2 - 2xy + y^2$, $x = 3$, $y = 1$, $\Delta x = dx = 0.1$, $\Delta y = dy = 0.2$

12. $z = f(x, y) = 2x^2 + xy - 3y^2$, $x = 2$, $y = 4$, $\Delta x = dx = 0.1$, $\Delta y = dy = 0.05$

13. $z = f(x, y) = 100\left(3 - \dfrac{x}{y}\right)$, $x = 2$, $y = 1$, $\Delta x = dx = 0.05$,

$\Delta y = dy = 0.1$

14. $z = f(x, y) = 50\left(1 + \dfrac{x^2}{y}\right)$, $x = 3$, $y = 9$, $\Delta x = dx = -0.1$,

$\Delta y = dy = 0.2$

In Problems 15–18 evaluate dw and Δw for each function at the indicated values.

15. $w = f(x, y, z) = x^2 + yz$, $x = 2$, $y = 3$, $z = 5$, $\Delta x = dx = 0.1$, $\Delta y = dy = 0.2$, $\Delta z = dz = 0.1$

16. $w = f(x, y, z) = 2xz + y^2 - z^2$, $x = 4$, $y = 2$, $z = 3$, $\Delta x = dx = 0.2$, $\Delta y = dy = 0.1$, $\Delta z = dz = -0.1$

17. $w = f(x, y, z) = \dfrac{10x + 20y}{z}$, $x = 4$, $y = 3$, $z = 5$,

$\Delta x = dx = 0.05$, $\Delta y = dy = -0.05$, $\Delta z = dz = 0.1$

18. $w = f(x, y, z) = 50\left(x + \dfrac{1}{y} + \dfrac{1}{z^2}\right)$, $x = 2$, $y = 2$, $z = 1$,

$\Delta x = dx = 0.2$, $\Delta y = dy = 0.1$, $\Delta z = dz = 0.1$

19. Approximate the hypotenuse of a right triangle with legs of length 3.1 and 3.9 inches.

20. Approximate the hypotenuse of a right triangle with legs of length 4.95 and 12.02 inches.

21. A can in the shape of a right circular cylinder with radius 5 inches and height 10 inches is coated with ice 0.1 inch thick. Use differentials to approximate the volume of the ice ($V = \pi r^2 h$).

22. A box with edges of length 10, 15, and 20 centimeters is covered with a 1 centimeter thick coat of fiberglass. Use differentials to approximate the volume of the fiberglass shell.

23. A plastic box is to be constructed with a square base and an open top. The plastic material used in construction is 0.1 centimeter thick. The inside dimensions of the box are 10 by 10 by 5 centimeters. Use differentials to approximate the volume of the plastic required for one box.

24. The surface area of a right circular cone with radius r and altitude h is given by

$$S = \pi r \sqrt{r^2 + h^2}$$

Use differentials to approximate the change in S when r changes from 6 to 6.1 inches and h changes from 8 to 8.05 inches.

C 25. Find dz if $z = xye^{x^2 + y^2}$.
26. Find dz if $z = x \ln(xy) + y \ln(xy)$.
27. Find dw if $w = xyze^{xyz}$.
28. Find dw if $w = xy \ln(xz) + yz \ln(xy)$.

APPLICATIONS

BUSINESS & ECONOMICS

29. *Cost function.* A microcomputer company manufactures two types of computers, model I and model II. The cost in thousands of dollars of producing x model I's and y model II's per month is given by

$$C(x, y) = x + 2y - \frac{1}{10} \sqrt{x^2 + y^2}$$

Currently, the company manufactures 30 model I computers and 40 model II computers each month. Use differentials to approximate the change in the cost function if the company decides to produce 5 more model I and 3 more model II computers each month.

30. *Advertising and sales.* For the company in Problem 30 in Exercise 6-1, use differentials to approximate the change in sales if the amount spent on newspaper advertising is increased from \$3,000 to \$3,100 per week and the amount spent on television advertising is increased from \$2,000 to \$2,200 per week.

31. *Revenue function.* For the supermarket in Problem 32 in Exercise 6-1, use differentials to approximate the change in revenue if the price of brand *A* is increased from $2.00 to $2.10 per pound and the price of brand *B* is decreased from $3.00 to $2.95 per pound.

32. *Productivity.* For the company in Problem 46 in Exercise 6-2, the current labor force is 5,000 workers. The current capital investment is $4 million. Use differentials to approximate the change in productivity if both the labor force and the capital investment are increased by 10%.

LIFE SCIENCES 33. *Blood flow.* For Poiseuille's law in Problem 34 in Exercise 6-1, use differentials to approximate the change in the resistance if the length of the vessel decreases from 8 to 7.5 and the radius decreases from 1 to 0.95.

34. *Drug concentration.* The concentration of a drug in the blood-stream after having been injected into a vein is given by

$$C(x, y) = \frac{1}{1 + \sqrt{x^2 + y^2}}$$

where x is the time passed since the injection and y is the distance from the point of injection. Use differentials to approximate $C(3.1, 4.1)$.

SOCIAL SCIENCES 35. *Safety research.* For Problem 36 in Exercise 6-1, use differentials to approximate the change in the length of the skid marks if the weight of the car is increased from 2,000 to 2,200 pounds and the speed is increased from 40 to 45 miles per hour.

36. *Psychology.* For Problem 37 in Exercise 6-1, use differentials to approximate the change in IQ as a person's mental age changes from 12 to 12.5 and chronological age changes from 10 to 11.

6-4   MAXIMA AND MINIMA

We are now ready to undertake a brief but useful analysis of local maxima and minima for functions of the type $z = f(x, y)$. Basically, we are going to extend the second-derivative test developed for functions of a single independent variable. To start, we assume that all higher-order partials exist for the function f in some circular region in the xy plane. This guarantees that the surface $z = f(x, y)$ has no sharp points, breaks, or ruptures. In other words, we are dealing only with smooth surfaces with no edges (like the edge of a box); or breaks (like an earthquake fault); or sharp points (like the bottom point of a golf tee). See Figure 6.

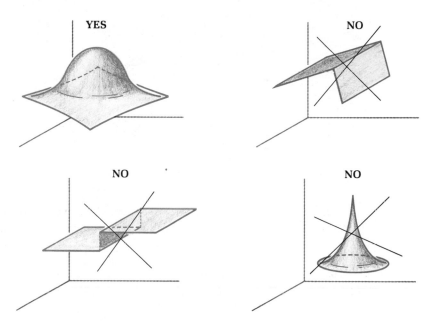

FIGURE 6

In addition, we will not concern ourselves with boundary points or absolue maxima—minima theory. In spite of these restrictions, the procedure we are now going to describe will help us solve a large number of useful problems.

What does it mean for $f(a, b)$ to be a local maximum or a local minimum? We say that **$f(a, b)$ is a local maximum** if there exists a circular region in the domain of f with (a, b) as the center, such that

$$f(a, b) \geq f(x, y)$$

for all (x, y) in the region. Similarly, we say that **$f(a, b)$ is a local minimum** if there exists a circular region in the domain of f with (a, b) as the center, such that

$$f(a, b) \leq f(x, y)$$

for all (x, y) in the region. In Section 6-1, Figure 2 illustrates a local minimum, Figure 3 illustrates a local maximum, and Figure 4 illustrates a saddle point, which is neither.

What happens to $f_x(a, b)$ and $f_y(a, b)$ if $f(a, b)$ is a local minimum or a local maximum and the partials of f exist in a circular region containing (a, b)? Figure 7 suggests that $f_x(a, b) = 0$ and $f_y(a, b) = 0$, since the tangents to the indicated curves are horizontal. Theorem 1 indicates that our intuitive reasoning is correct.

Theorem 1   Let $f(a, b)$ be an extreme (a local maximum or a local minimum) for the function f. If both f_x and f_y exist at (a, b), then

$$f_x(a, b) = 0 \qquad \text{and} \qquad f_y(a, b) = 0 \tag{1}$$

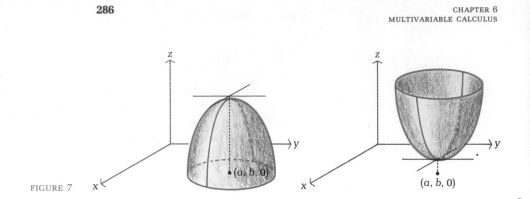

FIGURE 7

The converse of this theorem is false; that is, if $f_x(a, b) = 0$ and $f_y(a, b) = 0$, then $f(a, b)$ may or may not be a local extreme–the point $(a, b, f(a, b))$ may be a saddle point, for example.

Theorem 1 gives us what are called *necessary* (but not sufficient) conditions for $f(a, b)$ to be a local extreme. We thus find all points (a, b) such that $f_x(a, b) = 0$ and $f_y(a, b) = 0$ and test these further to determine whether $f(a, b)$ is a local extreme or a saddle point. Points (a, b) such that (1) holds are called **critical points.** The next theorem, using second-derivative tests, gives us sufficient conditions for a critical point to produce a local extreme or a saddle point. As was the case with Theorem 1, we state this theorem without proof.

Theorem 2

Second-derivative test for local extrema. If:

1. $z = f(x, y)$
2. $f_x(a, b) = 0$ and $f_y(a, b) = 0$ [(a, b) is a critical point]
3. All higher-order partials of f exist in some circular region containing (a, b) as a center
4. $A = f_{xx}(a, b), B = f_{xy}(a, b), C = f_{yy}(a, b)$

Then:

1. If $AC - B^2 > 0$ and $A < 0$, then $f(a, b)$ is a local maximum.
2. If $AC - B^2 > 0$ and $A > 0$, then $f(a, b)$ is a local minimum.
3. If $AC - B^2 < 0$, then f has a saddle point at (a, b).
4. If $AC - B^2 = 0$, the test fails.

Let us consider a few examples.

Example 12

Use Theorem 2 to find local extrema for

$$f(x, y) = -x^2 - y^2 + 6x + 8y - 21$$

Solution

Step 1. *Find critical points.* Find (x, y) such that $f_x(x, y) = 0$ and $f_y(x, y) = 0$, simultaneously.

$$f_x(x, y) = -2x + 6 = 0$$
$$x = 3$$
$$f_y(x, y) = -2y + 8 = 0$$
$$y = 4$$

The only critical point is $(a, b) = (3, 4)$.

Step 2. *Compute* $A = f_{xx}(a, b)$, $B = f_{xy}(a, b)$, *and* $C = f_{yy}(a, b)$.

$$f_{xx}(x, y) = -2, \quad \text{thus} \quad A = f_{xx}(3, 4) = -2$$
$$f_{xy}(x, y) = 0, \quad \text{thus} \quad B = f_{xy}(3, 4) = 0$$
$$f_{yy}(x, y) = -2, \quad \text{thus} \quad C = f_{yy}(3, 4) = -2$$

Step 3. *Evalute* $AC - B^2$ *and try to classify the critical point* (a, b) *using Theorem 2.*

$$AC - B^2 = (-2)(-2) - (0)^2 = 4 > 0 \quad \text{and} \quad A = -2 < 0$$

Therefore, case 1 in Theorem 2 holds. That is, $f(3, 4) = 4$ is a local maximum.

Problem 12 Use Theorem 2 to find local extrema for

$$f(x, y) = x^2 + y^2 - 10x - 2y + 36$$

Answer $f(5, 1) = 10$ is a local minimum

Example 13 Use Theorem 2 to find local extrema for

$$f(x, y) = xy + 8$$

Solution Step 1. *Find critical points.*

$$f_x(x, y) = y = 0$$
$$y = 0$$
$$f_y(x, y) = x = 0$$
$$x = 0$$

The only critical point is $(a, b) = (0, 0)$.

Step 2. *Compute* $A = f_{xx}(a, b)$, $B = f_{xy}(a, b)$, *and* $C = f_{yy}(a, b)$.

$$f_{xx}(x, y) = 0, \quad \text{thus} \quad A = f_{xx}(0, 0) = 0$$
$$f_{xy}(x, y) = 1, \quad \text{thus} \quad B = f_{xy}(0, 0) = 1$$
$$f_{yy}(x, y) = 0, \quad \text{thus} \quad C = f_{yy}(0, 0) = 0$$

Step 3. *Evaluate* $AC - B^2$ *and try to classify the critical point* (a, b) *using Theorem 2.*

$$AC - B^2 = (0)(0) - (1)^2 = -1 < 0$$

Therefore, case 3 in Theorem 2 applies. That is, f has a saddle point at $(0, 0)$.

Problem 13   Use Theorem 2 to find local extrema for

$$f(x, y) = x^2 - y^2$$

Answer f has a saddle point at $(0, 0)$

Example 14   Suppose the surfboard company discussed earlier has been able to develop the yearly profit equation

$$P(x, y) = -2x^2 + 2xy - y^2 + 10x - 4y + 107$$

where x is the number (in thousands) of standard surfboards produced per year, y is the number (in thousands) of competition surfboards produced per year, and P is profit (in thousands of dollars). How many of each type of board should be produced per year to realize a maximum profit? What is the maximum profit?

Solution   Step 1.   *Find critical points.*

$$P_x(x, y) = -4x + 2y + 10 = 0$$
$$P_y(x, y) = 2x - 2y - 4 = 0$$

Solving this system, we obtain $(3, 1)$ as the only critical point.

Step 2.   *Compute $A = P_{xx}(a, b), B = P_{xy}(a, b),$ and $C = P_{yy}(a, b)$.*

$$P_{xx}(x, y) = -4, \quad \text{thus} \quad A = P_{xx}(3, 1) = -4$$
$$P_{xy}(x, y) = 2, \quad \text{thus} \quad B = P_{xy}(3, 1) = 2$$
$$P_{yy}(x, y) = -2, \quad \text{thus} \quad C = P_{yy}(3, 1) = -2$$

Step 3.   *Evaluate $AC - B^2$ and try to classify the critical point (a, b) using Theorem 2.*

$$AC - B^2 = (-4)(-2) - (2)^2 = 8 - 4 = 4 > 0$$
$$A = -4 < 0$$

Therefore, case 1 in Theorem 2 applies. That is, $P(3, 1) = \$120,000$ is a local maximum. This is obtained by producing 3,000 standard boards and 1,000 competition boards per year.

Problem 14   Repeat Example 14 with

$$P(x, y) = -2x^2 + 4xy - 3y^2 + 4x - 2y + 77$$

Answer   Local maximum for $x = 2$ and $y = 1$; $P(2, 1) = \$80,000$

EXERCISE 6-4

Find local extrema using Theorem 2.

A   1. $f(x, y) = 6 - x^2 - 4x - y^2$
     2. $f(x, y) = 3 - x^2 - y^2 + 6y$

3. $f(x, y) = x^2 + y^2 + 2x - 6y + 14$
4. $f(x, y) = x^2 + y^2 - 4x + 6y + 23$

B 5. $f(x, y) = xy + 2x - 3y - 2$
6. $f(x, y) = x^2 - y^2 + 2x + 6y - 4$
7. $f(x, y) = -3x^2 + 2xy - 2y^2 + 14x + 2y + 10$
8. $f(x, y) = -x^2 + xy - 2y^2 + x + 10y - 5$
9. $f(x, y) = 2x^2 - 2xy + 3y^2 - 4x - 8y + 20$
10. $f(x, y) = 2x^2 - xy + y^2 - x - 5y + 8$

C 11. $f(x, y) = e^{xy}$ 12. $f(x, y) = x^2y - xy^2$

APPLICATIONS

BUSINESS & ECONOMICS

13. *Product mix for maximum profit.* A firm produces two types of calculators, x thousand of type A and y thousand of type B per year. If the revenue and cost equations for the year are (in millions of dollars)

$$R(x, y) = 2x + 3y$$
$$C(x, y) = x^2 - 2xy + 2y^2 + 6x - 9y + 5$$

find how many of each type of calculator should be produced per year to maximize profit. What is the maximum profit?

14. *Automation–labor mix for minimum cost.* The annual labor and automated equipment cost (in millions of dollars) for a company's production of television sets is given by

$$C(x, y) = 2x^2 + 2xy + 3y^2 - 16x - 18y + 54$$

where x is the amount spent per year on labor and y is the amount spent per year on automated equipment (both in millions of dollars). Determine how much should be spent on each per year to minimize this cost. What is the minimum cost?

15. *Research–advertising mix for maximum profit.* A pocket calculator company has developed the profit equation

$$P(x, y) = -3x^2 + 3xy - y^2 + 12x - 5y + 17$$

where x is the amount spent per year on research and development and y is the amount spent per year on advertising (all units are in millions of dollars). How much should be spent in each area per year to maximize profit? What is the maximum profit for this budget?

16. *Minimum material.* A rectangular box with no top is to be made to hold 32 cubic inches. What should its dimensions be in order to use the least amount of material in its construction?

17. *Maximum volume.* A mailing service states that a rectangular package shall have the sum of the length and girth not to exceed 120 inches (see the figure). What are the dimensions of the largest (in volume) mailing carton that can be constructed meeting these restrictions?

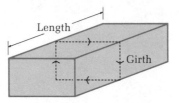

6-5 MAXIMA AND MINIMA USING LAGRANGE MULTIPLIERS

FUNCTIONS OF TWO INDEPENDENT VARIABLES

We will now consider a method of solving a certain class of maxima—minima problems that is particularly powerful. The method is due to Joseph Louis Lagrange (1736–1813), an eminent eighteenth century French mathematician, and it is called the **method of Lagrange multipliers.** We introduce the method through an example; then we will formalize the discussion in the form of a theorem.

A rancher wants to construct two feeding pens of the same size along an existing fence (see Fig. 8). If 720 feet of fencing are available, how long should x and y be in order to obtain the maximum total area? What is the maximum area?

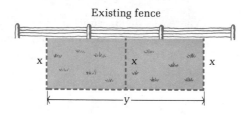

FIGURE 8

The total area is given by

$$f(x, y) = xy$$

which can be made as large as we like providing there are no restrictions on x and y. But there are restrictions on x and y, since we only have 720 feet of fencing. That is, x and y must be chosen so that

$$3x + y = 720$$

This restriction on x and y, also called a **constraint,** leads to the following maxima—minima problem:

$$\text{Maximize} \quad f(x, y) = xy \tag{1}$$
$$\text{Subject to} \quad 3x + y = 720 \tag{2}$$
$$\text{or} \quad 3x + y - 720 = 0$$

This problem is a special case of a general class of problems of the form

$$\text{Maximize (or minimize)} \quad z = f(x, y) \tag{3}$$
$$\text{Subject to} \quad g(x, y) = 0 \tag{4}$$

Of course, we could try to solve (4) for y in terms of x, or x in terms of y, then substitute the result into (3), and use methods developed in Section 3-6 for functions of a single variable. But what if (4) were more complicated than (2), and solving for one variable in terms of the other was either very difficult or impossible? In the method of Lagrange multipliers we work with $g(x, y)$ directly and avoid having to solve (4) for one variable in terms of the other. In addition, the method generalizes to functions of arbitrarily many variables subject to one or more constraints.

Now, to the method. We form a new function F, using functions f and g in (3) and (4), as follows:

$$F(x, y, L) = f(x, y) + Lg(x, y) \tag{5}$$

where L is called a **Lagrange multiplier.** Theorem 3 forms the basis for the method.

Theorem 3   The relative maxima and minima of the function $z = f(x, y)$ subject to the constraint $g(x, y) = 0$ will be among those points (x_0, y_0) for which (x_0, y_0, L_0) is a solution to the system

$$F_x(x, y, L) = 0$$
$$F_y(x, y, L) = 0$$
$$F_L(x, y, L) = 0$$

where $F(x, y, L) = f(x, y) + Lg(x, y)$, provided all the partial derivatives exist.

We now solve the fence problem using the method of Lagrange multipliers.

Step 1.   *Formulate the problem in the form of equations (3) and (4).*

$$\text{Maximize} \quad f(x, y) = xy$$
$$\text{Subject to} \quad g(x, y) = 3x + y - 720 = 0$$

Step 2.   *Form the function F, introducing the Lagrange multiplier L.*

$$F(x, y, L) = f(x, y) + Lg(x, y)$$
$$= xy + L(3x + y - 720)$$

Step 3.   *Solve the system $F_x = 0$, $F_y = 0$, $F_L = 0$.* (Solutions are called **critical points** for F.)

$$F_x = y + 3L = 0$$
$$F_y = x + L = 0$$
$$F_L = 3x + y - 720 = 0$$

From the first two equations, we see that

$$y = -3L$$
$$x = -L$$

Substitute these values for x and y into the third equation and solve for L.

$$-3L - 3L = 720$$
$$-6L = 720$$
$$L = -120$$

Thus,

$$y = -3(-120) = 360 \text{ feet}$$
$$x = -(-120) = 120 \text{ feet}$$

Step 4. *Test the critical points for maxima and minima.* The function F has only one critical point at $(120, 360, -120)$, and since $f(x, y) = xy$ has a minimum at $(0, 0)$, we conclude that $(120, 360)$ produces a maximum for f. Hence,

$$\text{Max } f(x, y) = f(120, 360)$$
$$= (120)(360)$$
$$= 43,200 \text{ square feet}$$

METHOD OF LAGRANGE MULTIPLIERS — KEY STEPS

1. Formulate the problem in the form

   Maximize (or minimize) $z = f(x, y)$
   Subject to $g(x, y) = 0$

2. Form the function F:

   $$F(x, y, L) = f(x, y) + Lg(x, y)$$

3. Find the critical points for F; that is, solve the system

   $$F_x(x, y, L) = 0$$
   $$F_y(x, y, L) = 0$$
   $$F_L(x, y, L) = 0$$

4. Evaluate $z = f(x, y)$ at each point (x_0, y_0) such that (x_0, y_0, L_0) satisfies the system in step 3. The maximum or minimum value of $f(x, y)$ will be among these values in the problems we consider.

Example 15   Minimize $f(x, y) = x^2 + y^2$ subject to $x + y = 10$.

Solution   Step 1.   Minimize $f(x, y) = x^2 + y^2$
                    Subject to $g(x, y) = x + y - 10 = 0$

         Step 2. $F(x, y, L) = x^2 + y^2 + L(x + y - 10)$

         Step 3. $F_x = 2x + L = 0$
                  $F_y = 2y + L = 0$
                  $F_L = x + y - 10 = 0$

From the first two equations,

$$x = -\frac{L}{2} \qquad y = -\frac{L}{2}$$

Substituting these into the third equation, we obtain

$$-\frac{L}{2} - \frac{L}{2} = 10$$

$$-L = 10$$

$$L = -10$$

The critical point is $(5, 5, -10)$.

         Step 4. $f(5, 5) = 5^2 + 5^2 = 50$
Checking other points on the line $x + y = 10$ near $(5, 5)$, we see that this is a minimum. (See Fig. 9.)

Problem 15   Maximize $f(x, y) = 25 - x^2 - y^2$ subject to $x + y = 4$.

Answer   Max $f(x, y) = f(2, 2) = 17$ (see Fig. 10)

FUNCTIONS OF THREE INDEPENDENT VARIABLES

As was indicated above, the method of Lagrange multipliers can be extended to functions with arbitrarily many independent variables with one or more constraints. We state a theorem for functions with three independent variables and one constraint.

Theorem 4   The relative maxima and minima of the function $w = f(x, y, z)$ subject to the constraint $g(x, y, z) = 0$ will be among those points (x_0, y_0, z_0) for which (x_0, y_0, z_0, L_0) is a solution to the system

$$F_x(x, y, z, L) = 0$$
$$F_y(x, y, z, L) = 0$$
$$F_z(x, y, z, L) = 0$$
$$F_L(x, y, z, L) = 0$$

where $F(x, y, z, L) = f(x, y, z) + Lg(x, y, z)$, provided all the partial derivatives exist.

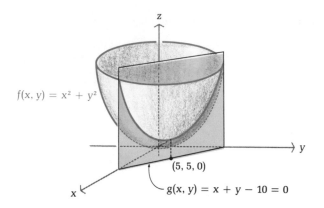

$f(x, y) = x^2 + y^2$

$(5, 5, 0)$

$g(x, y) = x + y - 10 = 0$

FIGURE 9

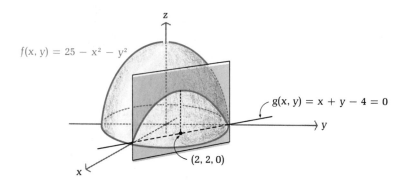

$f(x, y) = 25 - x^2 - y^2$

$g(x, y) = x + y - 4 = 0$

$(2, 2, 0)$

FIGURE 10

EXERCISE 6-5

Use the method of Lagrange multipliers in the following problems:

A 1. Maximize $f(x, y) = 2xy$
      Subject to $x + y = 6$

2. Minimize $f(x, y) = 6xy$
      Subject to $y - x = 6$

3. Minimize $f(x, y) = x^2 + y^2$
      Subject to $3x + 4y = 25$

4. Maximize $f(x, y) = 25 - x^2 - y^2$
      Subject to $2x + y = 10$

B 5. Find the maximum and minimum of $f(x, y) = 2xy$ subject to $x^2 + y^2 = 18$.

6. Find the maximum and minimum of $f(x, y) = x^2 - y^2$ subject to $x^2 + y^2 = 25$.

7. Maximize the product of two numbers if their sum must be 10.

8. Minimize the product of two numbers if their difference must be 10.

C   9.   Minimize $f(x, y, z) = x^2 + y^2 + z^2$
              Subject to $2x - y + 3z = -28$

       10.   Maximize $f(x, y, z) = xyz$
              Subject to $2x + y + 2z = 120$

APPLICATIONS

BUSINESS & ECONOMICS

11.   *Budgeting for least cost.*   A manufacturing company produces two models of a television set, x units of model A and y units of model B per week, at a cost in dollars of

$$C(x, y) = 6x^2 + 12y^2$$

If it is necessary (because of shipping considerations) that

$$x + y = 90$$

how many of each type of set should be manufactured per week to minimize cost? What is the minimum cost?

12.   *Budgeting for maximum production.*   A manufacturing firm has budgeted $60,000 per month for labor and materials. If x thousand dollars is spent on labor and y thousand dollars is spent on materials, and if the monthly output in units is given by

$$N(x, y) = 4xy - 8x$$

how should the $60,000 be allocated to labor and materials in order to maximize N? What is the maximum N?

13.   *Shipping problem.*   Rework Problem 17 in Exercise 6-4 using Lagrange multipliers. [*Hint:*   See Problem 10 above.]

LIFE SCIENCES

14.   *Agriculture.*   Three pens of the same size are to be built along an existing fence (see the figure). If 400 feet of fencing are available, what length should x and y be to produce the maximum total area? What is the maximum area?

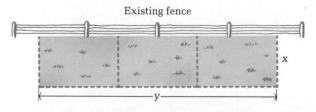

Existing fence

15. *Diet and minimum cost.* A group of guinea pigs is to receive 25,600 calories per week. Two available foods produce 200xy calories for a mixture of x kilograms of type M food and y kilograms of type N food. If type M costs $1 per kilogram and type N costs $2 per kilogram, how much of each type of food should be used to minimize weekly food costs? What is the minimum cost? [*Note:* $x \geq 0, y \geq 0$]

6-6   METHOD OF LEAST SQUARES

In this section we will use the optimization techniques discussed in Section 6-4 to find the equation of a line which is a "best" approximation to a set of points in a rectangular coordinate system. This very popular method is known as **least squares approximation** or **linear regression.** Let us begin by considering a specific case.

    A manufacturer wants to approximate the cost function for a product. The value of the cost function has been determined for certain levels of production, as listed in the table:

NUMBER OF UNITS x, in hundreds	COST y, in thousands of dollars
2	4
5	6
6	7
9	8

Although these points do not all lie on a line (see Fig. 11), they are very close to being linear. The manufacturer would like to approximate

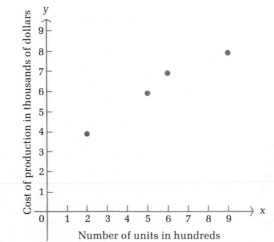

FIGURE 11

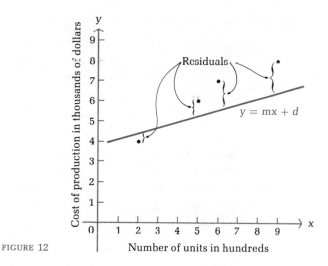

FIGURE 12

the cost function by a linear function; that is, determine values m and d so that the line

$$y = mx + d$$

is, in some sense, the "best" approximation to the cost function.

What do we mean by "best"? Since the line $y = mx + d$ will not go through all four points, it is reasonable to examine the differences between the y coordinates of the points listed in the table and the y coordinates of the corresponding points on the line. Each of these differences is called the **residual** at that point (see Fig. 12). For example, at $x = 2$ the point from the table is $(2, 4)$ and the point on the line is $(2, 2m + d)$, so the residual is

$$4 - (2m + d) = 4 - 2m - d$$

All the residuals are listed in the table below:

x	y	$mx + d$	Residual
2	4	$2m + d$	$4 - 2m - d$
5	6	$5m + d$	$6 - 5m - d$
6	7	$6m + d$	$7 - 6m - d$
9	8	$9m + d$	$8 - 9m - d$

Our criterion for the "best" approximation is the following: Determine the values of m and d that *minimize* the sum of the squares of the residuals. The resulting line is called the **least squares line** or the **regression line.** To this end, we minimize

$$F(m, d) = (4 - 2m - d)^2 + (6 - 5m - d)^2 + (7 - 6m - d)^2 + (8 - 9m - d)^2$$

Step 1. *Find critical points.*

$$F_m(m, d) = 2(4 - 2m - d)(-2) + 2(6 - 5m - d)(-5)$$
$$+ 2(7 - 6m - d)(-6) + 2(8 - 9m - d)(-9)$$
$$= -304 + 292m + 44d = 0$$

$$F_d(m, d) = 2(4 - 2m - d)(-1) + 2(6 - 5m - d)(-1)$$
$$+ 2(7 - 6m - d)(-1) + 2(8 - 9m - d)(-1)$$
$$= -50 + 44m + 8d = 0$$

Solving this system, we obtain $(m, d) = (0.58, 3.06)$ as the only critical point.

Step 2. *Compute $A = F_{mm}(m, d)$, $B = F_{md}(m, d)$, and $C = F_{dd}(m, d)$.*

$$F_{mm}(m, d) = 292, \quad \text{thus} \quad A = F_{mm}(0.58, 3.06) = 292$$
$$F_{md}(m, d) = 44, \quad \text{thus} \quad B = F_{md}(0.58, 3.06) = 44$$
$$F_{dd}(m, d) = 8, \quad \text{thus} \quad C = F_{dd}(0.58, 3.06) = 8$$

Step 3. *Evaluate $AC - B^2$ and try to classify the critical point (m, d) using Theorem 2 in Section 6-4.*

$$AC - B^2 = (292)(8) - (44)^2 = 400 > 0$$
$$A = 292 > 0$$

Therefore, case 2 in Theorem 2 applies, and $F(m, d)$ has a local minimum at the critical point $(0.58, 3.06)$.

Thus, the least squares line for the given data is

$$y = 0.58x + 3.06 \qquad \text{Least squares line}$$

Note that the sum of the squares of the residuals is minimized for this choice of m and d (see Fig. 13).

This linear function can now be used by the manufacturer to estimate any of the quantities normally associated with the cost function—such as costs, marginal costs, average costs, and so on. For example, the cost of producing 2,000 units is approximately

$$y = (0.58)(20) + 3.06 = 14.66 \quad \text{or} \quad \$14,660$$

The marginal cost function is

$$\frac{dy}{dx} = 0.58$$

The average cost function is

$$\bar{y} = \frac{0.58x + 3.06}{x}$$

Normally, least squares approximations are used with large sets of data points. We have used a very small data set here to keep the computations simple. Fortunately, even with a large number of points, m and

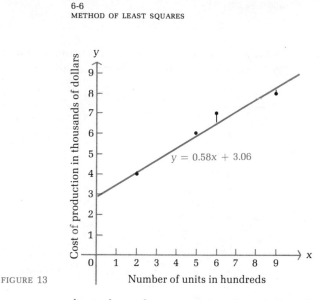

FIGURE 13

d can always be expressed as the solution of a system of linear equations. We will now state this system for a general set of data. These equations can be derived in a way similar to the procedure followed in the preceding discussion.

LEAST SQUARES APPROXIMATION

For a set of n points $(x_1, y_1), (x_2, y_2), \ldots, (x_n, y_n)$, the coefficients m and d of the least squares line

$$y = mx + d$$

are given by the solution of the system of equations

$$\left(\sum_{k=1}^{n} x_k \right) m + nd = \sum_{k=1}^{n} y_k \qquad (1)$$

$$\left(\sum_{k=1}^{n} x_k^2 \right) m + \left(\sum_{k=1}^{n} x_k \right) d = \sum_{k=1}^{n} x_k y_k \qquad (2)$$

Example 16   Find the least squares line for the data in the table. Graph the data and the least squares line.

x	y
−2	4
1	−1
2	1
4	0
5	−1

Solution   A table is a convenient way to compute all the necessary sums in the system of equations that determines m and d:

x_k	y_k	$x_k y_k$	x_k^2
-2	4	-8	4
1	-1	-1	1
2	1	2	4
4	0	0	16
5	-1	-5	25
Totals 10	3	-12	50

Thus,

$$\sum_{k=1}^{5} x_k = 10 \qquad \sum_{k=1}^{5} y_k = 3 \qquad \sum_{k=1}^{5} x_k y_k = -12 \qquad \sum_{k=1}^{5} x_k^2 = 50$$

Substituting these sums in equations (1) and (2) produces the following system:

$$10m + 5d = 3$$
$$50m + 10d = -12$$

The solution to this system is $(m, d) = (-0.6, 1.8)$. The least squares line is

$$y = -0.6x + 1.8$$

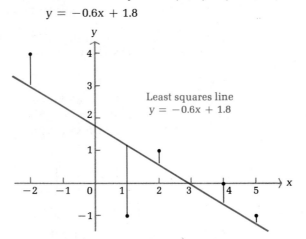

Least squares line
$y = -0.6x + 1.8$

Problem 16   Repeat Example 16 for the following data:

x	y
-3	-1
0	0
1	4
4	1
8	3

Answer $y = 0.3x + 0.8$

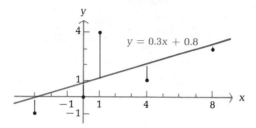

Example 17   The table lists the midterm and final examination scores for ten students in a calculus course.

MIDTERM	FINAL
49	61
53	47
67	72
71	76
74	68
78	77
83	81
85	79
91	93
99	99

(A) What is the least squares line?

(B) If a student scored 95 on the midterm examination, what would that student be expected to score on the final examination?

Solutions   (A)

x_k	y_k	$x_k y_k$	x_k^2
49	61	2,989	2,401
53	47	2,491	2,809
67	72	4,824	4,489
71	76	5,396	5,041
74	68	5,032	5,476
78	77	6,006	6,084
83	81	6,723	6,889
85	79	6,715	7,225
91	93	8,463	8,281
99	99	9,801	9,801
Totals 750	753	58,440	58,496

Substituting the totals in equations (1) and (2) produces the following system of equations:

$$750m + 10d = 753$$
$$58,496m + 750d = 58,440$$

The solution to this system is approximately $(m, d) \approx (0.875, 9.68)$. The least squares line is approximately

$$y = 0.875x + 9.68$$

(B) If a student scored 95 on the midterm examination, then, using the least squares line, the score on the final examination would be estimated to be

$$y = (0.875)(95) + 9.68 \approx 92.8$$

Problem 17   Repeat Example 17 for the following scores:

MIDTERM	FINAL
54	50
60	66
75	80
76	68
78	71
84	80
88	95
89	85
97	94
99	86

Answers   (A) $y = 0.85x + 9.47$ (B) 90.2

EXERCISE 6-6

A   *Find the least squares line. Graph the data and the least squares line.*

1.

x	y
1	1
2	3
3	4
4	3

2.

x	y
1	-2
2	-1
3	3
4	5

3.

x	y
1	8
2	5
3	4
4	0

4.

x	y
1	20
2	14
3	11
4	3

5.

x	y
1	3
2	4
3	5
4	6

6.

x	y
1	2
2	3
3	3
4	2

B   Find the least squares line and use it to estimate y for the indicated value of x.

7.

x	y
0	10
5	22
10	31
15	46
20	51

Estimate y when x = 25.

8.

x	y
-5	60
0	50
5	30
10	20
15	15

Estimate y when x = 20.

9.

x	y
-1	14
1	12
3	8
5	6
7	5

Estimate y when x = 2.

10.

x	v
2	-4
6	0
10	8
14	12
18	14

Estimate y for x = 15.

11.

x	y
0.5	25
2	22
3.5	21
5	21
6.5	18
9.5	12
11	11
12.5	8
14	5
15.5	1

Estimate y for x = 8.

12.

x	y
0	-15
2	-9
4	-7
6	-7
8	-1
12	11
14	13
16	19
18	25
20	33

Estimate y for $x = 10$.

C 13. The method of least squares can be generalized to curves other than straight lines. To find the coefficients of the parabola

$$y = ax^2 + bx + c$$

that is the "best" fit for the points $(1, 2)$, $(2, 1)$, $(3, 1)$, and $(4, 3)$, minimize the sum of the squares of the residuals

$$F(a, b, c) = (a + b + c - 2)^2 + (4a + 2b + c - 1)^2$$
$$+ (9a + 3b + c - 1)^2 + (16a + 4b + c - 3)^2$$

by solving the system

$$F_a(a, b, c) = 0 \qquad F_b(a, b, c) = 0 \qquad F_c(a, b, c) = 0$$

for a, b, and c. Graph the points and the parabola.

14. Repeat Problem 13 for the points $(-1, -2)$, $(0, 1)$, $(1, 2)$, and $(2, 0)$.

APPLICATIONS

BUSINESS & ECONOMICS

15. *Cost.* The cost y in thousands of dollars for producing x units of a product at various times in the past is given in the table.

x	y
10	5
12	6
15	7
18	8
20	9

(A) Find the least squares line for the data.
(B) Use the least squares line to estimate the cost of producing 25 units.

16. *Advertising and sales.* A company spends x thousand dollars on advertising each month and has y thousand dollars in monthly sales. The data in the table were obtained by examining the past history of the company.

x	y
4	100
5	120
6	150
7	190
8	240

(A) Find the least squares line for the data.

(b) Use the least squares line to estimate the sales if $10,000 is spent on advertising.

17. *Price–demand.* The price x in cents and the demand y in thousands of units for a certain item at various times in the past are given below:

x	y
10	120
15	120
20	130
25	125
30	135

(A) Find the least squares line for the data.

(B) Use the least squares line to estimate the demand and the revenue if the price is 40¢.

18. *Profit.* A company's annual profits in millions of dollars from 1965 to 1975 are listed in the table.

YEAR	PROFIT
1965	1.2
1966	1.4
1967	1.6
1968	1.8
1969	2.1
1970	2.4
1971	2.9
1972	3.3
1973	3.4
1974	3.5
1975	3.6

(A) Find the least squares line for the data.

(B) Use the least squares line to estimate the profit in 1980.

LIFE SCIENCES 19. *Air pollution.* The amounts of air pollution in parts per million in a large city at certain times of day are listed in the table. [*Note:* Count 12 noon as 0; then 8 AM is −4, 3 PM is 3, and so on.]

TIME	POLLUTION
8 AM	20
10 AM	47
1 PM	82
3 PM	107
4 PM	114

(A) Find the least squares line for the data.

(B) Use the least squares line to estimate the pollution at noon.

20. *Spread of disease.* A virus that affects dogs and other small mammals is spreading through a community. The table lists the number of cases (in thousands) each year from 1978 to 1982.

YEAR	CASES
1978	10
1979	14
1980	17
1981	19
1982	20

(A) Find the least squares line for the data.

(B) Use the least squares line to estimate the number of cases expected in 1987.

SOCIAL SCIENCES 21. *Learning.* The table lists the number of weeks of instruction in typing and the average number of words per minute typed for a group of students.

WEEKS OF PRACTICE	WORDS PER MINUTE
1	20
2	28
3	50
4	45
5	62

(A) Find the least squares line for the data.

(B) Estimate the number of weeks of practice required to be able to type 100 words per minute.

22. *Education.* The table lists the high school grade-point averages of ten students and their college grade-point averages after one semester of college.

HIGH SCHOOL GPA	COLLEGE GPA
2.0	1.5
2.2	1.5
2.4	1.6
2.7	1.8
2.9	2.1
3.0	2.3
3.1	2.5
3.3	2.9
3.4	3.2
3.7	3.5

(A) Find the least squares line for the data.

(B) Estimate the college GPA for a student with a high school GPA of 3.5.

(C) Estimate the high school GPA necessary for a college GPA of 2.7.

6-7   DOUBLE INTEGRALS OVER RECTANGULAR REGIONS

INTRODUCTION

We have generalized the concept of differentiation to functions with two or more independent variables. How can we do the same with integration and how can we interpret the results? Let us first look at the operation of antidifferentiation. We can antidifferentiate a function of two or more variables with respect to one of the variables by treating all the other variables as though they were constants. Thus, this operation is the reverse operation of partial differentiation, just as ordinary antidifferentiation is the reverse operation of ordinary differentiation.

Example 18

Evaluate:

(A) $\int (6xy^2 + 3x^2)\,dy$ (B) $\int (6xy^2 + 3x^2)\,dx$

Solutions

(A) Treating x as a constant and using the properties of antidifferentiation from Section 4-1,

$$\int (6xy^2 + 3x^2)\,dy = \int 6xy^2\,dy + \int 3x^2\,dy$$

$$= 6x \int y^2\,dy + 3x^2 \int dy$$

$$= 6x \left(\frac{y^3}{3}\right) + 3x^2(y) + C(x)$$

$$= 2xy^3 + 3x^2y + C(x)$$

The dy tells us we are looking for the anti-derivative of $(6xy^2 + 3x^2)$ with respect to y only, holding x constant.

Notice that the constant of integration can actually be *any function of x alone*, since, for any such function, $\dfrac{\partial}{\partial y}C(x) = 0$. We can verify that our answer is correct by using partial differentiation:

$$\frac{\partial}{\partial y}[2xy^3 + 3x^2y + C(x)] = 6xy^2 + 3x^2 + 0$$

$$= 6xy^2 + 3x^2$$

(B)   Now we treat y as a constant:

$$\int (6xy^2 + 3x^2)\,dx = \int 6xy^2\,dx + \int 3x^2\,dx$$

$$= 6y^2 \int x\,dx + 3 \int x^2\,dx$$

$$= 6y^2 \left(\frac{x^2}{2}\right) + 3\left(\frac{x^3}{3}\right) + E(y)$$

$$= 3x^2y^2 + x^3 + E(y)$$

This time the antiderivative contains an arbitrary function $E(y)$ of y alone.

Check: $\dfrac{\partial}{\partial x}[3x^2y^2 + x^3 + E(y)] = 6xy^2 + 3x^2 + 0$

$$= 6xy^2 + 3x^2$$

Problem 18   Evaluate:

(A) $\displaystyle\int (4xy + 12x^2y^3)\,dy$ (B) $\displaystyle\int (4xy + 12x^2y^3)\,dx$

Answers   (A) $2xy^2 + 3x^2y^4 + C(x)$ (B) $2x^2y + 4x^3y^3 + E(y)$

We will see applications of finding antiderivatives of the above types in Chapter 8. For now, we will use the process to evaluate definite integrals of functions with two independent variables.

Example 19   Evaluate, substituting the limits of integration in y if dy is used and in x if dx is used:

(A) $\displaystyle\int_0^2 (6xy^2 + 3x^2)\,dy$ (B) $\displaystyle\int_0^1 (6xy^2 + 3x^2)\,dx$

Solutions (A) From Example 18A, we know that $\int (6xy^2 + 3x^2)\,dy = 2xy^3 + 3x^2y + C(x)$. According to the definition of the definite integral for a function of one variable, we can use any antiderivative to evaluate the definite integral. Thus, choosing $C(x) = 0$, we have

$$\int_0^2 (6xy^2 + 3x^2)\,dy = (2xy^3 + 3x^2y)\Big|_{y=0}^{y=2}$$

$$= [2x(2)^3 + 3x^2(2)] - [2x(0)^3 + 3x^2(0)]$$

$$= 16x + 6x^2$$

(B) From Example 18B, we know that $\int (6xy^2 + 3x^2)\,dx = 3x^2y^2 + x^3 + E(y)$. Thus, choosing $E(y) = 0$, we have

$$\int_0^1 (6xy^2 + 3x^2)\,dx = (3x^2y^2 + x^3)\Big|_{x=0}^{x=1}$$

$$= [3y^2(1)^2 + (1)^3] - [3y^2(0)^2 + (0)^3]$$

$$= 3y^2 + 1$$

Problem 19 Evaluate:

(A) $\displaystyle\int_0^1 (4xy + 12x^2y^3)\,dy$ (B) $\displaystyle\int_0^3 (4xy + 12x^2y^3)\,dx$

Answers (A) $2x + 3x^2$ (B) $18y + 108y^3$

Notice that integrating and evaluating a definite integral, with integrand $f(x, y)$, with respect to y produces a function of x alone (or a constant). Likewise, integrating and evaluating a definite integral, with integrand $f(x, y)$, with respect to x produces a function of y alone (or a constant). Each of these results, involving at most one variable, can now be used as an integrand in a second definite integral.

Example 20 Evaluate:

(A) $\displaystyle\int_0^1 \left[\int_0^2 (6xy^2 + 3x^2)\,dy\right] dx$ (B) $\displaystyle\int_0^2 \left[\int_0^1 (6xy^2 + 3x^2)\,dx\right] dy$

Solutions (A) Example 19A showed that

$$\int_0^2 (6xy^2 + 3x^2)\,dy = 16x + 6x^2$$

Thus,

$$\int_0^1 \left[\int_0^2 (6xy^2 + 3x^2)\,dy\right] dx = \int_0^1 (16x + 6x^2)\,dx$$

$$= (8x^2 + 2x^3)\Big|_{x=0}^{x=1}$$

$$= [8(1)^2 + 2(1)^3] - [8(0)^2 + 2(0)^3)$$

$$= 10$$

(B)   Example 19B showed that

$$\int_0^1 (6xy^2 + 3x^2)\, dx = 3y^2 + 1$$

Thus,

$$\int_0^2 \left[\int_0^1 (6xy^2 + 3x^2)\, dx \right] dy = \int_0^2 (3y^2 + 1)\, dy$$

$$= (y^3 + y) \Big|_{y=0}^{y=2}$$

$$= [(2)^3 + 2] - [(0)^3 + 0]$$

$$= 10$$

Problem 20   Evaluate:

(A) $\displaystyle\int_0^3 \left[\int_0^1 (4xy + 12x^2y^3)\, dy \right] dx$ (B) $\displaystyle\int_0^1 \left[\int_0^3 (4xy + 12x^2y^3)\, dx \right] dy$

Answers   (A) 36   (B) 36

DOUBLE INTEGRAL   Notice that the answers in Example 20A and 20B are identical. This is not an accident. In fact, it is this property that enables us to define the concept of a double integral.

DOUBLE INTEGRAL

The double integral of a function $f(x, y)$ over a rectangle $R = \{(x, y) \mid a \le x \le b, \quad c \le y \le d\}$ is

$$\iint_R f(x, y)\, dy\, dx = \int_a^b \left[\int_c^d f(x, y)\, dy \right] dx$$

$$= \int_c^d \left[\int_a^b f(x, y)\, dx \right] dy$$

In the double integral $\iint_R f(x, y)\, dy\, dx$, $f(x, y)$ is called the **integrand** and R is called the **region of integration.** The expression $dy\, dx$ merely indicates that this is an integral over a two-dimensional region, and the order in which dx and dy are written is not meant to imply a particular order of integration. Thus, it is always correct to write either $\iint_R f(x, y)\, dy\, dx$ or $\iint_R f(x, y)\, dx\, dy$. The integrals

$$\int_a^b \left[\int_c^d f(x, y)\, dy \right] dx \qquad \text{and} \qquad \int_c^d \left[\int_a^b f(x, y)\, dx \right] dy$$

are referred to as **iterated integrals** (the brackets are often omitted), and the order in which dx and dy are written does indicate the order of integration. This is not the most general definition of the double integral over a rectangular region, however, it is equivalent to the general definition for all the functions we will consider. In the next section we will consider double integrals over regions that are not rectangles.

Example 21 Evaluate $\iint_R (x + y)\, dx\, dy$ over $R = \{(x, y)\,|\,1 \le x \le 3, \quad -1 \le y \le 2\}$.

Solution We can choose either order of iteration. As a check, we will evaluate the integral both ways:

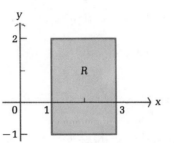

$$\iint_R (x + y)\, dx\, dy = \int_1^3 \int_{-1}^2 (x + y)\, dy\, dx$$

$$= \int_1^3 \left[\left(xy + \frac{y^2}{2} \right) \Big|_{y=-1}^{y=2} \right] dx$$

$$= \int_1^3 \left[(2x + 2) - \left(-x + \frac{1}{2} \right) \right] dx$$

$$= \int_1^3 \left(3x + \frac{3}{2} \right) dx$$

$$= \left(\frac{3}{2} x^2 + \frac{3}{2} x \right) \Big|_{x=1}^{x=3}$$

$$= \left(\frac{27}{2} + \frac{9}{2} \right) - \left(\frac{3}{2} + \frac{3}{2} \right)$$

$$= (18) - (3)$$

$$= 15$$

$$\iint\limits_{R} (x + y)\, dx\, dy = \int_{-1}^{2} \int_{1}^{3} (x + y)\, dx\, dy$$

$$= \int_{-1}^{2} \left[\left(\frac{x^2}{2} + xy \right) \Big|_{x=1}^{x=3} \right] dy$$

$$= \int_{-1}^{2} \left[\left(\frac{9}{2} + 3y \right) - \left(\frac{1}{2} + y \right) \right] dy$$

$$= \int_{-1}^{2} (4 + 2y)\, dy$$

$$= (4y + y^2) \Big|_{y=-1}^{y=2}$$

$$= (8 + 4) - (-4 + 1)$$

$$= (12) - (-3)$$

$$= 15$$

Problem 21 Evaluate both ways:

$$\iint\limits_{R} (2x - y)\, dx\, dy \text{ over } R = \{(x, y)|-1 \le x \le 5, 2 \le y \le 4\}.$$

Answer 12

Example 22 Evaluate: $\iint\limits_{R} 2xe^{x^2 + y}\, dx\, dy$ over $R = \{(x, y)|0 \le x \le 1, \ -1 \le y \le 1\}.$

Solution

$$\iint\limits_{R} 2xe^{x^2 + y}\, dx\, dy = \int_{-1}^{1} \int_{0}^{1} 2xe^{x^2 + y}\, dx\, dy$$

$$= \int_{-1}^{1} \left[(e^{x^2 + y}) \Big|_{x=0}^{x=1} \right] dy$$

$$= \int_{-1}^{1} (e^{1 + y} - e^{y})\, dy$$

$$= (e^{1 + y} - e^{y}) \Big|_{y=-1}^{y=1}$$

$$= (e^2 - e) - (e^0 - e^{-1})$$

$$= e^2 - e - 1 + e^{-1}$$

Problem 22   Evaluate: $\displaystyle\iint\limits_{R} \frac{x}{y^2}e^{x/y}\,dy\,dx$ over $R = \{(x, y)|0 \le x \le 1,\; 1 \le y \le 2\}.$

Answer $e - 2e^{1/2} + 1$

VOLUME AND DOUBLE INTEGRALS   An important application of the definite integral of a function with one variable is the calculation of areas, so it is not surprising that the definite integral of a function of two variables can be used to calculate volumes of solids.

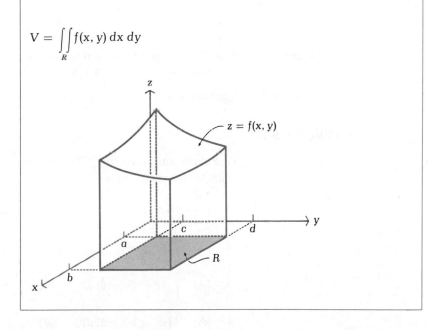

VOLUME UNDER A SURFACE

If $f(x, y) \ge 0$ over a rectangle R, $R = \{(x, y)|a \le x \le b,\;\; c \le y \le d\}$, then the volume of the solid formed by graphing f over the rectangle R is given by

$$V = \iint\limits_{R} f(x, y)\,dx\,dy$$

A proof of the statement in the box is left to a more advanced text.

Example 23   Find the volume of the solid under the graph of $f(x, y) = 1 + x^2 + y^2$ over the rectangle $R = \{(x, y)|0 \le x \le 1,\;\; 0 \le y \le 1\}.$

Solution

$$V = \iint\limits_{R} (1 + x^2 + y^2)\, dx\, dy$$

$$= \int_0^1 \int_0^1 (1 + x^2 + y^2)\, dx\, dy$$

$$= \int_0^1 \left[\left(x + \frac{1}{3}x^3 + xy^2 \right)\Big|_{x=0}^{x=1} \right] dy$$

$$= \int_0^1 \left(\frac{4}{3} + y^2 \right) dy$$

$$= \left(\frac{4}{3}y + \frac{1}{3}y^3 \right)\Big|_{y=0}^{y=1}$$

$$= \frac{5}{3}\ \text{cubic units}$$

Problem 23   Find the volume of the solid under the graph of $f(x, y) = 1 + x + y$ over the rectangle $R = \{(x, y)|0 \le x \le 1,\ \ 0 \le y \le 2\}$.

Answer   5 cubic units

EXERCISE 6-7

A   *Find each antiderivative. Then use the antiderivative to evaluate the definite integral.*

1. (A) $\displaystyle\int 12x^2y^3\, dy$ (B) $\displaystyle\int_0^1 12x^2y^3\, dy$

2. (A) $\displaystyle\int 12x^2y^3\, dx$ (B) $\displaystyle\int_{-1}^2 12x^2y^3\, dx$

3. (A) $\displaystyle\int (4x + 6y + 5)\, dx$ (B) $\displaystyle\int_{-2}^3 (4x + 6y + 5)\, dx$

4. (A) $\displaystyle\int (4x + 6y + 5)\, dy$ (B) $\displaystyle\int_1^4 (4x + 6y + 5)\, dy$

5. (A) $\displaystyle\int \frac{1 + 2y}{\sqrt{x}}\, dy$ (B) $\displaystyle\int_0^2 \frac{1 + 2y}{\sqrt{x}}\, dy$

6. (A) $\displaystyle\int \frac{1 + 2y}{\sqrt{x}}\, dx$ (B) $\displaystyle\int_1^9 \frac{1 + 2y}{\sqrt{x}}\, dx$

7. (A) $\displaystyle\int \frac{x}{\sqrt{y+x^2}}\,dx$ (B) $\displaystyle\int_0^2 \frac{x}{\sqrt{y+x^2}}\,dx$

8. (A) $\displaystyle\int \frac{x}{\sqrt{y+x^2}}\,dy$ (B) $\displaystyle\int_1^5 \frac{x}{\sqrt{y+x^2}}\,dy$

9. (A) $\displaystyle\int e^{x-y}\,dx$ (B) $\displaystyle\int_{-1}^1 e^{x-y}\,dx$

10. (A) $\displaystyle\int e^{x-y}\,dy$ (B) $\displaystyle\int_0^2 e^{x-y}\,dy$

B   *Evaluate each iterated integral. (See the indicated problem for the evaluation of the inner integral.)*

11. $\displaystyle\int_{-1}^2 \int_0^1 12x^2y^3 \,dy\,dx$
(see Problem 1)

12. $\displaystyle\int_0^1 \int_{-1}^2 12x^2y^3 \,dx\,dy$
(see Problem 2)

13. $\displaystyle\int_1^4 \int_{-2}^3 (4x+6y+5)\,dx\,dy$
(see Problem 3)

14. $\displaystyle\int_{-2}^3 \int_1^4 (4x+6y+5)\,dy\,dx$
(see Problem 4)

15. $\displaystyle\int_1^9 \int_0^2 \frac{1+2y}{\sqrt{x}}\,dy\,dx$
(see Problem 5)

16. $\displaystyle\int_0^2 \int_1^9 \frac{1+2y}{\sqrt{x}}\,dx\,dy$
(see Problem 6)

17. $\displaystyle\int_1^5 \int_0^2 \frac{x}{\sqrt{y+x^2}}\,dx\,dy$
(see Problem 7)

18. $\displaystyle\int_0^2 \int_1^5 \frac{x}{\sqrt{y+x^2}}\,dy\,dx$
(see Problem 8)

19. $\displaystyle\int_0^2 \int_{-1}^1 e^{x-y}\,dx\,dy$
(see Problem 9)

20. $\displaystyle\int_{-1}^1 \int_0^2 e^{x-y}\,dy\,dx$
(see Problem 10)

Use both orders of iteration to evaluate each double integral.

21. $\displaystyle\iint_R xy\,dy\,dx, \quad R=\{(x,y)|0\le x\le 2,\ \ 0\le y\le 4\}$

22. $\displaystyle\iint_R \sqrt{xy}\,dy\,dx, \quad R=\{(x,y)|1\le x\le 4,\ \ 1\le y\le 9\}$

23. $\displaystyle\iint_R (x+y)^5\,dy\,dx, \quad R=\{(x,y)|-1\le x\le 1,\ \ 1\le y\le 2\}$

24. $\displaystyle\iint_R xe^y\,dy\,dx, \quad R=\{(x,y)|-2\le x\le 3,\ \ 0\le y\le 2\}$

25. $\displaystyle\iint_R \frac{x}{1+x^2}\,dy\,dx, \quad R=\{(x,y)|0\le x\le 2,\ \ 0\le y\le 2\}$

26. $\displaystyle\iint\limits_R \frac{xy}{\sqrt{x^2 + y^2}}\, dy\, dx, \quad R = \{(x, y)|3 \le x \le 6, \quad 4 \le y \le 8\}$

Find the volume of the solid under the graph of each function over the given rectangle.

27. $f(x, y) = 1, \quad R = \{(x, y)|0 \le x \le 3, \quad 0 \le y \le 4\}$
28. $f(x, y) = 8 - x - y, \quad R = \{(x, y)|0 \le x \le 4, \quad 0 \le y \le 4\}$
29. $f(x, y) = 2 - x^2 - y^2, \quad R = \{(x, y)|0 \le x \le 1, \quad 0 \le y \le 1\}$
30. $f(x, y) = 5 - x, \quad R = \{(x, y)|0 \le x \le 5, \quad 0 \le y \le 5\}$
31. $f(x, y) = 4 - y^2, \quad R = \{(x, y)|0 \le x \le 2, \quad 0 \le y \le 2\}$
32. $f(x, y) = e^{-x-y}, \quad R = \{(x, y)|0 \le x \le 1, \quad 0 \le y \le 1\}$

C *Evaluate each double integral. Select the order of integration carefully—each problem is easy to do one way and difficult the other.*

33. $\displaystyle\iint\limits_R xe^{xy}\, dy\, dx, \quad R = \{(x, y)|0 \le x \le 1, \quad 1 \le y \le 2\}$

34. $\displaystyle\iint\limits_R xye^{x^2y}\, dy\, dx, \quad R = \{(x, y)|0 \le x \le 1, \quad 1 \le y \le 2\}$

35. $\displaystyle\iint\limits_R \frac{2y + 3xy^2}{1 + x^2}\, dy\, dx, \quad R = \{(x, y)|0 \le x \le 1, \quad -1 \le y \le 1\}$

36. $\displaystyle\iint\limits_R \frac{2x + 2y}{1 + 4y + y^2}\, dy\, dx, \quad R = \{(x, y)|1 \le x \le 3, \quad 0 \le y \le 1\}$

6-8   DOUBLE INTEGRALS OVER MORE GENERAL REGIONS

In this section we will extend the concept of double integration to nonrectangular regions. We begin with an example and some new terminology.

REGULAR REGIONS
    Let R be the region graphed in Figure 14. We can describe R with the following inequalities:

$$R = \{(x, y)|x \le y \le 6x - x^2, \quad 0 \le x \le 5\}$$

The region R can be viewed as a union of vertical line segments. For each x in the interval $[0, 5]$, the line segment from the point $(x, g(x))$ to the point $(x, f(x))$ lies in the region R. Any region that can be covered by vertical line segments in this manner is called a **regular x region.**
    Now consider the region S in Figure 15. S is *not* a regular x region, but it can be described with inequalities:

$$S = \{(x, y)|y^2 \le x \le y + 2, \quad -1 \le y \le 2\}$$

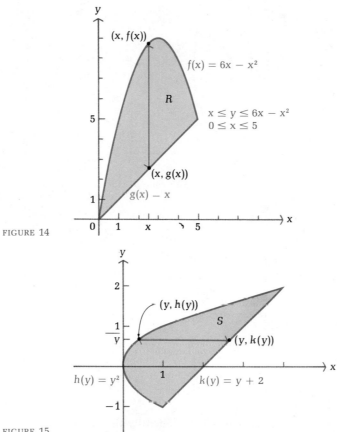

FIGURE 14

FIGURE 15

The region S can be viewed as a union of horizontal line segments going from the graph of $h(y) = y^2$ to the graph of $k(y) = y + 2$ on the interval $[-1, 2]$. Regions that can be described in this manner are called **regular y regions.** In general, **regular regions** are defined as follows:

REGULAR REGIONS

A region R in the xy plane is a **regular x region** if there exist functions $f(x)$ and $g(x)$ and numbers a and b so that

$$R = \{(x, y)|g(x) \leq y \leq f(x), \quad a \leq x \leq b\}$$

A region R is a **regular y region** if there exist functions $h(y)$ and $k(y)$ and numbers c and d so that

$$R = \{(x, y)|h(y) \leq x \leq k(y), \quad c \leq y \leq d\}$$

See Figure 16 (p. 318) for a geometric interpretation.

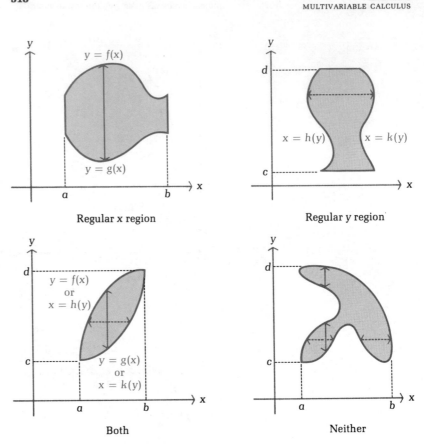

Regular x region

Regular y region

Both

Neither

FIGURE 16

Example 24 The region R is bounded by the graphs of $y = 4 - x^2$ and $y = x - 2$, $x \geq 0$, and the y axis. Graph R and describe R as a regular x region, a regular y region, both, or neither. If possible, represent R in terms of set notation and double inequalities.

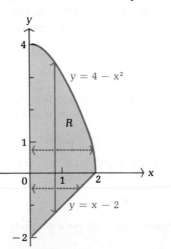

Solution As the solid line in the figure indicates, R can be covered by vertical line segments which go from the graph of $y = x - 2$ to the graph of $y = 4 - x^2$. Thus, R is a regular x region. In terms of set notation and double inequalities, we can write

$$R = \{(x, y)| x - 2 \le y \le 4 - x^2, \quad 0 \le x \le 2\}$$

On the other hand, a horizontal line passing through a point in the interval $[-2, 0]$ on the y axis will intersect R in a line segment which goes from the y axis to the graph of $y = x - 2$, while one that passes through a point in the interval $[0, 4]$ on the y axis goes from the y axis to the graph of $y = 4 - x^2$. Two such segments are shown as dashed lines in the figure. Thus, the region is not a regular y region.

Problem 24 Repeat Example 24 for the region R bounded by the graphs of $x = 6 - y$, $x = y^2$, $y \ge 0$, and the x axis.

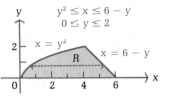

Answer $R = \{(x, y)| y^2 \le x \le 6 - y, \quad 0 \le y \le 2\}$ is a regular y region; R is not a regular x region.

Example 25 The region R is bounded by the graphs of $x + y^2 = 9$ and $x + 3y = 9$. Graph R and describe R as a regular x region, a regular y region, both, or neither. If possible, represent R using set notation and double inequalities.

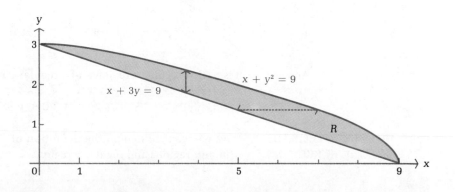

Solution   *Test for x region.*   Region R can be covered by vertical line segments which go from the graph of $x + 3y = 9$ to the graph of $x + y^2 = 9$. Thus, R is a regular x region. In order to describe R with inequalities, we must solve each equation for y in terms of x:

$$x + 3y = 9$$
$$3y = 9 - x$$
$$y = 3 - \tfrac{1}{3}x$$

$$x + y^2 = 9$$
$$y^2 = 9 - x$$
$$y = \sqrt{9 - x} \qquad \text{We use the positive square root, since}$$
$$\text{the graph is in the first quadrant.}$$

Thus,

$$R = \{(x, y) \mid 3 - \tfrac{1}{3}x \leq y \leq \sqrt{9 - x}, \quad 0 \leq x \leq 9\}$$

Test for y region.   Since region R can also be covered by horizontal line segments (dashed line in the figure) which go from the graph of $x + 3y = 9$ to the graph of $x + y^2 = 9$, it is a regular y region. Now we must solve each equation for x in terms of y:

$$x + 3y = 9 \qquad\qquad x + y^2 = 9$$
$$x = 9 - 3y \qquad\qquad x = 9 - y^2$$

Thus,

$$R = \{(x, y) \mid 9 - 3y \leq x \leq 9 - y^2, \quad 0 \leq y \leq 3\}$$

Problem 25   Repeat Example 25 for the region bounded by the graphs of $2y - x = 4$ and $y^2 - x = 4$.

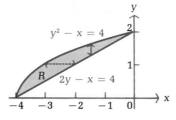

Answer   R is both a regular x region and a regular y region;
$$R = \{(x, y) \mid \tfrac{1}{2}x + 2 \leq y \leq \sqrt{x + 4}, \quad -4 \leq x \leq 0\}$$
$$= \{(x, y) \mid y^2 - 4 \leq x \leq 2y - 4, \quad 0 \leq y \leq 2\}$$

DOUBLE INTEGRALS   Now we want to extend the definition of double integration to include
OVER REGULAR REGIONS   regular x regions and regular y regions.

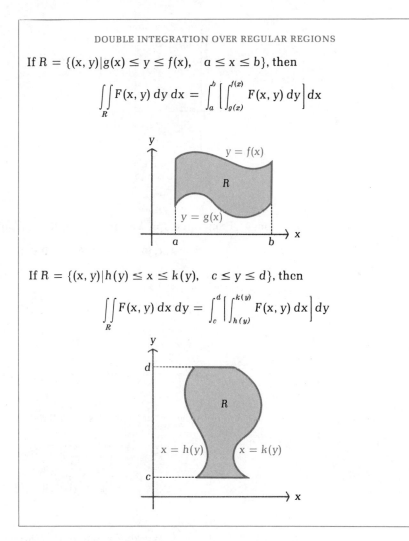

DOUBLE INTEGRATION OVER REGULAR REGIONS

If $R = \{(x, y)|g(x) \leq y \leq f(x), \quad a \leq x \leq b\}$, then

$$\iint\limits_{R} F(x, y) \, dy \, dx = \int_a^b \left[\int_{g(x)}^{f(x)} F(x, y) \, dy \right] dx$$

If $R = \{(x, y)|h(y) \leq x \leq k(y), \quad c \leq y \leq d\}$, then

$$\iint\limits_{R} F(x, y) \, dx \, dy = \int_c^d \left[\int_{h(y)}^{k(y)} F(x, y) \, dx \right] dy$$

Notice that the order of integration now depends on the nature of the region R. If R is a regular x region, we integrate with respect to y first, while if R is a regular y region, we integrate with respect to x first.

It is also important to note that the variable limits of integration (when present) are always on the inner integral, and the constant limits of integration are always on the outer integral.

Example 26   Evaluate $\iint_R 2xy \, dy \, dx$, where R is the region bounded by the graphs of $y = -x$ and $y = x^2$, $x \geq 0$, and the graph of $x = 1$.

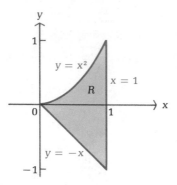

Solution   From the graph we can see that R is a regular x region described by

$$R = \{(x, y) \mid -x \le y \le x^2, \quad 0 \le x \le 1\}$$

Thus,

$$\iint_R 2xy \, dy \, dx = \int_0^1 \left[\int_{-x}^{x^2} 2xy \, dy \right] dx$$

$$= \int_0^1 \left[xy^2 \Big|_{y=-x}^{y=x^2} \right] dx$$

$$= \int_0^1 [x(x^2)^2 - x(-x)^2] \, dx$$

$$= \int_0^1 (x^5 - x^3) \, dx$$

$$= \left(\frac{x^6}{6} - \frac{x^4}{4} \right) \Big|_{x=0}^{x=1}$$

$$= (\tfrac{1}{6} - \tfrac{1}{4}) - (0 - 0)$$

$$= -\tfrac{1}{12}$$

Problem 26   Evaluate $\iint_R 3xy^2 \, dy \, dx$, where R is the region in Example 26.

Answer $\tfrac{13}{40}$

Example 27   Evaluate $\iint_R (2x + y) \, dy \, dx$, where R is the region bounded by the graphs of $y = \sqrt{x}$, $x + y = 2$, and $y = 0$.

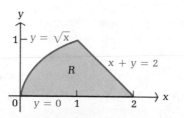

Solution   From the graph we can see that R is a regular y region. After solving each equation for x, we can write

$$R = \{(x, y)|y^2 \leq x \leq 2 - y, \quad 0 \leq y \leq 1\}$$

Thus,

$$\iint\limits_R (2x + y) \, dy \, dx = \int_0^1 \left[\int_{y^2}^{2-y} (2x + y) \, dx \right] dy$$

$$= \int_0^1 \left[(x^2 + yx) \Big|_{x=y^2}^{x=2-y} \right] dy$$

$$= \int_0^1 \{[(2 - y)^2 + y(2 - y)] - [(y^2)^2 + y(y^2)]\} \, dy$$

$$= \int_0^1 (4 - 2y - y^3 - y^4) \, dy$$

$$= (4y - y^2 - \tfrac{1}{4}y^4 - \tfrac{1}{5}y^5) \Big|_{y=0}^{y=1}$$

$$= (4 - 1 - \tfrac{1}{4} - \tfrac{1}{5}) - (0)$$

$$= {}^{51}\!/_{20}$$

Problem 27   Evaluate $\iint_R (y - 4x) \, dy \, dx$, where R is the region in Example 27.

Answer $-{}^{11}\!/_{20}$

Example 28   The region R is bounded by the graphs of $y = \sqrt{x}$ and $y = \tfrac{1}{2}x$. Evaluate $\iint_R 4xy^3 \, dy \, dx$ two different ways.

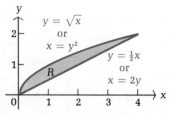

Solution   Region R is both a regular x region and a regular y region:

$$R = \{(x, y)|\tfrac{1}{2}x \leq y \leq \sqrt{x}, \quad 0 \leq x \leq 4\} \qquad \text{Regular x region}$$
$$R = \{(x, y)|y^2 \leq x \leq 2y, \quad 0 \leq y \leq 2\} \qquad \text{Regular y region}$$

Using the first representation (a regular x region), we obtain

$$\iint_R 4xy^3 \, dy \, dx = \int_0^4 \left[\int_{(1/2)x}^{\sqrt{x}} 4xy^3 \, dy \right] dx$$

$$= \int_0^4 \left[xy^4 \Big|_{y=(1/2)x}^{y=\sqrt{x}} \right] dx$$

$$= \int_0^4 [x(\sqrt{x})^4 - x(\tfrac{1}{2}x)^4] \, dx$$

$$= \int_0^4 (x^3 - \tfrac{1}{16}x^5) \, dx$$

$$= (\tfrac{1}{4}x^4 - \tfrac{1}{96}x^6) \Big|_{x=0}^{x=4}$$

$$= (64 - {}^{128}\!/_3) - 0$$

$$= {}^{64}\!/_3$$

Using the second representation (a regular y region), we obtain

$$\iint_R 4xy^3 \, dy \, dx = \int_0^2 \left[\int_{y^2}^{2y} 4xy^3 \, dx \right] dy$$

$$= \int_0^2 \left[2x^2y^3 \Big|_{x=y^2}^{x=2y} \right] dy$$

$$= \int_0^2 [2(2y)^2y^3 - 2(y^2)^2y^3] \, dy$$

$$= \int_0^2 (8y^5 - 2y^7) \, dy$$

$$= (\tfrac{4}{3}y^6 - \tfrac{1}{4}y^8) \Big|_{y=0}^{y=2}$$

$$= ({}^{256}\!/_3 - 64) - 0$$

$$= {}^{64}\!/_3$$

Problem 28   The region R is bounded by the graphs of $y = x$ and $y = \tfrac{1}{2}x^2$. Evaluate $\iint_R 4x^3y \, dy \, dx$ two different ways.

Answer ${}^{16}\!/_3$

REVERSING ORDER   Example 28 shows that
OF INTEGRATION

$$\iint_R 4xy^3 \, dy \, dx = \int_0^4 \left[\int_{(1/2)x}^{\sqrt{x}} 4xy^3 \, dy \right] dx = \int_0^2 \left[\int_{y^2}^{2y} 4xy^3 \, dx \right] dy$$

In general, if R is both a regular x region and a regular y region, the two iterated integrals are equal. In rectangular regions, reversing the order of integration in an iterated integral was a simple matter. As Example 28 illustrates, the process is more complicated in nonrectangular regions. The next example illustrates how to start with an iterated integral and reverse the order of integration. Since we are interested in the reversal process and not in the value of either integral, the integrand will not be specified.

Example 29   Reverse the order of integration in $\int_1^3 \left[\int_0^{x-1} f(x, y)\, dy \right] dx$.

Solution   The order of integration indicates that the region of integration is a regular x region:

$$R = \{(x, y) | 0 \le y \le x - 1, \quad 1 \le x \le 3\}$$

Graph region R to determine whether it is also a regular y region. The graph shows that R is also a regular y region, and we can write

$$R = \{(x, y) | y + 1 \le x \le 3, \quad 0 \le y \le 2\}$$

Thus,

$$\int_1^3 \left[\int_0^{x-1} f(x, y)\, dy \right] dx = \int_0^2 \left[\int_{y+1}^3 f(x, y)\, dx \right] dy$$

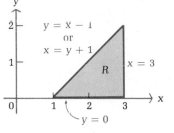

Problem 29   Reverse the order of integration in $\int_2^4 \left[\int_0^{4-x} f(x, y)\, dy \right] dx$.

Answer $\int_0^2 \int_2^{4-y} f(x, y)\, dx\, dy$

VOLUMES AND DOUBLE INTEGRALS   In Section 6-7 we used the double integral to calculate the volume of a solid with a rectangular base. In general, if a solid can be described by the graph of a positive function $f(x, y)$ over a regular region R (not necessarily a rectangle), then the double integral of the function f over the region R still represents the volume of the corresponding solid.

Example 30   The region R is bounded by the graphs of $x + y = 1$, $y = 0$, and $x = 0$. Find the volume of the solid under the graph of $z = 1 - x - y$ over the region R.

Solution

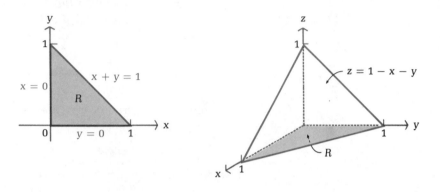

The graph of R indicates that R is a regular x region and can be described by

$$R = \{(x, y) | 0 \le y \le 1 - x, \quad 0 \le x \le 1\}$$

Thus, the volume of the solid is

$$V = \iint\limits_{R} (1 - x - y)\, dy\, dx = \int_{0}^{1} \left[\int_{0}^{1-x} (1 - x - y)\, dy \right] dx$$

$$= \int_{0}^{1} \left[(y - xy - \tfrac{1}{2}y^2) \Big|_{y=0}^{y=1-x} \right] dx$$

$$= \int_{0}^{1} [(1 - x) - x(1 - x) - \tfrac{1}{2}(1 - x)^2]\, dx$$

$$= \int_{0}^{1} (\tfrac{1}{2} - x + \tfrac{1}{2}x^2)\, dx$$

$$= (\tfrac{1}{2}x - \tfrac{1}{2}x^2 + \tfrac{1}{6}x^3) \Big|_{x=0}^{x=1}$$

$$= (\tfrac{1}{2} - \tfrac{1}{2} + \tfrac{1}{6}) - 0$$

$$= \tfrac{1}{6}$$

Problem 30   The region R is bounded by the graphs of $y + 2x = 2$, $y = 0$, and $x = 0$. Find the volume of the solid under the graph of $z = 2 - 2x - y$ over the region R.   [*Hint:* Sketch the region first—the solid does not have to be sketched.]

Answer $\tfrac{2}{3}$

EXERCISE 6-8

A   *Graph the region R bounded by the graphs of the equations. Express R in terms of set notation and double inequalities that describe R as a regular x region, a regular y region, or both.*

1. $y = 4 - x^2$, $y = 0$, $0 \le x \le 2$ 2. $y = x^2$, $y = 9$, $0 \le x \le 3$
3. $y = x^3$, $y = 12 - 2x$, $x = 0$ 4. $y = 5 - x$, $y = 1 + x$, $y = 0$
5. $y^2 = 2x$, $y = x - 4$ 6. $y = 4 + 3x - x^2$, $x + y = 4$

Evaluate each integral.

7. $\displaystyle\int_0^1 \int_0^x (x + y)\, dy\, dx$ 8. $\displaystyle\int_0^2 \int_0^y xy\, dx\, dy$

9. $\displaystyle\int_0^1 \int_{y^3}^{\sqrt{y}} (2x + y)\, dx\, dy$ 10. $\displaystyle\int_1^4 \int_x^{x^2} (x^2 + 2y)\, dy\, dx$

B   *Use the description of the region R to evaluate the indicated integral.*

11. $\displaystyle\iint_R (x^2 + y^2)\, dy\, dx$, $R = \{(x, y) | 0 \le y \le 2x, \ 0 < x \le 2\}$

12. $\displaystyle\iint_R 2x^2 y\, dy\, dx$, $R = \{(x, y) | 0 \le y \le 9 - x^2, \ -3 \le x \le 3\}$

13. $\displaystyle\iint_R (x + y - 2)^3\, dy\, dx$, $R = \{(x, y) | 0 \le x \le y + 2, \ 0 \le y \le 1\}$

14. $\displaystyle\iint_R (2x + 3y)\, dy\, dx$, $R = \{(x, y) | y^2 - 4 \le x \le 4 - 2y, \ 0 \le y \le 2\}$

15. $\displaystyle\iint_R e^{x+y}\, dy\, dx$, $R = \{(x, y) | -x \le y \le x, \ 0 \le x \le 2\}$

16. $\displaystyle\iint_R \frac{x}{\sqrt{x^2 + y^2}}\, dy\, dx$, $R = \{(x, y) | 0 \le x \le \sqrt{4y - y^2}, \ 0 \le y \le 2\}$

Graph the region R bounded by the graphs of the indicated equations. Describe R in set notation with double inequalities and evaluate the indicated integral.

17. $y = x + 1$, $y = 0$, $x = 1$; $\displaystyle\iint_R \sqrt{1 + x + y}\, dx\, dy$

18. $y = x^2$, $y = \sqrt{x}$; $\displaystyle\iint_R 12xy\, dy\, dx$

19. $y = 4x - x^2$, $y = 0$; $\displaystyle\iint_R \sqrt{y + x^2}\, dy\, dx$

20. $x = 1 + 3y$, $x - 1 = y$, $y = 1$; $\displaystyle\iint_R (x + y + 1)^3 \, dy \, dx$

21. $y = 1 - \sqrt{x}$, $y = 1 + \sqrt{x}$, $x = 4$; $\displaystyle\iint_R x(y - 1)^2 \, dy \, dx$

22. $y = \tfrac{1}{2}x$, $y = 6 - x$, $y = 1$; $\displaystyle\iint_R \frac{1}{x + y} \, dy \, dx$

Evaluate each integral. Graph the region of integration, reverse the order of integration, and then evaluate the integral with the order reversed.

23. $\displaystyle\int_0^3 \int_0^{3-x} (x + 2y) \, dy \, dx$

24. $\displaystyle\int_0^2 \int_0^y (y - x)^4 \, dx \, dy$

25. $\displaystyle\int_0^1 \int_0^{1-x^2} x\sqrt{y} \, dy \, dx$

26. $\displaystyle\int_0^2 \int_{x^3}^{4x} (1 + 2y) \, dy \, dx$

27. $\displaystyle\int_0^4 \int_{x/4}^{\sqrt{x}/2} x \, dy \, dx$

28. $\displaystyle\int_0^4 \int_{y^2/4}^{2\sqrt{y}} (1 + 2xy) \, dx \, dy$

Find the volume of the solid under the graph of $f(x, y)$ over the region R bounded by the graphs of the indicated equations. Sketch the region R—the solid does not have to be sketched.

29. $f(x, y) = 4 - x - y$; R is bounded by the graphs of $x + y = 4$, $y = 0$, $x = 0$

30. $f(x, y) = (x - y)^2$; R is the region bounded by the graphs of $y = x$, $y = 2$, $x = 0$

31. $f(x, y) = 4$; R is the region bounded by the graphs of $y = 1 - x^2$ and $y = 0$ for $0 \leq x \leq 1$

32. $f(x, y) = 4xy$; R is the region bounded by the graphs of $y = \sqrt{1 - x^2}$ and $y = 0$ for $0 \leq x \leq 1$

C *Reverse the order of integration for each integral. Evaluate the integral with the order reversed. Do not attempt to evaluate the integral in the original form.*

33. $\displaystyle\int_0^2 \int_{x^2}^4 \frac{4x}{1 + y^2} \, dy \, dx$

34. $\displaystyle\int_0^1 \int_y^1 \sqrt{1 - x^2} \, dx \, dy$

35. $\displaystyle\int_0^1 \int_{y^2}^1 4ye^{x^2} \, dx \, dy$

36. $\displaystyle\int_0^4 \int_{\sqrt{x}}^2 \sqrt{3x + y^2} \, dy \, dx$

6-9   CHAPTER REVIEW

6-1 Functions of several variables. functions of two independent variables, functions of several independent variables, surface, paraboloid, saddle point, $z = f(x, y)$, $w = f(x, y, z)$

6-2 Partial derivatives. partial derivative of f with respect to x, partial derivative of f with respect to y, second-order partials,

$$\frac{\partial z}{\partial x}, \quad \frac{\partial z}{\partial y}, \quad f_x(x, y), \quad f_y(x, y), \quad \frac{\partial^2 z}{\partial x^2} = f_{xx}(x, y), \quad \frac{\partial^2 z}{\partial x\,\partial y} = f_{yx}(x, y),$$

$$\frac{\partial^2 z}{\partial y\,\partial x} = f_{xy}(x, y), \quad \frac{\partial^2 z}{\partial y^2} = f_{yy}(x, y)$$

6-3 Total differentials and their applications. total differential of $z = f(x, y)$, $dz = f_x(x, y)\,dx + f_y(x, y)\,dy$, total differential of $w = f(x, y, z)$, $dw = f_x(x, y, z)\,dx + f_y(x, y, z)\,dy + f_z(x, y, z)\,dz$, differential approximation, $\Delta z = f(x + \Delta x, y + \Delta y) - f(x, y) \approx f_x(x, y)\,dx + f_y(x, y)\,dy = dz$

6-4 Maxima and minima. local maximum, local minimum, critical point, second-derivative test

6-5 Maxima and minima using Lagrange multipliers. method of Lagrange multipliers, constraints, critical points

6-6 Method of least squares. least squares approximation, linear regression, residual, least squares line, regression line, estimation, approximation, $y = mx + d$ where m and d satisfy

$$\left(\sum_{k=1}^{n} x_k\right) m + nd = \sum_{k=1}^{n} y_k, \quad \left(\sum_{k=1}^{n} x_k^2\right) m + \left(\sum_{k=1}^{n} x_k\right) d = \sum_{k=1}^{n} x_k y_k$$

6-7 Double integrals over rectangular regions. double integral, iterated integral,

$$\iint\limits_{R} f(x, y)\,dy\,dx = \int_a^b \left[\int_c^d f(x, y)\,dy\right] dx = \int_c^d \left[\int_a^b f(x, y)\,dx\right] dy,$$

volume under a surface, $V = \iint\limits_{R} f(x, y)\,dx\,dy$

6-8 Double integrals over more general regions. regular x region, regular y region, $\int_a^b \left[\int_{g(x)}^{f(x)} F(x, y)\,dy\right] dx$, $\int_c^d \left[\int_{h(y)}^{k(y)} F(x, y)\,dx\right] dy$, reversing the order of integration, volume under a surface

EXERCISE 6-9   CHAPTER REVIEW

Work through all the problems in this chapter review and check your answers in the back of the book. (Answers to all review problems are there.) Where weaknesses show up, review appropriate sections in the text. When you are satisfied that you know the material, take the practice test following this review.

A
1. For $f(x, y) = 2,000 + 40x + 70y$, find $f(5, 10)$, $f_x(x, y)$, and $f_y(x, y)$.
2. For $z = x^3y^2$, find $\partial^2z/\partial x^2$ and $\partial^2z/\partial x\,\partial y$.
3. For $z = 2x + 3y$, find dz.
4. For $z = x^4y^3$, find dz.

5. Evaluate: $\displaystyle\int (6xy^2 + 4y)\,dy$
6. Evaluate: $\displaystyle\int (6xy^2 + 4y)\,dx$

7. Evaluate: $\displaystyle\int_0^1\int_0^1 4xy\,dy\,dx$
8. Evaluate: $\displaystyle\int_0^1\int_0^x 4xy\,dy\,dx$

B
9. For $f(x, y) = 3x^2 - 2xy + y^2 - 2x + 3y - 7$, find $f(2, 3)$, $f_y(x, y)$, and $f_y(2, 3)$.
10. For $f(x, y) = -4x^2 + 4xy - 3y^2 + 4x + 10y + 81$, find $[f_{xx}(2, 3)]\,[f_{yy}(2, 3)] - [f_{xy}(2, 3)]^2$.
11. Find Δz and dz for $z = f(x, y) = x^4 + y^4$, $x = 1$, $y = 2$, $\Delta x = dx = 0.1$, and $\Delta y = dy = 0.2$.
12. Use the least squares line for the data in the table to estimate y when $x = 10$.

x	y
2	12
4	10
6	7
8	3

13. For $R = \{(x, y)|-1 \le x \le 1, \ 1 \le y \le 2\}$, evaluate
$$\iint_R (4x + 6y)\,dy\,dx \text{ two ways.}$$

14. Evaluate $\displaystyle\iint_R (x + y)^3\,dy\,dx$ for
$R = \{(x, y)|0 \le x \le y + 1, 0 \le y \le 3\}$.

C
15. For $f(x, y) = e^{x^2 + 2y}$, find f_x, f_y, and f_{xy}.
16. For $f(x, y) = (x^2 + y^2)^5$, find f_x and f_{xy}.
17. Use differentials to approximate the hypotenuse of a right triangle with legs 7.1 and 24.05 inches long, respectively.

18. Find the least squares line for the data in the table.

x	y
10	50
20	45
30	50
40	55
50	65
60	80
70	85
80	90
90	90
100	110

19. Find the volume of the solid under the graph of $z = x + y$ over the region bounded by the graphs of $y = \sqrt{1 - x^2}$ and $y = 0$ for $0 \le x \le 1$.

20. Evaluate $\int_0^1 \int_0^{\sqrt{1-x^2}} y \, dy \, dx$. Then evaluate the integral obtained by reversing the order of integration.

APPLICATIONS

BUSINESS & ECONOMICS

21. *Maximizing profit.* A company produces x units of product A and y units of product B (both in hundreds per month). The monthly profit equation (in thousands of dollars) is found to be

$$P(x, y) = -4x^2 + 4xy - 3y^2 + 4x + 10y + 81$$

(A) Find $P_x(1, 3)$ and interpret.

(B) How many of each product should be produced each month to maximize profit? What is the maximum profit?

22. *Profit.* A company's annual profit (in millions of dollars) over a 5 year period is given in the table. Use the least squares line to estimate the profit for the sixth year.

YEAR	PROFIT
1	2
2	2.5
3	3.1
4	4.2
5	4.3

LIFE SCIENCES

23. *Marine biology.* The function used for timing dives with scuba gear is

$$T(V, x) = \frac{33V}{x + 33}$$

where T is the time of the dive in minutes, V is the volume of air (at sea level pressure) compressed into tanks, and x is the depth of the dive in feet. Find $T_x(70, 17)$ and interpret.

24. *Blood flow.* In Poiseuille's law,

$$R(L, r) = k \frac{L}{r^4}$$

where R is the resistance for blood flow, L is the length of the blood vessel, r is the radius of the blood vessel, and k is a constant. Use differentials to approximate the change in R if L increases from 10 to 10.1 and r decreases from 0.5 to 0.45.

SOCIAL SCIENCES

25. *Sociology.* Joseph Cavanaugh, a sociologist, found that the number of long-distance telephone calls, n, between two cities in a given period of time varied (approximately) jointly as the populations P_1 and P_2 of the two cities, and varied inversely as the distance, d, between the two cities. In terms of an equation for a time period of 1 week,

$$n(P_1, P_2, d) = 0.001 \frac{P_1 P_2}{d}$$

Find $n(100,000, 50,000, 100)$.

26. *Education.* At the beginning of the semester, students in a foreign language course are given a proficiency exam. The same exam is given at the end of the semester. The results for five students are given in the table. Use the least squares line to estimate the score on the second exam for a student who scored 40 on the first exam.

FIRST EXAM	SECOND EXAM
30	60
50	75
60	80
70	85
90	90

PRACTICE TEST: CHAPTER 6

1. For $f(x, y) = 2x^2y + y + 1$, find:
   (A) $f(1, 2)$ (B) $f_x(1, 2)$
2. For $f(x, y) = (x^2y^3 - 2x)^4$, find $f_x(x, y)$ and $f_y(x, y)$.
3. For $z = x^3y^4$, find dz.
4. Find dz and Δz for $z = f(x, y) = x^2 + y^2, x = 1, y = 2, \Delta x = dx = 0.2$, and $\Delta y = dy = 0.1$.
5. For $z = x^3 - 2x^2y + y^2$, find $\partial^2z/\partial y \, \partial x$.
6. For $f(x, y) = e^{x^2y}$, find $f_{yx}(x, y)$.

7. Evaluate: $\displaystyle\int_0^2 \int_0^3 (6x + 4y) \, dy \, dx$

8. The daily revenue equation for two commodities is

$$R(x, y) = 10(-3x^2 + 2xy - 5y^2 + 250x + 200y)$$

where x and y are the unit prices of the commodities in dollars. Find $R_x(30, 20)$ and $R_y(20, 30)$.
9. Find all local extrema for $f(x, y) = x^4 + 8x^2 + y^2 - 4y$.
10. Find the volume of the solid under the graph of $f(x, y) = 1 - x - y$ over the region bounded by the graphs of $y = 1 - x, y = 0,$ and $x = 0$.
11. The cost y in thousands of dollars of producing x units of a certain product is given in the table for various levels of production. Use the least squares line to estimate the cost of producing 40 units.

x	y
20	40
25	45
30	51
35	56

CHAPTER 7
TRIGONOMETRIC FUNCTIONS

CONTENTS

7 TRIGONOMETRIC FUNCTIONS

7-1   INTRODUCTION

Up until now we have restricted our attention to algebraic, logarithmic, and exponential functions. These functions were used to model many real-life situations from business, economics, and the life and social sciences. Now we turn our attention to another important class of functions, called the **trigonometric functions.** These functions are particularly useful in describing periodic phenomena; that is, phenomena that repeat in cycles. Consider the sunrise times for a 2 year period starting January 1, as pictured in Figure 1. We see that the cycle repeats after 1 year. Business cycles, blood pressure in the aorta, seasonal growth, water waves, and amounts of pollution in the atmosphere are often periodic and can be illustrated with similar types of graphs.

We assume the reader has had a course in trigonometry. The next section provides a brief review of those topics that are most important for our purposes.

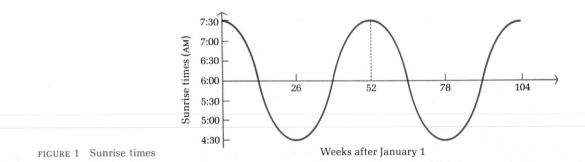

FIGURE 1   Sunrise times

336

7-2   TRIGONOMETRIC FUNCTIONS REVIEW

We start our discussion of trigonometry with the concept of **angle.** A point P on a line divides the line into two parts. Either part together with the end point is called a **half-line.** A geometric figure consisting of two half-lines with a common end point is called an **angle.** One of the half-lines is called the **initial side** and the other half-line is called the **terminal side.** The common end point is called the **vertex.** Figure 2 illustrates these concepts.

FIGURE 2   Angle θ

There are two widely used measures of angles—the **degree** and the **radian.** A central angle of a circle subtended by an arc $\frac{1}{360}$ of the circumference of the circle is said to have **degree measure 1,** written $1°$ (see Fig. 3). It follows that a central angle subtended by an arc $\frac{1}{4}$ the circumference has degree measure 90; $\frac{1}{2}$ the circumference has degree measure 180; and the whole circumference of a circle has degree measure 360.

FIGURE 3   Degree and radian measure

The other measure of angles, which we will use extensively when we discuss the calculus of the trigonometric functions in the next two sections, is radian measure. A central angle subtended by an arc of length equal to the radius (R) of the circle is said to have **radian measure 1,** written **1 radian** (see Fig. 3). In general, a central angle subtended by an arc of length s has radian measure determined as follows:

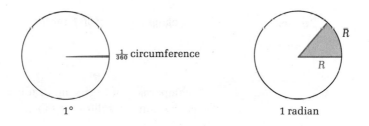

$$\theta^{\mathrm{r}} = \text{Radian measure of } \theta = \frac{\text{Arc length}}{\text{Radius}} = \frac{s}{R}$$

[*Note:* If $R = 1$, then $\theta^{\mathrm{r}} = s$.]

What is the radian measure of an angle of 180°? A central angle of 180° is subtended by an arc of ½ the circumference of a circle. Thus,

$$s = \frac{C}{2} = \frac{2\pi R}{2} = \pi R$$

and

$$\theta^r = \frac{s}{R} = \frac{\pi R}{R} = \pi \text{ radians}$$

The following proportion can be used to convert degree measure to radian measure and vice versa:

DEGREE–RADIAN CONVERSION FORMULA

$$\frac{\theta°}{180°} = \frac{\theta^r}{\pi \text{ radians}}$$

Example 1   Find the radian measure of 1°.

Solution
$$\frac{1°}{180°} = \frac{\theta^r}{\pi \text{ radians}}$$

$$\theta^r = \frac{\pi}{180} \text{radians} \approx 0.0175 \text{ radian}$$

Problem 1   Find the degree measure of 1 radian.

Answer
$$\frac{180}{\pi} \approx 57.3°$$

A comparison of degree and radian measures for a few important angles is given in the following table:

RADIAN	0	π	2π	$\pi/2$	$\pi/4$	$\pi/6$	$\pi/3$
DEGREE	0	180°	360°	90°	45°	30°	60°

GENERALIZED ANGLE   Preliminary to defining trigonometric functions, we will generalize the notion of angle defined above. Starting with a rectangular coordinate system and two half-lines coinciding with the nonnegative x axis, the initial side of the angle remains fixed and the terminal side rotates until it reaches its terminal position. When the terminal side is rotated counterclockwise, the angle formed is considered positive (see Figs. 4A and 4C). When it is rotated clockwise, the angle formed is considered negative (see Fig. 4B). Angles located in a coordinate system in this manner are said to be in a **standard position.**

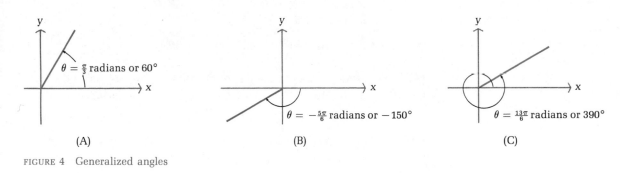

(A)   (B)   (C)

FIGURE 4   Generalized angles

TRIGONOMETRIC FUNCTIONS

Let us locate a unit circle (radius 1) in a coordinate system with center at the origin (Fig. 5). The terminal side of any angle will pass through this circle at some point P. The abscissa of this point P is called the **cosine of** θ (abbreviated **cos θ**), and the ordinate of the point is the **sine of** θ (abbreviated **sin θ**). Thus, the set of all ordered pairs of the form $(\theta, \cos \theta)$, and the set of all ordered pairs of the form $(\theta, \sin \theta)$ constitute, respectively, the **cosine and sine functions.** The **domain** of these two functions is the set of all angles, positive or negative, with measure either in degrees or radians. The **range** is a subset of the set of real numbers.

It is desirable, and necessary for our work in calculus, to define these two trigonometric functions in terms of real number domains. This is easily done as follows:

SINE AND COSINE FUNCTIONS WITH REAL NUMBER DOMAINS

For x any real number,

$$\sin x = \sin(x \text{ radians})$$
$$\cos x = \cos(x \text{ radians})$$

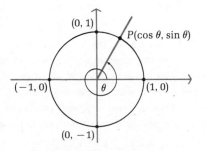

FIGURE 5

Example 2   Referring to Figure 5, find:

(A) $\cos 90°$ (B) $\sin\left(-\dfrac{\pi}{2} \text{ radians}\right)$ (C) $\cos \pi$

Solutions   (A)   The terminal side of an angle of degree measure 90 passes through (0, 1) on the unit circle. This point has abscissa zero. Thus,

$$\cos 90° = 0$$

(B)   The terminal side of an angle of radian measure $-\pi/2$ ($-90°$) passes through (0, -1) on the unit circle. This point has ordinate -1. Thus,

$$\sin\left(-\dfrac{\pi}{2} \text{ radians}\right) = -1$$

(C) $\cos \pi = \cos(\pi \text{ radians}) = -1$, since the terminal side of an angle of radian measure π (180°) passes through ($-1, 0$) on the unit circle and this point has abscissa -1.

Problem 2   Referring to Figure 5, find:
(A) $\sin 180°$ (B) $\cos(2\pi \text{ radians})$ (C) $\sin(-\pi)$

Answers   (A)   0   (B)   1   (C)   0

To find the value of either the sine or the cosine function for any angle or any real number by direct use of the definition is not easy. Tables are available, but hand calculators with SIN and COS buttons are even more convenient. Calculators generally have degree and radian options, so we can use the calculator to evaluate these functions for most of the real numbers in which we might have an interest. The following table includes a few values produced by a hand calculator. The x value is entered and then the SIN or COS button is pushed to obtain the desired value in display.

x	1	-7	35.26	-105.9
sin x	0.8415	-0.6570	-0.6461	0.7920
cos x	0.5403	0.7539	-0.7632	0.6105

Exact values of the sine and cosine functions can be obtained for multiples of the special angles in the triangles in Figure 6, since the triangles can be used to find the coordinate of the intersection of the terminal side of each angle with the unit circle. [*Note:* We now drop the word "radian" after $\pi/4$ and interpret $\pi/4$ as the radian measure of an angle or simply as a real number, depending on the context.]

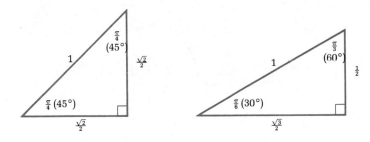

FIGURE 6

Example 3   Find the exact value of each of the following using Figure 6:
(A)   cos π/4   (B)   sin π/6   (C)   sin(−π/6)

Solutions   (A)   cos π/4 = √2/2   (B)   sin π/6 = ½

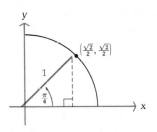   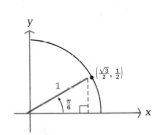

(C)   sin(−π/6) = −½

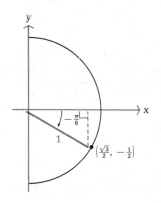

Problem 3   Find the exact value of each of the following using Figure 6:
(A)   sin π/4   (B)   cos π/3   (C)   cos(−π/3)

Answers   (A)   √2/2   (B)   ½   (C)   ½

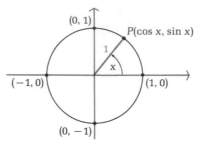

FIGURE 7

GRAPHS OF THE SINE AND COSINE FUNCTIONS

To graph $y = \sin x$ or $y = \cos x$ for x a real number, we could use a calculator to produce a table, and then plot the ordered pairs from the table in a coordinate system. However, we can speed the process up by returning to basic definitions. Referring to Figure 7, we see that as x increases and P moves around the unit circle in a counterclockwise (positive) direction, both $\sin x$ and $\cos x$ behave in uniform ways.

As x increases from	y = sin x	y = cos x
0 to $\pi/2$	Increases from 0 to 1	Decreases from 1 to 0
$\pi/2$ to π	Decreases from 1 to 0	Decreases from 0 to -1
π to $3\pi/2$	Decreases from 0 to -1	Increases from -1 to 0
$3\pi/2$ to 2π	Increases from -1 to 0	Increases from 0 to 1

Note that P has completed one revolution and is back at its starting place. If we let x continue to increase, then the second and third columns in the table will be repeated every 2π units. In general, it can be shown that

$$\sin(x + 2\pi) = \sin x$$
$$\cos(x + 2\pi) = \cos x$$

for all real numbers x. Functions such that

$$f(x + p) = f(x)$$

for some constant p and all real numbers x for which the functions are defined are said to be **periodic.** The smallest such value of p is called the **period** of the function. Thus, both the sine and cosine functions are periodic (a very important property) with period 2π.

Putting all this information together, and, perhaps, adding a few values obtained from a calculator or Figure 6, we obtain the graphs of the sine and cosine functions illustrated in Figure 8.

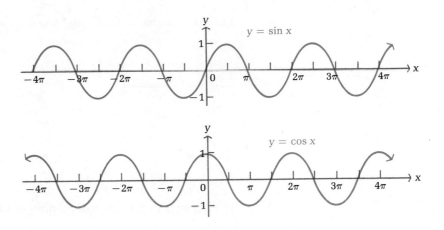

FIGURE 8

FOUR OTHER TRIGONOMETRIC FUNCTIONS

The sine and cosine functions are only two of six trigonometric functions. They are, however, the most important of the six for many applications. We define the other four trigonometric functions below. Exercises pertaining to these functions may be found in Exercise 7-2, part C.

FOUR OTHER TRIGONOMETRIC FUNCTIONS

$$\tan x = \frac{\sin x}{\cos x} \qquad \cos x \neq 0$$

$$\cot x = \frac{\cos x}{\sin x} \qquad \sin x \neq 0$$

$$\sec x = \frac{1}{\cos x} \qquad \cos x \neq 0$$

$$\csc x = \frac{1}{\sin x} \qquad \sin x \neq 0$$

EXERCISE 7-2

A   *Recall that 180° corresponds to π radians. Mentally convert each degree measure to radian measure in terms of π.*

1. 60°   2. 90°   3. 45°

4. 360°   5. 30°   6. 120°

Indicate the quadrant in which the terminal side of each angle lies.

7. 150°	8. −190°	9. −⅔ radians
10. ⁷π/₆ radians	11. 400°	12. −250°

Use Figure 5 to find the exact value of each of the following:

13. cos 0°	14. sin 90°	15. sin π
16. cos π/₂	17. cos(−π)	18. sin ³π/₂

B   *Recall that π radians corresponds to 180°. Mentally convert each radian measure to degree measure.*

19. π/₃ radians	20. 2π radians	21. π/₄ radian
22. π/₂ radians	23. π/₆ radian	24. ⁵π/₆ radians

Use Figure 6 to find the exact value of each of the following:

25. cos 30°	26. sin(−45°)	27. sin(π/₆ radian)
28. sin(−π/₃)	29. cos ⁵π/₆	30. cos(−120°)

Use a hand calculator to find the value (to four decimal places) of each of the following:

31. sin 3	32. cos 13	33. cos 33.74
34. sin 325.9	35. sin(−43.06)	36. cos(−502.3)

C   *Convert to radian measure.*

37. 27°   38. 18°

Convert to degree measure.

39. π/₁₂ radian   40. π/₆₀ radian

Use Figure 6 to find the exact value of each of the following:

41. tan 45°	42. cot 45°	43. sec π/₃
44. csc π/₆	45. cot π/₃	46. tan π/₆

47. Refer to Figure 5 and use the Pythagorean theorem to show that

$$(\sin x)^2 + (\cos x)^2 = 1$$

for all x.

48. Use the results of Problem 47 and basic definitions to show that
(A) $(\tan x)^2 + 1 = (\sec x)^2$ (B) $1 + (\cot x)^2 = (\csc x)^2$

APPLICATIONS

BUSINESS & ECONOMICS

49. *Seasonal business cycle.* Suppose profit in the sale of swimming suits in a chain department store over a 2 year period is given approximately by

$$P(t) = 5 - 5\cos\left(\frac{\pi}{26}\right)t$$

where P is profit in hundreds of dollars for a week of sales t weeks after January 1. The graph of the profit function is shown in the figure.

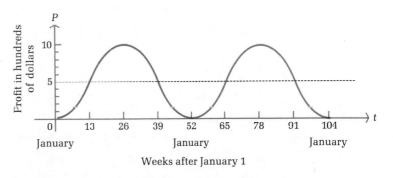

Weeks after January 1

(A) Find the exact values of P(13), P(26), P(39), and P(52) by evaluating $P(t) = 5 - 5\cos(\pi/_{26})t$ without tables or a calculator.
(B) Use a calculator to find P(30) and P(100).

LIFE SCIENCES

50. *Pollution.* In a large city the amount of sulfur dioxide pollutant released into the atmosphere due to the burning of coal and oil for heating purposes varies seasonally. Suppose the number of tons of pollutant released into the atmosphere during the nth week after January 1 is given approximately by

$$P(n) = 1 + \cos\left(\frac{\pi}{26}\right)n$$

The graph of the pollution function is shown in the figure on page 346.

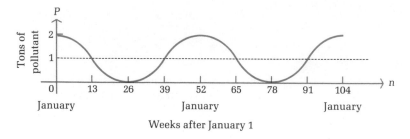

(A) Find the exact value of $P(0)$, $P(39)$, $P(52)$, and $P(65)$ by evaluating $P(n) = 1 + \cos(\frac{\pi}{26})n$ without tables or a calculator.

(B) Use a calculator to find $P(10)$ and $P(95)$.

SOCIAL SCIENCES 51. *Psychology–perception.* An important area of study in psychology is perception. Individuals perceive objects differently in different settings. Consider the well-known illusions shown in the figure.

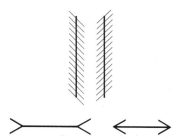

Lines that appear parallel in one setting may appear to be curved in another (the two vertical lines are actually parallel). Lines of the same length may appear to be of different legnths in two different settings (the two horizontal lines are actually the same length). An interesting experiment in visual perception was conducted by psychologists Berliner and Berliner (*American Journal of Psychology*, 1952, 65: 271–277). They reported that when subjects were presented with a large tilted field of parallel lines and were asked to estimate the position of a horizontal line in the field, most of the subjects were consistently off. They found that the difference in degrees, d, between their estimate and the actual horizontal could be approximated by the equation

$$d = a + b \sin 4\theta$$

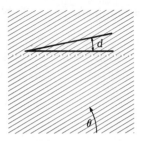

where a and b were constants associated with a particular individual and θ was the angle of tilt of the visual field in degrees. Suppose for a given individual $a = -2.1$ and $b = -4$, find d if:

(A) $\theta = 30°$ (B) $\theta = 10°$

7-3   DERIVATIVES OF SINE AND COSINE FUNCTIONS

In this section we will develop derivative formulas for the sine and cosine functions. Once we have these formulas, we will automatically have integral formulas for the same functions, which we will discuss in the next section.

To develop the derivative formulas for the sine and cosine functions, we will need the five properties of the trigonometric functions listed below, which we state without proof. For all real numbers a and b,

$$\sin(a + b) = \sin a \cos b + \cos a \sin b \tag{1}$$

$$\sin\left(\frac{\pi}{2} - a\right) = \cos a \tag{2}$$

$$\cos\left(\frac{\pi}{2} - a\right) = \sin a \tag{3}$$

In addition to these properties, we need the following special limits:

$$\lim_{\Delta x \to 0} \frac{\sin \Delta x}{\Delta x} = 1 \tag{4}$$

$$\lim_{\Delta x \to 0} \frac{\cos \Delta x - 1}{\Delta x} = 0 \tag{5}$$

(These limits are discussed intuitively in Problems 21 and 22, Exercise 7-3.)

Now let us show that

$$D_x \sin x = \cos x$$

From the definition of derivative (Section 2-4),

$$D_x \sin x = \lim_{\Delta x \to 0} \frac{\sin(x + \Delta x) - \sin x}{\Delta x}$$

Using property (1) we see that

$$\sin(x + \Delta x) = \sin x \cos \Delta x + \cos x \sin \Delta x$$

Thus,

$$\frac{\sin(x + \Delta x) - \sin x}{\Delta x} = \frac{\sin x \cos \Delta x + \cos x \sin \Delta x - \sin x}{\Delta x}$$

$$= \frac{\sin x (\cos \Delta x - 1) + \cos x \sin \Delta x}{\Delta x}$$

$$= \sin x \frac{\cos \Delta x - 1}{\Delta x} + \cos x \frac{\sin \Delta x}{\Delta x}$$

Hence,

$$\lim_{\Delta x \to 0} \frac{\sin(x + \Delta x) - \sin x}{\Delta x} = \lim_{\Delta x \to 0} \left(\sin x \frac{\cos \Delta x - 1}{\Delta x} + \cos x \frac{\sin \Delta x}{\Delta x} \right)$$

$$= \sin x \lim_{\Delta x \to 0} \frac{\cos \Delta x - 1}{\Delta x} + \cos x \lim_{\Delta x \to 0} \frac{\sin \Delta x}{\Delta x}$$

$$= (\sin x)(0) + (\cos x)(1) \qquad \text{Using (4) and (5)}$$

$$= \cos x$$

Thus,

$$D_x \sin x = \cos x$$

If $u = u(x)$, then we can use the chain rule for derivatives discussed in Section 2-7 to obtain the more general formula

$$\frac{d \sin u}{dx} = \cos u \frac{du}{dx} \tag{6}$$

We can now use (2), (3), and (6) to show that

$$D_x \cos x = -\sin x$$

We proceed as follows:

$$D_x \cos x = D_x \sin\left(\frac{\pi}{2} - x\right) \qquad \text{Property (2)}$$

$$= \cos\left(\frac{\pi}{2} - x\right) D_x\left(\frac{\pi}{2} - x\right) \qquad \text{Formula (6)}$$

$$= -\cos\left(\frac{\pi}{2} - x\right)$$

$$= -\sin x \qquad \text{Property (3)}$$

Thus,

$$D_x \cos x = -\sin x$$

We can now add the following two important derivative formulas to our list of derivative formulas from preceding chapters:

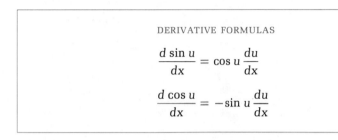

DERIVATIVE FORMULAS

$$\frac{d \sin u}{dx} = \cos u \frac{du}{dx}$$

$$\frac{d \cos u}{dx} = -\sin u \frac{du}{dx}$$

Example 4 (A) $D_x \sin x^2 = \cos x^2 \, D_x x^2 = 2x \cos x^2$

(B) $D_x \cos(2x - 5) = -\sin(2x - 5) D_x(2x - 5) = -2 \sin(2x - 5)$

(C) $D_x(3x^2 - x)\cos x = (3x^2 - x) D_x \cos x + (\cos x) D_x(3x^2 - x)$
$$= -(3x^2 - x) \sin x + (6x - 1) \cos x$$
$$= (x - 3x^2) \sin x + (6x - 1) \cos x$$

Problem 4 Find each of the following derivatives:

(A) $D_x \cos x^3$ (B) $D_x \sin(5 - 3x)$ (C) $D_x \dfrac{\sin x}{x}$

Answers (A) $-3x^2 \sin x^3$ (B) $-3 \cos(5 - 3x)$ (C) $\dfrac{x \cos x - \sin x}{x^2}$

Example 5 Find the slope of the graph of $f(x) = \sin x$ at $(\frac{\pi}{2}, 1)$, and sketch in the tangent line to the graph at this point.

Solution Slope at $(\frac{\pi}{2}, 1) = f'(\frac{\pi}{2}) = \cos \frac{\pi}{2} = 0$.

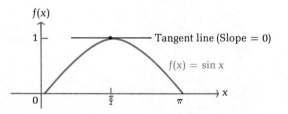

Problem 5 Find the slope of the graph of $f(x) = \cos x$ at $(\frac{\pi}{6}, \frac{\sqrt{3}}{2})$.

Answer $-\frac{1}{2}$

Example 6   Find $D_x \sec x$.

Solution

$$D_x \sec x = D_x \frac{1}{\cos x} \qquad \text{Since } \sec x = \frac{1}{\cos x}$$

$$= D_x(\cos x)^{-1}$$

$$= -(\cos x)^{-2} D_x \cos x$$

$$= -(\cos x)^{-2}(-\sin x)$$

$$= \frac{\sin x}{(\cos x)^2} = \left(\frac{\sin x}{\cos x}\right)\left(\frac{1}{\cos x}\right)$$

$$= \tan x \sec x \qquad \text{Since } \tan x = \frac{\sin x}{\cos x}$$

Problem 6   Find $D_x \csc x$.

Answer $-\cot x \csc x$

EXERCISE 7-3

Find the following derivatives:

A   1. $D_t \cos t$ 2. $D_w \sin w$
     3. $D_x \sin x^3$ 4. $D_x \cos(x^2 - 1)$

B   5. $D_t t \sin t$ 6. $D_u u \cos u$

     7. $D_x \sin x \cos x$ 8. $D_x \dfrac{\sin x}{\cos x}$

     9. $D_x(\sin x)^5$ 10. $D_x(\cos x)^8$
    11. $D_x \sqrt{\sin x}$ 12. $D_x \sqrt{\cos x}$
    13. $D_x \cos \sqrt{x}$ 14. $D_x \sin \sqrt{x}$

    15. Find the slope of the graph of $f(x) = \sin x$ at $x = \frac{\pi}{6}$.
    16. Find the slope of the graph of $f(x) = \cos x$ at $x = \frac{\pi}{4}$.

C   *Find the following derivatives:*

    17. $D_x \tan x$ 18. $D_x \cot x$
    19. $D_x \sin \sqrt{x^2 - 1}$ 20. $D_x \cos \sqrt{x^4 - 1}$

    21. Use a hand calculator to complete the following table:

Δx	-0.5	-0.05	-0.005	0.005	0.05	0.5
$\dfrac{\cos \Delta x - 1}{\Delta x}$						

[The results of this exercise suggest that $\lim_{\Delta x \to 0}[(\cos \Delta x - 1)/\Delta x] = 0$ is correct. A formal proof of this fact can be found in a more advanced text on calculus.]

22. Use a hand calculator to complete the following table:

Δx	-0.5	-0.05	-0.005	0.005	0.05	0.5
$\dfrac{\sin \Delta x}{\Delta x}$						

[The results of this exercise suggest that $\lim_{\Delta x \to 0}[(\sin \Delta x)/\Delta x] = 1$ is correct. A formal proof of this fact can be found in a more advanced text on calculus.]

23. The figure shows a piston connected to a wheel turning at 10 revolutions per second (rps). It can be shown that the point x on the piston moves according to

$$x(t) = \cos 20\pi t + \sqrt{25 - (\sin 20\pi t)^2}$$

if P is at $(1, 0)$ when counting of time is started ($t = 0$). The x units are inches.

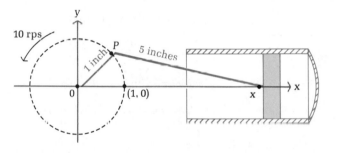

(A) Locate x when $t = 0, \frac{1}{40}, \frac{1}{20}$, and $\frac{1}{10}$ second.
(B) Find $x'(t)$.
(C) How fast is the piston moving when $t = \frac{1}{40}$ second? When $t = \frac{1}{20}$ second? [*Hint:* Find $x'(\frac{1}{40})$ and $x'(\frac{1}{20})$.]

7-4 INTEGRATION INVOLVING SINE AND COSINE FUNCTIONS

Now that we know that

$$D_x \sin x = \cos x \qquad \text{and} \qquad D_x \cos x = -\sin x$$

then from the definition of the indefinite integral of a function (Section 4-1), we automatically have the two integral formulas

$$\int \cos x \, dx = \sin x + C \qquad \text{and} \qquad \int \sin x \, dx = -\cos x + C$$

Example 7   Find the area under the sine curve $y = \sin x$ from 0 to π.

Solution

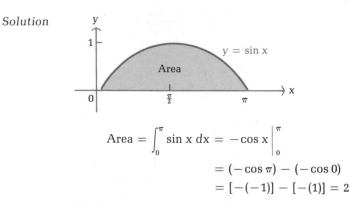

$$\text{Area} = \int_0^\pi \sin x \, dx = -\cos x \Big|_0^\pi$$

$$= (-\cos \pi) - (-\cos 0)$$

$$= [-(-1)] - [-(1)] = 2$$

Problem 7   Find the area under the cosine curve $y = \cos x$ from 0 to $\pi/2$.

Answer   1

From the fact that

$$D_x \sin u = \cos u \frac{du}{dx} \quad \text{and} \quad D_x \cos u = -\sin u \frac{du}{dx}.$$

we obtain the more general integral formulas below.

INTEGRAL FORMULAS

$$\int \sin u \, du = -\cos u + C$$

$$\int \cos u \, du = \sin u + C$$

where $u = u(x)$

Example 8   Integrate $\int x \sin x^2 \, dx$.

Solution

$$\int x \sin x^2 \, dx = \frac{1}{2} \int 2x \sin x^2 \, dx$$

$$= \frac{1}{2} \int (\sin x^2) \, 2x \, dx$$

$$= \frac{1}{2} \int \sin u \, du \qquad \text{Let } u = x^2; \text{ then } du = 2x \, dx$$

$$= -\frac{1}{2} \cos x^2 + C$$

To check, we differentiate the result to obtain the original integrand:

$$D_x\left(-\frac{1}{2}\cos x^2\right) = -\frac{1}{2}D_x\cos x^2$$

$$= -\frac{1}{2}(-\sin x^2)D_x x^2$$

$$= -\frac{1}{2}(-\sin x^2)(2x)$$

$$= x\sin x^2$$

Problem 8 Integrate $\int \cos 20\pi t\ dt$.

Answer $\dfrac{1}{20\pi}\sin 20\pi t + C$

Example 9 Integrate $\int (\sin x)^5 \cos x\ dx$.

Solution This is of the form $\int u^p\ du$, where $u = \sin x$ and $du = \cos x\ dx$. Thus,

$$\int (\sin x)^5 \cos x\ dx = \frac{(\sin x)^6}{6} + C$$

Problem 9 Integrate $\int \sqrt{\sin x}\ \cos x\ dx$.

Answer $\dfrac{2}{3}(\sin x)^{3/2} + C$

Example 10 Evaluate $\int_2^{3.5} \cos x\ dx$ using a hand calculator.

Solution $$\int_2^{3.5} \cos x\ dx = \sin x\ \Big|_2^{3.5}$$

$$= \sin 3.5 - \sin 2$$

$$= -0.3508 - 0.9093$$

$$= -1.2601$$

Problem 10 Evaluate $\int_1^{1.5} \sin x\ dx$ using a hand calculator.

Answer 0.4696

EXERCISE 7-4

Find each of the following indefinite integrals:

A   1. $\int \sin t\ dt$ 2. $\int \cos w\ dw$

    3. $\int \cos 3x\ dx$ 4. $\int \sin 2x\ dx$

5. $\displaystyle\int (\sin x)^{12} \cos x \, dx$

6. $\displaystyle\int \sin x \cos x \, dx$

B 7. $\displaystyle\int \sqrt[3]{\cos x}\, \sin x \, dx$

8. $\displaystyle\int \frac{\cos x}{\sqrt{\sin x}} \, dx$

9. $\displaystyle\int x^2 \cos x^3 \, dx$

10. $\displaystyle\int (x + 1) \sin(x^2 + 2x) \, dx$

Evaluate each of the following definite integrals:

11. $\displaystyle\int_0^{\pi/2} \cos x \, dx$

12. $\displaystyle\int_0^{\pi/4} \cos x \, dx$

13. $\displaystyle\int_{\pi/2}^{\pi} \sin x \, dx$

14. $\displaystyle\int_{\pi/6}^{\pi/3} \sin x \, dx$

15. Find the shaded area under the cosine curve in the figure.

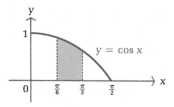

16. Find the shaded area under the sine curve in the figure.

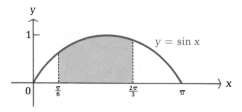

Use a hand calculator to evaluate the definite integrals below after performing the indefinite integration. (Remember that the limits are real numbers, hence the radian mode must be used on the calculator.)

17. $\displaystyle\int_0^2 \sin x \, dx$

18. $\displaystyle\int_0^{0.5} \cos x \, dx$

19. $\displaystyle\int_1^2 \cos x \, dx$

20. $\displaystyle\int_1^3 \sin x \, dx$

C   *Find each of the following indefinite integrals:*

21. $\int e^{\sin x} \cos x \, dx$ 22. $\int e^{\cos x} \sin x \, dx$

23. $\int \dfrac{\cos x}{\sin x} \, dx$ 24. $\int \dfrac{\sin x}{\cos x} \, dx$

25. $\int \tan x \, dx$ 26. $\int \cot x \, dx$

APPLICATIONS

BUSINESS & ECONOMICS   27. *Seasonal business cycle.*   In Problem 49 in Exercise 7-2 we were given the following profit equation for the sale of swimming suits:

$$P(t) = 5 - 5 \cos\left(\frac{\pi}{26}\right)t$$

where P is profit in hundreds of dollars for a week of sales t weeks after January 1. The area under the graph of this equation for a 2 year period represents the total profit on sales of swimming suits for that period.

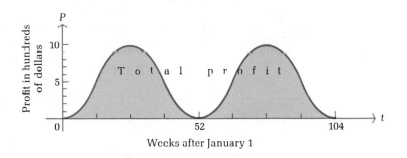

Weeks after January 1

(A) Set up a definite integral that will give the total profit for 2 years of sales (104 weeks).

(B) Find the total profit.

LIFE SCIENCES   28. *Pollution.*   In Problem 50 in Exercise 7-2 we were given the pollution equation

$$P(n) = 1 + \cos\left(\frac{\pi}{26}\right)n$$

where P is the number of tons of pollutant released into the atmosphere during the nth week (see the figure on the next page).

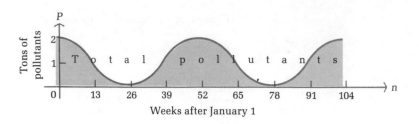

Weeks after January 1

(A) Set up a definite integral that will give the total number of tons of pollutants emitted into the atmosphere over the 2 year period (104 weeks).

(B) Evaluate the integral in part A.

7-5 OTHER TRIGONOMETRIC FUNCTIONS

In Example 6 and Problem 6 in Section 7-3, the formulas

$$D_x \sec x = \tan x \sec x \quad \text{and} \quad D_x \csc x = -\cot x \csc x$$

were derived. In this section, we will derive differentiation formulas for tan x and cot x and integral formulas for a variety of trigonometric functions. We will need the following property of trigonometric functions (see Problem 47 in Exercise 7-2):

$$(\sin x)^2 + (\cos x)^2 = 1 \tag{1}$$

Example 11 Find $D_x \tan x$.

Solution

$$D_x \tan x = D_x \frac{\sin x}{\cos x} \qquad \text{Since } \tan x = \frac{\sin x}{\cos x}$$

$$= \frac{\cos x \, D_x \sin x - \sin x \, D_x \cos x}{(\cos x)^2}$$

$$= \frac{\cos x \cos x - \sin x(-\sin x)}{(\cos x)^2}$$

$$= \frac{(\cos x)^2 + (\sin x)^2}{(\cos x)^2}$$

$$= \frac{1}{(\cos x)^2} \qquad \text{Use equation (1)}$$

$$= (\sec x)^2 \qquad \text{Since } \frac{1}{\cos x} = \sec x$$

Problem 11   Find $D_x \cot x$.

Answer $-(\csc x)^2$

If $u = u(x)$, then we can use the chain rule to state more general formulas for the derivatives of these functions.

DERIVATIVE FORMULAS

$$D_x \tan u = (\sec u)^2 D_x u \qquad D_x \sec u = \tan u \sec u \, D_x u$$

$$D_x \cot u = -(\csc u)^2 D_x u \qquad D_x \csc u = -\cot u \csc u \, D_x u$$

Example 12   (A) $D_x \tan 5x = (\sec 5x)^2 D_x 5x = 5(\sec 5x)^2$

(B) $D_x \cot x^3 = -(\csc x^3)^2 D_x x^3 = -3x^2(\csc x^3)^2$

(C) $D_x x^2 \sec x = x^2 D_x \sec x + \sec x \, D_x x^2$

$$= x^2 \tan x \sec x + 2x \sec x$$

(D) $D_x \dfrac{\csc(x + 1)}{x + 1} = \dfrac{(x + 1) D_x \csc(x + 1) - \csc(x + 1) D_x(x + 1)}{(x + 1)^2}$

$$= \dfrac{(x + 1)[-\cot(x + 1) \csc(x + 1)] - \csc(x + 1)}{(x + 1)^2}$$

$$= -\dfrac{\csc(x + 1)}{(x + 1)^2}[(x + 1) \cot(x + 1) + 1]$$

Problem 12   Find each of the following derivatives:

(A) $D_x \cot 3x$ (B) $D_x \tan \sqrt{x}$ (C) $D_x \dfrac{\sec x}{x}$ (D) $D_x \cot x \csc x$

Answers   (A) $-3(\csc 3x)^2$ (B) $\dfrac{1}{2\sqrt{x}}(\sec \sqrt{x})^2$

(C) $\dfrac{\sec x(x \tan x - 1)}{x^2}$ (D) $-(\cot x)^2 \csc x - (\csc x)^3$

As before, these derivative formulas also provide us with some new integration formulas.

Example 13   Evaluate: $\displaystyle\int_0^{\pi/4} (\sec x)^2 \, dx$

Solution $\displaystyle\int_0^{\pi/4} (\sec x)^2 \, dx = \tan x \Big|_0^{\pi/4} = \tan \dfrac{\pi}{4} - \tan 0$

$$= 1 - 0 = 1$$

Problem 13   Evaluate: $\displaystyle\int_0^{\pi/6} \tan x \sec x \, dx$

Answer $\frac{2}{3}\sqrt{3} - 1$

The new integral formulas are summarized in the box.

INTEGRAL FORMULAS

$$\int (\sec u)^2 \, du = \tan u + C \qquad \int \tan u \sec u \, du = \sec u + C$$

$$\int (\csc u)^2 \, du = -\cot u + C \qquad \int \cot u \csc u \, du = -\csc u + C$$

Example 14   Integrate: $\displaystyle\int \cot 4x \csc 4x \, dx$

Solution $\displaystyle\int \cot 4x \csc 4x \, dx = \frac{1}{4}\int \cot 4x \csc 4x \, 4dx$

$$= \frac{1}{4}\int \cot u \csc u \, du \qquad \text{Let } u = 4x; \text{ then } du = 4dx$$

$$= -\frac{1}{4}\csc 4x + C$$

Problem 14   Integrate: $\displaystyle\int (\csc 2\pi x)^2 \, dx$

Answer $-\dfrac{1}{2\pi}\cot 2\pi x + C$

Most hand calculators have SIN, COS, and TAN buttons. These can be combined with the reciprocal button (1/X) to evaluate the other three trigonometric functions by means of the following relationships:

$$\sec x = \frac{1}{\cos x} \qquad \csc x = \frac{1}{\sin x} \qquad \cot x = \frac{1}{\tan x}$$

Example 15   Evaluate $\displaystyle\int_{0.5}^{1.5} \tan x \sec x \, dx$ using a hand calculator.   [*Note:*   Make sure the calculator is in the radian mode.]

Solution $\displaystyle\int_{0.5}^{1.5} \tan x \sec x \, dx = \sec x \Big|_{0.5}^{1.5} = \sec 1.5 - \sec 0.5$

$$= 14.137 - 1.139$$
$$= 12.998$$

Problem 15 Evaluate $\displaystyle\int_{0.25}^{0.75} (\csc x)^2 \, dx$ using a hand calculator.

Answer 2.843

      The integral formulas on page 358 are of interest primarily because they were easily derived from the derivative formulas. But how can we find integral formulas for the functions tan x, cot x, sec x, and csc x? It turns out that each of these functions can be integrated by first expressing the function in a different form and then making a substitution. Two of these integrals are evaluated in Examples 16 and 17; the other two are left as matching problems.

Example 16 Integrate: $\displaystyle\int \tan x \, dx$

Solution $\displaystyle\int \tan x \, dx = \int \frac{\sin x}{\cos x} \, dx$ Since $\tan x = \dfrac{\sin x}{\cos x}$

$$= \int \frac{1}{\cos x} \sin x \, dx$$

$$= -\int \frac{1}{\cos x} (-\sin x) \, dx$$

$$= -\int \frac{1}{u} \, du \qquad \text{Let } u = \cos x; \text{ then } du = -\sin x \, dx$$

$$= -\ln|u| + C$$

$$= -\ln|\cos x| + C$$

Problem 16 Integrate: $\displaystyle\int \cot x \, dx$

Answer $\ln|\sin x| + C$

Example 17 Integrate: $\displaystyle\int \sec x \, dx$

Solution $\displaystyle\int \sec x \, dx = \int \sec x \, \frac{\sec x + \tan x}{\sec x + \tan x} \, dx$ Since $\displaystyle\frac{\sec x + \tan x}{\sec x + \tan x} = 1$

$\displaystyle = \int \frac{(\sec x)^2 + \tan x \sec x}{\sec x + \tan x} \, dx$

$\displaystyle = \int \frac{1}{\sec x + \tan x} \, [\tan x \sec x + (\sec x)^2] \, dx$

$\displaystyle = \int \frac{1}{u} \, du$ Let $u = \sec x + \tan x$; then

$du = [\tan x \sec x + (\sec x)^2] \, dx$

$= \ln|u| + C$

$= \ln|\sec x + \tan x| + C$

Problem 17   Integrate: $\displaystyle\int \csc x \, dx$

Answer $-\ln|\csc x + \cot x| + C$

The general integral formulas obtained by the methods described above are summarized in the box. It is generally easier to remember the technique used to derive these formulas than it is to commit each formula to memory.

INTEGRAL FORMULAS

$\displaystyle\int \tan u \, du = -\ln|\cos u| + C$ $\displaystyle\int \sec u \, du = \ln|\sec u + \tan u| + C$

$\displaystyle\int \cot u \, du = \ln|\sin u| + C$ $\displaystyle\int \csc u \, du = -\ln|\csc u + \cot u| + C$

Example 18   Integrate: $\displaystyle\int x \tan x^2 \, dx$

Solution $\displaystyle\int x \tan x^2 \, dx = \frac{1}{2} \int (\tan x^2) 2x \, dx$

$\displaystyle = \frac{1}{2} \int \tan u \, du$ Let $u = x^2$; then $du = 2x \, dx$

$\displaystyle = -\frac{1}{2} \ln|\cos u| + C$

$\displaystyle = -\frac{1}{2} \ln|\cos x^2| + C$

Problem 18   Integrate: $\displaystyle\int \csc(2x + 3)\,dx$

Answer $-\tfrac{1}{2}\ln|\csc(2x + 3) + \cot(2x + 3)| + C$

Example 19   Evaluate $\displaystyle\int_{-0.5}^{0.75} \sec x\,dx$ using a hand calculator.

Solution $\displaystyle\int_{-0.5}^{0.75} \sec x\,dx = \ln|\sec x + \tan x|\,\Big|_{-0.5}^{0.75}$

$$= \ln|\sec 0.75 + \tan 0.75| - \ln|\sec(-0.5) + \tan(-0.5)|$$
$$\approx \ln(1.3667 + 0.9316) - \ln(1.1395 - 0.5463)$$
$$= \ln 2.2983 - \ln 0.5932$$
$$\approx 0.8322 + 0.5222 = 1.3544$$

Problem 19   Evaluate $\displaystyle\int_{0.75}^{1.25} \csc x\,dx$ using a hand calculator.

Answer   0.6059

EXERCISE 7-5

A   Find each of the following derivatives:

1. $D_t \tan t$
2. $D_w \cot w$
3. $D_u \sec u$
4. $D_y \csc y$
5. $D_x \tan 2x$
6. $D_t \cot t^2$
7. $D_u \sec u^3$
8. $D_w \csc(2w + 1)$

B
9. $D_x \tan x \sec x$
10. $D_x \cot x \csc x$
11. $D_x(\sec x)^2$
12. $D_x \sqrt{\cot x}$
13. $D_x \tan e^x$
14. $D_x \csc(\ln x)$

Find each of the following indefinite integrals:

15. $\displaystyle\int (\sec t)^2\,dt$
16. $\displaystyle\int \cot w \csc w\,dw$

17. $\displaystyle\int (\csc \tfrac{1}{2} x)^2\,dx$
18. $\displaystyle\int \tan \pi x \sec \pi x\,dx$

19. $\displaystyle\int \tan w\,dw$
20. $\displaystyle\int \csc t\,dt$

21. $\displaystyle\int \cot 2\pi x\,dx$
22. $\displaystyle\int \sec\left(\tfrac{1}{3}x + \tfrac{\pi}{3}\right)dx$

23. $\displaystyle\int x^2(\cot x^3)(\csc x^3)\,dx$

24. $\displaystyle\int \frac{1}{x^2}\tan\frac{1}{x}\sec\frac{1}{x}\,dx$

25. $\displaystyle\int \sqrt{\tan x}\,(\sec x)^2\,dx$

26. $\displaystyle\int (\cot x)^{3/2}\,(\csc x)^2\,dx$

27. $\displaystyle\int e^x \tan(1 + e^x)\,dx$

28. $\displaystyle\int \frac{1}{x}\sec(\ln x)\,dx$

Find the slope of the graph of y = f(x) at the indicate value of x.

29. $f(x) = \tan x$ at $x = \frac{\pi}{4}$

30. $f(x) = \cot x$ at $x = \frac{\pi}{3}$

31. $f(x) = \sec \pi x$ at $x = 1$

32. $f(x) = \csc \pi x$ at $x = \frac{1}{3}$

Evaluate each of the following definite integrals:

33. $\displaystyle\int_0^{\pi/6} \tan x\,dx$

34. $\displaystyle\int_{\pi/3}^{\pi/2} \cot x\,dx$

35. $\displaystyle\int_{\pi/4}^{\pi/2} \csc x\,dx$

36. $\displaystyle\int_{-\pi/4}^{\pi/4} \sec x\,dx$

Use a hand calculator to evaluate each definite integral after finding the indefinite integral. (Remember that the limits are real numbers, so the radian mode must be used on the calculator.)

37. $\displaystyle\int_0^{0.5} (\sec x)^2\,dx$

38. $\displaystyle\int_0^1 \tan x \sec x\,dx$

39. $\displaystyle\int_{0.1}^{0.9} (\csc x)^2\,dx$

40. $\displaystyle\int_1^2 \cot x \csc x\,dx$

C *Find each indefinite integral. You will need the identities $(\tan x)^2 + 1 = (\sec x)^2$ and $1 + (\cot x)^2 = (\csc x)^2$. (See Problem 48 in Exercise 7-2.)*

41. $\displaystyle\int (\tan x)^2\,dx$

42. $\displaystyle\int (\cot x)^2\,dx$

7-6 CHAPTER REVIEW

IMPORTANT TERMS AND SYMBOLS

7-2 Trigonometric functions review. angle, initial side, terminal side, vertex, degree measure, radian measure, standard position, generalized angle, sin x, cos x, tan x, sec x, csc x, cot x

7-3 Derivatives of trigonometric functions. $D_x \sin u = \cos u\,D_x u$, $D_x \cos u = -\sin u\,D_x u$

7-4 *Integration of trigonometric functions.* $\int \cos u \, du = \sin u + C,$

$\int \sin u \, du = -\cos u + C$

7-5 *Other trigonometric functions.* $D_x \tan u = (\sec u)^2 \, D_x u,$
$D_x \cot u = -(\csc u)^2 \, D_x u, \quad D_x \sec u = \tan u \sec u \, D_x u,$
$D_x \csc u = -\cot u \csc u \, D_x u,$

$\int (\sec u)^2 \, du = \tan u + C, \quad \int (\csc u)^2 \, du = -\cot u + C,$

$\int \tan u \sec u \, du = \sec u + C, \quad \int \cot u \csc u \, du = -\csc u + C,$

$\int \tan u \, du = -\ln|\cos u| + C, \quad \int \cot u \, du = \ln|\sin u| + C,$

$\int \sec u \, du = \ln|\sec u + \tan u| + C, \quad \int \csc u \, du = -\ln|\csc u + \cot u| + C$

EXERCISE 7-6   CHAPTER REVIEW

Work through all the problems in this chapter review and check your answers in the back of the book. (Answers to all review problems are there.) Where weaknesses show up, review appropriate sections in the text. When you are satisfied that you know the material, take the practice test following this review

A

1. Convert to radian measure in terms of π:
   (A) 30°   (B) 45°   (C) 60°   (D) 90°
2. Evaluate exactly without using a table or calculator:
   (A) $\cos \pi$ (B) $\sin 0$ (C) $\sin \frac{\pi}{2}$

Find:

3. $D_m \cos m$

4. $D_u \sin u$

5. $D_z \tan z$

6. $D_y \cot y$

7. $D_m \sec m$

8. $D_w \csc w$

9. $D_x \sin(x^2 - 2x + 1)$

10. $D_t \cos(1 - t^3)$

11. $D_z \tan 2z$

12. $D_w \csc \pi w$

13. $D_x \sec \dfrac{1}{x}$

14. $D_t \cot \sqrt{t}$

15. $\int \sin 3t \, dt$

16. $\int \cos 2x \, dx$

17. $\int \tan\left(\frac{1}{2} y\right) dy$

18. $\int \cot\left(\frac{w}{3}\right) dw$

B　19. Convert to degree measure:
　　　(A) $\frac{\pi}{6}$　(B) $\frac{\pi}{4}$　(C) $\frac{\pi}{3}$　(D) $\frac{\pi}{2}$
　　20. Evaluate exactly without using a table or calculator:
　　　(A) $\sin \frac{\pi}{6}$　(B) $\cos \frac{\pi}{4}$　(C) $\sin \frac{\pi}{3}$
　　21. Evaluate to four decimal places using a hand calculator:
　　　(A) $\cos 2.367$　(B) $\sin(-0.04312)$　(C) $\tan 8$
　　22. Evaluate to four decimal places using a hand calculator:
　　　(A) $\sec 1.0132$　(B) $\csc 3$　(C) $\cot(-2)$

Find:

23. $D_x(x^2 - 1) \sin x$

24. $D_x(\sin x)^6$

25. $D_x \sqrt[3]{\sin x}$

26. $D_x \, x \sec 2x$

27. $D_x \sqrt{\cot x}$

28. $D_u \, u \tan u$

29. $\displaystyle\int t \cos(t^2 - 1)\, dt$

30. $\displaystyle\int_0^\pi \sin u \, du$

31. $\displaystyle\int_0^{\pi/3} \cos x \, dx$

32. $\displaystyle\int \frac{1}{\sqrt{x}} (\sec \sqrt{x})^2 \, dx$

33. $\displaystyle\int w \cot(w^2 + 1) \csc(w^2 + 1)\, dw$

34. $\displaystyle\int_0^{\pi/3} \sec x \, dx$

35. $\displaystyle\int_0^{\pi/4} \tan x \sec x \, dx$

36. $\displaystyle\int_1^{2.5} \cos x \, dx$

37. $\displaystyle\int_{0.5}^{1.5} \cot x \, dx$

38. Find the slope of the cosine curve $(y = \cos x)$ at $x = \frac{\pi}{4}$.
39. Find the area under the sine curve $(y = \sin x)$ above the x axis
　　from $x = \frac{\pi}{4}$ to $x = \frac{3\pi}{4}$.

C　40. Convert 15° to radian measure.
　　41. Evaluate exactly without using a table or calculator:
　　　(A) $\sin \frac{3\pi}{2}$　(B) $\cos \frac{5\pi}{6}$　(C) $\sin(-\frac{\pi}{6})$

Find:

42. $D_x \, e^{\cos x^2}$

43. $D_u \sin u \tan u$

44. $\displaystyle\int e^{\sin x} \cos x \, dx$

45. $\displaystyle\int e^{\tan x} (\sec x)^2 \, dx$

46. $\displaystyle\int \sqrt[3]{\tan x} \, (\sec x)^2 \, dx$

47. $\displaystyle\int (\sec x)^5 \tan x \, dx$

48. $\displaystyle\int_{\pi/4}^{\pi/3} [(\sec x)^2 + (\csc x)^2] \, dx$

49. $\displaystyle\int_2^5 (5 + 2 \cos 2x) \, dx$

50. $\displaystyle\int \frac{\cot x \csc x}{2 + \csc x} \, dx$

PRACTICE TEST: CHAPTER 7

Find:

1. $D_x \sin(x^2 + x)$

2. $D_x \sin e^{x^2}$

3. $D_x e^x \cos x^2$

4. $D_x \sqrt[3]{(\sin x)^2}$

5. $D_x (\tan x)^2$

6. $D_x \sin x \sec x$

7. $D_x[\tfrac{1}{2}(\cot x)^2 + \ln|\sin x|]$

8. $\displaystyle\int (x - 1) \cos(x^2 - 2x)\,dx$

9. $\displaystyle\int e^{\cos x} \sin x\,dx$

10. $\displaystyle\int \frac{\cos x}{\sin x}\,dx$

11. $\displaystyle\int_{-\pi/2}^{\pi} \sin x\,dx$

12. $\displaystyle\int e^x \sin e^x\,dx$

13. $\displaystyle\int x \cot x^2\,dx$

14. $\displaystyle\int (\tan x)^{4/3} (\sec x)^2\,dx$

15. $\displaystyle\int_{\pi/2}^{2\pi/3} \csc x\,dx$

CHAPTER 8
DIFFERENTIAL EQUATIONS

CONTENTS

8 DIFFERENTIAL EQUATIONS

The use of differential equations represents one of the most important applications in mathematics. Many problems in business, economics, and the sciences can be stated in terms of differential equations. Unfortunately, there is no single method that will solve all the differential equations that arise—even in very simple applications. In this chapter we will study several different types of differential equations that are frequently encountered. Each section will introduce a method for solving one of these types of equations. A wide variety of applications are included to illustrate the importance of differential equations as a problem-solving tool.

8-1 SEPARATION OF VARIABLES

INTRODUCTION

A **differential equation** is an equation involving an unknown function, usually denoted by y, and one or more of its derivatives. For example,

$$y' = 2xy \tag{1}$$

is a differential equation. Since only the first derivative of the unknown function y appears in this equation, it is called a **first-order** differential equation. In general, the **order** of a differential equation is the highest derivative present in the equation.

Now consider the function

$$y = 4e^{x^2}$$

whose derivative is

$$y' = 4e^{x^2}2x$$

Substituting for y and y' in equation (1) gives

$$4e^{x^2}2x = 2x4e^{x^2}$$

which is certainly true for all values of x. This shows that the function

$y = 4e^{x^2}$ is a **solution** of differential equation (1). But this function is not the only solution. In fact, if C is any constant, substituting $y = Ce^{x^2}$ and $y' = 2xCe^{x^2}$ in equation (1) yields the identity

$$Ce^{x^2}2x = 2xCe^{x^2}$$

It turns out that all solutions of $y' = 2xy$ can be obtained from $y = Ce^{x^2}$ by assigning C appropriate values; hence, $y = Ce^{x^2}$ is called the **general solution** of equation (1). The function $y = 4e^{x^2}$, obtained by letting $C = 4$ in the general solution, is called a **particular solution** of the equation.

METHOD OF SEPARATION OF VARIABLES

How was the general solution of equation (1) found? It was found by using a method called **separation of variables.** We will now go through the steps that lead to the general solution. We begin by writing the equation in the form

$$\frac{dy}{dx} = 2xy$$

Treating dy and dx as differentials and multiplying both sides by dx/y yields

$$\frac{dy}{y} = 2x\,dx$$

Thus, we have "separated" the variables. Now we find the indefinite integral of both sides:

$$\int \frac{dy}{y} = \int 2x\,dx$$
$$\ln|y| = x^2 + A$$

where A is an arbitrary constant. Converting this last expression to exponential form, we have

$$|y| = e^{x^2 + A} = e^{x^2}e^A = Ke^{x^2}$$

where $K = e^A$ is any positive constant. Removing the absolute value sign, we have

$$y = \pm Ke^{x^2} = Ce^{x^2} \qquad \text{where} \quad C = \pm K$$

At this point, $C = \pm K$ could be any *nonzero* constant. If $C = 0$, then $y = 0$ for all values of x. Since the constant function $y = 0$ obviously satisfies the original differential equation, we can say that the function $y = Ce^{x^2}$ is a solution for any constant C.

If we specify that the function y must satisfy the **initial condition** $y(0) = 4$, then substituting $x = 0$ and $y = 4$ in the general solution gives

$$4 = Ce^0 = C$$

Thus, the function $y = 4e^{x^2}$ is the particular solution satisfying the initial condition $y(0) = 4$.

In summary, if the variables can be separated in a differential equation, then we may be able to solve the differential equation by integration.

Example 1 Find the general solution of $y' = y^2$ and then find the particular solution that satisfies $y(0) = \frac{1}{2}$.

Solution **Step 1.** *Find the general solution.*

$$\frac{dy}{dx} = y^2$$

$$\frac{dy}{y^2} = dx \qquad \text{Multiply by } dx/y^2 \text{ to separate the variables}$$

$$\int \frac{dy}{y^2} = \int dx \qquad \text{Find indefinite integrals for both sides}$$

$$-\frac{1}{y} = x + C$$

$$y = -\frac{1}{x + C} \qquad \text{Solve for } y \text{ to find the general solution}$$

Check: $\qquad y' = y^2$

$$\left(\frac{-1}{x + C}\right)' \overset{?}{=} \left(-\frac{1}{x + C}\right)^2 \qquad \begin{array}{l} \text{Substitute } y = -1/(x + C) \text{ into} \\ y' = y^2 \text{ to obtain an identity} \end{array}$$

$$\frac{1}{(x + C)^2} = \frac{1}{(x + C)^2} \qquad \text{An identity}$$

This verifies that our general solution is correct. (You should develop the habit of checking the solution of each differential equation that you solve, as we have done here. From now on, we will leave it up to you to check most of the examples worked in the text.)

Step 2. *Find the particular solution that meets the initial condition.*
We are given that $y = \frac{1}{2}$ when $x = 0$:

$$\frac{1}{2} = -\frac{1}{0 + C} \qquad \begin{array}{l} \text{Substitute initial conditions in} \\ \text{general solution} \end{array}$$

$$C = -2 \qquad \text{Solve for } C$$

$$y = -\frac{1}{x + (-2)} \qquad \text{Substitute for } C \text{ in general solution}$$

$$y = \frac{1}{2 - x} \qquad \begin{array}{l} \text{Simplify to obtain the particular} \\ \text{solution} \end{array}$$

Problem 1   Find the general solution of $y' = y^2/x$ and then find the particular solution that satisfies $y(1) = 1$. Check your answer.

Answer   General solution: $y = -\dfrac{1}{\ln|x| + C}$

Particular solution: $y = \dfrac{1}{1 - \ln|x|}$

APPLICATIONS   In Section 5-5 we discussed the unlimited growth law related to

$$\frac{dy}{dt} = ky \qquad k > 0$$

the decay law related to

$$\frac{dy}{dt} = -ky \qquad k > 0$$

and the limited growth law related to

$$\frac{dy}{dt} = k(M - y)$$

In all three cases, the equations are solved by separating the variables. We will now look at some additional applications that lead to equations of this type.

Example 2   A certain brand of mothballs evaporate, losing half their volume every 4 weeks. If the volume of each mothball is initially 15 cubic centimeters and a mothball becomes ineffective when its volume reaches 1 cubic centimeter, how long will these mothballs be effective?

Solution   The volume of each mothball is decaying at a rate proportional to its volume. If V is the volume of a mothball after t weeks, then

$$\frac{dV}{dt} = kV$$

Since the initial volume is 15 cubic centimeters, we know that $V(0) = 15$. After 4 weeks, the volume will be half the original volume, so $V(4) = 7.5$. Summarizing these requirements, we have the following mathematical model:

$$\frac{dV}{dt} = kV$$

$$V(0) = 15$$

$$V(4) = 7.5$$

We want to determine the value of t that satisfies the equation $V(t) = 1$. First, we use separation of variables to find the general solution of the differential equation:

$$\frac{dV}{dt} = kV$$

$$\frac{dV}{V} = k\,dt$$

$$\ln V = kt + C$$ We can write $\ln V$ and not $\ln|V|$, since $V > 0$

$$V = e^{kt+C} = e^{C}e^{kt} = Ae^{kt}$$ General solution

Now we use the initial condition to determine the value of the constant A:

$$V(0) = Ae^0 = A = 15$$
$$V(t) = 15e^{kt}$$

Next, we apply the condition $V(4) = 7.5$ to determine the constant k:

$$V(4) = 15e^{4k} = 7.5$$

$$e^{4k} = \frac{7.5}{15} = 0.5$$

$$4k = \ln 0.5$$

$$k = \frac{\ln 0.5}{4} \approx -0.1733$$

$$V(t) = 15e^{-0.1733t}$$ Particular solution

Notice that k is negative. This is to be expected in any exponential decay problem. Finally, to find how long the mothballs will be effective, we find t when $V = 1$:

$$V(t) = 1$$
$$15e^{-0.1733t} = 1$$
$$e^{-0.1733t} = \frac{1}{15}$$
$$-0.1733t = \ln \frac{1}{15}$$
$$t = -\frac{\ln \frac{1}{15}}{0.1733} \approx 15.6 \text{ weeks}$$

Problem 2   Repeat Example 2 if the mothballs lose half their volume every 5 weeks.

Answer   19.5 weeks

Example 3   The annual sales of a new company are expected to grow at a rate proportional to the difference between the sales and an upper limit of

$20 million. The sales are 0 initially and $4 million for the second year of operation.

(A) What will the sales be during the tenth year?

(B) In what year will the sales be $15 million?

Solutions

If S is the annual sales in millions of dollars during year t, then the model for this problem is

$$\frac{dS}{dt} = k(20 - S) \qquad (2)$$

$$S(0) = 0$$

$$S(2) = 4$$

For part A we want to find $S(10)$, and for part B we want to solve $S(t) = 15$ for t. First, separating the variables in (2) and integrating both sides, we obtain

$$\frac{dS}{20 - S} = k\,dt$$

$$-\ln(20 - S) = kt + C \qquad \text{We can write } -\ln(20 - S) \text{ in place}$$
$$\text{of } -\ln|20 - S|, \text{ since } 0 < S < 20$$

$$\ln(20 - S) = -kt - C$$

$$20 - S = e^{-kt - C} = e^{-C}e^{-kt} = Ae^{-kt} \qquad A = e^{-C}$$

$$S = 20 - Ae^{-kt} \qquad \text{General solution}$$

Now we use the conditions $S(0) = 0$ and $S(2) = 4$ to determine the constants A and k:

$$S(0) = 20 - Ae^0 = 20 - A = 0$$

$$A = 20$$

$$S(t) = 20 - 20e^{-kt}$$

$$S(2) = 20 - 20e^{-2k} = 4$$

$$20e^{-2k} = 16$$

$$e^{-2k} = \frac{16}{20} = 0.8$$

$$-2k = \ln 0.8$$

$$k = -\frac{\ln 0.8}{2} \approx 0.1116$$

$$S(t) = 20 - 20e^{-0.1116t} \qquad \text{Particular solution}$$

(Compare this with the limited growth equation in Section 5-5.)

(A) $S(10) = 20 - 20e^{-0.1116(10)} \approx \$13.45$ million

(B) $S(t) = 20 - 20e^{-0.1116t} = 15$

$$20e^{-0.1116t} = 5$$

$$e^{-0.1116t} = \frac{5}{20} = 0.25$$

$$-0.1116t = \ln 0.25$$

$$t = -\frac{\ln 0.25}{0.1116} \approx 12.42 \text{ years}$$

The annual sales will exceed \$15 million in the thirteenth year.

Problem 3   Repeat Example 3 if the sales during the second year are \$3 million.

Answers   (A)   \$11.13 million   (B)   Seventeenth year ($t \approx 17.05$ years)

EXERCISE 8-1

A   *Find the general solution for each differential equation. Then find the particular solution satisfying the initial condition.*

1. $y' = 1, \quad y(0) = 2$

2. $y' = 2x, \quad y(0) = 4$

3. $y' = \frac{1}{x}, \quad y(1) = -2$

4. $y' = 3x^{1/2}, \quad y(0) = 7$

5. $y' = y, \quad y(0) = 10$

6. $y' = y - 10, \quad y(0) = 15$

7. $y' = 25 - y, \quad y(0) = 5$

8. $y' = 3x^2y, \quad y(0) = \frac{1}{2}$

9. $y' = \frac{y}{x}, \quad y(1) = 5, \quad x > 0$

10. $y' = \frac{y}{x^2}, \quad y(-1) = 2e$

B   *Find the general solution for each differential equation. Then find the particular solution satisfying the initial condition.*

11. $y' = \frac{1}{y^2}, \quad y(1) = 3$

12. $y' = \frac{x^2}{y^2}, \quad y(0) = 2$

13. $y' = ye^x, \quad y(0) = 3e$

14. $y' = -y^2e^x, \quad y(0) = \frac{1}{2}$

15. $y' = \frac{e^x}{e^y}, \quad y(0) = \ln 2$

16. $y' = y^2(2x + 1), \quad y(0) = -\frac{1}{5}$

17. $y' = xy + x, \quad y(0) = 2$

18. $y' = (2x + 4)(y - 3), \quad y(0) = 2$

19. $y' = (2 - y)^2e^x, \quad y(0) = 1$

20. $y' = \frac{x^{1/2}}{(y - 5)^2}, \quad y(1) = 7$

Find the general solution for each differential equation. Do not attempt to find an explicit expression for the solution.

21. $y' = \dfrac{1 + x^2}{1 + y^2}$

22. $y' = \dfrac{6x - 9x^2}{2y + 4}$

23. $xyy' = (1 + x^2)(1 + y^2)$

24. $(xy^2 - x)y' = x^2y - x^2$

25. $x^2e^yy' = x^3 + x^3e^y$

26. $y' = \dfrac{xe^x}{\ln y}$

C   Find an explicit expression for the particular solution for each differential equation.

27. $xyy' = \ln x, \quad y(1) = 1$

28. $y' = \dfrac{xe^{x^2}}{y}, \quad y(0) = 2$

29. $xy' = x\sqrt{y} + 2\sqrt{y}, \quad y(1) = 4$

30. $y' = x(x - 1)^{1/2}(y - 1)^{1/2}, \quad y(1) = 1$

31. $yy' = xe^{-y^2}, \quad y(0) = 0$

32. $yy' = x(1 + y^2), \quad y(0) = 1$

APPLICATIONS

BUSINESS & ECONOMICS

33. *Interest.* If $5,000 is invested at 12% compounded continuously, how much will be in the account at the end of 10 years?

34. *Interest.* If a sum of money is invested at 9% compounded continuously, how long will it take for the sum to double?

35. *Advertising.* A company is using radio advertising to introduce a new product to a community of 100,000 people. Suppose the rate at which people learn about the new product is proportional to the number who have not yet heard of it. If no one is aware of the product at the start of the advertising campaign and after 7 days 20,000 people are aware of the product, how long will it take for 50,000 people to become aware of the product?

36. *Advertising.* Prior to the beginning of an advertising campaign, 10% of the potential consumers of a product are aware of the product. After the first week of the campaign, 20% of the consumers are aware of the product. If the percentage of informed consumers is growing at a rate proportional to the product of the percentage of informed consumers and the percentage of uninformed consumers, what percentage of consumers will be aware of

the product after 5 weeks of advertising? Note that you will need the algebraic identity

$$\frac{1}{y(1-y)} = \frac{1}{y} + \frac{1}{1-y}$$

to perform the necessary integration.

37. *Product analysis.* A company wishes to analyze a new room deodorizer. The active ingredient evaporates at a rate proportional to the amount present. Half of the ingredient evaporates in the first 30 days after the deodorizer is installed. If the deodorizer becomes ineffective after 90% of the active ingredient has evaporated, how long will one of these deodorizers remain effective?

38. *Natural resources.* According to the United States Department of Agriculture Forest Service, the total consumption of wood in the United States was 11.6 billion cubic feet in 1960 and 13.4 billion cubic feet in 1970. If the consumption is increasing at a rate proportional to the total consumption, how much wood will be consumed in the year 2000?

39. *Sales growth.* The annual sales of a new company are expected to grow at a rate proportional to the difference between the sales and an upper limit of $5 million. If the sales are 0 initially and $1 million during the fourth year of operation, when will the sales reach $4 million?

40. *Sales growth—Gompertz growth law.* When a new owner took over a company, the annual sales were $2 million. After 1 year, the sales reached $3 million. Suppose the amount of sales y satisfies the Gompertz growth law,

$$\frac{dy}{dt} = kye^{-0.4t}$$

(A) When will the sales reach $5 million?
(B) What is the long-range sales limit?

LIFE SCIENCES

41. *Population growth.* A culture of bacteria is growing at a rate proportional to the number present. The culture initially contains 100 bacteria. After 1 hour there are 140 bacteria in the culture.
(A) How many bacteria will be present after 5 hours?
(B) When will the culture contain 1,000 bacteria?

42. *Population growth.* A culture of bacteria is growing in a medium that can support a maximum of 1,100 bacteria. The rate of change

of the number of bacteria is proportional to the product of the number present and the difference between 1,100 and the number present. The culture initially contains 100 bacteria. After 1 hour there are 140 bacteria.

(A) How many bacteria are present after 5 hours?

(B) When will the culture contain 1,000 bacteria?

Note that you will need the algebraic identity

$$\frac{1}{y(1,100 - y)} = \frac{1}{1,100}\left(\frac{1}{y} + \frac{1}{1,100 - y}\right)$$

to perform the necessary integration.

43. *Simple epidemic.* An influenza epidemic has spread throughout a community of 50,000 people at a rate proportional to the product of the number of people who have been infected and the number who have not been infected. If 100 individuals were infected initially and 500 were infected 10 days later:

(A) How many people will be infected after 20 days?

(B) When will half the community be infected?

Note that you will need the algebraic identity

$$\frac{1}{y(50,000 - y)} = \frac{1}{50,000}\left(\frac{1}{y} + \frac{1}{50,000 - y}\right)$$

to perform the necessary integration.

44. *Epidemic–Gompertz growth law.* An influenza epidemic has spread throughout a community of 50,000 people. Let $y(t)$ be the number of people infected t days after the epidemic began, and assume that $y(t)$ satisfies the Gompertz growth law,

$$\frac{dy}{dt} = kye^{-0.03t}$$

If 100 individuals were infected initially and 500 were infected 10 days later:

(A) How many people will be infected after 20 days?

(B) When will half the community be infected?

SOCIAL SCIENCES

45. *Sensory perception.* A person is subjected to a physical stimulus that has a measurable magnitude, but the intensity of the resulting sensation is difficult to measure. If s is the magnitude of the stimulus and $I(s)$ is the intensity of sensation, experimental evidence suggests that

$$\frac{dI}{ds} = k\frac{I}{s}$$

for some constant k. Express I as a function of s.

46. *Learning.* The number of words per minute, N, a person can type
    increases with practice. Suppose the rate of change of N is propor-
    tional to the difference between N and an upper limit of 140. It is
    reasonable to assume that a beginner cannot type at all. Thus, N = 0
    when t = 0. If a person can type 35 words per minute after 10 hours
    of practice:
    (A)   How many words per minute can that individual type after
          20 hours of practice?
    (B)   How many hours must that individual practice to be able to
          type 105 words per minute?

47. *Rumor spread.* A rumor spreads through a population of 1,000
    people at a rate proportional to the product of the number that have
    heard it and the number that have not heard it. If five people
    initiated a rumor and ten people had heard it after 1 day:
    (A)   How many people will have heard the rumor after 7 days?
    (B)   How long will it take for 850 people to hear the rumor?
    Note that you will need the algebraic identity

    $$\frac{1}{y(1,000 - y)} = \frac{1}{1,000}\left(\frac{1}{y} + \frac{1}{1,000 - y}\right)$$

    to perform the necessary integration.

48. *Rumor spread—Gompertz growth law.* The rate of propagation of
    a rumor in a group of 1,000 individuals satisfies the Gompertz
    growth law,

    $$\frac{dy}{dt} = kye^{-(73/520)t}$$

    where y represents the number of people who have heard the rumor
    after t days. If five people initiate a rumor and ten people have
    heard it after 1 day:
    (A)   How many people will have heard the rumor after 7 days?
    (B)   How long will it take for 850 people to hear the rumor?

8-2   EXACT EQUATIONS

METHOD OF   The differential equation
EXACT EQUATIONS

$$\frac{dy}{dx} = \frac{2x - y}{x} \tag{1}$$

cannot be solved by separating the variables. (Try it!) Instead, we will
multiply both sides of the equation by x dx to obtain

$$x \, dy = (2x - y) \, dx$$

which can also be written as

$$(2x - y) \, dx - x \, dy = 0 \tag{2}$$

When a differential equation is written in this form, it resembles the differential of a function $F(x, y)$ and is called a **differential form** (see Section 6-3). We shall use the fact that the differential of F is

$$dF = \frac{\partial F}{\partial x} dx + \frac{\partial F}{\partial y} dy$$

to solve this equation. Consider the function

$$F(x, y) = x^2 - xy$$

with differential

$$dF = (2x - y) \, dx - x \, dy$$

From (2), we have

$$dF = 0$$

Since the only function of two variables that has a differential equal to zero for all values of x and y is a constant function, F must satisfy

$$F(x, y) = C$$

Thus, we must have

$$x^2 - xy = C \qquad \text{Since } F(x, y) = x^2 - xy$$

where C is an arbitrary constant. Solving for y determines the general solution of equation (1):

$$y = \frac{x^2 - C}{x}$$

In general, a differential form

$$P(x, y) \, dx + Q(x, y) \, dy$$

is **exact** if there is a function $F(x, y)$ that satisfies

$$dF = P(x, y) \, dx + Q(x, y) \, dy$$

that is, if

$$\frac{\partial F}{\partial x} = P \qquad \text{and} \qquad \frac{\partial F}{\partial y} = Q$$

If such a function exists, then the solution of the equation

$$P(x, y)\, dx + Q(x, y)\, dy = 0$$

is defined implicitly by the relation

$$F(x, y) = C$$

In order to use these ideas to solve differential equations, we must be able to do two things:

1. Recognize that a differential form is exact (most are not).
2. Find the function $F(x, y)$ if the differential form is exact.

The test given in the box shows how to recognize an exact differential form.

TEST FOR EXACTNESS

The differential form

$$P(x, y)\, dx + Q(x, y)\, dy$$

is exact if

$$\frac{\partial P}{\partial y} = \frac{\partial Q}{\partial x}$$

The following example shows how to find the function $F(x, y)$ when it exists.

Example 4   Solve $(2xy^2 + 3x^2)\, dx + (2x^2y + 1)\, dy = 0$.

Solution $P(x, y) = 2xy^2 + 3x^2$ and $Q(x, y) = 2x^2y + 1$

Since

$$\frac{\partial P}{\partial y} = 4xy = \frac{\partial Q}{\partial x}$$

the original differential form is exact. Thus, there must be a function $F(x, y)$ such that

$$dF = (2xy^2 + 3x^2)\, dx + (2x^2y + 1)\, dy$$

or, equivalently,

$$\frac{\partial F}{\partial x} = 2xy^2 + 3x^2 \quad \text{and} \quad \frac{\partial F}{\partial y} = 2x^2y + 1 \tag{3}$$

Integrating the first of these two equations with respect to x gives

$$F(x, y) = \int (2xy^2 + 3x^2)\, dx = x^2y^2 + x^3 + g(y) \tag{4}$$

Notice that we have added a *function* g(y) instead of a *constant of integration*. (This step is essential!) Since

$$\frac{\partial}{\partial x} g(y) = 0$$

for any function g(y), F will satisfy the first equation in (3) for any choice of g(y). However, F must also satisfy the second equation in (3). This will determine a specific function g(y). Computing $\partial F / \partial y$ using (4) and substituting the result in the second equation in (3), we obtain

$$\frac{\partial F}{\partial y} = 2x^2y + g'(y) = 2x^2y + 1$$

$$g'(y) = 1$$

This implies that $g(y) = y + C_1$, where C_1 is an ordinary constant of integration. The most general function F satisfying

$$dF = (2xy^2 + 3x^3)\,dx + (2x^2y + 1)\,dy$$

results from substituting $g(y) = y + C_1$ into (4) to obtain

$$F(x, y) = x^2y^2 + x^3 + y + C_1$$

Thus, the solution of the original differential equation is defined implicitly by

$$F(x, y) = x^2y^2 + x^3 + y + C_1 = C_2$$

or

$$x^2y^2 + x^3 + y = C$$

where $C = C_2 - C_1$ is a constant. Usually, we will not try to solve for y explicitly.

Problem 4   Solve $(3x^2y^3 + 1)\,dx + (3x^3y^2 + 2y)\,dy = 0$.

Answer $x^3y^3 + x + y^2 = C$

Example 5   Solve $ye^x\,dx + (e^x + y^{-1})\,dy = 0$, and then find the particular solution that satisfies $y(0) = 1$.

Solution   Step 1. *Test for exactness.*

$$P(x, y) = ye^x \quad \text{and} \quad Q(x, y) = e^x + y^{-1}$$

$$\frac{\partial P}{\partial y} = e^x = \frac{\partial Q}{\partial x} \qquad \begin{matrix} P(x, y)\,dx + Q(x, y)\,dy \text{ is exact (hence,} \\ F(x, y) \text{ exists)} \end{matrix}$$

Step 2. *Find F(x, y) and the general solution.*

$$\frac{\partial F}{\partial x} = P(x, y) = ye^x$$

$F(x, y) = ye^x + g(y)$	Integrate with respect to x
$\frac{\partial F}{\partial y} = e^x + g'(y)$	Differentiate $F(x, y)$ with respect to y
$e^x + g'(y) = e^x + y^{-1}$	Equate $\partial F/\partial y$ with $Q(x, y)$
$g'(y) = y^{-1}$	Solve for $g'(y)$
$g(y) = \ln|y| + C_1$	Solve for $g(y)$
$F(x, y) = ye^x + \ln|y| + C_1$	Substitute g in $F(x, y)$ above
$ye^x + \ln|y| = C$	General solution

Step 3. *Find the particular solution.*

$1e^0 + \ln|1| = C$	Substitue initial values
$1 + 0 = C$	$(y = 1$ when $x = 0)$ to find C
$C = 1$	
$ye^x + \ln|y| = 1$	Particular solution

Problem 5   Find the solution of $(2x - 2y^2) dx + (4y^3 - 4xy) dy = 0$ that satisfies $y(0) = 2$.

Answer $x^2 - 2xy^2 + y^4 = 16$

APPLICATIONS   Let us consider an example in which we want to find the equation of a curve given its slope function.

Example 6   The slope of a curve at any point (x, y) is given by

$$\frac{dy}{dx} = \frac{x - y}{x + y}$$

If the curve passes through the point $(2, 4)$, find the equation of the curve.

Solution   Rewriting the equation as $(y - x) dx + (x + y) dy = 0$, we see that it is exact, since

$$\frac{\partial P}{\partial y} = 1 = \frac{\partial Q}{\partial x}$$

Then, since $F(x, y)$ exists, we can proceed as in Example 4:

$$\frac{\partial F}{\partial x} = y - x$$

$$F(x, y) = xy - \frac{1}{2}x^2 + g(y)$$

$$\frac{\partial F}{\partial y} = x + g'(y) = x + y$$

$$g'(y) = y$$

$$g(y) = \frac{1}{2}y^2 + C_1$$

Thus,

$$F(x, y) = xy - \frac{1}{2}x^2 + \frac{1}{2}y^2 + C_1$$

and the general solution is given implicitly by

$$xy - \frac{1}{2}x^2 + \frac{1}{2}y^2 - C$$

We now find C so that the graph passes through the point $(2, 4)$:

$$8 - 2 + 8 = 14 = C$$

The equation of the curve is

$$xy - \frac{1}{2}x^2 + \frac{1}{2}y^2 = 14$$

or

$$2xy - x^2 + y^2 = 28$$

Problem 6   The slope of a curve at any point is given by

$$\frac{dy}{dx} = \frac{y}{y - x}$$

If the curve passes through the point $(3, 2)$, find its equation.

Answer $2xy - y^2 = 8$

In economics the **elasticity of a cost function** is defined to be the marginal cost divided by the average cost. Thus, if $y = C(x)$ is a cost function, the elasticity function is

$$E(x) = \frac{C'(x)}{C(x)/x} = \frac{dy/dx}{y/x} = \frac{x}{y}\frac{dy}{dx}$$

Example 7   Find the cost function $y = C(x)$ if the elasticity function is given by

$$E(x) = \frac{6x - y}{2y - 3x}$$

and $C(5) = 20$.

Solution   The cost function is the solution of the differential equation

$$E(x) = \frac{x}{y}\frac{dy}{dx} = \frac{6x - y}{2y - 3x}$$

Multiplying both sides by $(2y - 3x)y\,dx$ and regrouping, we can express this differential equation as a differential form:

$$(2y - 3x)x\,dy = (6x - y)y\,dx$$
$$(y^2 - 6xy)\,dx + (2xy - 3x^2)\,dy = 0$$

Now we must test for exactness:

$$\frac{\partial}{\partial y}(y^2 - 6xy) = 2y - 6x = \frac{\partial}{\partial x}(2xy - 3x^2)$$

The test indicates that $F(x, y)$ exists, so we can proceed as before:

$$\frac{\partial F}{\partial x} = y^2 - 6xy$$

$$F(x, y) = xy^2 - 3x^2y + g(y)$$

$$\frac{\partial F}{\partial y} = 2xy - 3x^2 + g'(y) = 2xy - 3x^2$$

$$g'(y) = 0$$

$$g(y) = C_1$$

$$F(x, y) = xy^2 - 3x^2y + C_1$$

$$xy^2 - 3x^2y = K \qquad \text{General solution (implicitly defined);}$$
$$\text{note that we use } K, \text{ since } C \text{ is used above}$$
$$\text{for } C(x)$$

We now find the particular solution that satisfies $C(5) = 20$; that is, $y = 20$ when $x = 5$:

$$5(20)^2 - 3(5)^2 20 = 2{,}000 - 1{,}500 = 500 = K$$
$$xy^2 - 3x^2y = 500 \qquad \text{Particular solution}$$
$$\text{(implicitly defined)} \tag{5}$$

Equation (5) provides an implicit definition of the cost function $y = C(x)$. What do we do if we want an explicit representation of the cost function? Since (5) is a quadratic equation in y, we can use the quadratic formula to solve for y. First, we rewrite the equation:

$$xy^2 - 3x^2y - 500 = 0$$

Next, we apply the quadratic formula with $a = x$, $b = -3x^2$, and $c = -500$ (see Section A-7):

$$y = \frac{-(-3x^2) \pm \sqrt{(-3x^2)^2 - 4(x)(-500)}}{2x}$$

$$= \frac{3x^2 \pm \sqrt{9x^4 + 2{,}000x}}{2x}$$

Now, we must decide which sign to use in front of the square root. Since we know that $C(5) = 20$, we will evaluate y at $x = 5$:

$$y = \frac{75 \pm \sqrt{15{,}625}}{10} = \frac{75 \pm 125}{10}$$

We see that we will obtain the value 20 for y if we use the plus sign in front of the square root. Thus,

$$y = C(x) = \frac{3x^2 + \sqrt{9x^4 + 2{,}000x}}{2x}$$

is the desired cost function.

Problem 7   Find the cost function if

$$E(x) = \frac{3x^2 - y}{2y - x^2}$$

and $C(5) = 30$.

Answer $C(x) = \dfrac{x^3 + \sqrt{x^6 + 3{,}000x}}{2x}$

EXERCISE 8-2

A   *Find the general solution, and when an initial condition is given, find the particular solution. Leave your answer in implicit form unless an explicit form is obvious.*

1. $y\,dx + x\,dy = 0$ 2. $2\,dx + 5\,dy = 0$
3. $2x\,dx + 2y\,dy = 0$ 4. $(2x + y)\,dx + (x + 2y)\,dy = 0$
5. $(2 - 3y)\,dx + (4 - 3x)\,dy = 0$

6. $\dfrac{1}{y}\,dx - \dfrac{x}{y^2}\,dy = 0$

7. $2xy^2\,dx + 2x^2y\,dy = 0$, $y(1) = 2$
8. $(3 + 4y)\,dx + (4x - 2)\,dy = 0$, $y(0) = -3$

9. $\dfrac{xy^2}{\sqrt{x^2 + 9}}\,dx + 2y\sqrt{x^2 + 9}\,dy = 0$, $y(4) = 2$

10. $\dfrac{y^2}{(1-x)^2}\,dx + \dfrac{2y}{1-x}\,dy = 0, \quad y(-1) = 4$

B   *Find the general solution, and when an initial condition is given, find the particular solution. Leave your answer in implicit form unless an explicit form is obvious.*

11. $(3x^2 + 2xy^2)\,dx + (2x^2y + 4y^3)\,dy = 0$

12. $(2x - 3x^2y^3)\,dx + (2y - 3x^3y^2)\,dy = 0$

13. $\left(2x + \dfrac{1}{x+y}\right)dx + \left(3y^2 + \dfrac{1}{x+y}\right)dy = 0$

14. $[2x(1+xy)^3 + 3x^2y(1+xy)^2]\,dx + 3x^3(1+xy)^2\,dy = 0$

15. $(xye^x + ye^x)\,dx + (xe^x + e^y)\,dy = 0$

16. $(1 + \ln x + \ln y)\,dx + \dfrac{x}{y}\,dy = 0$

17. $(e^y + ye^x)\,dx + (xe^y + e^x)\,dy = 0, \quad y(1) = 1$

18. $\left(1 + \dfrac{2x}{x^2+y^2}\right)dx + \left(1 + \dfrac{2y}{x^2+y^2}\right)dy = 0, \quad y(1) = 0$

19. $\dfrac{(1+y)^2}{(2-x)^2}\,dx + \dfrac{2(1+y)}{(2-x)}\,dy = 0, \quad y(1) = 3$

20. $(e^{xy} + xye^{xy})\,dx + x^2e^{xy}\,dy = 0, \quad y(4) = 0$

Find the equation for the curve that has the indicated slope and that passes through the indicated point.

21. $\dfrac{dy}{dx} = \dfrac{y-x}{2y-x}, \quad (2,1)$

22. $\dfrac{dy}{dx} = \dfrac{3y+4x}{2y-3x}, \quad (1,3)$

23. $\dfrac{dy}{dx} = \dfrac{y}{y^2-x}, \quad (1,1)$

24. $\dfrac{dy}{dx} = \dfrac{x^2-y}{x+y^2}, \quad (3,3)$

25. $\dfrac{dy}{dx} = -\dfrac{2xy+y^2}{2xy+x^2}, \quad (2,1)$

26. $\dfrac{dy}{dx} = \dfrac{y-2xy^2}{2x^2y-x}, \quad (2,1)$

C   *Problems 27–32 are mixed. Some are separable, some are exact, and some are both. Solve each by the most convenient method. Express each solution in explicit form.*

27. $y' = \dfrac{1-2y}{2x+3}, \quad y(0) = 1$

28. $y' = \dfrac{y}{1+x}, \quad y(0) = 4$

29. $y' = \dfrac{x-y}{x+1}, \quad y(2) = 1$

30. $y' = \dfrac{1-y^2}{xy}, \quad y(1) = 2$

31. $y' = \dfrac{1-x^2}{2x^2y}, \quad y(1) = 2$

32. $y' = \dfrac{1-y^2}{2xy+1}, \quad y(1) = 2$

APPLICATIONS

BUSINESS & ECONOMICS

33. *Marginal revenue.* If the marginal revenue $y' = R'(x)$ (in dollars) satisfies

$$\frac{dy}{dx} = \frac{2x - y}{x + 10}$$

where x is the number of units sold and $R(0) = 0$, find the revenue received from the sale of 100 units.

34. *Advertising.* The amount x (in thousands of dollars) that a company spends on advertising a product and the profit $y = P(x)$ (in thousands of dollars) that results from the advertising expenditures satisfy the equation

$$\frac{dy}{dx} = \frac{y - 10}{y - 10 - x}$$

If spending $1,000 on advertising results in a profit of $6,000, what will the profit be if $5,000 is spent on advertising?

35. *Demand.* The price p (in dollars) of a commodity and the number of units x demanded by consumers at that price are related by the equation

$$\frac{dp}{dx} = - \frac{p}{p + x}$$

If $p = 2$ when $x = 1$, find p when $x = 5$.

36. *Elasticity.* Find an explicit representation for the cost function $y = C(x)$ if the elasticity function is

$$E(x) = \frac{y - 4x}{2x - 2y}$$

and $C(1) = 10$.

37. *Elasticity.* Find an explicit relationship for the cost function $y = C(x)$ if the elasticity function is

$$E(x) = \frac{6x^2 - y}{2y - 2x^2}$$

and $C(2) = 10$.

8-3   INTEGRATING FACTORS

SOLUTION USING AN INTEGRATING FACTOR

Consider the differential equation

$$y' + \frac{2}{x}y = x$$

This equation cannot be solved by either of the two previous methods. (Try them to convince yourself that this is true.) Instead, we will change the form of the equation by multiplying both sides by x^2:

$$x^2y' + 2xy = x^3$$

How was x^2 chosen? We will discuss that below. Let us first see how this choice leads to a solution of the problem.

Recall that the product rule for differentiation can be written as

$$uv' + u'v = (uv)'$$

Notice the similarity between the left-hand sides of the last two equations. In fact, if we equate u with x^2 and v with y, then the two expressions are identical:

$$x^2y' + 2xy = uv' + u'v = (uv)' = (x^2y)'$$

Thus, making use of the product rule, we can write the differential equation as

$$(x^2y)' = x^3$$

Now we can integrate both sides:

$$\int (x^2y)' \, dx = \int x^3 \, dx$$

$$x^2y = \frac{x^4}{4} + C$$

Solving for y, we obtain the general solution

$$y = \frac{x^2}{4} + \frac{C}{x^2}$$

The function x^2 which we used to transform the original equation into one that we could solve as illustrated is called an **integrating factor.** In certain cases, there is an explicit formula for determining the integrating factor for a differential equation. We will consider only one type of equation that can be solved this way.

> **FIRST-ORDER LINEAR DIFFERENTIAL EQUATIONS**
>
> An equation that can be written in the form
> $$y' + p(x)y = q(x)$$
> is a **first-order linear differential equation.** The function
> $$I(x) = e^{\int p(x)\,dx}$$
> is an **integrating factor** for this equation.

Using the formula in the box, we can compute the integrating factor $I(x) = x^2$ that was used to help us solve

$$y' + \frac{2}{x}y = x$$

at the beginning of this section (since the equation is a first-order linear equation):

$$I(x) = e^{\int (2/x)\,dx} = e^{2\ln|x|} = e^{\ln|x|^2} = |x^2| = x^2$$

The identity $e^{\ln f(x)} = f(x)$ will be used frequently in this section. Since $\ln f(x)$ is only defined when $f(x) > 0$, we will assume that the domain of f has been restricted so that this is true.

Example 8   Solve $2xy' + y = 10x^2$ where $y(1) = 4$.

Solution

$y' + \dfrac{1}{2x}y = 5x$	Multiply both sides by $1/2x$ to obtain the form $y' + p(x)y = q(x)$
$I(x) = e^{\int (1/2x)\,dx}$	Compute the integrating factor
$\quad = e^{(1/2)\ln x}$	
$\quad = e^{\ln x^{1/2}}$	Property of logs: $r\ln t = \ln t^r$
$\quad = x^{1/2}$	Property of logs: $e^{\ln r} = r, r > 0$
$x^{1/2}y' + \dfrac{1}{2}x^{-1/2}y = 5x^{3/2}$	Multiply both sides by the integrating factor $x^{1/2}$
$(x^{1/2}y)' = 5x^{3/2}$	Write the left side as the derivative of a product
$x^{1/2}y = 2x^{5/2} + C$	Integrate both sides to obtain the general solution
$y = 2x^2 + Cx^{-1/2}$	Multiply both sides by $x^{-1/2}$ to find y
$4 = 2 + C$	Substitute initial condition $y = 4$ when $x = 1$
$C = 2$	Determine C
$y = 2x^2 + 2x^{-1/2}$	Substitute for C to obtain the particular solution

Problem 8   Solve $xy' + 4y = x^5$ where $y(1) = 1$.

Answer $I(x) = x^4$, $y = \frac{8}{9}x^{-4} + \frac{1}{9}x^5$

APPLICATIONS   Let us now consider applications that make use of the discussion above.

Example 9   Earlier, we studied population growth governed by the exponential growth law $dy/dt = ky$. External changes, such as food supply, can modify this growth law. Suppose that the population of a certain species grows according to the growth law

$$\frac{dy}{dt} = 0.1y + 0.1t$$

where y is measured in thousands and t is measured in years. If the current population is 100,000, what will the population be in 2 years? In 10 years?

Solution $\dfrac{dy}{dt} - 0.1y = 0.1t$

$I(t) = e^{\int -0.1\, dt} = e^{-0.1t}$ Integrating factor

$e^{-0.1t}\dfrac{dy}{dt} - 0.1e^{-0.1t}y = 0.1te^{-0.1t}$

$\dfrac{d}{dt}(e^{-0.1t}y) = 0.1te^{-0.1t}$

$e^{-0.1t}y = -te^{-0.1t} - 10e^{-0.1t} + C$ Use integration by parts on the right

$y = -t - 10 + Ce^{0.1t}$

This is the general solution. Applying the initial conditions, $y = 100$ when $t = 0$, gives

$100 = -10 + C$
$C = 110$

Thus, the particular solution is

$y = -t - 10 + 110e^{0.1t}$

To find the population in 2 years we must evaluate $y(2)$:

$y(2) = -2 - 10 + 110e^{(0.1)2} \approx 122.35$

Since y is measured in thousands, the population after 2 years is approximately 122,350. In the same manner, $y(10) \approx 279.01$, so the population after 10 years is approximately 279,010.

Problem 9 The growth equation for a certain species is

$$\frac{dy}{dt} = 0.1(100 - y) - 2e^{0.1t}$$

where y is measured in thousands and t is measured in years. If the current population is 20,000, what will the population be in 10 years? In 20 years?

Answer $y = 100 - 10e^{0.1t} - 70e^{-0.1}t$; the population after 10 years is approximately 47,100; the population after 20 years is approximately 16,600

Example 10 In economics, the supply S and the demand D for a commodity can often be considered as functions of both the price, $p(t)$, and the rate of change of the price, $p'(t)$. (Thus, S and D are ultimately functions of time t.) The **equilibrium price*** $p_e(t)$ is the solution of the equation $S = D$. For example, if

$$D = 50 - 2p(t) + 2p'(t)$$
$$S = 20 + 4p(t) + 5p'(t)$$

and $p(0) = 15$, then the equilibrium price is the solution of the equation

$$50 - 2p(t) + 2p'(t) = 20 + 4p(t) + 5p'(t)$$

This simplifies to

$$p'(t) + 2p(t) = 10$$

which is a first-order linear equation with integrating factor

$$I(t) = e^{\int 2\,dt} = e^{2t}$$

Proceeding as before,

$$e^{2t}p'(t) + 2e^{2t}p(t) = 10e^{2t}$$
$$[e^{2t}p(t)]' = 10e^{2t}$$
$$e^{2t}p(t) = 5e^{2t} + C$$
$$p(t) = 5 + Ce^{-2t}$$
$$15 = p(0) = 5 + C$$
$$C = 10$$
$$p_e(t) = 5 + 10e^{-2t} \qquad \text{Equilibrium price at time } t$$

Problem 10 If $D = 70 + 2p(t) + 2p'(t)$, $S = 30 + 6p(t) + 3p'(t)$, and $p(0) = 25$, find the equilibrium price as a function of time t.

Answer $p_e(t) = 10 + 15e^{-4t}$

*The e in $p_e(t)$ does not represent a partial derivative; $p_e(t)$ represents "equilibrium price at time t."

EXERCISE 8-3

A Find the integrating factor $I(x)$ for each equation, and then find the general solution. Find a particular solution when an initial condition is given. Assume $x > 0$ when the natural logarithm function is involved.

1. $y' + 2y = 4$
2. $y' - 3y = 3$, $y(0) = -1$
3. $y' + y = e^{-2x}$
4. $y' - 2y = e^{3x}$, $y(0) = 2$
5. $y' - y = 2e^x$
6. $y' + 4y = 3e^{-4x}$, $y(0) = 5$
7. $y' + y = 9x^2 e^{-x}$
8. $y' - 3y = 6\sqrt{x}\, e^{3x}$, $y(0) = -2$
9. $y' + \dfrac{1}{x}y = 2$
10. $y' - \dfrac{3}{x}y = 4$, $y(1) = 1$
11. $y' + \dfrac{2}{x}y = 10x^2$
12. $y' - \dfrac{1}{x}y = \dfrac{9}{x^3}$, $y(2) = 3$

B Find the integrating factor $I(x)$ for each equation, and then find the general solution. Assume $x > 0$ when the natural logarithm function is involved.

13. $y' + xy = 5x$
14. $y' - 2xy = 6x$
15. $y' - 2y = 4x$
16. $y' + y = x^2$
17. $y' + \dfrac{1}{x}y = e^x$
18. $y' + \dfrac{2}{x}y = e^{3x}$
19. $y' + \dfrac{1}{x}y = \ln x$
20. $y' - \dfrac{1}{x}y = \ln x$

C Problems 21–26 are mixed. Each one can be solved by one or more of the three methods we have discussed. Find the general solution.

21. $y' = \dfrac{1 - y}{x}$
22. $y' = \dfrac{y + 2}{x + 1}$
23. $y' = \dfrac{2x - y}{1 + x}$
24. $y' = \dfrac{3x^2 - 2xy}{1 + x^2}$
25. $y' = \dfrac{2x + 2xy}{1 + x^2}$
26. $y' = \dfrac{2x - y^3}{3y^2 + 3xy^2}$

APPLICATIONS

BUSINESS & ECONOMICS

27. *Sales growth.* The annual sales y (in millions of dollars) of a company satisfy the growth equation

$$\frac{dy}{dt} = 0.05(100 - y) + 0.2t$$

where t is time in years. If the sales are $60 million initially, find the annual sales in 10 years.

28. *Supply–demand.* The supply S and demand D for a certain commodity satisfy the equations

$$S = 70 - 3p + 2p' - 20e^{-3t} \quad \text{and} \quad D = 100 - 5p + p'$$

If $p(0) = 5$, find the equilibrium price $p_e(t)$. What is $\lim_{t \to \infty} p_e(t)$?

LIFE SCIENCES

29. *Pollution.* If $W(t)$ is the amount of natural waste (in parts per million) in a body of water, $D(t)$ is the rate of injection of natural pollutants into the body, and u is the waste composition coefficient for natural wastes, then W, D, and u are related by the equation

$$\frac{dW}{dt} + uW = D$$

Suppose $D(t) = e^{-t}$, $u = 1$, and $W(0) = 100$. Find $W(t)$.

30. *Population growth.* Normally, a culture of bacteria grows at a rate proportional to the size of the culture. If a scientist adds a drug to the culture that inhibits the growth of the bacteria, the modified growth rate is

$$\frac{dy}{dt} = 0.01y - 0.5$$

where y is the size of the culture at time t (in hours). If the initial size of the culture is 75, what is the size when $t = 100$ hours?

8-4   SECOND-ORDER DIFFERENTIAL EQUATIONS

HOMOGENEOUS
EQUATIONS

In this section we will study equations of the form

$$ay'' + by' + cy = 0 \tag{1}$$

where a, b, and c are constants. Equations of this type are called **second-order linear homogeneous equations with constant coefficients.** (If the 0 in equation (1) is replaced with a nonzero real constant, then the equation is called a **second-order linear nonhomogeneous equation.**)

We begin by considering the specific differential equation

$$y'' - 2y' - 3y = 0 \tag{2}$$

We will try to determine whether this second-order equation has any solutions of the form $y = Ce^{mx}$. Substituting $y = Ce^{mx}$, $y' = Cme^{mx}$, and $y'' = Cm^2 e^{mx}$ in equation (2) yields

$$Cm^2 e^{mx} - 2Cme^{mx} - 3Ce^{mx} = 0$$
$$Ce^{mx}(m^2 - 2m - 3) = 0$$

Remember that for $y = Ce^{mx}$ to be a solution to the differential equation, the last equation must be satisfied for *all* values of x. This will happen if $C = 0$ (which implies $y = 0$ for all x) or if

$$m^2 - 2m - 3 = 0 \tag{3}$$
$$(m - 3)(m + 1) = 0$$
$$m_1 = 3 \qquad m_2 = -1$$

Thus, $y = C_1 e^{3x}$ and $y = C_2 e^{-x}$ will both satisfy the original differential equation for any arbitrary constants C_1 and C_2. The general solution is obtained by combining these two solutions:

$$y = C_1 e^{3x} + C_2 e^{-x}$$

To verify that this is the solution, we compute y' and y'' and substitute in equation (2):

$$y' = 3C_1 e^{3x} - C_2 e^{-x}$$
$$y'' = 9C_1 e^{3x} + C_2 e^{-x}$$
$$y'' - 2y' - 3y = 9C_1 e^{3x} + C_2 e^{-x} - 2(3C_1 e^{3x} - C_2 e^{-x}) - 3(C_1 e^{3x} + C_2 e^{-x})$$
$$= (9 - 6 - 3)C_1 e^{3x} + (1 + 2 - 3)C_2 e^{-x}$$
$$= 0$$

The quadratic equation $m^2 - 2m - 3 = 0$ [equation (3) above] is called the **characteristic equation** for equation (2). Comparing equations (2) and (3), we see that the characteristic equation can be obtained from the original differential equation (2) by simply substituting m^2 for y'', m for y', and 1 for y. In general, it can be shown that for a second-order differential equation

$$ay'' + by' + cy = 0$$

its characteristic equation is given by

$$am^2 + bm + c = 0$$

Recognizing this enables us to write the characteristic equation directly without going through the above type of calculations.

Example 11   Solve $y'' - 16y = 0$.

Solution
$$m^2 - 16 = 0 \qquad \text{Characteristic equation}$$
$$(m - 4)(m + 4) = 0$$
$$m_1 = 4 \qquad m_2 = -4 \qquad \text{Roots of the characteristic equation}$$
$$y = C_1 e^{4x} + C_2 e^{-4x} \qquad \text{General solution}$$

Problem 11   Solve $y'' - y' - 2y = 0$.

Answer $y = C_1 e^{2x} + C_2 e^{-x}$

Since the general solution of a second-order equation involves two arbitrary constants, two conditions are required to determine a particular solution. If the value of y and the value of y' are both given for the same value of x, then both conditions are called initial conditions.

Example 12 Find the particular solution of $2y'' + 3y' - 2y = 0$ that satisfies the initial conditions $y(0) = 2$ and $y'(0) = -1$.

Solution

$2m^2 + 3m - 2 = 0$	Characteristic equation
$(2m - 1)(m + 2) = 0$	
$m_1 = \tfrac{1}{2} \qquad m_2 = -2$	Roots of characteristic equation
$y = C_1 e^{x/2} + C_2 e^{-2x}$	General solution
$y(0) = C_1 + C_2 = 2$	First initial condition
$y' = \tfrac{1}{2} C_1 e^{x/2} - 2C_2 e^{-2x}$	
$y'(0) = \tfrac{1}{2} C_1 - 2C_2 = -1$	Second initial condition

In order to determine the values of C_1 and C_2, we must solve the system of equations

$C_1 + C_2 = 2$	
$\tfrac{1}{2} C_1 - 2C_2 = -1$	
$C_1 = 2 - C_2$	Solve the first equation for C_1
$\tfrac{1}{2}(2 - C_2) - 2C_2 = -1$	Substitute for C_1 in the second
$C_2 = \tfrac{4}{5}$	equation and solve for C_2
$C_1 = 2 - \tfrac{4}{5}$	Substitute the value for C_2 and
$\quad = \tfrac{6}{5}$	solve for C_1

Thus, the particular solution satisfying the given initial conditions is

$$y = \tfrac{6}{5} e^{x/2} + \tfrac{4}{5} e^{-2x}$$

Problem 12 Find the particular solution of $3y'' - 7y' - 6y = 0$ that satisfies the initial conditions $y(0) = 1$ and $y'(0) = 2$.

Answer $y = \tfrac{8}{11} e^{3x} + \tfrac{3}{11} e^{-2x/3}$

Example 13 Solve $y'' - 2y' + y = 0$.

Solution

$m^2 - 2m + 1 = 0$	Characteristic equation
$(m - 1)(m - 1) = 0$	
$m = 1$	Single, repeated root of the characteristic equation

Since this characteristic equation has only one root, we conclude that $y = C_1 e^x$ is a solution. However, the general solution of a second-order equation must be the sum of two different functions involving two arbitrary constants. It can be shown that when the characteristic equation has a single root m, then the second function in the solution is $y = C_2 xe^{mx}$. Thus, for this example, the second function is $y = C_2 xe^x$ and the general solution is

$$y = C_1 e^x + C_2 xe^x$$

You should check that this function does satisfy the original differential equation.

Problem 13   Solve $y'' + 6y' + 9y = 0$.

Answer $y = C_1 e^{-3x} + C_2 xe^{-3x}$

We now know how to find the solution of a second-order equation if the characteristic equation has two distinct real roots or one repeated real root. There is a third possibility. If the discriminant $b^2 - 4ac$ of the characteristic equation is negative, the characteristic equation has two complex roots. In this case, the general solution involves both exponential functions and trigonometric functions. For completeness, the complex roots case is included in the box (also, see Problems 19–22 in Exercise 8-4).

SECOND-ORDER HOMOGENEOUS DIFFERENTIAL EQUATIONS
WITH CONSTANT COEFFICIENTS

Characteristic equation	*Differential equation*
$am^2 + bm + c = 0$	$ay'' + by' + cy = 0$

Solution	*Nature of roots*
$y = C_1 e^{m_1 x} + C_2 e^{m_2 x}$	Two real distinct roots, m_1 and m_2
$y = C_1 e^{mx} + C_2 xe^{mx}$	A single repeated root, m
$y = e^{ax}(C_1 \cos bx + C_2 \sin bx)$	Two distinct complex roots, $m_1 = a + bi$ and $m_2 = a - bi$

NONHOMOGENEOUS EQUATIONS

Several applications of second-order equations are included in the exercises. Our primary application will be to the solution of systems of differential equations in the next section. In one of the systems in the next section, we will have to solve a nonhomogeneous second-order differential equation (an equation of form (1) with 0 replaced by a nonzero real constant). The following example shows how to solve such an equation.

Example 14   Solve $y'' - 2y' - 8y = 16$.

Solution   Since we only know how to solve homogeneous equations, we will make a substitution that changes this equation into a homogeneous equation. Let

$$y = u - 2 \qquad \text{See the box below}$$

Then

$$y' = u'$$
$$y'' = u''$$

Substituting these expressions in the original equation produces the following homogeneous equation:

$$u'' - 2u' - 8(u - 2) = 16$$
$$u'' - 2u' - 8u + 16 = 16$$
$$u'' - 2u' - 8u = 0$$

$$m^2 - 2m - 8 = 0 \qquad \text{Characteristic equation}$$
$$(m - 4)(m + 2) = 0$$

$$m_1 = 4 \qquad m_2 = -2 \qquad \text{Roots of characteristic equation}$$

$$u = C_1 e^{4x} + C_2 e^{-2x} \qquad \text{Solution of homogeneous equation}$$
$$y = C_1 e^{4x} + C_2 e^{-2x} - 2 \qquad \text{Solution of nonhomogeneous equation}$$

How did we select the substitution $y = u - 2$ in Example 14? In general, we proceed as indicated in the box. (You should verify the results in the box.)

FROM NONHOMOGENEOUS TO HOMOGENEOUS

The nonhomogeneous equation

$$ay'' + by' + cy = d$$

can be transformed into a homogeneous equation

$$au'' + bu' + cu = 0$$

by the substitution

$$y = u + \frac{d}{c} \qquad c \neq 0$$

Problem 14 Solve $y'' + y' - 12y = 6$.

Answer $y = C_1 e^{-4x} + C_2 e^{3x} - \frac{1}{2}$

EXERCISE 8-4

A Find the general solution for each equation.

1. $y'' + 3y' + 2y = 0$
2. $y'' - 6y' + 8y = 0$
3. $y'' + 2y' - 15y = 0$
4. $y'' - 25y = 0$
5. $y'' + 6y' = 0$
6. $y'' - 3y' = 0$
7. $y'' - 4y' + 4y = 0$
8. $y'' + 10y' + 25y = 0$

B Find the particular solution for each equation that satisfies the initial conditions.

9. $y'' - y = 0$; $y(0) = 3$, $y'(0) = 1$
10. $y'' - y' - 2y = 0$; $y(0) = 1$, $y'(0) = 2$
11. $3y'' - 10y' + 3y = 0$; $y(0) = 1$, $y'(0) = -1$
12. $y'' - 4y' = 0$; $y(0) = 0$, $y'(0) = 3$
13. $y'' + 2y' + y = 0$; $y(0) = 2$, $y'(0) = 4$
14. $y'' - 2y = 0$; $y(0) = 0$, $y'(0) = 1$

C Find the general solution for each equation.

15. $y'' - 3y' - 4y = 12$
16. $y'' + 2y' + y = 5$

Find the particular solution that satisfies the initial conditions.

17. $y'' + y' - 2y = 6$; $y(0) = 0$, $y'(0) = 0$
18. $y'' - 3y' - 10y = 100$; $y(0) = -10$, $y'(0) = 0$

Problems $19 - 22$ are optional (trigonometric functions are involved). Find the general solution for each equation.

19. $y'' + y = 0$
20. $y'' + 4y = 0$
21. $y'' - 4y' + 13y = 0$
22. $y'' + 2y' + 3y = 0$

APPLICATIONS

BUSINESS & ECONOMICS

23. *Supply–demand.* In earlier exercises, supply and demand were considered as functions of the price, $p(t)$, and the rate of change of the price, $p'(t)$. In studying certain markets, economists include the second derivative $p''(t)$ in the differential equation to reflect whether

the rate of change of $p(t)$ is increasing or decreasing. Suppose that S and D satisfy the equations

$$S = 3 + 0.2p' - 0.05p - p'' \qquad \text{and} \qquad D = 2 + 0.8p' - 0.01p + p''$$

and that $p(0) = 75$ and $p'(0) = -15$. The equilibrium price $p_e(t)$ is the solution of the equation $S = D$. Find $p_e(t)$ and evaluate $\lim_{t \to \infty} p_e(t)$.

24. *Public debt.* According to the Domar burden-of-debt model, the total public debt $D(t)$ can be modeled by the equation

$$D''(t) - \beta D(t) = 0$$

where β is the constant relative growth rate of income ($0 < \beta < 1$).
(A) Find the general solution of this equation for any constant β.
(B) Find the particular solution satisfying $D(0) = 1$ and $D'(0) = -\sqrt{\beta}$.
(C) Find the limit of this particular solution as $t \to \infty$.

SOCIAL SCIENCES
25. *Learning theory.* The differential equation

$$y'' + 5y' + 4y = 8$$

is typical of the equations that occur in the study of learning curves of rats in certain types of psychological experiments. Find the particular solution satisfying $y(0) = 1$ and $y'(0) = 1$. Find the limit of this particular solution as $t \to \infty$.

8-5 SYSTEMS OF DIFFERENTIAL EQUATIONS

SYSTEMS
In many applications we are interested in the relationship between two quantities, both of which are changing with respect to time. This often leads to a system of differential equations of the type illustrated in Example 15.

Example 15 Solve:

$$\frac{dx}{dt} = x + y \tag{1}$$

$$\frac{dy}{dt} = 4x - 2y \tag{2}$$

Solution Equations (1) and (2) form a **first-order linear system of differential equations.** Systems of this form can be solved by eliminating one of the variables, in the same way systems of linear algebraic equations are solved. We start by differentiating equation (1) to obtain

$$\frac{d^2x}{dt^2} = \frac{dx}{dt} + \frac{dy}{dt}$$

Solve for dy/dt

$$\frac{dy}{dt} = \frac{d^2x}{dt^2} - \frac{dx}{dt}$$

Substitute for dy/dt in equation (2)

$$\frac{d^2x}{dt^2} - \frac{dx}{dt} = 4x - 2y$$

Now, y must be eliminated; solve equation (1) for y

$$y = \frac{dx}{dt} - x$$

Substitute for y in the last equation

$$\frac{d^2x}{dt^2} - \frac{dx}{dt} = 4x - 2\left(\frac{dx}{dt} - x\right)$$

Simplify and combine like terms to obtain a second-order homogeneous equation

$$\frac{d^2x}{dt^2} + \frac{dx}{dt} - 6x = 0$$

Thus, we have completely eliminated y and dy/dt, so we now have a second-order differential equation involving x alone. The characteristic equation is $m^2 + m - 6 = 0$, which has roots $m_1 = -3$ and $m_2 = 2$. The general solution is

$$x = C_1e^{-3t} + C_2e^{2t}$$

To determine y, we must first compute dx/dt:

$$\frac{dx}{dt} = -3C_1e^{-3t} + 2C_2e^{2t}$$

Substituting for dx/dt and x in the equation $y = (dx/dt) - x$ above gives

$$y = -3C_1e^{-3t} + 2C_2e^{2t} - C_1e^{-3t} - C_2e^{2t}$$
$$= -4C_1e^{-3t} + C_2e^{2t} \quad .$$

Thus, the solution to this system is

$$x = C_1e^{-3t} + C_2e^{2t} \quad \text{and} \quad y = -4C_1e^{-3t} + C_2e^{2t}$$

You should check this solution.

Problem 15   Solve: $\dfrac{dx}{dt} = x + y$

$$\frac{dy}{dt} = 3x - y$$

Answer $x = C_1e^{2t} + C_2e^{-2t} \quad \text{and} \quad y = C_1e^{2t} - 3C_2e^{-2t}$

Now, suppose that we want to find a particular solution of the system in Example 15. Since there are two arbitrary constants in the general solution, this will require two initial conditions, one for x and one for y.

Example 16   Find the particular solution of the system

$$\frac{dx}{dt} = x + y$$

$$\frac{dy}{dt} = 4x - 2y$$

that satisfies the initial conditions $x(0) = 3$ and $y(0) = -2$. (See Example 15 above for the general solution.)

Solution   Substituting $x = 3$, $y = -2$, and $t = 0$ in the general solution for Example 15,

$$x = C_1 e^{-3t} + C_2 e^{2t}$$
$$y = -4C_1 e^{-3t} + C_2 e^{2t}$$

we obtain the following system of equations:

$$C_1 + C_2 = 3$$
$$-4C_1 + C_2 = -2$$

The solution to this system is $C_1 = 1$ and $C_2 = 2$. Thus, the particular solution we are seeking is

$$x = e^{-3t} + 2e^{2t} \qquad \text{and} \qquad y = -4e^{-3t} + 2e^{2t}$$

Problem 16   Find the particular solution of the system

$$\frac{dx}{dt} = x + y$$

$$\frac{dy}{dt} = 3x - y$$

that satisfies the initial conditions $x(0) = 5$ and $y(0) = -3$. (See Problem 15 above for the general solution.)

Answer $x = 3e^{2t} + 2e^{-2t} \qquad \text{and} \qquad y = 3e^{2t} - 6e^{-2t}$

APPLICATION: INTERRELATED MARKETS   In studying the relationship between the prices of two commodities in an interrelated market, economists often assume that the rate of change of each price is proportional to a linear combination of both prices. This leads to a system of first-order linear differential equations.

Example 17 The prices p and q of two commodities in an interrelated market satisfy the following system of differential equations, where p and q are functions of time t:

$$p' = \quad p + 2q - 300 \qquad p(0) = 150$$
$$q' = -4p - 5q + 900 \qquad q(0) = \quad 25$$

Find the solution to this system and analyze the long-term behavior of p and q; that is, the behavior of p and q as t gets larger and larger.

Solution
$$p'' = p' + 2q' \qquad \qquad \text{Differentiate the first equation}$$
$$\qquad \qquad \qquad \qquad \text{with respect to } t$$
$$p'' = p' + 2(-4p - 5q + 900) \quad \text{Use the second equation to}$$
$$\quad = p' - 8p - 10q + 1{,}800 \quad \text{eliminate } q'$$
$$q = \tfrac{1}{2}p' - \tfrac{1}{2}p + 150 \qquad \qquad \text{Solve the first equation for } q \qquad (3)$$

$$p'' = p' - 8p - 10(\tfrac{1}{2}p' - \tfrac{1}{2}p + 150) + 1{,}800 \quad \text{Eliminate}$$
$$\qquad = -4p' - 3p + 300 \qquad \qquad \qquad \qquad q \text{ in the}$$
$$p'' + 4p' + 3p = 300 \qquad \qquad \qquad \qquad \text{equation}$$
$$\qquad \qquad \qquad \qquad \qquad \qquad \qquad \qquad \text{for } p''$$

We have eliminated q and q', but the resulting equation is nonhomogeneous. Using the method illustrated in Example 13, we find (details omitted) the general solution of this nonhomogeneous equation to be

$$p = C_1 e^{-t} + C_2 e^{-3t} + 100$$

Substituting p and p' in (3) determines q:

$$q = -C_1 e^{-t} - 2C_2 e^{-3t} + 100$$

Applying the initial conditions $p = 150$ and $q = 25$ when $t = 0$ produces the following system of equations:

$$C_1 + \quad C_2 = 50$$
$$C_1 + 2C_2 = 75$$

The solution to this system is $C_1 = 25$ and $C_2 = 25$. Thus,

$$p = 25e^{-t} + 25e^{-3t} + 100$$
$$q = -25e^{-t} - 50e^{-3t} + 100$$

To determine the behavior of p and q for large values of t, first note that

$$p' = -25e^{-t} - 75e^{-3t} < 0 \qquad \text{for all } t$$

and

$$q' = 25e^{-t} + 150e^{-3t} > 0 \qquad \text{for all } t$$

Thus, p is always decreasing and q is always increasing. Furthermore,

$$\lim_{t \to \infty} p = \lim_{t \to \infty} (25e^{-t} + 25e^{-3t} + 100)$$
$$= 0 + 0 + 100 = 100$$

and

$$\lim_{t \to \infty} q = \lim_{t \to \infty} (-25e^{-t} - 50e^{-3t} + 100)$$
$$= 0 + 0 + 100 = 100$$

The graph of $y = p(t)$ is always falling and approaches the line $y = 100$ from above, while the graph of $y = q(t)$ is always rising and approaches the line $y = 100$ from below (see the figure).

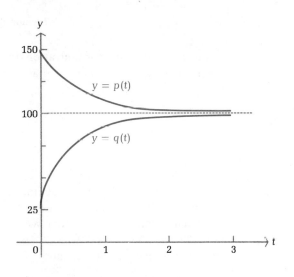

Problem 17   Repeat Example 16 for the system

$$p' = -3p + q + 225 \qquad p(0) = 150$$
$$q' = 2p - 4q + 100 \qquad q(0) = 35$$

Answer $p = 20e^{-2t} + 30e^{-5t} + 100$ and $q = 20e^{-2t} - 60e^{-5t} + 75$

APPLICATION:
COMPARTMENT
ANALYSIS

In studying organs in the body, scientists often must deal with the relationship between different organs or different parts of the same organ. Each organ or part of an organ is considered to be a compartment. A given substance, such as a drug, may be able to enter or leave a compartment at certain rates and may be able to pass back and forth between adjacent compartments.

Example 18   Suppose an organ has two compartments separated by a membrane, and assume that a drug injected into compartment 1 can move back and forth between the compartments and can also leave the organ from compartment 2 (see the figure). If x is the drug concentration in compartment

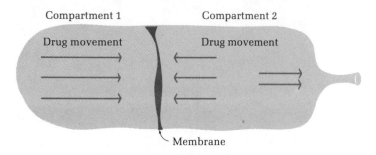

Compartment 1 — Drug movement

Compartment 2 — Drug movement

Membrane

1 and y is the drug concentration in compartment 2 at any time t, experimental evidence suggests that x and y satisfy the following system of differential equations:

$$\frac{dx}{dt} = y - 6x$$

$$\frac{dy}{dt} = 6x - 7y$$

where x and y are functions of time t in hours. If there was no drug present in either compartment prior to the injection and if 40 units are injected into compartment 1, then we have the initial conditions $x(0) = 40$ and $y(0) = 0$. Find the solution of the system, and find the amount of the drug present in each compartment after 6 minutes.

Solution   Proceeding as before (details omitted), we arrive at the second-order equation

$$\frac{d^2x}{dt^2} + 13\frac{dx}{dt} + 36x = 0$$

which has the solution

$$x = C_1 e^{-9t} + C_2 e^{-4t}$$

This gives

$$y = -3C_1 e^{-9t} + 2C_2 e^{-4t}$$

Applying the initial conditions produces the following system of equations:

$$C_1 + C_2 = 40$$
$$-3C_1 + 2C_2 = 0$$

which has the solution $C_1 = 16$ and $C_2 = 24$. Thus, the amount of the drug present in each compartment at time t is

$$x = \quad 16e^{-9t} + 24e^{-4t} \qquad \text{Compartment 1}$$
$$y = -48e^{-9t} + 48e^{-4t} \qquad \text{Compartment 2}$$

If t is measured in hours, then after 6 minutes, or 0.1 hour, the amount of the drug present in each compartment is

$$x(0.1) = \quad 16e^{-0.9} + 24e^{-0.4} \approx 22.6 \text{ units}$$
$$y(0.1) = -48e^{-0.9} + 48e^{-0.4} \approx 12.7 \text{ units}$$

Of the original 40 units injected into compartment 1, 22.6 units are still present, 12.7 units are now in compartment 2, and so, 4.7 units have left the organ completely.

Problem 18

If the drug concentrations x and y are governed by the system of equations

$$\frac{dx}{dt} = y - 2x$$

$$\frac{dy}{dt} = 2x - 3y$$

and the initial concentration is 30 units in compartment 1 and none in compartment 2, find the concentration in each compartment after 6 minutes. After 30 minutes. After 1 hour.

Answer

The solution to the system is $x = 10e^{-4t} + 20e^{-t}$ and $y = -20e^{-4t} + 20e^{-t}$. The concentrations are (approximately) 24.8 units in compartment 1 and 4.7 units in compartment 2 after 6 minutes, 13.5 units in compartment 1 and 9.4 units in compartment 2 after 30 minutes, 7.5 units in compartment 1 and 7.0 units in compartment 2 after 1 hour.

APPLICATION: GROWTH OF INTERACTING SPECIES

In Section 5-5 we saw that the population growth of many different species over short periods of time is often governed by the exponential growth law $dx/dt = kx$. If we consider the situation where two species compete for the same food supply, it is reasonable to assume that the rate of growth of each species will be affected by the size of the populations of both species. If x and y are the populations of the two species, then the following system of differential equations can be used as a model for this situation:

$$\frac{dx}{dt} = ax - by$$

$$\frac{dy}{dt} = -cx + dy$$

where a, b, c, and d are positive constants. Notice that the first equation indicates that the rate of growth of x increases as x increases but decreases as y increases. The term $-by$ introduces the competition between the two species. The second equation may be interpreted in a similar manner.

Example 19 The populations x and y of two species satisfy the system of differential equations

$$\frac{dx}{dt} = 0.125x - 0.05y$$

$$\frac{dy}{dt} = -0.05x + 0.05y$$

and the initial conditions $x(0) = 90$ and $y(0) = 130$. Find the solution to this system and analyze the long-term growth of each species.

Solution Eliminating y and dy/dt (details omitted) leads to the second-order differential equation

$$\frac{d^2x}{dt^2} - 0.175\frac{dx}{dt} + 0.00375x = 0$$

which has the solution

$$x = C_1 e^{0.15t} + C_2 e^{0.025t}$$

Substituting for dx/dt and x in the first equation and solving for y gives

$$y = -0.5C_1 e^{0.15t} + 2C_2 e^{0.025t}$$

Applying the initial conditions gives the system

$$\begin{aligned} C_1 + C_2 &= 90 \\ -\tfrac{1}{2}C_1 + 2C_2 &= 130 \end{aligned}$$

which has the solution $C_1 = 20$ and $C_2 = 70$. Thus,

$$x = 20e^{0.15t} + 70e^{0.025t} \qquad \text{and} \qquad y = -10e^{0.15t} + 140e^{0.025t}$$

For example, the population of each species after 10 years is

$$x(10) = 20e^{1.5} + 70e^{0.25} \approx 180$$
$$y(10) = -10e^{1.5} + 140e^{0.25} \approx 135$$

We see that x has increased from 90 to 180 thousand and y has increased from 130 to 135 thousand. Since x is the sum of the two increasing functions, the first species will continue to increase. On the other hand, y is the difference between two increasing functions. Is it possible that the second species will die out? That is, is y ever 0? To find out, we set the solution for y equal to 0 and try to solve for t:

$$y = -10e^{0.15t} + 140e^{0.025t} = 0$$

$$10e^{0.15}t = 140e^{0.025t}$$

$$\frac{e^{0.15t}}{e^{0.025t}} = \frac{140}{10}$$

$$e^{0.125t} = 14$$

$$0.125t = \ln 14$$

$$t = \frac{\ln 14}{0.125} \approx 21$$

This indicates that the second species will die out after 21 years, and since negative populations do not make any sense, the solution to this system should not be used for larger values of t.

Problem 19   Repeat Example 19 for the system

$$\frac{dx}{dt} = 0.04x - 0.02y$$

$$\frac{dy}{dt} = 0.01x + 0.03y$$

$$x(0) = 50 \qquad y(0) = 35$$

Answer $x(t) = 40e^{0.02t} + 10e^{0.05t}$ and $y(t) = 40e^{0.02t} - 5e^{0.05t}$; the first species grows without bound; the second species will die out after approximately 69 years

EXERCISE 8-5

A   *Find the general solution for each system of differential equations, and find the particular solution when initial conditions are given.*

1. $x' = -x + y, \quad y' = 2x$
2. $x' = 3x + y, \quad y' = -2x$
3. $x' = 2x - y, \quad y' = 3x - 2y; \quad x(0) = 1, \quad y(0) = -1$
4. $x' = 5x + 2y, \quad y' = 2x + 2y; \quad x(0) = 3, \quad y(0) = -1$
5. $x' = 2x + y, \quad y' = 2x + y; \quad x(0) = 2, \quad y(0) = -1$
6. $x' = x - y, \quad y' = x - y; \quad x(0) = 1, \quad y(0) = 2$

B   7. $x' = -2x + y, \quad y' = -3x + 2y$
8. $x' = 3x - y, \quad y' = 4x - y$
9. $x' = 5x - 3y + 2, \quad y' = 6x - 4y + 4$
10. $x' = y - 3, \quad y' = 4x - 16; \quad x(0) = 3, \quad y(0) = 1$

APPLICATIONS

BUSINESS & ECONOMICS

11. *Interrelated markets.* The prices p and q of two commodities in an interrelated market satisfy the following system of equations:

$$p' = -4p + q + 260$$
$$q' = -2p - q + 250$$
$$p(0) = 100 \qquad q(0) = 100$$

Find the solution to this system and analyze the long-term behavior of p and q.

12. *Interrelated markets.* Repeat Problem 11 above for the following system:

$$p' = -26p - 40q + 1{,}320$$
$$q' = 15p + 23q - \phantom{1{,}}760$$
$$p(0) = 33 \qquad q(0) = 12$$

LIFE SCIENCES

13. *Compartment analysis.* An organ has two compartments separated by a membrane. The drug concentration in the first compartment is represented by $x(t)$ and in the second compartment by $y(t)$ at any time t (measured in hours). Initially, there are no traces of a drug in either compartment. Fifty units of a drug are injected into compartment 1. The flow of the drug is governed by the equations

$$x' = y - 5x$$
$$y' = x - 5y$$

Find the concentration in each compartment after 6 minutes. After 30 minutes. After 1 hour.

14. *Population growth.* The populations x and y of two competing species satisfy the system of differential equations

$$\frac{dx}{dt} = 0.09x - 0.02y$$

$$\frac{dy}{dt} = -0.02x + 0.06y$$

$$x(0) = 200 \qquad y(0) = 150$$

Find the solution of this system and analyze the long-term growth of each species.

8-6   CHAPTER REVIEW

EXERCISE 8-6   CHAPTER REVIEW

Work through all the problems in this chapter review and check your answers in the back of the book. (Answers to all review problems are there.) Where weaknesses show up, review appropriate sections in the text. When you are satisfied that you know the material, take the practice test following this review.

A   *Find the general solution.*

1. $y' = -\dfrac{4y}{x}$

2. $y' = -\dfrac{4y}{x} + x$

3. $2x\,dx + 3y^2\,dy = 0$

4. $y' = 3x^2y^2$

5. $y' = 2y - e^x$

6. $(3y^3 - 2)\,dx + 3y^2(3x - 2)\,dy = 0$

7. $y' = \dfrac{5}{x}y + x^6$

8. $y' = \dfrac{3 + y}{2 + x}, \quad x > -2$

9. $y'' - 4y' - 21y = 0$

10. $y'' + 12y' + 36y = 0$

B   *Find the particular solution that satisfies the initial condition(s).*

11. $y' = 10 - y; \quad y(0) = 0$

12. $y' + y = x; \quad y(0) = 0$

13. $y' = 2ye^{-x}; \quad y(0) = 1$

14. $y' = \dfrac{2x - y}{x + 4}$; $y(0) = 1$

15. $y' = \dfrac{2x - y}{x + 4y}$; $y(1) = 1$

16. $y' + \dfrac{2}{x}y = \ln x$; $y(1) = 2$

17. $yy' = \dfrac{x(1 + y^2)}{1 + x^2}$; $y(0) = 1$

18. $y' = -\dfrac{y^2}{1 + 2xy}$; $y(1) = 1$

19. $y'' + 4y' = 0$; $y(0) = 1$, $y'(0) = -2$
20. $y'' - 16y = 16$; $y(0) = 1$, $y'(0) = 0$

C   *Find the particular solution that satisfies the initial condition(s).*

21. $x' = -2x + y$
     $y' = -4x + 2y$
     $x(0) = 1$, $y(0) = 1$

22. $x' = -2x + 2y + 6$
     $y' = 4x - 8$
     $x(0) = 2$, $y(0) = 2$

APPLICATIONS

BUSINESS & ECONOMICS

23. *Present value.*   A note will pay $100,000 at maturity in 20 years. How much should an investor be willing to pay for the note if money is worth 12% compounded continuously?

24. *Interrelated markets.*   The prices p and q of two commodities of an interrelated market satisfy the following system of equations:

$$p' = -2p + q + 125$$
$$q' = -2p - 5q + 575$$
$$p(0) = 50 \qquad q(0) = 150$$

Find the solution of this system and analyze the long-term behavior of *p and q.*

LIFE SCIENCES

25. *Crop yield.*   The yield per acre, y(t), of a corn crop satisfies the equation

$$\frac{dy}{dt} = 100 + e^{-t} - y$$

If $y(0) = 0$, find y at any time *t.*

26. *Population growth.* The populations x and y of the two competing species satisfy the following systems of differential equations:

$$\frac{dx}{dt} = 0.03x - 0.01y$$

$$\frac{dy}{dt} = -0.02x + 0.02y$$

$$x(0) = 75 \qquad y(0) = 75$$

Find the solution of this system. Analyze the long-term behavior of each species.

SOCIAL SCIENCES

27. *Rumor spread.* A single individual starts a rumor in a community of 200 people. The rumor spreads at a rate proportional to the number of people who have not yet heard the rumor. After 2 days, 10 people have heard the rumor.

(A) How many people will have heard the rumor after 5 days?

(B) How long will it take for the rumor to spread to 100 people?

PRACTICE TEST: **CHAPTER 8**

Find the general solution in Problems 1–3.

1. $y' = 4x^3y$ 2. $y' + \dfrac{3}{x}y = x$ 3. $y'' - 49y = 0$

Find the particular solution in Problems 4–6.

4. $y' - y = e^{2x}; \quad y(0) = 0$

5. $(2x - y)\,dx + (y - x)\,dy = 0; \quad y(0) = 2$

6. $y'' - 4y' + 3y = 0; \quad y(0) = 0, \quad y'(0) = -4$

7. Find the solution of the following system:

$$x' = -3x + y$$
$$y' = -2x$$
$$x(0) = 3 \qquad y(0) = 4$$

8. *Sales growth.* A new company has sales of $50,000 during the first year of operation. The rate of growth of the annual sales s is proportional to the difference between s and an upper limit of $200,000. Assuming $s = 0$ at $t = 0$, how long will it take for the annual sales to reach $150,000?

9. *Archaeology.* A piece of human bone discovered at an archaeological site contains 15% of the original amount of radioactive carbon-14. The half-life of carbon-14 is 5,730 years. How old is the bone?

10. *Supply–demand.* The supply S and demand D for a certain commodity satisfy the equations

$$S = 100 + p - 15e^{-t} \quad \text{and} \quad D = 200 - p' - p$$

If $p = 75$ when $t = 0$, find the equilibrium price p_e at any time t.

CHAPTER 9
TAYLOR POLYNOMIALS AND SERIES;
L'HÔPITAL'S RULE

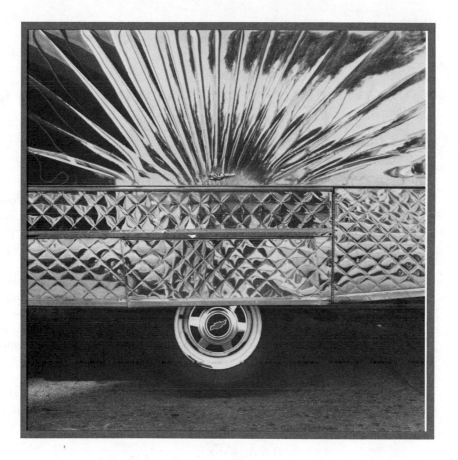

CONTENTS

9 TAYLOR POLYNOMIALS AND SERIES; L'HÔPITAL'S RULE

The circuits inside a hand-held electronic calculator are only capable of performing the basic operations of addition, subtraction, multiplication, and division. Yet, many calculators have keys that allow you to evaluate functions such as e^x, ln x, and sin x. How is this done? In most cases, the values of these functions are *approximated* by using a carefully selected polynomial. Of course, polynomials can be evaluated by using the basic arithmetic operations. Thus, the approximating polynomials give the calculator the capability of evaluating nonpolynomial functions. In this chapter we will develop several methods of determining these polynomials and we will examine some of their applications.

9-1 TAYLOR POLYNOMIALS AT 0

We begin by approximating the function $f(x) = e^x$ for values of x near 0 with a first-degree polynomial:

$$p_1(x) = a_0 + a_1 x \tag{1}$$

We want to place conditions on p_1 that will enable us to determine the unknown coefficients a_0 and a_1. Since we want to approximate f for values near 0, it is reasonable to require that f and p_1 agree at 0. Thus,

$$a_0 = p_1(0) = f(0) = e^0 = 1$$

This determines the value of a_0. To determine the value of a_1, we require that both functions have the same slope at 0. Since $p_1'(x) = a_1$ and $f'(x) = e^x$, this implies that

$$a_1 = p_1'(0) = f'(0) = e^0 = 1$$

Thus, after substituting $a_0 = 1$ and $a_1 = 1$ into (1), we obtain $p_1(x) = 1 + x$, which is a first-degree polynomial satisfying

$$p_1(0) = f(0) \qquad \text{and} \qquad p_1'(0) = f'(0)$$

414

Since we want to use $1 + x$ to approximate the values of e^x near $x = 0$, we compare these two functions in Table 1.

TABLE 1

x	$p_1(x) = 1 + x$	$f(x) = e^x$
0.000 1	1.000 1	1.000 1
0.001	1.001	1.001 000 5
0.01	1.01	1.010 050 2
0.1	1.1	1.105 170 9
1.0	2.0	2.718 281 8
10.0	11.0	22,026.466

The first two rows in Table 1 indicate that $1 + x$ is a good approximation to e^x for x very close to 0. The next three rows show that the accuracy of the approximation decreases as x gets larger. The last row indicates that $1 + x$ cannot be used at all to approximate e^x for large values of x. This is not surprising, since the graph of $1 + x$ is a straight line and the graph of e^x is certainly not straight.

Now we will try to improve this approximation by using a second-degree polynomial,

$$p_2(x) = a_0 + a_1x + a_2x^2 \tag{2}$$

We still require that $p_2(0) = f(0)$ and $p_2'(0) = f'(0)$. From the requirement $p_2(0) = f(0)$, we have

$$a_0 = p_2(0) = f(0) = e^0 = 1$$

and from the requirement $p_2'(0) = f'(0)$, since $p_2'(x) = a_1 + 2a_2x$, we have

$$a_1 = p_2'(0) = f'(0) = e^0 = 1$$

as before. The value of a_2 will be determined by requiring that

$$p_2''(0) = f''(0)$$

Now, $p_2''(x) = 2a_2$ and $f''(x) = e^x$. Thus,

$$2a_2 = p_2''(0) = f''(0) = e^0 = 1$$

and

$$a_2 = \tfrac{1}{2}$$

We now substitute $a_0 = 1$, $a_1 = 1$, and $a_2 = \tfrac{1}{2}$ into (2) to obtain

$$p_2(x) = 1 + x + \tfrac{1}{2}x^2$$

Now we compare the values of $1 + x$, $1 + x + \tfrac{1}{2}x^2$, and e^x (Table 2), and we see that $1 + x + \tfrac{1}{2}x^2$ is a better approximation to e^x than $1 + x$.

TABLE 2

x	$p_1(x) = 1 + x$	$p_2(x) = 1 + x + \frac{1}{2}x^2$	$f(x) = e^x$
0.000 1	1.000 1	1.000 1	1.000 1
0.001	1.001	1.001 000 5	1.001 000 5
0.01	1.01	1.010 05	1.010 050 2
0.1	1.1	1.105	1.105 170 9
1.0	2.0	2.5	2.718 281 8
10.0	11.0	61.0	22,026.466

It seems reasonable to assume that a third-degree polynomial would yield a still better approximation. If we adopt the convention that $f^{(0)}(x) = f(x)$ (that is, the **zeroth derivative of a function** is the function itself), then we can state the required conditions as:

$$\text{Find} \quad p_3(x) = a_0 + a_1 x + a_2 x^2 + a_3 x^3 \quad \text{satisfying} \tag{3}$$
$$p_3^{(k)}(0) = f^{(k)}(0) \qquad k = 0, 1, 2, 3$$

As before, we obtain the value of the additional coefficient a_3 by adding the requirement that

$$p_3^{(3)}(0) = f^{(3)}(0)$$

Thus,

$$
\begin{array}{l|l}
p_3(x) = a_0 + a_1 x + a_2 x^2 + a_3 x^3 & a_0 = p_3(0) = f(0) = e^0 = 1 \\
p_3'(x) = \quad\quad a_1 + 2a_2 x + 3a_3 x^2 & a_1 = p_3'(0) = f'(0) = e^0 = 1 \\
p_3''(x) = \quad\quad\quad\quad\quad 2a_2 + 6a_3 x & 2a_2 = p_3''(0) = f''(0) = e^0 = 1 \\
& a_2 = \frac{1}{2} \\
p_3^{(3)}(x) = \quad\quad\quad\quad\quad\quad\quad 6a_3 & 6a_3 = p_3^{(3)}(0) = f^{(3)}(0) = e^0 = 1 \\
& a_3 = \frac{1}{6}
\end{array}
$$

Thus, substituting the coefficients a_0, a_1, a_2, a_3 found in the second column into (3), we obtain

$$p_3(x) = 1 + x + \frac{1}{2}x^2 + \frac{1}{6}x^3$$

Table 3 shows that the approximations provided by p_3 are an improvement over those provided by p_2 and p_1.

TABLE 3

x	$p_1(x) = 1 + x$	$p_2(x) = 1 + x + \frac{1}{2}x^2$	$p_3(x) = 1 + x + \frac{1}{2}x^2 + \frac{1}{6}x^3$	$f(x) = e^x$
0.000 1	1.000 1	1.000 1	1.000 1	1.000 1
0.001	1.001	1.001 000 5	1.001 000 5	1.001 000 5
0.01	1.01	1.010 05	1.010 050 2	1.010 050 2
0.1	1.1	1.105	1.105 166 7	1.105 170 9
1.0	2.0	2.5	2.666 666 7	2.718 281 8
10.0	11.0	61.0	227.666 67	22,026.466

Obviously, this process can be continued. Given any positive integer n, we define

$$p_n(x) = a_0 + a_1x + a_2x^2 + \cdots + a_nx^n$$

and require that

$$p_n^{(k)}(0) = f^{(k)}(0) \qquad k = 0, 1, 2, \ldots, n$$

The polynomial p_n is called a *Taylor polynomial*. Before determining p_n for $f(x) = e^x$, it will be convenient to make some general statements concerning the relationship between a_k, $p_n^{(k)}(0)$, and $f^{(k)}(0)$ for an arbitrary function f. We differentiate $p_n(x)$ n times to obtain the following relationships:

$$
\begin{aligned}
p_n(x) &= a_0 + a_1x + a_2x^2 + a_3x^3 + \cdots + a_nx^n \\
p_n'(x) &= a_1 + 2a_2x + 3a_3x^2 + \cdots + na_nx^{n-1} \\
p_n''(x) &= 2a_2 + (3)(2)a_3x + \cdots + n(n-1)a_nx^{n-2} \\
p_n^{(3)}(x) &= (3)(2)a_3 + \cdots + n(n-1)(n-2)a_nx^{n-3} \\
&\vdots \\
p_n^{(n)}(x) &= n(n-1)(n-2)\cdots(1)a_n = n!a_n
\end{aligned}
$$

$$
\begin{aligned}
a_0 &= p_n(0) = f(0) \\
a_1 &= p_n'(0) = f'(0) \\
2a_2 &= p_n''(0) = f''(0) \\
(3)(2)a_3 &= p_n^{(3)}(0) = f^{(3)}(0) \\
&\vdots \\
n!a_n &= p_n^{(n)}(0) = f^{(n)}(0)
\end{aligned}
$$

Using the facts that $0! = 1$, $1! = 1$, and $k! = k(k-1)\cdots 1$ for k any positive integer, each of the equations on the right-hand side of the above can be rewritten as

$$a_k = \frac{p_n^{(k)}(0)}{k!} = \frac{f^{(k)}(0)}{k!}$$

This relationship enables us to state the general definition of a **Taylor polynomial.**

TAYLOR POLYNOMIAL AT 0

If f has n derivatives at 0, then the nth-degree Taylor polynomial for f at 0 is

$$p_n(x) = a_0 + a_1x + a_2x^2 + \cdots + a_nx^n$$

where

$$p_n^{(k)}(0) = f^{(k)}(0) \qquad \text{and} \qquad a_k = \frac{f^{(k)}(0)}{k!} \qquad k = 0, 1, 2, \ldots, n$$

This result can be stated in a form that is more readily remembered, as follows:

TAYLOR POLYNOMIAL AT 0: CONCISE FORM

The Taylor polynomial of degree n for f at 0 is

$$p_n(0) = f(0) + f'(0)x + \frac{f''(0)}{2!}x^2 + \cdots + \frac{f^{(n)}(0)}{n!}x^n$$

Returning to our original function $f(x) = e^x$, it is now an easy matter to find the nth degree Taylor polynomial for this function. Since $D_x e^x = e^x$, it follows that

$$f^{(k)}(x) = e^x \qquad f^{(k)}(0) = e^0 = 1 \qquad a_k = \frac{f^{(k)}(0)}{k!} = \frac{1}{k!}$$

for all values of k. Thus, for any n,

$$p_n(x) = 1 + x + \frac{1}{2!}x^2 + \frac{1}{3!}x^3 + \cdots + \frac{1}{n!}x^n$$

is the nth-degree Taylor polynomial for e^x.

Example 1 Find the third-degree Taylor polynomial at 0 for $f(x) = \sqrt{x + 4}$. Use p_3 to approximate $\sqrt{5}$.

Solution Step 1. *Find the derivatives.*

$$f(x) = (x + 4)^{1/2}$$
$$f'(x) = \tfrac{1}{2}(x + 4)^{-1/2}$$
$$f''(x) = -\tfrac{1}{4}(x + 4)^{-3/2}$$
$$f^{(3)}(x) = -\tfrac{3}{8}(x + 4)^{-5/2}$$

Step 2. *Evaluate the derivatives at 0.*

$$f(0) = 4^{1/2} = 2$$
$$f'(0) = \tfrac{1}{2}\,4^{-1/2} = \tfrac{1}{4}$$
$$f''(0) = -\tfrac{1}{4}\,4^{-3/2} = -\tfrac{1}{32}$$
$$f^{(3)}(0) = \tfrac{3}{8}\,4^{-5/2} = \tfrac{3}{256}$$

Step 3. *Find the coefficients of the Taylor polynomial.*

$$a_0 = \frac{f(0)}{0!} = f(0) = 2$$

$$a_1 = \frac{f'(0)}{1!} = f'(0) = \tfrac{1}{4}$$

$$a_2 = \frac{f''(0)}{2!} = \frac{-\tfrac{1}{32}}{2} = -\tfrac{1}{64}$$

$$a_3 = \frac{f^{(3)}(0)}{3!} = \frac{\tfrac{3}{256}}{6} = \tfrac{1}{512}$$

Step 4. *Write down the Taylor polynomial.*

$$p_3(x) = 2 + \tfrac{1}{4}x - \tfrac{1}{64}x^2 + \tfrac{1}{512}x^3 \qquad \text{Taylor polynomial}$$

Thus,

$$\sqrt{5} = f(1) \approx p_3(1) = 2 + \tfrac{1}{4} - \tfrac{1}{64} + \tfrac{1}{512} \approx 2.2363281$$

[*Note:* The value obtained by using a calculator is $\sqrt{5} = 2.236068$.]

Problem 1   Find the second-degree Taylor polynomial at 0 for $f(x) = \sqrt{x + 9}$. Use $p_2(x)$ to approximate $\sqrt{10}$.

Answer $p_2(x) = 3 + \tfrac{1}{6}x - \tfrac{1}{216}x^2, \quad \sqrt{10} \approx 3.162037$

Example 2   Find the nth-degree Taylor polynomial at 0 for $f(x) = e^{2x}$.

Solution   **Step 1**   **Step 2**   **Step 3**

$$f(x) = \ e^{2x} \qquad\qquad f(0) = 1 \qquad\qquad a_0 = f(0) \ = 1$$
$$f'(x) = 2e^{2x} \qquad\qquad f'(0) = 2 \qquad\qquad a_1 = f'(0) \ = 2$$

$$f''(x) = 2^2 e^{2x} \qquad\quad f''(0) = 2^2 \qquad\quad a_2 = \frac{f''(0)}{2!} = \frac{2^2}{2!}$$

$$f^{(3)}(x) = 2^3 e^{2x} \qquad\quad f^{(3)}(0) = 2^3 \qquad\quad a_3 = \frac{f^{(3)}(0)}{3!} = \frac{2^3}{3!}$$

$$\vdots \qquad\qquad\qquad \vdots \qquad\qquad\qquad \vdots$$

$$f^{(n)}(x) = 2^n e^{2x} \qquad f^{(n)}(0) = 2^n \qquad a_n = \frac{f^{(n)}(0)}{n!} = \frac{2^n}{n!}$$

Step 4. *Write down the Taylor polynomial.*

$$p_n(x) = 1 + 2x + \frac{2^2}{2!}x^2 + \frac{2^3}{3!}x^3 + \cdots + \frac{2^n}{n!}x^n$$

Problem 2   Find the nth-degree Taylor polynomial at 0 for $f(x) = e^{x/3}$.

Answer $p_n(x) = 1 + \dfrac{1}{3}x + \dfrac{1}{3^2}\dfrac{1}{2!}x^2 + \dfrac{1}{3^3}\dfrac{1}{3!}x^3 + \cdots + \dfrac{1}{3^n}\dfrac{1}{n!}x^n$

Example 3   Find the nth-degree Taylor polynomial at 0 for $f(x) = \ln(1 + x)$.

Solution   **Step 1.**   *Find the derivatives.*

$$f(x) = \ln(1 + x)$$
$$f'(x) = (1 + x)^{-1} = 0!(1 + x)^{-1}$$
$$f''(x) = (-1)(1 + x)^{-2} = -1!(1 + x)^{-2}$$
$$f^{(3)}(x) = (-2)(-1)(1 + x)^{-3} = 2!(1 + x)^{-3}$$
$$f^{(4)}(x) = (-3)(-2)(-1)(1 + x)^{-4} = -3!(1 + x)^{-4}$$
$$f^{(5)}(x) = (-4)(-3)(-2)(-1)(1 + x)^{-5} = 4!(1 + x)^{-5}$$
$$\vdots$$
$$f^{(n)}(x) = (-1)^{n-1}(n - 1)!(1 + x)^{-n}$$

Step 2.   *Evaluate the derivatives at 0.*

$$f(0) = \ln 1 = 0$$
$$f'(0) = 1$$
$$f''(0) = (-1)$$
$$f^{(3)}(0) = (-1)^2 2!$$
$$f^{(4)}(0) = (-1)^3 3!$$
$$f^{(5)}(0) = (-1)^4 4!$$
$$\vdots$$
$$f^{(n)}(0) = (-1)^{n-1}(n - 1)!$$

Step 3.   *Find the coefficients of the Taylor polynomial.*

$$a_0 = f(0) \quad = 0$$
$$a_1 = f'(0) \quad = 1$$
$$a_2 = \frac{f''(0)}{2!} = \frac{(-1)}{2!} = -\frac{1}{2}$$
$$a_3 = \frac{f^{(3)}(0)}{3!} = \frac{(-1)^2 2!}{3!} = \frac{(-1)^2}{3} = \frac{1}{3}$$
$$a_4 = \frac{f^{(4)}(0)}{4!} = \frac{(-1)^3 3!}{4!} = \frac{(-1)^3}{4} = -\frac{1}{4}$$
$$a_5 = \frac{f^{(5)}(0)}{5!} = \frac{(-1)^4 4!}{5!} = \frac{(-1)^4}{5} = \frac{1}{5}$$
$$\vdots$$
$$a_n = \frac{f^{(n)}(0)}{n!} = \frac{(-1)^{n-1}(n - 1)!}{n!} = \frac{(-1)^{n-1}}{n}$$

Step 4.   *Write down the Taylor polynomial.*

$$p_n(x) = x - \frac{1}{2}x^2 + \frac{1}{3}x^3 - \frac{1}{4}x^4 + \frac{1}{5}x^5 - \cdots + \frac{(-1)^{n-1}}{n}x^n$$

Problem 3   Find the nth-degree Taylor polynomial at 0 for $f(x) = \dfrac{1}{2 + x}$.

Answer $p_n(x) = \dfrac{1}{2} - \dfrac{1}{2^2}x + \dfrac{1}{2^3}x^2 - \dfrac{1}{2^4}x^3 + \cdots + \dfrac{(-1)^n}{2^{n+1}}x^n$

EXERCISE 9-1

A   *In Problems 1–10 find the indicated Taylor polynomial at 0.*

1. $f(x) = e^{-x}, \quad p_4(x)$
2. $f(x) = e^{4x}, \quad p_3(x)$
3. $f(x) = e^{-2x}, \quad p_3(x)$
4. $f(x) = \ln(1 - x), \quad p_4(x)$
5. $f(x) = \ln(1 + 2x), \quad p_3(x)$
6. $f(x) = \ln(1 + \frac{1}{2}x), \quad p_4(x)$
7. $f(x) = \sqrt{x + 16}, \quad p_2(x)$
8. $f(x) = \sqrt{(x + 4)^3}, \quad p_3(x)$
9. $f(x) = \sqrt[3]{x + 1}, \quad p_3(x)$
10. $f(x) = \sqrt[4]{x + 16}, \quad p_2(x)$
11. Use p_4 from Problem 1 and $x = 0.1$ to approximate $e^{-0.1}$.
12. Use p_3 from Problem 2 and $x = 0.01$ to approximate $e^{0.04}$.
13. Use p_3 from Problem 3 and $x = 0.25$ to approximate $e^{-0.5}$.
14. Use p_4 from Problem 4 and $x = 0.1$ to approximate $\ln(0.9)$.
15. Use p_3 from Problem 5 and $x = 0.1$ to approximate $\ln(1.2)$.
16. Use p_4 from Problem 6 and $x = -0.04$ to approximate $\ln(0.98)$.
17. Use p_2 from Problem 7 and $x = 1$ to approximate $\sqrt{17}$.
18. Use p_3 from Problem 8 and $x = 1$ to approximate $\sqrt{125}$.
19. Use p_3 from Problem 9 and $x = 0.03$ to approximate $\sqrt[3]{1.03}$.
20. Use p_2 from Problem 10 and $x = -0.5$ to approximate $\sqrt[4]{15.5}$.

B   *Find the nth-degree Taylor polynomial at 0.*

21. $f(x) = e^{2x}$
22. $f(x) = e^{-x}$

23. $f(x) = \dfrac{1}{1 - x}$
24. $f(x) = \ln(1 - x)$

25. $f(x) = \dfrac{2}{1 + 2x}$
26. $f(x) = \ln(1 + 2x)$

27. $f(x) = \ln(3 + 4x)$
28. $f(x) = \dfrac{x}{2 - x}$

29. $f(x) = \dfrac{1}{(1 - x)^2}$
30. $f(x) = \dfrac{1 + x}{1 - x}$

Find the indicated Taylor polynomial at 0.

31. $f(x) = xe^x, \quad p_4(x)$
32. $f(x) = x\ln(1 + x), \quad p_5(x)$
33. $f(x) = e^{x^2}, \quad p_4(x)$
34. $f(x) = \sqrt{1 + x^2}, \quad p_4(x)$

35. $f(x) = \dfrac{1}{1 + x^2}, \quad p_4(x)$
36. $f(x) = \ln(1 + e^x), \quad p_2(x)$

C   *In Problems 37–40 t is a constant. Find the second-degree Taylor polynomial at 0.*

37. $f(x) = (1 + x)^t$

38. $f(x) = \ln(1 + tx)$

39. $f(x) = e^{tx}$

40. $f(x) = t^x, \quad t > 0$

[*Hint:*   Use $t^x = e^{x \ln t}$ and Problem 39.]

41.   Let $f(x) = (x + 1)^4$.

(A)   Find p_0, p_1, p_2, p_3, p_4, and p_5 at 0.

(B)   What is the relationship between f and p_4? Between p_4 and p_5? Between p_4 and p_n for any $n > 4$?

42.   If $f(x) = \ln(1 + x)$, then the first three Taylor polynomials are (see Example 3)

$$p_1(x) = x$$
$$p_2(x) = x - \tfrac{1}{2}x^2$$
$$p_3(x) = x - \tfrac{1}{2}x^2 + \tfrac{1}{3}x^3$$

Construct a table similar to Table 3 to compare the values of these functions.

APPLICATIONS

BUSINESS & ECONOMICS

43.   *Cost.*   For a company that manufactures beach balls, the cost of producing x beach balls is given by

$$C(x) = 5 + \sqrt{100 + x^2}$$

Use the second-degree Taylor polynomial at 0 to estimate the cost of producing ten beach balls.

44.   *Profit.*   The profit from the sale of x radios is given by

$$P(x) = x \sqrt[3]{1,000 + x} - 400$$

Use the third-degree Taylor polynomial at 0 to approximate the profit from the sale of 100 radios.

45.   *Demand equation.*   The daily demand (in pounds) for jelly beans at x dollars per pound is given by

$$D(x) = 1,000 e^{-x/10}$$

Use the second-degree Taylor polynomial at 0 to approximate the demand when the price is $1.00.

46.   *Advertising.*   Using past records, it is estimated that a company will sell N units of a product after spending x thousand dollars in advertising, as given by

$$N = 600 \ln\left(1 + \frac{x}{10}\right)$$

Use the second-degree Taylor polynomial at 0 to estimate the number of units that will be sold if $2,000 is spent on advertising.

LIFE SCIENCES

47. *Medicine.* The amount of a drug (in milligrams) present in the bloodstream of a patient x minutes after the drug is administered is given by

$$D(x) = 100e^{-x/100}$$

Use the second-degree Taylor polynomial at 0 to approximate the amount of the drug in the bloodstream after 10 minutes.

48. *Ecology.* The atmospheric pressure P (in pounds per square inch) is given approximately by

$$P = 14.7e^{-0.21h}$$

where h is the altitude above sea level in miles. Use the second-degree Taylor polynomial at 0 to approximate the pressure $\frac{1}{2}$ mile above sea level.

SOCIAL SCIENCES

49. *Learning theory.* A particular person's history of learning to type is given by

$$N = 100(1 - e^{-0.1t})$$

where N is the number of words per minute typed after t weeks of training. Use the second-degree Taylor polynomial at 0 to approximate N when t = 2.

50. *Learning theory.* If a person learns y items in x hours, where y is given by

$$y = 90(1 + x)^{1/3} - 90$$

use the second-degree Taylor polynomial at 0 to approximate the number of items learned after $\frac{1}{2}$ hour.

9-2   TAYLOR POLYNOMIALS AT a

There are two reasons why we need to be able to find Taylor polynomials at points other than 0. First, if we want to approximate the value of a function at a point that is not close to 0, the Taylor polynomial at 0 may require too many terms to handle easily. For example, to approximate $f(x) = e^x$ for x = 10 to three decimal places, the Taylor polynomial at 0 needs more than forty terms. Obviously, computations with polynomials this long are impractical. Second, functions such as ln x and $\sqrt{x}$ do not have derivatives at 0, and hence do not have Taylor polynomials at 0. On the other hand, both functions have derivatives of all orders at any $x > 0$; consequently, it is possible to find some type of approximating polynomial.

Before we define Taylor polynomials at an arbitrary point a, we need to review a basic property of polynomials. We are used to expressing polynomials in powers of x. However, it is also possible to express any polynomial in powers of $(x - a)$ for an arbitrary number a. By **constructing a Taylor polynomial at a point a,** we mean to express the Taylor polynomial in powers of $(x - a)$.

We will proceed as we did in the previous section. Given a function f that has n derivatives at a point a, we want to find a polynomial p_n with the property that

$$p_n^{(k)}(a) = f^{(k)}(a) \qquad k = 0, 1, \ldots, n$$

First, we express p_n as a polynomial in powers of $(x - a)$ and compute the derivatives:

$$
\begin{aligned}
p_n(x) &= a_0 + a_1(x - a) + a_2(x - a)^2 + a_3(x - a)^3 + \cdots + a_n(x - a)^n \\
p_n'(x) &= \qquad\quad a_1 \quad + 2a_2(x - a) + 3a_3(x - a)^2 + \cdots + na_n(x - a)^{n-1} \\
p_n''(x) &= \qquad\qquad\qquad\quad 2a_2 \quad + (3)(2)a_3(x - a) + \cdots + n(n - 1)a_n(x - a)^{n-2} \\
p_n^{(3)}(x) &= \qquad\qquad\qquad\qquad\qquad\quad (3)(2)(1)a_3 \quad + \cdots + n(n - 1)(n - 2)a_n(x - a)^{n-3}
\end{aligned}
$$

$$\vdots$$

$$p_n^{(n)}(x) = n(n - 1) \cdots (1)a_n = n!a_n$$

Now we evaluate each derivative at a:

$$
\begin{aligned}
a_0 &= p_n(a) = f(a) \\
a_1 &= p_n'(a) = f'(a) \\
2a_2 &= p_n''(a) = f''(a) \\
(3)(2)(1)a_3 &= p_n^{(3)}(a) = f^{(3)}(a)
\end{aligned}
$$

$$\vdots \qquad\qquad \vdots$$

$$n!a_n = p_n^{(n)}(a) = f^{(n)}(a)$$

Thus, each coefficient of p_n satisfies

$$a_k = \frac{f^{(k)}(a)}{k!} \qquad k = 0, 1, 2, \ldots, n$$

TAYLOR POLYNOMIAL AT a

The nth-degree Taylor polynomial at a for a function f is

$$p_n(x) = f(a) + f'(a)(x - a) + \frac{f''(a)}{2!}(x - a)^2 + \cdots + \frac{f^{(n)}(a)}{n!}(x - a)^n$$

provided f has n derivatives at a.

Example 4 Find the third-degree Taylor polynomial at $a = 1$ for $f(x) = \sqrt[4]{x}$. Use $p_3(x)$ to approximate $\sqrt[4]{2}$.

Solution **Step 1.** *Find the derivatives.*

$$f(x) = x^{1/4}$$
$$f'(x) = \tfrac{1}{4}x^{-3/4}$$
$$f''(x) = -\tfrac{3}{16}x^{-7/4}$$
$$f^{(3)}(x) = \tfrac{21}{64}x^{-11/4}$$

Step 2. *Evaluate the derivatives at $a = 1$.*

$$f(1) = 1$$
$$f'(1) = \tfrac{1}{4}$$
$$f''(1) = -\tfrac{3}{16}$$
$$f^{(3)}(1) = \tfrac{21}{64}$$

Step 3. *Find the coefficients of the Taylor polynomial.*

$$a_0 = f(1) = 1$$
$$a_1 = f'(1) = \tfrac{1}{4}$$
$$a_2 = \frac{f''(1)}{2!} = \frac{-\tfrac{3}{16}}{2} = -\tfrac{3}{32}$$
$$a_3 = \frac{f^{(3)}(1)}{3!} = \frac{\tfrac{21}{64}}{6} = \tfrac{7}{128}$$

Step 4. *Write down the Taylor polynomial.*

$$p_3(x) = 1 + \tfrac{1}{4}(x - 1) - \tfrac{3}{32}(x - 1)^2 + \tfrac{7}{128}(x - 1)^3$$

Now we use the Taylor polynomial to approximate $\sqrt[4]{2}$.

$$\sqrt[4]{2} = f(2) \approx p_3(2) = 1 + \tfrac{1}{4} - \tfrac{3}{32} + \tfrac{7}{128} = 1.2109375$$

[*Note:* The value obtained by using a calculator is $\sqrt[4]{2} \approx 1.1892071$.]

Problem 4 Find the second-degree Taylor polynomial at $a = 8$ for $f(x) = \sqrt[3]{x}$. Use $p_2(x)$ to approximate $\sqrt[3]{9}$.

Answer $p_2(x) = 2 + \dfrac{1}{12}(x - 8) - \dfrac{1}{288}(x - 8)^2$, $\sqrt[3]{9} \approx 2.0798611$

Example 5 Find the nth-degree Taylor polynomial at $a = 3$ for $f(x) = 1/x$.

Solution

Step 1. *Find the derivatives.*

$$f(x) = x^{-1} = 0!x^{-1}$$
$$f'(x) = (-1)x^{-2} = -1!x^{-2}$$
$$f''(x) = (-2)(-1)x^{-3} = 2!x^{-3}$$
$$f^{(3)}(x) = (-3)(-2)(-1)x^{-4} = -3!x^{-4}$$
$$\vdots$$
$$f^{(n)}(x) = (-n)[-(n-1)] \cdots (-1)x^{-n-1} = (-1)^n n! x^{-n-1}$$

Step 2. *Evaluate the derivatives at $a = 3$.*

$$f(3) = \frac{1}{3}$$

$$f'(3) = \frac{(-1)}{3^2}$$

$$f''(3) = \frac{(-1)^2 2!}{3^3}$$

$$f^{(3)}(3) = \frac{(-1)^3 3!}{3^4}$$

$$\vdots$$

$$f^{(n)}(3) = \frac{(-1)^n n!}{3^{n+1}}$$

Step 3. *Find the coefficients of the Taylor polynomial.*

$$a_0 = f(3) = \frac{1}{3}$$

$$a_1 = f'(3) = \frac{(-1)}{3^2}$$

$$a_2 = \frac{f''(3)}{2!} = \frac{(-1)^2 2!}{3^3 2!} = \frac{(-1)^2}{3^3}$$

$$a_3 = \frac{f^{(3)}(3)}{3!} = \frac{(-1)^3 3!}{3^4 3!} = \frac{(-1)^3}{3^4}$$

$$\vdots$$

$$a_n = \frac{f^{(n)}(3)}{n!} = \frac{(-1)^n n!}{3^{n+1} n!} = \frac{(-1)^n}{3^{n+1}}$$

Step 4. *Write down the Taylor polynomial.*

$$p_n(x) = \frac{1}{3} - \frac{1}{3^2}(x-3) + \frac{1}{3^3}(x-3)^2 - \frac{1}{3^4}(x-3)^3 + \cdots + \frac{(-1)^n}{3^{n+1}}(x-3)^n$$

Problem 5 Find the nth-degree Taylor polynomial at $a = 1$ for $f(x) = e^{-x}$.

Answer $$p_n(x) = \frac{1}{e} - \frac{1}{e}(x - 1) + \frac{1}{e(2!)}(x - 1)^2 - \frac{1}{e(3!)}(x - 1)^3 + \cdots + \frac{(-1)^n}{e(n!)}(x - 1)^n$$

Although the primary reason for introducing Taylor polynomials was to approximate nonpolynomial functions, this procedure has an interesting application when f is a polynomial. We state the following theorem without proof.

Theorem 1 If $f(x)$ is an nth-degree polynomial and $p_n(x)$ is the nth-degree Taylor polynomial for f at any point a, then $f(x) = p_n(x)$.

　　　　　This theorem provides a simple way to express any polynomial in powers of $(x - a)$.

Example 6 Expand $f(x) = x^4 - 1$ in powers of $(x + 1)$.

Solution Choose $a = -1$ and find the fourth-degree Taylor polynomial:

Step 1	Step 2	Step 3
$f(x) = x^4 - 1$	$f(-1) = \quad 0$	$a_0 = \quad f(-1) = 0$
$f'(x) = 4x^3$	$f'(-1) = \quad -4$	$a_1 = \quad f'(-1) = -4$
$f''(x) = 12x^2$	$f''(-1) = \quad 12$	$a_2 = \dfrac{f''(-1)}{2!} = \dfrac{12}{2!} = 6$
$f^{(3)}(x) = 24x$	$f^{(3)}(-1) = -24$	$a_3 = \dfrac{f^{(3)}(-1)}{3!} = \dfrac{-24}{3!} = -4$
$f^{(4)}(x) = 24$	$f^{(4)}(-1) = \quad 24$	$a_4 = \dfrac{f^{(4)}(-1)}{4!} = \dfrac{24}{4!} = 1$

Step 4. *Write down the Taylor polynomial.*

$$x^4 - 1 = -4(x + 1) + 6(x + 1)^2 - 4(x + 1)^3 + (x + 1)^4$$

Problem 6 Expand $f(x) = x^4$ in powers of $(x - 1)$.

Answer $x^4 = 1 + 4(x - 1) + 6(x - 1)^2 + 4(x - 1)^3 + (x - 1)^4$

EXERCISE 9-2

A *In Problems 1–10 find the indicated Taylor polynomial at the given value of a.*

1. $f(x) = e^{x-1}$,　$p_4(x)$ at 1
2. $f(x) = e^{2x}$,　$p_3(x)$ at $\frac{1}{2}$
3. $f(x) = e^{-x}$,　$p_3(x)$ at -2
4. $f(x) = \ln(2 - x)$,　$p_4(x)$ at 1

5. $f(x) = \ln(3x)$, $p_3(x)$ at $\frac{1}{3}$
6. $f(x) = \ln(5 + x)$, $p_4(x)$ at -3
7. $f(x) = \sqrt{x}$, $p_4(x)$ at 1
8. $f(x) = \sqrt{x}$, $p_3(x)$ at 4
9. $f(x) = \sqrt[3]{x}$, $p_4(x)$ at -1
10. $f(x) = \sqrt{(2 + x)^3}$, $p_3(x)$ at 2
11. Use p_4 from Problem 7 and $x = 1.2$ to approximate $\sqrt{1.2}$.
12. Use p_3 from Problem 8 and $x = 3.95$ to approximate $\sqrt{3.95}$.
13. Use p_4 from Problem 9 and $x = -1.5$ to approximate $\sqrt[3]{-1.5}$.
14. Use p_3 from Problem 10 and $x = 1.8$ to approximate $(3.8)^{1.5}$.

B   *In Problems 15–22 find the nth-degree Taylor polynomial at the indicated value of a.*

15. $f(x) = e^x$ at 3
16. $f(x) = \dfrac{1}{3 + x}$ at -2

17. $f(x) = \ln x$ at $\dfrac{1}{2}$
18. $f(x) = \dfrac{1}{2 - x}$ at 1

19. $f(x) = \dfrac{1}{(2 - x)^2}$ at 1
20. $f(x) = \dfrac{2 + x}{x}$ at 2

21. $f(x) = e^{-x}$ at 10
22. $f(x) = x \ln x$ at 1
23. Expand $f(x) = x^2$ in powers of $(x - 1)$.
24. Expand $f(x) = 3x^3 - 2x^2 + 5x - 7$ in powers of $(x + 1)$.
25. Expand $f(x) = (x - 1)^5$ in powers of x.
26. Expand $f(x) = x^5$ in powers of $(x - 2)$.

C   27. Find the nth-degree Taylor polynomial for $f(x) = e^x$ at any point a.
28. Find the nth-degree Taylor polynomial for $f(x) = \ln x$ at any $a > 0$.
29. Find the nth-degree Taylor polynomial for $f(x) = 1/x$ at any $a \neq 0$.
30. Expand $f(x) = (x + a)^n$ in powers of x, where n is a positive integer.

APPLICATIONS

BUSINESS & ECONOMICS

31. *Revenue.*   A company estimates that the revenue R it will receive from selling x units of a product is given by

$$R(x) = x(1 + e^{1-0.1x})$$

Use the second-degree Taylor polynomial at $a = 10$ to approximate the revenue received from the sale of eleven units.

32. *Supply–demand.*   The supply and demand equations for a certain market are

$$p = S(x) = 0.4x + 15 \quad \text{and} \quad p = D(x) = 10(1 + e^{1-0.05x})$$

where p is price and x is number of units. The equilibrium point is determined by the equation

$$S(x) = D(x)$$

In this case, the equilibrium equation cannot be solved algebraically. To approximate the solution, replace D by its second-degree Taylor polynomial at $a = 20$ and solve the equation

$$S(x) = p_2(x)$$

LIFE SCIENCES

33. *Pulse rate.* The average pulse rate y (in beats per minute) of a healthy person x inches tall is given approximately by

$$y = \frac{590}{\sqrt{x}} \qquad 30 \le x \le 75$$

Use the second-degree Taylor polynomial at $a = 64$ to approximate y when $x = 70$.

34. *Air pollution.* On an average summer day in a particular large city the air pollution P (in parts per million) is given by

$$P(x) = 20\sqrt{3x^2 + 25} - 80$$

where x is the number of hours elapsed since 8:00 AM. Use the second-degree Taylor polynomial expanded at $a = 5$ to approximate the amount of pollution at 2:00 PM.

SOCIAL SCIENCES

35. *Small group analysis.* If ten members of a discussion group are ranked according to the number of times each participates in the discussion, then the number of times, $N(k)$, that the kth-ranked person participates is given approximately by

$$N(k) = 100e^{-0.11(k-1)} \qquad 1 \le k \le 10$$

Use the third-degree Taylor polynomial at $a = 1$ to approximate the number of times the third-ranked person participates in the discussion.

36. *Learning theory.* The time T (in seconds) it takes a person to learn x nonsense syllables is given approximately by

$$T(x) = 12x\sqrt{x - 4} \qquad 10 \le x \le 50$$

Use the fourth-degree Taylor polynomial expanded at $a = 40$ to approximate the time it will take this person to learn 43 nonsense syllables.

9-3 TAYLOR SERIES

If f is a function with derivatives of all order at a point a, then we can construct the Taylor polynomial p_n at a for any integer n. Now we are interested in the relationship between the original function f and the corresponding Taylor polynomial p_n as n assumes larger and larger values. If the Taylor polynomial is a useful tool for approximating functions, then the accuracy of the approximation should improve as we increase the size of n. Indeed, Tables 1, 2, and 3 in Section 9-1 indicate that this is the case for the function e^x.

For a given value of x, we would like to know whether we can make $p_n(x)$ arbitrarily close to $f(x)$ by making n sufficiently large. In other words, we want to know whether

$$\lim_{n \to \infty} p_n(x) = f(x) \tag{1}$$

For most functions, it turns out that there is a set of values of x for which equation (1) is valid. We will consider a specific example to illustrate this. Let

$$f(x) = \frac{1}{1 - x} \quad \text{and} \quad a = 0$$

First, we find p_n:

Step 1	Step 2	Step 3
$f(x) = (1 - x)^{-1}$	$f(0) = 1$	$a_0 = f(0) = 1$
$f'(x) = (1 - x)^{-2}$	$f'(0) = 1$	$a_1 = f'(0) = 1$
$f''(x) = 2(1 - x)^{-3}$	$f''(0) = 2$	$a_2 = \dfrac{f''(0)}{2!} = \dfrac{2}{2!} = 1$
$f^{(3)}(x) = 3!(1 - x)^{-4}$	$f^{(3)}(0) = 3!$	$a_3 = \dfrac{f^{(3)}(0)}{3!} = \dfrac{3!}{3!} = 1$
$\vdots$	$\vdots$	$\vdots$
$f^{(n)}(x) = n!(1 - x)^{-n-1}$	$f^{(n)}(0) = n!$	$a_n = \dfrac{f^{(n)}(1)}{n!} = 1$

Step 4. *Write down the Taylor polynomial.*

$$p_n(x) = 1 + x + x^2 + x^3 + \cdots + x^n$$

Now we want to evaluate

$$\lim_{n \to \infty} p_n(x) = \lim_{n \to \infty}(1 + x + x^2 + x^3 + \cdots + x^n)$$

It is not possible to evaluate the limit as $n \to \infty$ when p_n is expressed as a sum of n terms. The difficulty lies in the fact that the number of terms in the sum is increasing as n increases. However, it is possible to express p_n in a form that will allow us to evaluate the limit. You may recognize that p_n is a finite geometric series with common ratio x. We will now find a simpler expression for p_n.

$$p_n(x) = 1 + x + x^2 + \cdots + x^{n-1} + n^n$$
$$xp_n(x) = \qquad x + x^2 + \cdots + x^{n-1} + x^n + x^{n+1} \qquad \text{Multiply both}$$

sides by x

$$p_n(x) - xp_n(x) = 1 \qquad\qquad\qquad\qquad\qquad\qquad - x^{n+1} \qquad \text{Subtract the second equation from the first}$$

$$p_n(x)(1 - x) = 1 - x^{n+1} \qquad\qquad\qquad\qquad\qquad\qquad \text{Solve for } p_n(x)$$

$$p_n(x) = \frac{1 - x^{n+1}}{1 - x} \qquad x \neq 1$$

It is not difficult to see that

$$\lim_{n \to \infty} x^{n+1} = \begin{cases} 0 & -1 < x < 1 \\ 1 & x = 1 \end{cases}$$

and does not exist if $x \leq -1$ or $x > 1$. Now we can use the limit properties in Section 2-2 to evaluate the limit of $p_n(x)$ as $n \to \infty$:

$$\lim_{n \to \infty} p_n(x) = \lim_{n \to \infty} \frac{1 - x^{n+1}}{1 - x}$$

$$= \frac{\lim_{n \to \infty} (1 - x^{n+1})}{\lim_{n \to \infty} (1 - x)} = \frac{1 - \lim_{n \to \infty} x^{n+1}}{1 - x}$$

$$= \frac{1 - 0}{1 - x} = \frac{1}{1 - x} \qquad -1 < x < 1$$

This limit does not exist for $x \leq -1$ or $x \geq 1$. [Recall that at $x = 1$, $(1 - x^{n+1})/(1 - x)$ is not defined.] We can now conclude that

$$f(x) = \frac{1}{1 - x} = \lim_{n \to \infty} p_n(x) \qquad \text{for} \quad -1 < x < 1$$

This is often abbreviated by writing

$$\frac{1}{1 - x} = 1 + x + x^2 + \cdots + x^n + \cdots \qquad -1 < x < 1$$

or

$$\frac{1}{1-x} = \sum_{k=0}^{\infty} x^k \qquad -1 < x < 1$$

The expression

$$1 + x + x^2 + \cdots + x^n + \cdots$$

is called the **Taylor series** or the **infinite power series** for $1/(1-x)$ at $a = 0$. The set of values $\{x \mid -1 < x < 1\}$ where this representation is valid is called the **interval of convergence** of the Taylor series. Notice that the interval of convergence of the Taylor series and the domain of the function are not the same. The function $1/(1-x)$ is defined for all x except $x = 1$, but is represented by its Taylor series only for $-1 < x < 1$.

We now generalize the above discussion for an arbitrary function f at a point a.

TAYLOR SERIES

If f has derivatives of all order at a point a, then then **Taylor series** expansion for f at a is

$$a_0 + a_1(x-a) + a_2(x-a)^2 + \cdots + a_n(x-a)^n + \cdots$$
$$= \sum_{k=0}^{\infty} a_k(x-a)^k$$

where

$$a_k = \frac{f^{(k)}(a)}{k!}$$

The set of values of x for which

$$f(x) = a_0 + a_1(x-a) + a_2(x-a) + \cdots + a_n(x-a)^n + \cdots$$

is called the **interval of convergence.**

We must emphasize that we are not really adding up an infinite number of terms, nor is the Taylor series an "infinite" polynomial. Rather, for x in the interval of convergence, the expression

$$f(x) = \sum_{k=0}^{\infty} a_k(x-a)^k$$

is just shorthand for

$$f(x) = \lim_{n \to \infty} \sum_{k=0}^{n} a_k(x-a)^k$$

Thus, when we write

$$\frac{1}{1-x} = 1 + x + x^2 + \cdots + x^n + \cdots \qquad -1 < x < 1$$

we mean

$$\frac{1}{1-x} = \lim_{n \to \infty} (1 + x + x^2 + \cdots + x^n) \qquad -1 < x < 1$$

For most functions it is very difficult to evaluate $\lim_{n \to \infty} p_n(x)$. The techniques involved in evaluating this type of limit are beyond the scope of this book. The Taylor series expansions at 0 and the corresponding intervals of convergence for the most commonly used functions are listed in Table 4. We will be primarily interested in the first three expansions.

TABLE 4

$f(x)$	TAYLOR SERIES EXPANSION AT 0	INTERVAL OF CONVERGENCE
$\dfrac{1}{1-x}$	$1 + x + x^2 + \cdots + x^n + \cdots$	$-1 < x < 1$
e^x	$1 + x + \dfrac{1}{2!}x^2 + \cdots + \dfrac{1}{n!}x^n + \cdots$	$-\infty < x < \infty$
$\ln(1 + x)$	$x - \dfrac{1}{2}x^2 + \dfrac{1}{3}x^3 - \cdots + \dfrac{(-1)^{n+1}}{n}x^n + \cdots$	$-1 < x < 1$
$\sin x$	$x - \dfrac{1}{3!}x^3 + \dfrac{1}{5!}x^5 - \cdots + \dfrac{(-1)^k}{(2k+1)!}x^{2k+1} + \cdots$	$-\infty < x < \infty$
$\cos x$	$1 - \dfrac{1}{2!}x^2 + \dfrac{1}{4!}x^4 - \cdots + \dfrac{(-1)^k}{(2k)!}x^{2k} + \cdots$	$-\infty < x < \infty$

OPERATIONS ON TAYLOR SERIES

We like to use the Taylor series notation

$$f(x) = a_0 + a_1(x - a) + a_2(x - a)^2 + \cdots + a_n(x - a)^n + \cdots$$

since it resembles polynomial notation. We shall see that Taylor series and polynomials have many properties in common. Some of these properties are listed in the boxes below. Others will be discussed in the next section.

PROPERTY 1: ADDITION

If

$$f(x) = a_0 + a_1(x - a) + a_2(x - a)^2 + \cdots + a_n(x - a)^n + \cdots$$

and

$$g(x) = b_0 + b_1(x - a) + b_2(x - a)^2 + \cdots + b_n(x - a)^n + \cdots$$

are the Taylor series for f and g at a, then the Taylor series expansion for the function $(f + g)$ at a is

$$f(x) + g(x) = (a_0 + b_0) + (a_1 + b_1)(x - a) + (a_2 + b_2)(x - a)^2 + \\ \cdots + (a_n + b_n)(x - a)^n + \cdots$$

This representation is valid in the interval that is the intersection of the intervals of convergence of f and g.

Example 7   Use the Taylor series expansions in Table 4 and property 1 to find the Taylor series expansion at 0 for $f(x) = e^x + \ln(1 + x)$.

Solution   From Table 4,

$$e^x = 1 + x + \frac{1}{2!}x^2 + \frac{1}{3!}x^3 + \cdots + \frac{1}{n!}x^n + \cdots$$

$$-\infty < x < \infty$$

and

$$\ln(1 + x) = \quad x - \frac{1}{2}x^2 + \frac{1}{3}x^3 - \cdots + \frac{(-1)^{n+1}}{n}x^n + \cdots$$

$$-1 < x < 1$$

From property 1,

$$f(x) = e^x + \ln(1 + x)$$

$$= 1 + 2x + \frac{1}{2}x^3 + \cdots + \left(\frac{1}{n!} + \frac{(-1)^{n+1}}{n}\right)x^n + \cdots$$

This representation is valid for $-1 < x < 1$, since this is the intersection of the intervals of convergence of e^x and $\ln(1 + x)$.

Problem 7   Use the Taylor series expansions in Table 4 and property 1 to find the Taylor series expansion at 0 for

$$f(x) = e^x + \frac{1}{1 - x}$$

Answer $f(x) = 2 + 2x + \frac{3}{2}x^2 + \cdots + \left(\frac{1}{n!} + 1\right)x^n + \cdots, \quad -1 < x < 1$

PROPERTY 2: MULTIPLICATION BY A CONSTANT c

If

$$f(x) = a_0 + a_1(x - a) + a_2(x - a)^2 + \cdots + a_n(x - a)^n + \cdots$$

is the Taylor series for f at a and c is a constant, then the Taylor series expansion for the function cf at a is

$$cf(x) = ca_0 + ca_1(x - a) + ca_2(x - a)^2 + \cdots + ca_n(x - a)^n + \cdots$$

This representation is valid for the same inverval of convergence as f.

Example 8   Use a Taylor series expansion from Table 4 and property 2 to find the Taylor series expansion at 0 for

$$f(x) = \frac{3}{1 - x}$$

Solution   From Table 4,

$$\frac{1}{1 - x} = 1 + x + x^2 + \cdots + x^n + \cdots \qquad -1 < x < 1$$

Property 2 indicates that

$$\frac{3}{1 - x} = 3\left(\frac{1}{1 - x}\right) = 3 + 3x + 3x^2 + \cdots + 3x^n + \cdots$$

$$-1 < x < 1$$

Problem 8   Use a Taylor series expansion from Table 4 and property 2 to find the Taylor series expansion at 0 for $f(x) = 2 \ln(1 + x)$.

Answer $f(x) = 2x - x^2 + \dfrac{2}{3}x^3 - \cdots + \dfrac{2(-1)^{n+1}}{n}x^n + \cdots, \quad -1 < x < 1$

PROPERTY 3: MULTIPLICATION BY $(x - a)^r$

If

$$f(x) = a_0 + a_1(x - a) + a_2(x - a)^2 + \cdots + a_n(x - a)^n + \cdots$$

is the Taylor series for f at a and r is a positive integer, then the Taylor series expansion for $(x - a)^r f(x)$ at a is

$$(x - a)^r f(x) = a_0(x - a)^r + a_1(x - a)^{r+1} + a_2(x - a)^{r+2} + \cdots$$
$$+ a_n(x - a)^{r+n} + \cdots$$

This representation is valid for the same interval of convergence as f.

Example 9   Use a Taylor series expansion from Table 4 and property 3 to find the Taylor series expansion at 0 for $f(x) = x^2 e^x$.

Solution   From Table 4,

$$e^x = 1 + x + \frac{1}{2!}x^2 + \cdots + \frac{1}{n!}x^n + \cdots \qquad -\infty < x < \infty$$

Multiplying this by x^2, we get

$$x^2 e^x = x^2 + x^3 + \frac{1}{2!}x^4 + \cdots + \frac{1}{n!}x^{n+2} + \cdots \qquad -\infty < x < \infty$$

Problem 9   Use a Taylor series expansion from Table 4 and property 3 to find the Taylor series expansion at 0 for $f(x) = x^3 \ln(1 + x)$.

Answer $f(x) = x^4 - \frac{1}{2}x^5 + \frac{1}{3}x^6 - \cdots + \frac{(-1)^{n+1}}{n}x^{n+3} + \cdots, \qquad -1 < x < 1$

PROPERTY 4: COMPOSITION

A Taylor series may be formed by composition with other functions. Depending on the particular composition used, the interval of convergence may be changed.

Example 10   Use a Taylor series expansion from Table 4 and property 4 to find the Taylor series expansion at 0 for

$$f(x) = \frac{1}{1 - 2x}$$

Solution   If we let

$$g(x) = \frac{1}{1 - x} \qquad \text{then} \qquad f(x) = g(2x)$$

From Table 4,

$$g(x) = 1 + x + x^2 + \cdots + x^n + \cdots \qquad -1 < x < 1$$

Substituting $2x$ in the Taylor series expansion for $g(x)$, we obtain

$$f(x) = \frac{1}{1 - 2x} = g(2x) = 1 + 2x + (2x)^2 + \cdots + (2x)^n + \cdots$$

$$= 1 + 2x + 2^2 x^2 + \cdots + 2^n x^n + \cdots$$

This representation is valid for $-1 < 2x < 1$, or $-\frac{1}{2} < x < \frac{1}{2}$.

Problem 10   Use a Taylor series expansion from Table 4 and property 4 to find the Taylor series expansion at 0 for $f(x) = \ln(1 + 3x)$.

Answer $f(x) = 3x - \dfrac{9}{2}x^2 + 9x^3 - \cdots + \dfrac{(-1)^{n+1}3^n}{n}x^n + \cdots, \quad -\frac{1}{3} < x < \frac{1}{3}$

Example 11   Use a Taylor series expansion from Table 4 and property 4 to find the Taylor series expansion at 0 for $f(x) = e^{-x^2}$.

Solution   If we let $g(x) = e^x$, then $f(x) = g(-x^2)$. From Table 4,

$$g(x) = 1 + x + \frac{1}{2!}x^2 + \cdots + \frac{1}{n!}x^n + \cdots \qquad -\infty < x < \infty$$

Substituting $-x^2$ in the Taylor series expansion for $g(x)$, we obtain

$$f(x) = e^{-x^2} = g(-x^2) = 1 + (-x^2) + \frac{1}{2!}(-x^2)^2 + \cdots + \frac{1}{n!}(-x^2)^n + \cdots$$

$$= 1 - x^2 + \frac{1}{2!}x^4 - \cdots + \frac{(-1)^n}{n!}x^{2n} + \cdots$$

Since the Taylor series expansion for g was valid for all values of x, the Taylor series expansion for f is also valid for all values of x (that is, $-\infty < x < \infty$).

Problem 11   Use a Taylor series expansion from Table 4 and property 4 to find the Taylor series expansion at 0 for

$$f(x) = \frac{1}{1 + x^2}$$

Answer $f(x) = 1 - x^2 + x^4 - \cdots + (-1)^n x^{2n} + \cdots, \quad -1 < x < 1$

Example 12   Use a Taylor series expansion from Table 4 and property 4 to find the Taylor series expansion at 0 for

$$f(x) = \frac{1}{5 - x}$$

Solution   If we factor 5 out of the denominator of f, then f can be related to one of the functions in Table 4:

$$f(x) = \frac{1}{5 - x} = \frac{1}{5\left(1 - \dfrac{x}{5}\right)} = \frac{1}{5}\left(\frac{1}{1 - \dfrac{x}{5}}\right)$$

If we let

$$g(x) = \frac{1}{1 - x} \quad \text{then} \quad f(x) = \frac{1}{5}\left(\frac{1}{1 - \dfrac{x}{5}}\right) = \frac{1}{5}g\left(\frac{x}{5}\right)$$

Since

$$g(x) = 1 + x + x^2 + \cdots + x^n + \cdots \qquad -1 < x < 1$$

we have

$$f(x) = \frac{1}{5 - x} = \frac{1}{5}g\left(\frac{x}{5}\right) = \frac{1}{5}\left(1 + \frac{x}{5} + \left(\frac{x}{5}\right)^2 + \cdots + \left(\frac{x}{5}\right)^n + \cdots\right)$$

$$-1 < \frac{x}{5} < 1$$

$$= \frac{1}{5} + \frac{1}{5^2}x + \frac{1}{5^3}x^2 + \cdots + \frac{1}{5^{n+1}}x^n + \cdots$$

$$-5 < x < 5$$

Note that we also made use of property 2 in this example.

Problem 12   Use a Taylor series expansion from Table 4 and property 4 to find the Taylor series expansion at 0 for

$$f(x) = \frac{1}{3 - x}$$

Answer $f(x) = \dfrac{1}{3} + \dfrac{1}{3^2}x + \dfrac{1}{3^3}x^2 + \cdots + \dfrac{1}{3^{n+1}}x^n + \cdots, \quad -3 < x < 3$

Example 13   Use the Taylor series expansion from Example 12 and property 4 to find the Taylor series expansion at 2 for

$$F(x) = \frac{1}{7 - x}$$

Solution   Since we want to expand F in powers of $(x - 2)$, we first express the denominator of $F(x)$ in powers of $(x - 2)$:

$$7 - x = 7 - 2 - x + 2 = 5 - (x - 2)$$

If we let

$$f(x) = \frac{1}{5 - x} \quad \text{then} \quad F(x) = \frac{1}{7 - x} = \frac{1}{5 - (x - 2)} = f(x - 2)$$

From Example 12,

$$f(x) = \frac{1}{5 - x} = \frac{1}{5} + \frac{1}{5^2}x + \frac{1}{5^3}x^2 + \cdots + \frac{1}{5^{n+1}}x^n + \cdots \qquad -5 < x < 5$$

Substituting $(x - 2)$ in the expansion for $f(x)$, we obtain

$$F(x) = \frac{1}{7 - x} = f(x - 2) = \frac{1}{5} + \frac{1}{5^2}(x - 2) + \frac{1}{5^3}(x - 2)^2 + \cdots$$

$$+ \frac{1}{5^{n+1}}(x - 2)^n + \cdots$$

This representation is valid for $-5 < x - 2 < 5$, which can be written as $-3 < x < 7$.

Problem 13 Use Problem 12 and property 4 to find the Taylor series expansion at 1 for

$$F(x) = \frac{1}{4 - x}$$

Answer $F(x) - \frac{1}{3} + \frac{1}{3^2}(x - 1) + \frac{1}{3^3}(x - 1)^2 + \cdots + \frac{1}{3^{n+1}}(x - 1)^n + \cdots,$

$$-2 < x < 4$$

Example 14 Use a Taylor series expansion from Table 4 and property 4 to find the Taylor series expansion at 3 for $f(x) = \ln x$.

Solution Proceeding as in Example 13,

$$\ln x = \ln(3 + x - 3) = \ln\left[3\left(1 + \frac{x - 3}{3}\right)\right]$$

$$= \ln 3 + \ln\left(1 + \frac{x - 3}{3}\right) \qquad \text{Property of logs}$$

If we let

$$g(x) = \ln(1 + x)$$

$$= x - \frac{1}{2}x^2 + \frac{1}{3}x^3 - \cdots + \frac{(-1)^{n+1}}{n}x^n + \cdots \qquad -1 < x < 1$$

then

$$\ln x = \ln 3 + \ln\left(1 + \frac{x-3}{3}\right)$$

$$= \ln 3 + g\left(\frac{x-3}{3}\right)$$

$$= \ln 3 + \left(\frac{x-3}{3}\right) - \frac{1}{2}\left(\frac{x-3}{3}\right)^2 + \frac{1}{3}\left(\frac{x-3}{3}\right)^3 - \cdots$$

$$+ \frac{(-1)^{n+1}}{n}\left(\frac{x-3}{3}\right)^n + \cdots \qquad -1 < \frac{x-3}{3} < 1$$

$$= \ln 3 + \frac{1}{3}(x-3) - \frac{1}{2(3^2)}(x-3)^2 + \frac{1}{3(3^3)}(x-3)^3 - \cdots$$

$$+ \frac{(-1)^{n+1}}{n(3^n)}(x-3)^n + \cdots$$

This representation is valid when

$$-1 < \frac{x-3}{3} < 1$$
$$-3 < x - 3 < 3$$
$$0 < \quad x \quad < 6 \qquad \text{Interval of convergence}$$

Problem 14   Use a Taylor series expansion from Table 4 and property 4 to find the Taylor series expansion at 2 for $f(x) = e^x$.

Answer $f(x) = e^2 + e^2(x-2) + \dfrac{e^2}{2!}(x-2)^2 + \cdots + \dfrac{e^2}{n!}(x-2)^n + \cdots,$

$$-\infty < x < \infty$$

EXERCISE 9-3

Solve all the problems in this exercise by using the Taylor series in Table 4 and properties 1–4.

A   *Find the Taylor series at 0. Give the interval of convergence for each series.*

1. $f(x) = \dfrac{1}{1-x} + \ln(1+x)$ 2. $f(x) = 3e^x$

3. $f(x) = \dfrac{x^3}{1-x}$ 4. $f(x) = \ln\left(1 + \dfrac{1}{2}x\right)$

5. $f(x) = \dfrac{1}{1-x} + \ln(1-x)$ 6. $f(x) = x^3 e^x$

7. $f(x) = \dfrac{1}{1 - x^2}$

8. $f(x) = \ln(1 + x^3)$

9. $f(x) = \dfrac{e^x + e^{-x}}{2}$

10. $f(x) = \dfrac{e^x - e^{-x}}{2}$

B   *Find the Taylor series at 0. Give the interval of convergence for each series.*

11. $f(x) = xe^{x^2}$

12. $f(x) = \dfrac{1}{4 - 3x}$

13. $f(x) = \dfrac{x^2}{1 + x^4}$

14. $f(x) = xe^{-2x^2}$

15. $f(x) = \dfrac{1}{1 - 8x^3}$

16. $f(x) = \ln(4 + x^2)$

Find the Taylor series at the indicated value of a. Give the interval of convergence.

17. $f(x) = e^{2x}$ at -1

18. $f(x) = e^{x+1}$ at 2

19. $f(x) = \ln x$ at 2

20. $f(x) = \ln(1 + 3x)$ at 3

21. $f(x) = \dfrac{1}{x}$ at 1

22. $f(x) = \dfrac{1}{x}$ at -3

23. $f(x) = \dfrac{1}{4 - 3x}$ at 1

24. $f(x) = \dfrac{1}{4 - 3x}$ at -1

C   25.   Find the Taylor series at 0 for

$$f(x) = \frac{1 + x}{1 - x}$$

Note that

$$\frac{1 + x}{1 - x} = \frac{1}{1 - x} + \frac{x}{1 - x}$$

26.   Find the Taylor series at 2 for

$$f(x) = \frac{x}{x - 1}$$

Note that

$$\frac{x}{x - 1} = 1 + \frac{1}{1 + (x - 2)}$$

27. Find the Taylor series at 0 for

$$f(x) = \frac{1}{2} \ln \left(\frac{1 + x}{1 - x} \right)$$

Note that

$$\ln \left(\frac{1 + x}{1 - x} \right) = \ln(1 + x) - \ln(1 - x)$$

28. If a and b are constants ($a \neq b$), find the Taylor series at a for

$$f(x) = \frac{1}{b - x}$$

9-4   DIFFERENTIATION AND INTEGRATION OF TAYLOR SERIES

One of the reasons polynomials are convenient functions to use in calculus is that they are easy to differentiate and integrate. It is a very important fact that Taylor series share this property.

PROPERTY 5: DIFFERENTIATION AND INTEGRATION

If

$$f(x) = a_0 + a_1(x - a) + a_2(x - a)^2 + a_3(x - a)^3 + \cdots + a_n(x - a)^n + \cdots$$

is the Taylor series for f at a, then the Taylor series for $f'(x)$ at a is

$$f'(x) = a_1 + 2a_2(x - a) + 3a_3(x - a)^2 + \cdots + na_n(x - a)^{n-1} + \cdots$$

and the Taylor series for $\int f(x)\, dx$ at a is

$$\int f(x)\, dx = a_0(x - a) + \frac{1}{2}a_1(x - a)^2 + \frac{1}{3}a_2(x - a)^3 + \frac{1}{4}a_3(x - a)^4 + \cdots + \frac{1}{n + 1}a_n(x - a)^{n+1} + \cdots + C$$

Both operations are valid within the interval of convergence of f. (Thus, as long as we restrict x to be in the interval of convergence of a Taylor series, we can differentiate and integrate the series as if it were a polynomial.)

Example 15   Use the Taylor series for $\ln(1 + x)$ from Table 4 and property 5 to find the Taylor series at 0 for

$$f(x) = \frac{1}{1 + x}$$

Solution

$$\ln(1 + x) = x - \frac{1}{2}x^2 + \frac{1}{3}x^3 - \cdots + \frac{(-1)^{n\,|\,1}}{n}x^n + \cdots \qquad -1 < x < 1$$

$$D_x \ln(1 + x) = D_x\left[x - \frac{1}{2}x^2 + \frac{1}{3}x^3 - \cdots + \frac{(-1)^{n+1}}{n}x^n + \cdots\right]$$

$$-1 < x < 1$$

$$\frac{1}{1 + x} = 1 - x + x^2 - \cdots + (-1)^{n+1}x^{n-1} + \cdots \qquad -1 < x < 1$$

Problem 15   Use the Taylor series from Example 15 and property 5 to find a Taylor series at 0 for

$$f(x) = \frac{1}{(1 + x)^2}$$

Answer $$\frac{1}{(1 + x)^2} = 1 - 2x + 3x^2 - \cdots + (-1)^n(n - 1)x^{n-2} + \cdots,$$

$$-1 < x < 1$$

Example 16   If $f(x)$ is a function that satisfies $f'(x) = x^2e^x$ and $f(0) = 2$, find a Taylor series at 0 for f.

Solution   **Step 1.** *Find a Taylor series at 0 for f'.*

$$f'(x) = x^2e^x = x^2\left(1 + x + \frac{1}{2!}x^2 + \cdots + \frac{1}{n!}x^n + \cdots\right) \qquad -\infty < x < \infty$$

$$= x^2 + x^3 + \frac{1}{2!}x^4 + \cdots + \frac{1}{n!}x^{n+2} + \cdots \qquad -\infty < x < \infty$$

Step 2. *Integrate to find the series for f.*

$$f(x) = \int\left(x^2 + x^3 + \frac{1}{2!}x^4 + \cdots + \frac{1}{n!}x^{n+2} + \cdots\right)dx \qquad -\infty < x < \infty$$

$$= \left[\frac{1}{3}x^3 + \frac{1}{4}x^4 + \frac{1}{5(2!)}x^5 + \cdots + \frac{1}{(n + 3)n!}x^{n+3} + \cdots\right] + C$$

$$-\infty < x < \infty$$

Step 3. *Evaluate the constant of integration.*

$$2 = f(0) = [0 + 0 + \cdots + 0 + \cdots] + C = C$$

Thus,

$$f(x) = 2 + \frac{1}{3}x^3 + \frac{1}{4}x^4 + \frac{1}{5(2!)}x^5 + \cdots + \frac{1}{(n+3)n!}x^{n+3} + \cdots$$

$$-\infty < x < \infty$$

Problem 16

If $f(x)$ is a function that satisfies $f'(x) = x \ln(1 + x)$ and $f(0) = 4$, find a Taylor series at 0 for f.

Answer

$$f(x) = 4 + \frac{1}{3}x^3 - \frac{1}{8}x^4 + \frac{1}{15}x^5 - \cdots + \frac{(-1)^{n+1}}{n(n+2)}x^{n+2} + \cdots,$$

$$-1 < x < 1$$

APPROXIMATION WITH ALTERNATING SERIES

Now that we can determine the values of x for which a function may be represented by its Taylor series expansion, we return to our original goal: approximating the values of a function.

If x is in the interval of convergence of the Taylor series expansion for a function f, then

$$\lim_{n \to \infty} p_n(x) = f(x)$$

How large should n be in order to ensure sufficient accuracy in the approximation $p_n(x) \approx f(x)$? In general, this is a difficult question to answer, but there is one type of Taylor series expansion for which the answer is very easy. If x_0 is in the interval of convergence of the Taylor series expansion for f and the terms in the series for $f(x_0)$ $[a_0, a_1(x_0 - a),$ $a_2(x_0 - a)^2, \ldots]$ **alternate in sign,** then the error introduced by replacing $f(x_0)$ with the first n terms of its Taylor series expansion evaluated at x_0 is no larger than the absolute value of the $(n + 1)$st term in the Taylor series. Thus, when we approximate $f(x_0)$ by $p_n(x_0)$, we can estimate the error by examining the next term in the series. It is very important to understand that *this error estimation technique only works when the numbers in the series for $f(x_0)$ alternate in sign.* It is incorrect to apply this method to a series that has terms that do not alternate in sign.

ALTERNATING SERIES APPROXIMATION

If

$$f(x) = a_0 + a_1(x - a) + a_2(x - a)^2 + \cdots + a_n(x - a)^n + \cdots$$

is the Taylor series for f at a;

$$p_n(x) = a_0 + a_1(x - a) + a_2(x - a)^2 + \cdots + a_n(x - a)^n$$

is the nth-degree Taylor polynomial for f at a; x_0 is in the interval of convergence of the Taylor series for f; and the numbers a_0, $a_1(x_0 - a), a_2(x_0 - a)^2, \ldots, a_n(x_0 - a)^n, \ldots$ alternate in sign, then

$$|f(x_0) - p_n(x_0)| \le |a_{n+1}(x_0 - a)^{n+1}|$$

Example 17   Use a Taylor series from Table 4 to approximate $e^{-1/2}$ correct to four decimal places.

Solution   From Table 4,

$$e^x = 1 + x + \frac{1}{2!}x^2 + \cdots + \frac{1}{n!}x^n + \cdots \qquad -\infty < x < \infty$$

Thus,

$$e^{-1/2} = 1 + \left(-\frac{1}{2}\right) + \frac{1}{2!}\left(-\frac{1}{2}\right)^2 + \frac{1}{3!}\left(-\frac{1}{2}\right)^3 + \frac{1}{4!}\left(-\frac{1}{2}\right)^4$$

$$+ \frac{1}{5!}\left(-\frac{1}{2}\right)^5 + \frac{1}{6!}\left(-\frac{1}{2}\right)^6 + \cdots$$

$$= 1 - \frac{1}{2} + \frac{1}{8} - \frac{1}{48} + \frac{1}{384} - \frac{1}{3,840} + \frac{1}{46,080} + \cdots$$

$$= 1 - 0.5 + 0.125 - 0.0208333 + 0.0026042$$
$$- 0.0002604 + 0.0000217 + \cdots$$

Since these numbers alternate in sign, we can use the alternating series approximation criteria to estimate the error. If we use the first six terms in this series, then we can conclude that

$$|e^{-1/2} - (1 - 0.5 + 0.125 - 0.0208333 + 0.0026042 - 0.0002604)| \le |0.0000217| \le |0.00003|$$

Thus,

$$e^{-1/2} \approx 1 - 0.5 + 0.125 - 0.02083 + 0.00260 - 0.00026$$
$$\approx 0.6065$$

and we are assured that the maximum possible error in this approximation is less than 0.00003.

Problem 17   Use a Taylor series from Table 4 to approximate $e^{-0.1}$ correct to five decimal places.

Answer $1 - \frac{1}{10} + \frac{1}{200} - \frac{1}{6,000} + \frac{1}{240,000} = 0.90484$

APPLICATIONS OF TAYLOR SERIES   We will now consider two important applications of Taylor series: approximating definite integrals and solving differential equations. Our first example will involve the function $f(x) = e^{-x^2}$, which is related to an important probability density function (see Section 5-7). This function is nonintegrable in terms of elementary functions, but the integral can be approximated by Taylor series techniques.

Example 18   Approximate $\int_0^1 e^{-x^2}\,dx$ correct to three decimal places.

Solution   **Step 1.** *Find a Taylor series for the integrand.* **From Table 4,**

$$e^x = 1 + x + \frac{1}{2!}x^2 + \frac{1}{3!}x^3 + \cdots + \frac{1}{n!}x^n + \cdots \qquad -\infty < x < \infty$$

$$e^{-x^2} = 1 + (-x^2) + \frac{1}{2}(-x^2)^2 + \frac{1}{6}(-x^2)^3 + \cdots + \frac{1}{n!}(-x^2)^n + \cdots$$

$$= 1 - x^2 + \frac{1}{2}x^4 - \frac{1}{6}x^6 + \cdots + \frac{(-1)^n}{n!}x^{2n} + \cdots$$

Step 2. *Find the Taylor series for the antiderivative.* Using property 5,

$$\int e^{-x^2}\,dx = \int \left[1 - x^2 + \frac{1}{2}x^4 - \frac{1}{6}x^6 + \cdots + \frac{(-1)^n}{n!}x^{2n} + \cdots \right] dx$$

$$= x - \frac{1}{3}x^3 + \frac{1}{10}x^5 - \frac{1}{42}x^7 + \cdots + \frac{(-1)^n}{n!(2n+1)}x^{2n+1} + \cdots$$

Step 3. *Approximate the definite integral.*

$$\int_0^1 e^{-x^2}\,dx = \left[x - \frac{1}{3}x^3 + \frac{1}{10}x^5 - \frac{1}{42}x^7 + \cdots + \frac{(-1)^n}{n!(2n+1)}x^{2n+1} + \cdots \right]_0^1$$

$$= 1 - \frac{1}{3} + \frac{1}{10} - \frac{1}{42} + \frac{1}{216} - \frac{1}{1,320} + \frac{1}{9,360} + \cdots$$

$$= 1 - 0.3333333 + 0.1 - 0.0238095 + 0.0046296 - 0.0007576$$
$$+ 0.0001068 + \cdots$$

Since the seventh term is 0.0001068, the error introduced by approximating the definite integral with the first six terms of this series will be no more than 0.0002. Thus,

$$\int_0^1 e^{-x^2}\,dx \approx 1 - 0.3333333 + 0.1 - 0.0238095 + 0.0046296 - 0.0007576$$

$$\approx 0.747$$

Problem 18   Approximate $\int_0^{0.5} e^{-x^2}\,dx$ correct to four decimal places.

Answer $\int_0^{0.5} e^{-x^2}\,dx \approx 0.5 - 0.0416667 + 0.003125 - 0.000186 \approx 0.4613$

Example 19   Solve the differential equation

$$y'' - xy = 0 \qquad y(0) = 1 \qquad y'(0) = 0 \tag{1}$$

Solution   Since the coefficient of y is not a constant, this equation cannot be solved by the methods discussed in Section 8-4. However, we can find the solution by using Taylor series. Suppose we denote the (unknown) solution by $y(x)$. If we assume that $y(x)$ has a Taylor series at 0, then

$$y(x) = y(0) + y'(0)x + \frac{y''(0)}{2!}x^2 + \frac{y^{(3)}(0)}{3!}x^3 + \cdots + \frac{y^{(n)}(0)}{n!}x^n + \cdots$$

We chose the Taylor series at 0 because the initial conditions were given for $x = 0$. In fact, the initial conditions give us the first two coefficients in the Taylor series [$y(0) = 1$ and $y'(0) = 0$]. Thus,

$$y(x) = 1 + \frac{y''(0)}{2!}x^2 + \frac{y^{(3)}(0)}{3!}x^3 + \frac{y^{(4)}(0)}{4!}x^4 + \cdots$$

How can we evaluate the coefficient of x^2? Since $y(x)$ is assumed to satisfy the differential equation in (1), we can solve that equation for $y''(x)$:

$$y''(x) = xy(x) \tag{2}$$

Evaluating (2) at $x = 0$ implies

$$y''(0) = 0y(0) = 0$$

To evaluate $y^{(3)}(0)$, we first differentiate equation (2):

$$y^{(3)}(x) = xy'(x) + y(x)$$

This implies $y^{(3)}(0) = 0y'(0) + y(0) = 1$. We can continue this process indefinitely:

$$
\begin{aligned}
y^{(4)}(x) &= xy''(x) + 2y'(x) & y^{(4)}(0) &= 0 \\
y^{(5)}(x) &= xy^{(3)}(x) + 3y''(x) & y^{(5)}(0) &= 0 \\
y^{(6)}(x) &= xy^{(4)}(x) + 4y^{(3)}(x) & y^{(6)}(0) &= 4 \\
y^{(7)}(x) &= xy^{(5)}(x) + 5y^{(4)}(x) & y^{(7)}(0) &= 0 \\
y^{(8)}(x) &= xy^{(6)}(x) + 6y^{(5)}(x) & y^{(8)}(0) &= 0 \\
y^{(9)}(x) &= xy^{(7)}(x) + 7y^{(6)}(x) & y^{(9)}(0) &= 28
\end{aligned}
$$

Thus, we have determined the first four nonzero terms of the Taylor series for the solution $y(x)$,

$$y(x) = 1 + \frac{1}{3!}x^3 + \frac{4}{6!}x^6 + \frac{28}{9!}x^9 + \cdots$$

We will not attempt to find a formula for the nth term of this series, nor will we discuss the interval of convergence. These topics are left to a more advanced treatment of the subject.

Problem 19   Find the first four nonzero terms in the Taylor series at 0 for the solution of the differential equation

$$y'' - xy = 1 \qquad y(0) = 0 \qquad y'(0) = 1$$

Answer $y(x) = x + \dfrac{1}{2!}x^2 + \dfrac{2}{4!}x^4 + \dfrac{3}{5!}x^5 + \cdots$

EXERCISE 9-4

A   *Use the Taylor series from Table 4 to approximate each of the following correct to three decimal places. Include the minimal number of terms in the Taylor series that you must use to obtain this accuracy.*

    1. $e^{-0.002}$ 2. $e^{-0.02}$ 3. $e^{-0.2}$ 4. e^{-1}

    5. $\ln(1.001)$ 6. $\ln(1.01)$ 7. $\ln(1.1)$ 8. $\ln(1.4)$

B   9. (A) Use the Taylor series at 0 for $1/(1 - x)$ and property 5 to find the Taylor series at 0 for $1/(1 - x)^2$. Find the interval of convergence.

    (B) Use the series from part A and property 5 to find the Taylor series at 0 for $1/(1 - x)^3$. Find the interval of convergence.

  10. (A) Find the Taylor series at 1 for $1/x$. Find the interval of convergence.

    (B) Use the series from part A and property 5 to find the Taylor series at 1 for $1/x^2$. Find the interval of convergence.

    (C) Use the series from part B and property 5 to find the Taylor series at 1 for $1/x^3$. Find the interval of convergence.

  11. (A) Find the Taylor series at -1 for $1/(1 - x)$. Find the interval of convergence.

    (B) Use the series from part A and property 5 to find the Taylor series at -1 for $1/(1 - x)^2$. Find the interval of convergence.

    (C) Use the series from part B and property 5 to find the Taylor series at -1 for $1/(1 - x)^3$. Find the interval of convergence.

  12. Use the Taylor series at 0 for $\ln(1 + x)$ and property 5 to find the Taylor series at 0 for

$$f(x) = \int_0^x \ln(1 + t)\, dt$$

Find the interval of convergence.

  13. (A) Find the Taylor series at 1 for $\ln x$. Find the interval of convergence.

    (B) Use the series from part A and property 5 to find the Taylor series at 1 for

$$f(x) = \int_1^x \ln t\, dt$$

Find the interval of convergence.

  14. (A) Find the Taylor series at 1 for $\ln(1 + x)$. Find the interval of convergence.

    (B) Use the series from part A and property 5 to find the Taylor series at 1 for

$$f(x) = \int_1^x \ln(1 + t)\, dt$$

Find the interval of convergence.

Use a Taylor series at 0 to approximate each integral correct to three decimal places.

15. $\displaystyle\int_0^{0.2} \frac{1}{1+x^2}\,dx$
16. $\displaystyle\int_0^{0.5} \frac{x}{1+x^4}\,dx$

17. $\displaystyle\int_0^{0.6} \ln(1+x^2)\,dx$
18. $\displaystyle\int_0^{0.7} x\ln(1+x^4)\,dx$

19. $\displaystyle\int_0^{0.5} e^{-x^2}\,dx$
20. $\displaystyle\int_0^{0.8} x^4 e^{-x^2}\,dx$

Find the first four nonzero terms in the Taylor series at 0 for the solution of each differential equation.

21. $y'' - y = x;\quad y(0) = 1,\quad y'(0) = 0$
22. $y'' - y = e^x;\quad y(0) = 0,\quad y'(0) = 1$
23. $y'' + y = e^{-x};\quad y(0) = 0,\quad y'(0) = -1$
24. $y'' - xy = 1 + x;\quad y(0) = 0,\quad y'(0) = 0$
25. $y'' - 2y' + 3y = e^{-x};\quad y(0) = 1,\quad y'(0) = 1$
26. $y'' - 2y' + 3y = x^2;\quad y(0) = 0,\quad y'(0) = 0$

C **27.** Let $f(x) = \dfrac{e^{-x} - 1}{x}$.

   (A) Find a Taylor series at 0 that agrees with $f(x)$ when $x \neq 0$.

   (B) Use the Taylor series in part A to evaluate

$$\lim_{x \to 0} \frac{e^{-x} - 1}{x}$$

   (C) Use the series in part A to approximate

$$\int_0^{0.1} \frac{e^{-x} - 1}{x}$$

   correct to five decimal places.

28. Let $f(x) = \dfrac{\ln(1 + x)}{x}$.

   (A) Find the Taylor series at 0 that agrees with $f(x)$ when $x \neq 0$.

   (B) Use the Taylor series in part A to evaluate

$$\lim_{x \to 0} \frac{\ln(1 + x)}{x}$$

   (C) Use the series in part A to approximate

$$\int_0^{0.1} \frac{\ln(1 + x)}{x}\,dx$$

   correct to four decimal places.

29. Let $f(x) = \dfrac{e^x - e^2}{x - 2}$.

(A) Find the Taylor series at 2 that agrees with $f(x)$ when $x \neq 2$.

(B) Use the Taylor series in part A to evaluate

$$\lim_{x \to 2} \frac{e^x - e^2}{x - 2}$$

30. Let $f(x) = \dfrac{\ln x}{x - 1}$.

(A) Find the Taylor series at 1 that agrees with $f(x)$ when $x \neq 1$.

(B) Use the Taylor series in part A to evaluate

$$\lim_{x \to 1} \frac{\ln x}{x - 1}$$

31. How many terms of the series for e^{-x} at 0 are required to approximate $e^{-0.2}$ correct to five decimal places? To eight decimal places?

32. How many terms of the series for $\ln(1 + x)$ at 0 are required to approximate $\ln 1.2$ correct to five decimal places? To eight decimal places?

33. How many terms of the series for e^{-x} at 0 are required to approximate e^{-1} correct to five decimal places?

34. How many terms of the series for $\ln(1 + x)$ at 0 are required to approximate $\ln 2$ correct to five decimal places?

APPLICATIONS

BUSINESS & ECONOMICS

35. *Manufacturing.* A china manufacturer produces a bowl in the shape of the solid of revolution formed by revolving the region bounded by the graphs of

$$y = \frac{6}{\sqrt{4 + x^2}} \qquad y = 0 \qquad x = 0 \qquad x = 1$$

about the x axis. Express the volume of the bowl as a definite integral and use a Taylor series at 0 to approximate this integral correct to two decimal places.

36. *Supply–demand.* The supply and demand in a certain market satisfy the equations

$$S = 0.1p + 0.05p' + p'' \qquad \text{and} \qquad D = -0.3p - e^{-t}$$

where $p(0) = 10$ and $p'(0) = 0$. (Note that price p changes with time t.) The equilibrium price $p_e(t)$ is the solution to the equation

$$S(t) = D(t)$$

Find the first four nonzero terms in the Taylor series expansion for p_e at 0.

LIFE SCIENCES

37. *Temperature.* The temperature in an artificial habitat is made to change according to the equation

$$C(t) = 5 \ln[40 + 10(t - 1)^2] \qquad 0 \le t \le 2$$

(in degrees Celsius) over a 2 hour period. Use the Taylor series expansion for $C(t)$ at 1 to approximate the average temperature during this period correct to two decimal places.

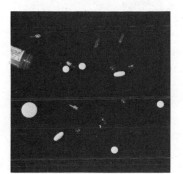

38. *Drug assimilation.* After a person takes a pill, the body begins to assimilate the drug contained in the pill into the bloodstream. The rate of assimilation t minutes after taking the pill is

$$R(t) = 10t^2 e^{-0.1t^2}$$

Use the Taylor series at 0 to approximate, correct to two decimal places, the total amount of the drug that has entered the bloodstream 1 minute after the pill is taken.

SOCIAL SCIENCES

39. *Learning.* A person learns N items at a rate given by

$$N'(t) = \frac{500}{100 + t^2}$$

where t is the number of minutes of continuous study. Express the total number of items learned in 5 minutes of study as a definite integral and use a Taylor series at 0 to approximate the integral.

9-5 INDETERMINATE FORMS AND L'HÔPITAL'S RULE

Our first application of the concept of limit involved the slope of the tangent line at a point on a graph (see Section 2-3). We immediately encountered limits that could not be evaluated by direct substitution. For example, if $f(x) = x^2$, then the slope of the tangent line at the point $(2, 4)$ is given by *

$$\lim_{x \to 0} \frac{f(2 + x) - f(2)}{x} = \lim_{x \to 0} \frac{(2 + x)^2 - 4}{x}$$

If we try to evaluate this limit by substituting $x = 0$ in the numerator and in the denominator, we obtain the **indeterminate form 0/0.** It is impossible to assign a value to the symbol 0/0 that is consistent with the

---

*The form $\displaystyle\lim_{x \to 0} \frac{f(2 + x) - f(2)}{x}$ is equivalent to the form $\displaystyle\lim_{\Delta x \to 0} \frac{f(2 + \Delta x) - f(2)}{\Delta x}$,

but has the advantage of making the work that follows simpler.

rules of arithmetic, so we generally say that 0/0 is meaningless. Nevertheless, as we saw in Section 2-3, limits that involve indeterminate forms can still be evaluated. Usually, we first perform some algebraic simplifications, such as

$$\lim_{x \to 0} \frac{(2 + x)^2 - 4}{x} = \lim_{x \to 0} \frac{4 + 4x + x^2 - 4}{x}$$

$$= \lim_{x \to 0} \frac{4x + x^2}{x} = \lim_{x \to 0} \frac{x(4 + x)}{x} = \lim_{x \to 0}(4 + x) = 4$$

This method works as long as we can perform the necessary algebraic manipulations, but what if we cannot simplify the function? Consider the limit

$$\lim_{x \to 1} \frac{e^x - e}{x - 1}$$

As before, substituting $x = 1$ produces the indeterminate form 0/0. However, this time there are no apparent algebraic simplifications that will allow us to evaluate this limit by direct substitution. Fortunately, there is a procedure that can be used in cases like this. The rule given in the box, known as **L'Hôpital's rule,** can be used whenever a limit involves the indeterminate form 0/0.

L'HÔPITAL'S RULE

If $\lim_{x \to c} f(x) = 0$ and $\lim_{x \to c} g(x) = 0$, and if

$$\lim_{x \to c} \frac{D_x f(x)}{D_x g(x)} \qquad \text{exists}$$

then

$$\lim_{x \to c} \frac{f(x)}{g(x)} \qquad \text{exists}$$

and

$$\lim_{x \to c} \frac{f(x)}{g(x)} = \lim_{x \to c} \frac{D_x f(x)}{D_x g(x)}$$

If

$$\lim_{x \to c} \frac{D_x f(x)}{D_x g(x)} \qquad \text{fails to exist}$$

because $|D_x f(x)/D_x g(x)|$ becomes arbitrarily large for x near c, then

$$\lim_{x \to c} \frac{f(x)}{g(x)} \qquad \text{also fails to exist}$$

Example 20   Use L'Hôpital's rule to evaluate: $\lim\limits_{x \to 0} \dfrac{(2 + x)^2 - 4}{x}$

Solution   **Step 1.**   *Check to see if L'Hôpital's rule applies.*

$$\lim_{x \to 0}[(2 + x)^2 - 4] = 0 \qquad \text{and} \qquad \lim_{x \to 0} x = 0$$

Step 2.   *Apply L'Hôpital's rule.*

$$\lim_{x \to 0} \frac{D_x[(2 + x)^2 - 4]}{D_x\, x} = \lim_{x \to 0} \frac{2(2 + x)}{1} = \frac{2(2)}{1} = 4$$

Thus,

$$\lim_{x \to 0} \frac{(2 + x)^2 - 4}{x} = 4$$

Problem 20   Use L'Hôpital's rule to evaluate: $\lim\limits_{x \to 0} \dfrac{(2 + x)^3 - 8}{x}$

Answer   12

Example 21   Evaluate: $\lim\limits_{x \to 1} \dfrac{e^x - e}{x - 1}$

Solution   **Step 1.**   *Check to see if L'Hôpital's rule applies.*

$$\lim_{x \to 1}(e^x - e) = e^1 - e = 0 \qquad \text{and} \qquad \lim_{x \to 1}(x - 1) = 0$$

Step 2.   *Apply L'Hôpital's rule.*

$$\lim_{x \to 1} \frac{D_x(e^x - e)}{D_x(x - 1)} = \lim_{x \to 1} \frac{e^x}{1} = \frac{e^1}{1} = e$$

Thus,

$$\lim_{x \to 1} \frac{e^x - e}{x - 1} = e$$

Problem 21   Evaluate: $\lim\limits_{x \to 4} \dfrac{e^x - e^4}{x - 4}$

Answer e^4

Example 22   Evaluate: $\lim\limits_{x \to 1} \dfrac{\ln x}{x}$

Solution   Step 1. *Check to see if L'Hôpital's rule applies.*

$$\lim_{x \to 1} \ln x = \ln 1 = 0 \qquad \text{but} \qquad \lim_{x \to 1} x = 1 \neq 0$$

So L'Hôpital's rule does not apply.

Step 2. *Evaluate by another method.* By direct substitution,

$$\lim_{x \to 1} \frac{\ln x}{x} = \frac{\ln 1}{1} = \frac{0}{1} = 0$$

Problem 22   Evaluate: $\displaystyle \lim_{x \to 0} \frac{x}{e^x}$

Answer   0

Example 22 points out that you must always check to see if L'Hôpital's rule applies before you use the rule. If you had applied L'Hôpital's rule in Example 22 without checking, you would have obtained

$$\lim_{x \to 1} \frac{D_x(\ln x)}{D_x x} = \lim_{x \to 1} \frac{1/x}{1} = 1$$

which is not the correct answer.

Example 23   Evaluate: $\displaystyle \lim_{x \to 1} \frac{e^x}{x - 1}$

Solution   Step 1. *Check to see if L'Hôpital's rule applies.*

$$\lim_{x \to 1} e^x = 1 \neq 0 \qquad \lim_{x \to 1} (x - 1) = 0$$

L'Hôpital's rule does not apply.

Step 2. *Evaluate by another method.* Since the numerator is approaching a nonzero number and the denominator is approaching 0, the value of $e^x/(x - 1)$ is becoming arbitrarily large and

$$\lim_{x \to 1} \frac{e^x}{x - 1} \qquad \text{does not exist}$$

Problem 23   Evaluate: $\displaystyle \lim_{x \to 2} \frac{\ln x}{x - 2}$

Answer   Does not exist.

Example 24   Evaluate: $\displaystyle \lim_{x \to 0} \frac{x^2}{e^x - 1 - x}$

Solution **Step 1.** *Check to see if L'Hôpital's rule applies.*

$$\lim_{x \to 0} x^2 = 0 \quad \text{and} \quad \lim_{x \to 0} (e^x - 1 - x) = 0$$

L'Hôpital's rule applies.

Step 2. *Apply L'Hôpital's rule.*

$$\lim_{x \to 0} \frac{D_x\, x^2}{D_x(e^x - 1 - x)} = \lim_{x \to 0} \frac{2x}{e^x - 1}$$

Since this limit cannot be evaluated by direct substitution, we try L'Hôpital's rule again.

Step 3. *Check to see if L'Hôpital's rule applies a second time.*

$$\lim_{x \to 0} 2x = 0 \quad \text{and} \quad \lim_{x \to 0} (e^x - 1) = 0$$

L'Hôpital's rule applies.

Step 4. *Apply L'Hôpital's rule.*

$$\lim_{x \to 0} \frac{D_x\, 2x}{D_x(e^x - 1)} = \lim_{x \to 0} \frac{2}{e^x} = \frac{2}{1} = 2$$

Thus,

$$\lim_{x \to 0} \frac{x^2}{e^x - 1 - x} = \lim_{x \to 0} \frac{2x}{e^x - 1} = 2$$

Problem 24 Evaluate: $\displaystyle \lim_{x \to 0} \frac{0^{2x} \quad 1 \quad 2x}{x^2}$

Answer 2

Example 25 Evaluate: $\displaystyle \lim_{x \to 1} \frac{\ln x}{(x - 1)^2}$

Solution **Step 1.** *Check to see if L'Hôpital's rule applies.*

$$\lim_{x \to 1} \ln x = \ln 1 = 0 \quad \text{and} \quad \lim_{x \to 1}(x - 1)^2 = 0$$

Step 2. *Apply L'Hôpital's rule.*

$$\lim_{x \to 1} \frac{D_x(\ln x)}{D_x(x - 1)^2} = \lim_{x \to 1} \frac{1/x}{2(x - 1)} = \lim_{x \to 1} \frac{1}{2x(x - 1)}$$

As $x \to 1$, the value of $|1/2x(x - 1)|$ becomes arbitrarily large and the limit does not exist. Thus,

$$\lim_{x \to 1} \frac{\ln x}{(x - 1)^2} \qquad \text{does not exist}$$

Problem 25   Evaluate: $\lim\limits_{x \to 0} \dfrac{x}{e^{x^2} - 1}$

Answer   Does not exist.

EXERCISE 9-5

A   *Find each limit using L'Hôpital's rule.*

1. $\lim\limits_{x \to 2} \dfrac{x^2 + x - 6}{x^2 + 6x - 16}$

2. $\lim\limits_{x \to 4} \dfrac{x^2 - 8x + 16}{x^2 - 5x + 4}$

3. $\lim\limits_{x \to 0} \dfrac{e^x - 1}{x}$

4. $\lim\limits_{x \to 1} \dfrac{\ln x}{x - 1}$

5. $\lim\limits_{x \to 0} \dfrac{e^{2x} - 1}{x}$

6. $\lim\limits_{x \to 0} \dfrac{\ln(1 + 4x)}{x}$

7. $\lim\limits_{x \to 1} \dfrac{x^2 + x - 2}{\ln x}$

8. $\lim\limits_{x \to 2} \dfrac{x^2 - 3x + 2}{e^x - e^2}$

9. $\lim\limits_{x \to 4} \dfrac{\ln(5 - x)}{4 - x}$

10. $\lim\limits_{x \to 4} \dfrac{\sqrt{x} - 2}{x - 4}$

B   *Problems 11–34 are mixed. L'Hôpital's rule does not apply to every problem. Some problems will require more than one application of L'Hôpital's rule.*

11. $\lim\limits_{x \to 0} \dfrac{\ln(1 + x^2)}{x}$

12. $\lim\limits_{x \to 0} \dfrac{\ln(1 + x^2)}{x^2}$

13. $\lim\limits_{x \to 0} \dfrac{\ln(1 + x^2)}{x^3}$

14. $\lim\limits_{x \to 0} \dfrac{\ln(1 + x^2)}{1 + x^2}$

15. $\lim\limits_{x \to -2} \dfrac{x^2 + 2x + 1}{x^2 + x + 1}$

16. $\lim\limits_{x \to 1} \dfrac{2x^3 - 3x^2 + 1}{x^3 - 3x + 2}$

17. $\lim\limits_{x \to 2} \dfrac{2 - x + \ln(x - 1)}{x^3 - 3x^2 + 4}$

18. $\lim\limits_{x \to 0} \dfrac{e^{-2x} - 1 + 2x}{e^x - 1 - x}$

19. $\lim\limits_{x \to 0} \dfrac{e^{x^2} - 1}{x^2}$

20. $\lim\limits_{x \to 0} \dfrac{xe^x - x}{\ln(1 + x)}$

21. $\lim\limits_{x \to 0} \dfrac{\sqrt{1 + x^4} - 1}{x^2}$

22. $\lim\limits_{x \to 3} \dfrac{2x^3 - 9x^2 + 27}{x^3 - 7x - 6}$

23. $\lim\limits_{x \to 1} \dfrac{x^3 + 3x - 4}{x^3 - 3x + 2}$

24. $\lim\limits_{x \to 4} \dfrac{\sqrt{x^2 + 9} - 5}{2x - 8}$

25. $\lim\limits_{x \to 1} \dfrac{x \ln x - x}{e^x - 1}$

26. $\lim\limits_{x \to 0} \dfrac{e^x - e^{-x} - 2x}{x^3}$

27. $\displaystyle\lim_{x \to 1} \frac{(x + 1)\ln x + 1 - x^2}{(x - 1)^2}$

28. $\displaystyle\lim_{x \to 3} \frac{x^2 + 2x - 3}{x^2 - 2x - 3}$

29. $\displaystyle\lim_{x \to -2} \frac{\ln(x^2 + 4x + 5)}{x + 2}$

30. $\displaystyle\lim_{x \to 2} \frac{\ln(x^2 - 4x + 5)}{x + 2}$

31. $\displaystyle\lim_{x \to -2} \frac{\ln(x^2 - 4x + 5)}{x + 2}$

32. $\displaystyle\lim_{x \to 1} \frac{\ln(1 + 2\ln x)}{\ln(2 - x)}$

33. $\displaystyle\lim_{x \to 1} \frac{(\ln x)^2}{x^3 - x^2 - x + 1}$

34. $\displaystyle\lim_{x \to 1} \frac{(\ln x)^2}{x^4 - 2x^3 + 2x - 1}$

9-6   CHAPTER REVIEW

**IMPORTANT TERMS
AND SYMBOLS**

9-1 Taylor polynomials at 0.   nth-degree Taylor polynomial at 0,

$$p_n(x) = f(0) + f'(0)x + \frac{f''(0)}{2!}x^2 + \cdots + \frac{f^{(n)}(0)}{n!}x^n$$

9-2 Taylor polynomials at a.   nth-degree Taylor polynomial at a,

$$p_n(x) = f(a) + f'(a)(x - a) + \frac{f''(a)}{2!}(x - a)^2 + \cdots + \frac{f^{(n)}(a)}{n!}(x - a)^n$$

9-3 Taylor series.   Taylor series,   interval of convergence

$$\frac{1}{1 - x} - 1 + x + x^2 \mid x^3 \mid \quad \mid x^n \mid \quad , \quad -1 < x < 1$$

$$e^x = 1 + x + \frac{1}{2!}x^2 + \frac{1}{3!}x^3 + \cdots + \frac{1}{n!}x^n + \cdots, \quad -\infty < x < \infty$$

$$\ln(1 + x) = x - \frac{1}{2}x^2 + \frac{1}{3}x^3 - \frac{1}{4}x^4 + \cdots + \frac{(-1)^{n+1}}{n}x^n + \cdots,$$

$$-1 < x < 1$$

If $f(x) = a_0 + a_1(x - a) + \cdots + a_n(x - a)^n + \cdots$,
$g(x) = b_0 + b_1(x - a) + \cdots + b_n(x - a)^n + \cdots$,
and $y = h(x)$, then:

Property 1: $f(x) + g(x) = (a_0 + b_0) + (a_1 + b_1)(x - a) + \cdots$
$\qquad\qquad\qquad\qquad\qquad + (a_n + b_n)(x - a)^n + \cdots$

Property 2: $cf(x) = ca_0 + ca_1(x - a) + \cdots + ca_n(x - a)^n + \cdots$,
$\qquad\qquad$ where c is a constant

Property 3: $(x - a)^r f(x) = a_0(x - a)^r + a_1(x - a)^{r+1} + \cdots$
$\qquad\qquad\qquad + a_n(x - a)^{r+n} + \cdots$, where r is a positive integer

Property 4: $f[h(x)] = a_0 + a_1[h(x) - a] + \cdots + a_n[h(x) - a]^n + \cdots$

9-4 Differentiation and integration of Taylor series.

Property 5:   If $f(x) = a_0 + a_1(x - a) + a_2(x - a)^2 + \cdots$
$$+ a_n(x - a)^n + \cdots,$$

then

$$f'(x) = a_1 + 2a_2(x - a) + \cdots + na_n(x - a)^{n-1} + \cdots$$

$$\int f(x)\, dx = a_0(x - a) + \frac{1}{2}a_1(x - a)^2 + \frac{1}{3}a_2(x - a)^3 + \cdots$$

$$+ \frac{1}{n + 1}a_n(x - a)^{n+1} + \cdots + C$$

alternating series,   error estimation, $|f(x_0) - p_n(x_0)| \le |a_{n+1}(x_0 - a)^{n+1}|$
when $a_0, a_1(x_0 - a), \ldots, a_n(x_0 - a)^n$ alternate in sign,   approximation
of definite integrals, differential equations

9-5 Indeterminate forms and L'Hôpital's rule.   indeterminate form,   0/0

$$\lim_{x \to c} \frac{f(x)}{g(x)} = \lim_{x \to c} \frac{D_x f(x)}{D_x g(x)} \quad \text{if} \quad \lim_{x \to c} f(x) = 0 \quad \text{and} \quad \lim_{x \to c} g(x) = 0$$

EXERCISE 9-6   **CHAPTER REVIEW**

*Work through all the problems in this chapter review and check your
answers in the back of the book. (Answers to all review problems are
there.) Where weaknesses show up, review appropriate sections in the
text. When you are satisfied that you know the material, take the
practice test following this review.*

A   *Find the Taylor polynomial $p_3(x)$ at the indicated value of a.*

1. $f(x) = \sqrt{1 + x}$ at 0   2. $f(x) = \sqrt{1 + x}$ at 3
3. $f(x) = \sqrt[3]{1 + x}$ at 0   4. $f(x) = \sqrt[3]{1 + x}$ at 7
5. $f(x) = \sqrt{9 + x^2}$ at 0   6. $f(x) = \sqrt{9 + x^2}$ at 4
7.   Use $p_3(x)$ in Problem 1 to approximate $\sqrt{1.01}$.
8.   Use $p_3(x)$ in Problem 2 to approximate $\sqrt{3.9}$.
9.   Use $p_3(x)$ in Problem 3 to approximate $\sqrt[3]{1.01}$.
10.   Use $p_3(x)$ in Problem 4 to approximate $\sqrt[3]{7.8}$.
11.   Use $p_3(x)$ in Problem 5 and $x = 0.1$ to approximate $\sqrt{9.01}$.
12.   Use $p_3(x)$ in Problem 6 and $x = 4.1$ to approximate $\sqrt{25.81}$.

B   *In Problems 13–20 use Table 4 and properties 1–5 to find each Taylor
series expansions at the indicated value of a. Find the interval of
convergence.*

13. $f(x) = \dfrac{1}{4 - x}$ at 0   14. $f(x) = \dfrac{1}{4 - x}$ at 2

15. $f(x) = \dfrac{1}{4 - x}$ at -4

16. $f(x) = \dfrac{x^2}{4 - x^2}$ at 0

17. $f(x) = x^2 e^{3x}$ at 0

18. $f(x) = e^{2x}$ at 2

19. $f(x) = x \ln(e + x)$ at 0

20. $f(x) = \ln(2x - 2)$ at 2

21. Expand $1/x$ at 2 and use property 5 to obtain expansions for $1/x^2$ and $1/x^3$ at 2.

22. Expand $1/x$ at -1 and use property 5 to obtain expansions for $1/x^2$ and $1/x^3$ at 2.

In Problems 23 and 24 find the Taylor series expansion at 0 and its interval of convergence.

23. $f(x) = \displaystyle\int_0^x \dfrac{t^2}{9 + t^2}\, dt$

24. $f(x) = \displaystyle\int_0^x \dfrac{t^4}{16 - t^2}\, dt$

In Problems 25–30 evaluate each limit.

25. $\displaystyle\lim_{x \to 0} \dfrac{e^{3x} - 1}{x}$

26. $\displaystyle\lim_{x \to 2} \dfrac{x^2 - 5x + 6}{x^2 + x - 6}$

27. $\displaystyle\lim_{x \to 0} \dfrac{\ln(1 + x)}{x^2}$

28. $\displaystyle\lim_{x \to 0} \dfrac{\ln(1 + x)}{1 + x}$

29. $\displaystyle\lim_{x \to 1} \dfrac{x^2 - x + 1}{x^2 + 2x - 3}$

30. $\displaystyle\lim_{x \to 0} \dfrac{e^x + e^{-x} - 2}{x^2}$

C   Approximate each integral correct to four decimal places.

31. $\displaystyle\int_0^1 \dfrac{1}{16 + x^2}\, dx$

32. $\displaystyle\int_0^1 x^2 e^{-0.1x^2}\, dx$

33. $\displaystyle\int_0^1 \ln\left(1 + \dfrac{1}{9}x^2\right) dx$

34. $\displaystyle\int_1^{1.1} \dfrac{\ln x}{x - 1}\, dx$

Find the first four nonzero terms in the Taylor series expansion at 0 for the solution of the indicated differential equation.

35. $y'' - 2y = x^3;$ $y(0) = 0,$ $y'(0) = 1$

36. $y'' - x^2 y = 1;$ $y(0) = 1,$ $y'(0) = 0$

APPLICATIONS

BUSINESS & ECONOMICS

37. *Marginal analysis.* A company has a vending machine with the following marginal cost and revenue equations (in thousands of dollars per year):

$$C'(t) = 2e^{0.005t^2} \qquad R'(t) = 12 - 1.95t$$

where $C(t)$ and $R(t)$ represent total accumulated costs and revenues, respectively, t years after the machine is put into use. The area between the graphs of the marginal equations for the time period such that $R'(t) \geq C'(t)$ represents the total accumulated profit for the useful life of the machine. Approximate $C'(t)$ with its second-degree Taylor polynomial and find the useful life of the machine and the total accumulated profit.

38. *Supply–demand.* The supply and demand (in thousands of units) for a certain market are given by the equations

$$S(p) = 6 + 41p \quad \text{and} \quad D(p) = 10 + 8e^{-p}$$

where p is the price in dollars. The equilibrium price is the solution of the equation

$$S(p) = D(p)$$

Use the second-degree Taylor polynomial at 0 for D to approximate the equilibrium price.

LIFE SCIENCES

39. *Insulin level.* A large injection of insulin is administered to a patient. The level of insulin in the bloodstream t minutes after the injection is given approximately by

$$L(t) = \frac{5{,}000t^2}{10{,}000 + t^4}$$

Express the average insulin level for the 5 minute interval immediately following the injection as a definite integral. Use a Taylor series at 0 to approximate the integral correct to two decimal places.

40. *Energy expenditure.* The rate at which a person uses energy during vigorous exercise is given by

$$E(t) = \begin{cases} 5 - 18 \ln\left[1 + \left(\dfrac{t-4}{2}\right)^2\right] & 3 \leq t \leq 5 \\ 1 & \text{otherwise} \end{cases}$$

Express the total energy used during the period from $t = 3$ to $t = 5$ as a definite integral. Use a Taylor series at 4 to approximate the integral correct to two decimal places.

SOCIAL SCIENCES

41. *Politics.* In a newly incorporated city the number of voters (in thousands) during the first 5 years was found to be given by

$$N(t) = 10 + 2t - 5e^{-0.01t^2} \qquad 0 \leq t \leq 5$$

Express the average number of voters during this 5 year period as a definite integral. Use a Taylor series at 0 to approximate the integral correct to one decimal place.

PRACTICE TEST: CHAPTER 9

1. Find the third-degree Taylor polynomial $p_3(x)$ at 1 for $f(x) = 1/\sqrt{x}$.
2. Find the Taylor series at 0 and the interval of convergence.

   (A) $f(x) = \ln(4 + x)$ (B) $f(x) = \dfrac{e^x - e^{-x}}{2}$

3. Use the Taylor series at 0 for e^x to approximate $e^{-0.3}$ correct to four decimal places. Indicate the minimum number of terms required to obtain this accuracy.
4. Approximate

$$\int_0^{0.6} \frac{t^5}{1 + t^3}\, dt$$

   correct to four decimal places.
5. Find each limit if it exists:

   (A) $\lim\limits_{x \to 1} \dfrac{(x - 1)^2}{\ln x}$ (B) $\lim\limits_{x \to 1} \dfrac{(x - 1)^2}{\ln x + 1 - x}$ (C) $\lim\limits_{x \to 1} \dfrac{(x - 1)}{\ln x + 1 - x}$

6. Find Taylor series for $1/(1 + x)$ and $1/(1 + x)^2$ at 2. In each case, find the interval of convergence.
7. Find the first four nonzero terms of the Taylor series at 0 for the solution of $y'' - 3y = x$, $y(0) = 1$, $y'(0) = 0$.
8. The revenue and cost functions for a company are $R(x) = xe^{-0.0001x}$ and $C(x) = 135 + 0.085x$, respectively. Use the second-degree Taylor polynomial at 0 for $R(x)$ in order to approximate the break-even point.

CHAPTER 10
NUMERICAL TECHNIQUES

CONTENTS

10 NUMERICAL TECHNIQUES

In earlier chapters we have developed techniques for evaluating definite integrals and solving differential equations. But many problems cannot be solved explicitly by using these techniques. In such cases, numerical methods are used to approximate the desired solutions. In this chapter we will develop methods for approximating the roots of an equation, the value of a definite integral, and the solution of a differential equation. In actual practice, these approximations are computed by using a digital computer. For the beginning student, it is preferable to perform the calculations on a hand-held calculator. We have used a scientific calculator to perform our calculations. You will need a calculator with similar capabilities in order to work the problems and exercises in this chapter.

CALCULATOR VARIATION

Do not be alarmed if your calculator results do not agree exactly with those printed in this chapter or in the answer section. Since there is a large variety of scientific calculators on the market and not all these perform operations in exactly the same way, you can expect slight variations in the last two or three decimal places when the calculators are used to their full capacities. Slight variations in results may even occur in the same calculator when a given calculation is performed in two different ways.

10-1   ROOT APPROXIMATION

A number r is a **zero** of a function f or a **root** of the equation $f(x) = 0$ if $f(r) = 0$. If $f(x) = ax^2 + bx + c$ is any second-degree polynomial, then its zeros are given by the **quadratic formula**

$$r = \frac{-b \pm \sqrt{b^2 - 4ac}}{2a}$$

Example 1   Find the zeros of $f(x) = x^2 - 8x + 13$.

Solution   Since f is a second-degree polynomial and $a = 1$, $b = -8$, $c = 13$, the zeros are

$$r_1 = \frac{8 - \sqrt{64 - 52}}{2} = \frac{8 - 2\sqrt{3}}{2} = 4 - \sqrt{3} \approx 2.2679492$$

$$r_2 = \frac{8 + \sqrt{64 - 52}}{2} = 4 + \sqrt{3} \approx 5.7320508$$

Problem 1   Use the quadratic formula to find the zeros of $f(x) = x^2 - 6x + 4$.

Answer $r_1 = 3 + \sqrt{5} \approx 5.236068$, $r_2 = 3 - \sqrt{5} \approx 0.76393202$

If we graph $y = f(x)$, then the zeros are often referred to as the **x intercepts,** since these are the points where the graph crosses the x axis. Figure 1 illustrates this for the function in Example 1.

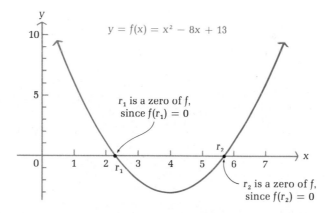

Using algebraic methods to find the exact zeros of more complicated functions can be very difficult, if not impossible. Many methods have been developed to approximate the zeros of a function. One of these—usually called **Newton's method**—is particularly well-suited to being used with a calculator.

**NEWTON'S METHOD:
AN INFORMAL
DEVELOPMENT**   Suppose we are given a function f and we wish to approximate an (unknown) zero r (that is, find a number r such that $f(r) = 0$). To begin, recall that the equation of the line tangent to the graph of $y = f(x)$ at

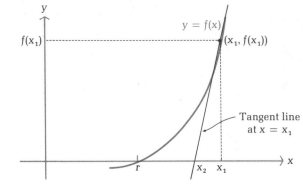

FIGURE 2

$x = x_1$ is

$$y - f(x_1) = f'(x_1)(x - x_1) \tag{1}$$

if $f'(x_1)$ exists (see Section 2-4). If we choose an initial value x_1 which we suspect is close to the zero r, then the x intercept of the tangent line at $x = x_1$ is generally even closer to r (see Fig. 2). Let x_2 denote the x intercept of the tangent line at $x = x_1$. Now we can substitute $x = x_2$ and $y = 0$ in (1) and solve for x_2:

$$0 - f(x_1) = f'(x_1)(x_2 - x_1)$$

$$x_2 - x_1 = -\frac{f(x_1)}{f'(x_1)} \qquad f'(x_1) \neq 0$$

$$x_2 = x_1 - \frac{f(x_1)}{f'(x_1)} \tag{2}$$

Notice that if $f'(x_1) = 0$, then the tangent line at $x = x_1$ is horizontal and no x intercept exists. To avoid this difficulty, we will assume that $f'(x) \neq 0$ for any values of x near the zero r. As Figure 2 indicates, x_2 is usually a better approximation to r than x_1. If we repeat this process, beginning now with the tangent line at $x = x_2$, we should obtain an even better approximation to r (see Fig. 3).

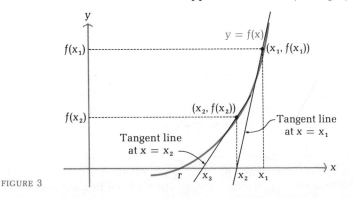

FIGURE 3

We let x_3 be the x intercept of the tangent line at $x = x_2$ and proceed exactly as before:

$$y - f(x_2) = f'(x_2)(x - x_2) \qquad \text{Tangent line at } x = x_2$$

$$0 - f(x_2) = f'(x_2)(x_3 - x_2) \qquad \text{Substitute } x = x_3 \text{ and } y = 0$$

$$x_3 = x_2 - \frac{f(x_2)}{f'(x_2)} \qquad \text{Solve for } x_3 \tag{3}$$

Notice the similarity between the expression for x_2 in equation (2) and the one for x_3 in equation (3). Continued repetition of this process will produce a sequence of numbers $x_1, x_2, \ldots, x_n, \ldots$, where each number in the sequence (after x_1) is obtained by using the formula

$$x_n = x_{n-1} - \frac{f(x_{n-1})}{f'(x_{n-1})} \qquad n > 1$$

The process described above is referred to as **Newton's method of root approximation.**

Example 2   Use Newton's method to approximate the smaller zero of $f(x) = x^2 - 8x + 13$. Compare the result with that obtained by use of the quadratic formula in Example 1.

Solution   **Step 1.**   *Find the formula for x_n.*

$$f(x) = x^2 - 8x + 13$$

$$f'(x) = 2x - 8$$

$$x_n = x_{n-1} - \frac{f(x_{n-1})}{f'(x_{n-1})}$$

$$= x_{n-1} - \frac{x_{n-1}^2 - 8x_{n-1} + 13}{2x_{n-1} - 8}$$

$$= \frac{x_{n-1}(2x_{n-1} - 8) - (x_{n-1}^2 - 8x_{n-1} + 13)}{2x_{n-1} - 8}$$

$$x_n = \frac{x_{n-1}^2 - 13}{2x_{n-1} - 8} \tag{4}$$

Step 2.   *Approximate the root using equation (4).* **Examining the graph of $y = f(x)$ in Figure 1 (and ignoring the fact that we know the exact root), we see that $x_1 = 2$ is a reasonable first approximation to the root r_1. Using equation (4) and $x_1 = 2$ as a first approximation, we obtain the following successive approximations for r_1:**

$$x_1 = 2$$

$$x_2 = \frac{(2)^2 - 13}{2(2) - 8} = 2.25$$

$$x_3 = \frac{(2.25)^2 - 13}{2(2.25) - 8} = 2.2678571$$

$$x_4 = \frac{(2.2678571)^2 - 13}{2\,(2.2678571) - 8} = 2.2679492$$

$$x_5 = \frac{(2.2679492)^2 - 13}{2(2.2679492) - 8} = 2.2679492$$

Since $x_4 = x_5$ in the display in our calculator, we have reached the limits of accuracy for our calculator and assume $r_1 \approx 2.2679492$. Notice that this is the *same approximation to the exact zero* that we obtained in Example 1 using the quadratic formula and our calculator directly.

Problem 2   Use Newton's method to approximate the larger zero of $f(x) = x^2 - 8x + 13$. First sketch a graph of the function f to locate its zeros approximately. Then use Newton's method to approximate the larger root to the limit of accuracy of your calculator; that is, until two successive approximations are the same in the display of your calculator.

Answer $r_2 \approx 5.7320508$

In Example 2, it took us only three calculations to approximate the zero correct to eight significant digits. We were able to obtain this much accuracy with so few computations because the initial approximation was reasonably close to the actual zero. It is always important to try to select a good first approximation. Figures 4 and 5 illustrate two situations that may occur if x_1 is too far away from r. In Figure 4, each x_n is larger than the preceding value and further away from r. This type of behavior is very common if the initial approximation x_1 is not selected carefully. In Figure 5, the approximating values oscillate between x_1 and x_2, never getting any closer to r. If either of these situations

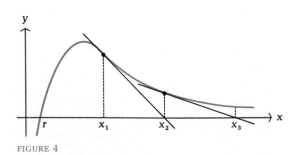

FIGURE 4

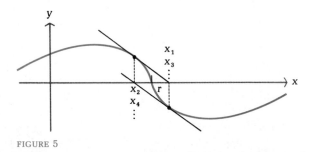

FIGURE 5

occur when you are using Newton's method, you must select a better initial value for x_1.

NEWTON'S METHOD

Given the function f and the initial approximation x_1, define

$$x_n = x_{n-1} - \frac{f(x_{n-1})}{f'(x_{n-1})} \qquad n > 1, \quad f'(x_{n-1}) \neq 0$$

If $\lim_{n \to \infty} x_n$ exists, then

$$r = \lim_{n \to \infty} x_n$$

is a zero of f.

Example 3   Sketch the graph of $f(x) = x^3 - 9x^2 + 15x + 10$ and approximate the x intercepts.

Solution   **Step 1.**   *Sketch the graph of f.*

$f'(x) = 3x^2 - 18x + 15 = 3(x - 5)(x - 1)$
Critical values: $x = 1, 5$

The graph of f rises for x on the intervals $(-\infty, 1)$ and $(5, \infty)$ and falls for x on the interval $(1, 5)$.

$f''(x) = 6x - 18 = 6(x - 3)$

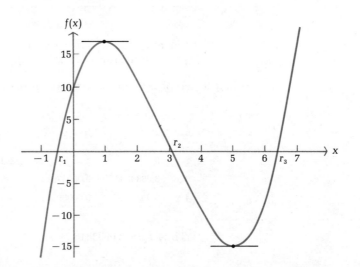

The graph of f is concave down for x on the interval $(-\infty, 3)$ and concave up for x on the interval $(3, \infty)$. The figure on page 469 indicates that f has three zeros: r_1 in $(-1, 0)$, r_2 in $(3, 4)$, and r_3 in $(6, 7)$.

Step 2. *Approximate r_1.* We will use $x_1 = -1$ for an initial approximation and tabulate partial and final results in a table. You may prefer to do the complete calculation in your calculator without writing down partial results. If you have a calculator with sufficient storage capacity, this procedure is recommended. To facilitate calculator computation, we write $f(x)$ and $f'(x)$ in the following "nested factored" forms:

$$f(x) = x^3 - 9x^2 + 15x + 10 = [(x - 9)x + 15]x + 10$$
$$f'(x) = 3x^2 - 18x + 15 = (3x - 18)x + 15$$

n	x_{n-1}	$f(x_{n-1})$	$f'(x_{n-1})$	$x_n = x_{n-1} - \dfrac{f(x_{n-1})}{f'(x_{n-1})}$
2	-1	-15	36	$-1 - \dfrac{-15}{36} = -0.58333333$
3	-0.58333333	-2.0109953	26.520833	$-0.58333333 - \dfrac{-2.0109953}{26.520833} = -0.50750633$
4	-0.50750633	-0.06137372	24.907803	$-0.50750632 - \dfrac{-0.06137372}{24.907802} = -0.50504229$
5	-0.50504229	-0.00006377	24.855964	$-0.50504229 - \dfrac{-0.00006377}{24.855964} = -0.50503972$
6	-0.50503972	0.00000011	19.855910	$-0.505003972 - \dfrac{0.00000011}{19.855910} = -0.50503973$

Since x_5 and x_6 are very nearly the same, we conclude that $r_1 \approx -0.5050397$ is a good approximation.

Step 3. *Approximate r_2.* Initial approximation: $x_1 = 3$.

n	x_{n-1}	$f(x_{n-1})$	$f'(x_{n-1})$	$x_n = x_{n-1} - \dfrac{f(x_{n-1})}{f'(x_{n-1})}$
2	3	1	-12	3.0833333
3	3.0833333	0.00057911	-11.979167	3.0833816
4	3.0833816	0.00000051	-11.979143	3.0833816

Thus, $r_2 \approx 3.0833816$.

Step 4. *Approximate* r_3. Initial approximation: $x_1 = 6$.

n	x_{n-1}	$f(x_{n-1})$	$f'(x_{n-1})$	$x_n = x_{n-1} - \dfrac{f(x_{n-1})}{f'(x_{n-1})}$
2	6	-8	15	6.5333333
3	6.5333333	2.7117036	25.453333	6.426797
4	6.426797	0.11910063	23.228814	6.4216698
5	6.4216698	0.00027004	23.123472	6.4216581
6	6.4216581	0.00000039	23.123232	6.4216581

Thus, $r_3 \approx 6.4216581$.

Problem 3 Repeat Example 3 for $f(x) = x^4 - 4x^3 + 10$.

Answer $r_1 \approx 1.6117934$, $r_2 \approx 3.8207044$

Example 4 Determine the points of intersection for the graphs of $f(x) = c^{-x}$ and $g(x) = 2x$.

Solution The figure indicates that the graphs of $f(x)$ and $g(x)$ have one point of intersection. Furthermore, $x_1 = 0.5$ appears to be a good initial approximation to the x coordinate of the point of intersection. Since Newton's method can only be used to find the zero of a *single* function, we define

$$F(x) = g(x) - f(x) = 2x - e^{-x}$$

Then the zero of F will be the x coordinate of the point of intersection of the graphs of f and g.

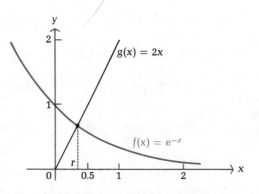

Step 1. *Find a formula for x_n.*

$$x_n = x_{n-1} - \frac{F(x_{n-1})}{F'(x_{n-1})}$$

$$= x_{n-1} - \frac{2x_{n-1} - e^{-x_{n-1}}}{2 + e^{-x_{n-1}}}$$

$$= \frac{2x_{n-1} + x_{n-1}e^{-x_{n-1}} - 2x_{n-1} + e^{-x_{n-1}}}{2 + e^{-x_{n-1}}}$$

$$= \frac{e^{-x_{n-1}}(x_{n-1} + 1)}{2 + e^{-x_{n-1}}} \frac{e^{x_{n-1}}}{e^{x_{n-1}}}$$

$$x_n = \frac{x_{n-1} + 1}{2e^{x_{n-1}} + 1}$$

Step 2. *Find successive approximations for r.*

$x_1 = 0.5$

$x_2 = 0.34904481$ Notice how the first four approximations

$x_3 = 0.35173277$ alternate from left to right, but finally

settle on 0.3517331

$x_4 = 0.35173371$

$x_5 = 0.35173371$

Since (to the accuracy of our calculator)

$$f(0.35173371) = 0.70346742 = g(0.35173371)$$

the point of intersection is approximately (0.35173371, 0.70346742).

Problem 4 Find the points of intersection of the graphs of $f(x) = e^{-x}$ and $g(x) = \frac{1}{5}x$.

Answer The formula for x_n is

$$x_n = \frac{5(x_{n-1} + 1)}{e^{x_{n-1}} + 5}$$

and the point of intersection is approximately (1.3267247, 0.26534993).

Example 5 Find the equilibrium point for the supply–demand equations

$$p = D(x) = 100 - 10\sqrt{x} \quad \text{and} \quad p = S(x) = 10 + \frac{1}{50}x^2$$

Solution Let

$$F(x) = D(x) - S(x) = 90 - 10x^{1/2} - \frac{1}{50}x^2$$

$$F'(x) = -5x^{-1/2} - \frac{1}{25}x$$

The figure shows that $x_1 = 40$ is a good first approximation to the

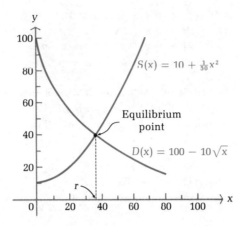

x coordinate of the point of intersection, so we construct the following table:

n	x_{n-1}	$F(x_{n-1})$	$F'(x_{n-1})$	$x_n = x_{n-1} - \dfrac{F(x_{n-1})}{F'(x_{n-1})}$
2	40	-5.2455532	-2.3905694	37.805731
3	37.805731	-0.07183014	-2.3254176	37.774842
4	37.774842	-0.00001403	-2.3245145	37.774835
5	37.774835	0.00000107	-2.3245143	37.774835

Since $D(37.774835) \approx 38.538764 \approx S(37.774835)$, the equilibrium point is approximately $(37.774835, 38.538764)$.

Problem 5 Find the equilibrium point for the supply–demand equations

$$p = D(x) = 36 - \frac{1}{16}x^2 \quad \text{and} \quad p = S(x) = \sqrt{100 + x^2}$$

Answer (16.393686, 19.202941)

EXERCISE 10-1

A In Problems 1–10 find and simplify the formula for x_n. Use this formula and the given value of x_1 to compute x_n for the indicated value of n.

1. $f(x) = x^2 - 4$; $x_1 = 1$; $n = 5$
2. $f(x) = x^2 - 2$; $x_1 = 1$; $n = 5$
3. $f(x) = x^3 - 8$; $x_1 = 3$; $n = 5$
4. $f(x) = x^3 - 2$; $x_1 = 4$; $n = 7$
5. $f(x) = e^x + x$; $x_1 = 0$; $n = 5$
6. $f(x) = e^x + 2x$; $x_1 = 1$; $n = 5$

7. $f(x) = \ln x + x$; $\quad x_1 = 1$; $\quad n = 5$
8. $f(x) = \ln x + 2x - 8$; $\quad x_1 = 10$; $\quad n = 4$
9. $f(x) = \ln x + x^2$; $\quad x_1 = 2$; $\quad n = 5$
10. $f(x) = \ln x - e^{-x}$; $\quad x_1 = 3$; $\quad n = 7$

B In Problems 11–18 use Newton's method to find all the zeros of each function. Sketch the graph of the function to obtain initial approximations and to be certain that all the zeros have been located.

11. $f(x) = x^2 - 7x + 2$ 12. $f(x) = x^2 + 5x + 3$
13. $f(x) = x^3 + 4x + 10$ 14. $f(x) = x^3 - 6x^2 + 12x - 4$
15. $f(x) = x^3 - 12x^2 + 22$ 16. $f(x) = x^3 - 18x^2 + 60x - 11$
17. $f(x) = x^4 - 4x^3 - 8x^2 + 4$ 18. $f(x) = x^4 - 4x^3 + 4x^2 + 1$

In Problems 19–24 graph f and g on the same set of axes and use Newton's method to find the x coordinate of all points of intersection of the two graphs.

19. $f(x) = x^3$, $\quad g(x) = x + 4$ 20. $f(x) = \sqrt{x + 1}$, $\quad g(x) = x^2 - 4$

21. $f(x) = x^{1/3}$, $\quad g(x) = 12 - x$ 22. $f(x) = x^3$, $\quad g(x) = \dfrac{1}{1 + x^2}$

23. $f(x) = e^x$, $\quad g(x) = x + 2$ 24. $f(x) = \ln x$, $\quad g(x) = x^2 - 4$

C 25. Newton's algorithm for approximating the square root of a positive number A is often stated as

$$x_n = \frac{x_{n-1}}{2} + \frac{A}{2x_{n-1}}$$

Show that this formula can be derived by applying Newton's method to the function $f(x) = x^2 - A$.

26. Apply Newton's method to the function $f(x) = x^3 - A$ and derive a formula for approximating cube roots.

27. Apply Newton's method to the function $f(x) = x^p - A$, where p is a positive integer, and derive a formula for approximating the pth root of A.

28. Apply Newton's method to the function

$$f(x) = \frac{1}{x} - A$$

and derive a formula for approximating reciprocals without the use of division.

29. Apply Newton's method to

$$f(x) = \frac{x}{\sqrt{1 + x^2}}$$

with $x_1 = 1$. Why does Newton's method fail in this case?

30. Apply Newton's method to $f(x) = 17 + 8x^2 - x^4$ with $x_1 = 1$. Why does Newton's method fail in this case?

APPLICATIONS

BUSINESS & ECONOMICS

31. *Break-even point.* A company has determined that its revenue and cost functions are

$$R(x) = x\sqrt{16 + x^2} \quad \text{and} \quad C(x) = x + 10$$

Find the break-even point correct to two decimal places.

32. *Supply–demand.* Find the equilibrium point correct to two decimal places if supply and demand are given by

$$p = S(x) = e^{0.1x} \quad \text{and} \quad p = D(x) = 10 - e^{0.05x}$$

33. *Marginal analysis.* The marginal revenue and cost functions for a commercial electronic game are given by

$$R'(t) = \frac{1,000}{100 + t^2} \quad \text{and} \quad C'(t) = \frac{1}{2}t + 4 \qquad 0 \le t \le 10$$

where $R(t)$ and $C(t)$ represent total accumulated revenues and costs, respectively, t years after the game has been put on the market. The company wants to continue to use the game as long as $R'(t) \ge C'(t)$. Find the useful life of the game correct to one decimal place.

34. *Sales growth.* The annual sales y (in millions of dollars) of a company satisfy the growth equation

$$\frac{dy}{dt} = 0.05(100 - y) + 0.2t$$

where t is time in years. This year the sales were $60 million. The solution to this differential equation is

$$y = 20 + 4t + 60e^{-0.05t}$$

(see Problem 27 in Exercise 8-3). How long will it take for the sales to reach $90 million?

LIFE SCIENCES

35. *Pollution.* The amount of natural waste $w(t)$ (in parts per million) in a body of water satisfies the differential equation

$$\frac{dw}{dt} + w = e^{-t} \qquad w(0) = 100$$

where t is measured in years. The solution to this equation is

$$w(t) = te^{-t} + 100e^{-t}$$

(see Problem 29 in Exercise 8-3). When will the amount of waste reach 50 parts per million?

36. *Population growth.* A population grows according to the growth law

$$\frac{dy}{dt} = ky + 0.1$$

The general solution is

$$y = \frac{1}{10k}(Ce^{kt} - 1)$$

If $y(0) = 0$ and $y(1) = 20$, determine C and k.

SOCIAL SCIENCES 37. *Learning.* The relationship between the number of units $N(t)$ a worker can assemble in 1 day and the number of hours t of training the worker has received is given by

$$N(t) = t^3 - 6t^2 + 25t$$

How many hours of training are required in order for a worker to be able to assemble 500 units a day?

10-2 NUMERICAL INTEGRATION

In Section 4-5 we saw that the definite integral of a continuous function can be approximated by using the rectangle rule. If the interval $[a, b]$ is divided into n equal subintervals of length $\Delta x = (b - a)/n$ and if t_k is any point on the kth subinterval, then

$$\int_a^b f(x)\,dx \approx [f(t_1) + f(t_2) + \cdots + f(t_n)]\Delta x$$

In order to carry out this calculation, it is necessary to specify how the points $t_1, t_2, \ldots, t_n$ are chosen. We will use three different methods for selecting these points. It will simplify matters if we first introduce some new notation for the end points of each subinterval. Let

$$x_0 = a$$
$$x_1 = a + \Delta x$$
$$x_2 = a + 2\Delta x$$
$$\vdots$$
$$x_{n-1} = a + (n - 1)\Delta x$$
$$x_n = a + n\Delta x = a + n\left(\frac{b - a}{n}\right) = b$$

**METHOD 1:
THE LEFT-HAND
END POINT RULE**

Choose the left-hand end point of each subinterval for the point in that interval and apply the rectangle rule. Thus,

$$t_1 = x_0, t_2 = x_1, t_3 = x_2, \ldots, t_n = x_{n-1}$$

and

$$L = [f(x_0) + f(x_1) + \cdots + f(x_{n-1})]\Delta x$$

is an approximation of the definite integral; that is,

$$L \approx \int_a^b f(x)\, dx$$

Example 6

Use the left-hand end point rule and $n = 10$ to approximate $\int_0^1 3x^{1/2}\, dx$.

Then compare the value you obtain with the exact value obtained by using the fundamental theorem of calculus.

Solution

Part 1. *Use the left-hand end point rule.* We are given $f(x) = 3x^{1/2}$, $a = 0, b = 1$, and $\Delta x = 0.1$, so we have

$$
\begin{aligned}
L &= [f(0) + f(0.1) + f(0.2) + \cdots + f(0.9)](0.1) \\
&= [0 + 0.9486833 + 1.3416408 + 1.6431677 + 1.8973666 \\
&\quad + 2.1213203 + 2.32379 + 2.5099001 + 2.6832816 \\
&\quad + 2.8460499](0.1) \\
&= [18.31528](0.1) = 1.831528
\end{aligned}
$$

Part 2. *Use the fundamental theorem.*

$$A = \int_0^1 3x^{1/2}\, dx = 2x^{3/2}\Big|_0^1 = 2(1)^{3/2} - 2(0)^{3/2} = 2$$

COMMENTS

The area A in part 2 of Example 6 is the shaded region in Figure 6A (on the next page) and the area L in part 1 of the example is the shaded region in Figure 6B. The difference

$$|A - L| = |2 - 1.831528| = 0.168472$$

is the **absolute error** in the approximation. Normally, numerical methods are used to approximate definite integrals that *cannot* be evaluated by the fundamental theorem. In such cases, the absolute error cannot be determined exactly, but can often be estimated. We will not consider techniques for estimating the absolute error in this text.

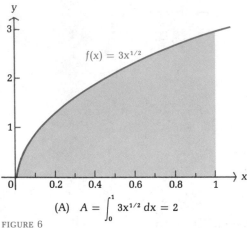

(A) $A = \int_0^1 3x^{1/2}\,dx = 2$

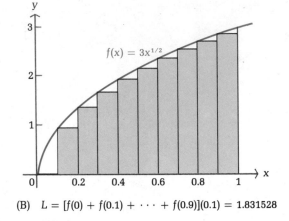

(B) $L = [f(0) + f(0.1) + \cdots + f(0.9)](0.1) = 1.831528$

FIGURE 6

Problem 6   Use the left-hand end point rule with $n = 10$ to approximate $\int_1^2 x^4\,dx$.

Then use the fundamental theorem to find the exact value, and find the absolute error.

Answer $L = 5.47333$, exact value is 6.2, absolute error is 0.72667

METHOD 2:
THE RIGHT-HAND
END POINT RULE

Choose the right-hand end point of each subinterval for the point in that interval and apply the rectangle rule. Thus,

$$t_1 = x_1, t_2 = x_2, t_3 = x_3, \ldots, t_n = x_n$$

and

$$R = [f(x_1) + f(x_2) + \cdots + f(x_n)]\Delta x$$

is an approximation of the definite integral; that is,

$$R \approx \int_a^b f(x)\,dx$$

Example 7   Use the right-hand end point rule with $n = 10$ to approximate $\int_0^1 3x^{1/2}\,dx$. Then find the absolute error.

Solution $R = [f(0.1) + f(0.2) + \cdots + f(1)](0.1)$
$= [0.9486833 + 1.3416408 + 1.6431677 + 1.8973666$
$\quad + 2.1213203 + 2.32379 + 2.5099801 + 2.6832816$
$\quad + 2.8460499 + 3](0.1)$
$= [21.31528](0.1) = 2.131528$
Absolute error $= |2 - 2.131528| = 0.131528$

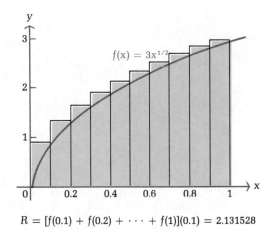

$$R = [f(0.1) + f(0.2) + \cdots + f(1)](0.1) = 2.131528$$

The area R is the shaded region in the figure.

Problem 7   Use the right-hand end point rule with $n = 10$ to approximate $\int_1^2 x^4 \, dx$. Then find the absolute error.

Answer $R = 6.97333$, Absolute error $= 0.77333$

METHOD 3:   Choose the midpoint of each subinterval for the point in that interval
THE MIDPOINT RULE   and apply the rectangle rule. Thus,

$$t_1 = \frac{x_0 + x_1}{2}, \, t_2 = \frac{x_1 + x_2}{2}, \ldots, t_n = \frac{x_{n-1} + x_n}{2}$$

and

$$M = \left[f\left(\frac{x_0 + x_1}{2}\right) + f\left(\frac{x_1 + x_2}{2}\right) + \cdots + f\left(\frac{x_{n-1} + x_n}{2}\right) \right] \Delta x$$

is an approximation of the definite integral; that is,

$$M \approx \int_a^b f(x) \, dx$$

Example 8   Use the midpoint rule and $n = 10$ to approximate $\int_0^1 3x^{1/2} \, dx$. Then find the absolute error.

Solution $M = [f(0.05) + f(0.15) + \cdots + f(0.95)](0.1)$
$= [0.67082039 + 1.161895 + 1.5 + 1.7748239 + 2.0124612 + 2.2248595$
$\quad + 2.4186773 + 2.5980762 + 2.7658633 + 2.9240383](0.1)$
$= [20.051515](0.1) = 2.0051515$
Absolute error $= |2 - 2.0051515| = 0.0051515$

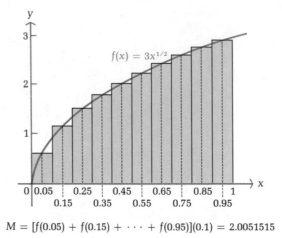

$$M = [f(0.05) + f(0.15) + \cdots + f(0.95)](0.1) = 2.0051515$$

The area M is the shaded region in the figure.

Problem 8   Use the midpoint rule and $n = 10$ to approximate $\int_{1}^{2} x^4 \, dx$. Then find the absolute error.

Answer $M = 6.1883363$, Absolute error $= 0.0116637$

      Notice that the absolute error obtained by using the midpoint rule is much smaller than the error in either of the first two cases. This will happen with most functions, because the rectangle tops are usually above part of the graph and below part of the graph instead of entirely above or entirely below the graph (compare Fig. 6B and the figures that accompany Examples 7 and 8). This tends to reduce the amount of error that occurs.

      Now we will examine some other popular methods for approximating the definite integral. These methods are not based on the rectangle rule, but they turn out to be related to the first three methods we have discussed.

METHOD 4:
THE TRAPEZOID RULE

Another way to approximate a definite integral is to use trapezoids in place of rectangles. The area of a trapezoid with base b and altitudes h_1 and h_2 is (see Fig. 7)

$$A = b \left(\frac{h_1 + h_2}{2} \right)$$

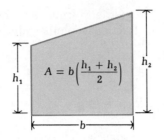

FIGURE 7

The area of the trapezoid with vertices $(x_{k-1}, 0)$, $(x_{k-1}, f(x_{k-1}))$, $(x_k, f(x_k))$, and $(x_k, 0)$ is (see Fig. 8)

$$A = \left[\frac{f(x_{k-1}) + f(x_k)}{2}\right]\Delta x$$

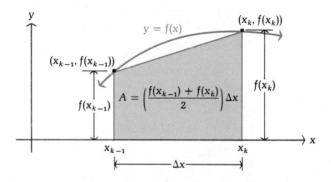

FIGURE 8

If we sum the areas of the n trapezoids formed by connecting the points $(x_0, f(x_0))$, $(x_1, f(x_1))$, . . . , $(x_n, f(x_n))$, as shown in Figure 9, then we have another approximation of the definite integral:

$$T = \left[\frac{f(x_0) + f(x_1)}{2}\right]\Delta x + \left[\frac{f(x_1) + f(x_2)}{2}\right]\Delta x + \cdots + \left[\frac{f(x_{n-1}) + f(x_n)}{2}\right]\Delta x$$

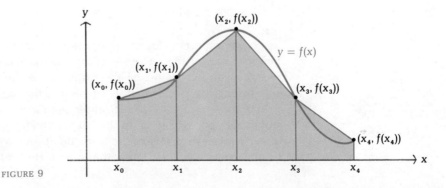

FIGURE 9

If we group the first term from each fraction in a sum and the second term from each fraction in another sum, we can express T as

$$T = \frac{1}{2}[f(x_0) + f(x_1) + \cdots + f(x_{n-1})]\Delta x$$

$$+ \frac{1}{2}[f(x_1) + f(x_2) + \cdots + f(x_n)]\Delta x$$

Comparing these sums with the left- and right-hand end point rules, we see that

$$T = \frac{1}{2}L + \frac{1}{2}R = \frac{1}{2}(L + R)$$

Thus, the trapezoid rule is simply the average of the left- and right-hand end point rules.

Example 9   Use the trapezoid rule with $n = 10$ to approximate $\int_0^1 3x^{1/2}\, dx$. Then find the absolute error.

Solution   From Examples 6 and 7, we have

$$L = 1.831528 \quad \text{and} \quad R = 2.131528$$

Thus,

$$T = \frac{1}{2}(1.831528 + 2.131528) = 1.981528$$

$$\text{Absolute error} = |2 - 1.981528| = 0.018472$$

Problem 9   Use the trapezoid rule with $n = 10$ to approximate $\int_1^2 x^4\, dx$. Then find the absolute error.

Answer $T = 6.22333$, Absolute error $= 0.02333$

The trapezoid rule usually produces a more accurate approximation than any of the rectangle rules, because the top of each trapezoid goes through two points on the graph, while the top of each rectangle goes through only one point. If we replace the top edge of the trapezoid with a curve that goes through three points on the graph, we expect a still better approximation. In the next method, we will use the midpoint of each interval to obtain the third point and we will use a parabola to approximate the area under the graph instead of a straight line.

METHOD 5:
SIMPSON'S RULE

On each interval $[x_{k-1}, x_k]$ we consider a parabola P that passes through the points

$$(x_{k-1}, f(x_{k-1})), \qquad \left(\frac{x_{k-1} + x_k}{2}, f\left(\frac{x_{k-1} + x_k}{2}\right)\right), \qquad (x_k, f(x_k))$$

Omitting the calculatons, it can be shown that the area under the parabola P is given exactly by

$$A = \left[f(x_{k-1}) + 4f\left(\frac{x_{k-1} + x_k}{2}\right) + f(x_k)\right]\frac{\Delta x}{6}$$

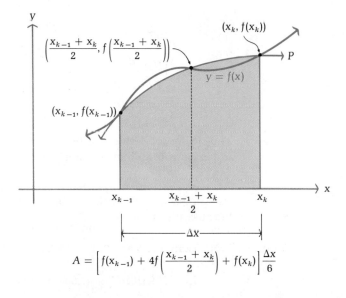

FIGURE 10

$$A = \left[f(x_{k-1}) + 4f\left(\frac{x_{k-1} + x_k}{2}\right) + f(x_k)\right]\frac{\Delta x}{6}$$

This area is the shaded region in Figure 10. Summing these areas, we obtain another approximation to the definite integral, usually referred to as **Simpson's rule:**

$$S = \left[f(x_0) + 4f\left(\frac{x_0 + x_1}{2}\right) + f(x_1)\right]\frac{\Delta x}{6}$$

$$+ \left[f(x_1) + 4f\left(\frac{x_1 + x_2}{2}\right) + f(x_2)\right]\frac{\Delta x}{6}$$

$$+ \cdots + \left[f(x_{n-1}) + 4f\left(\frac{x_{n-1} + x_n}{2}\right) + f(x_n)\right]\frac{\Delta x}{6}$$

Grouping all the first terms from each square bracket, all the second terms, and all the third terms, this sum can be expressed using the trapezoid and midpoint formulas:

$$S = \frac{1}{6}[f(x_0) + f(x_1) + \cdots + f(x_{n-1})]\Delta x$$

$$\underbrace{\phantom{S = \frac{1}{6}[f(x_0) + f(x_1) + \cdots + f(x_{n-1})]\Delta x}}_{L}$$

$$+ \frac{4}{6}\left[f\left(\frac{x_0 + x_1}{2}\right) + f\left(\frac{x_1 + x_2}{2}\right) + \cdots + f\left(\frac{x_{n-1} + x_n}{2}\right)\right]\Delta x$$

$$\underbrace{\phantom{+ \frac{4}{6}\left[f\left(\frac{x_0 + x_1}{2}\right) + f\left(\frac{x_1 + x_2}{2}\right) + \cdots + f\left(\frac{x_{n-1} + x_n}{2}\right)\right]\Delta x}}_{M}$$

$$+ \frac{1}{6}[f(x_1) + f(x_2) + \cdots + f(x_n)]\Delta x$$

$$\underbrace{\phantom{+ \frac{1}{6}[f(x_1) + f(x_2) + \cdots + f(x_n)]\Delta x}}_{R}$$

$$= \frac{1}{6}L + \frac{2}{3}M + \frac{1}{6}R$$

$$= \frac{1}{3}\left(\frac{L + R}{2}\right) + \frac{2}{3}M$$

$$= \frac{1}{3}T + \frac{2}{3}M$$

Thus, Simpson's rule can be considered as a weighted average of the trapezoid rule and the midpoint rule.

Example 10   Use Simpson's rule with $n = 10$ to approximate $\int_0^1 3x^{1/2}\, dx$. Then find the absolute error.

Solution   From Examples 8 and 9, $T = 1.981528$ and $M = 2.0051515$. Thus,

$$S = \frac{1}{3}(1.981528) + \frac{2}{3}(2.0051515) = 1.997277$$

Absolute error $= |2 - 1.997277| = 0.002723$

Problem 10   Use Simpson's rule with $n = 10$ to approximate $\int_1^2 x^4\, dx$. Then find the absolute error.

Answer $S = 6.2000009$, Absolute error $= 0.0000009$

ORGANIZING THE COMPUTATIONS   In most situations, Simpson's rule is the most accurate of the five methods discussed and requires hardly any more work. We have expressed it as a weighted average of the midpoint and trapezoid rules because it is more convenient to calculate T and M separately when using a hand-held calculator. Table 1 shows how to organize all the

TABLE 1   Approximating sums for $\int_0^1 3x^{1/2}\,dx$ with $n = 10$

x_k	$f(x_k)$	$\dfrac{x_{k-1} + x_k}{2}$	$f\!\left(\dfrac{x_{k-1} + x_k}{2}\right)$
0	0		
		0.05	0.67082039
0.1	0.9486833		
		0.15	1.161895
0.2	1.3416408		
		0.25	1.5
0.3	1.6431677		
		0.35	1.7748239
0.4	1.8973666		
		0.45	2.0124612
0.5	2.1213203		
		0.55	2.2248595
0.6	2.32379		
		0.65	2.4186773
0.7	2.5099801		
		0.75	2.5980762
0.8	2.6832816		
		0.85	2.7658633
0.9	2.8460499		
		0.95	2.9240383
1	3		

↑
End points
of the
intervals

↑
Midpoints of
the intervals
in column 1

data that are necessary for the convenient computation of the five approximations we have discussed. Procedures for computing each approximation using the table are outlined below.

Left-hand end point rule.   To compute L, add all entries in column 2, *except the last one,* and multiply by $\Delta x = 0.1$.

   Result: $L = 1.831528$

Right-hand end point rule.   To compute R, add all entries in column 2, *except the first one,* and multiply by $\Delta x = 0.1$.

   Result: $R = 2.131528$

Midpoint rule.   To compute M, add *all* entries in column 4 and multiply by $\Delta x = 0.1$.

   Result: $M = 2.0051515$

Trapezoid rule.   To compute T, use $T = \frac{1}{2}(L + R)$.

   Result: $T = 1.981528$

Simpson's rule. To compute S, use $S = \frac{1}{3}T + \frac{2}{3}M$.

Result: $S = 1.997277$

Example 11   Find all five approximating sums with $n = 10$ for

$$\int_0^2 \frac{8}{4 + x^2}\, dx$$

Solution   Approximating sums for $\int_0^2 \dfrac{8}{4 + x^2}\, dx$ with $n = 10$

x_k	$f(x_k)$	$\dfrac{x_{k-1} + x_k}{2}$	$f\!\left(\dfrac{x_{k-1} + x_k}{2}\right)$
0	2		
		0.1	1.9950125
0.2	1.980198		
		0.3	1.9559902
0.4	1.9230769		
		0.5	1.8823529
0.6	1.8348624		
		0.7	1.7817372
0.8	1.7241379		
		0.9	1.6632017
1	1.6		
		1.1	1.5355086
1.2	1.4705882		
		1.3	1.4059754
1.4	1.3422819		
		1.5	1.28
1.6	1.2195122		
		1.7	1.161103
1.8	1.1049724		
		1.9	1.0512484
2	1		

$$L = (16.19963)(0.2) = 3.239926$$
$$R = (15.19963)(0.2) = 3.039926$$
$$M = (15.71213)(0.2) = 3.142426$$
$$T = \tfrac{1}{2}(L + R) = \tfrac{1}{2}(3.239926 + 3.039926) = 3.139926$$
$$S = \tfrac{1}{3}T + \tfrac{2}{3}M = \tfrac{1}{3}(3.139926) + \tfrac{2}{3}(3.142426) = 3.1415927$$

Problem 11   Find all five approximating sums with $n = 10$ for $\int_1^3 (1/x)\, dx$ by first forming a table similar to that used in Example 11.

Answer $L = 1.168229,\ R = 1.0348957,\ M = 1.0971421,\ T = 1.1015624,\ S = 1.0986155$

SUMMARY OF THE FIVE METHODS   We now summarize the formulas for each approximating sum. We have collected similar terms in the sums for T and S in order to simplify the formulas and represent them in forms you are likely to encounter in other texts and courses.

APPROXIMATING SUMS FOR $\displaystyle\int_a^b f(x)\,dx$

Left-hand end point rule

$$L = [f(x_0) + f(x_1) + \cdots + f(x_{n-1})]\Delta x$$

Right-hand end point rule

$$R = [f(x_1) + f(x_2) + \cdots + f(x_n)]\Delta x$$

Midpoint rule

$$M = \left[f\left(\frac{x_0 + x_1}{2}\right) + f\left(\frac{x_1 + x_2}{2}\right) + \cdots + f\left(\frac{x_{n-1} + x_n}{2}\right)\right]\Delta x$$

Trapezoid rule

$$T = [f(x_0) + 2f(x_1) + 2f(x_2) + \cdots + 2f(x_{n-1}) + f(x_n)]\frac{\Delta x}{2}$$

Simpson's rule

$$S = \left[f(x_0) + 4f\left(\frac{x_0 + x_1}{2}\right) + 2f(x_1) + 4f\left(\frac{x_1 + x_2}{2}\right) + \cdots \right.$$

$$\left. + 2f(x_{n-1}) + 4f\left(\frac{x_{n-1} + x_n}{2}\right) + f(x_n)\right]\frac{\Delta x}{6}$$

Example 12 Use the midpoint rule with $n = 4$ to approximate the area of the region bounded by the graphs of

$$y = \frac{1}{\sqrt{1 + x^2}} \qquad y = 0 \qquad x = 0 \qquad x = 1$$

Solution From the graph shown in the figure, we see that the desired area is given by

$$A = \int_0^1 \frac{1}{\sqrt{1 + x^2}}\,dx$$

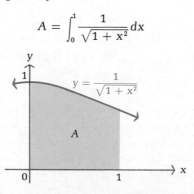

Step 1. *Find the length of each subinterval.*

$$\Delta x = \frac{b - a}{n} = \frac{1 - 0}{4} = 0.25$$

Step 2. *Find the midpoint of each subinterval.*

Subintervals: [0, 0.25], [0.25, 0.5], [0.5, 0.75], [0.75, 1]
Midpoints: 0.125,   0.375,   0.625,   0.875

Step 3. *Compute M.*

$$M = [f(0.125) + f(0.375) + f(0.625) + f(0.875)](0.25)$$
$$= [0.99227788 + 0.93632918 + 0.8479983 + 0.7525767](0.25)$$
$$= [3.5291821](0.25)$$
$$= 0.88229551$$

Thus, $A \approx 0.88229551$.

Problem 12

Use the midpoint rule with $n = 8$ to approximate the area of the region bounded by the graphs of

$$y = \frac{1}{1 + x^4} \qquad y = 0 \qquad x = 1 \qquad x = 3$$

Answer $A \approx 0.22887607$

Example 13

Use the trapezoid rule with $n = 10$ to approximate the average value (see Section 4-5) of $f(x) = \sqrt{8 + x^3}$ on the interval [1, 6].

Solution   Step 1. $\Delta x = \dfrac{b - a}{n} = \dfrac{6 - 1}{10} = \dfrac{5}{10} = 0.5$

Step 2. Intervals: [1, 1.5], [1.5, 2], [2, 2.5], [2.5, 3], [3, 3.5],
                         [3.5, 4], [4, 4.5], [4.5, 5], [5, 5.5], [5.5, 6]

Step 3. $T = [f(1) + 2f(1.5) + 2f(2) + 2f(2.5) + 2f(3) + 2f(3.5)$

$$+ 2f(4) + 2f(4.5) + 2f(5) + 2f(5.5) + f(6)]\left(\frac{\Delta x}{2}\right)$$

$$= [3 + 6.7453688 + 8 + 9.721111 + 11.83216 + 14.265343$$
$$+ 16.970563 + 19.912308 + 23.065125 + 26.410225$$
$$+ 14.96663]\left(\frac{0.5}{2}\right)$$

$$= [154.88883](0.25) = 38.722208$$

The average value is

$$\frac{1}{5}\int_{1}^{6} \sqrt{8 + x^3}\, dx \approx \frac{1}{5}T = \frac{1}{5}(38.722208) = 7.7444416$$

Problem 13 Use the trapezoid rule with $n = 12$ to approximate the average value (see Section 4-5) of $f(x) = \ln(1 + \sqrt{x})$ on the interval $[1, 4]$.

Answer 0.93165681

Example 14 Use Simpson's rule with $n = 4$ to approximate the consumers' surplus and the producers' surplus for

$$p = D(x) = \frac{12}{1 + \sqrt{x}} \quad \text{and} \quad p = S(x) = 2 + \sqrt{x}$$

(See Section 4-4 for a discussion of consumers' and producers' surplus.)

Solution Step 1. *Sketch the graph.*

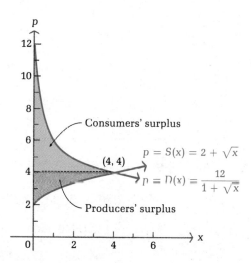

Step 2. *Find the equilibrium point.*

$$D(x) = S(x)$$

$$\frac{12}{1 + \sqrt{x}} = 2 + \sqrt{x}$$

$$12 = (2 + \sqrt{x})(1 + \sqrt{x})$$

$$= 2 + 3\sqrt{x} + x$$

$$10 - x = 3\sqrt{x}$$

$$(10 - x)^2 = (3\sqrt{x})^2$$

$$100 - 20x + x^2 = 9x$$

$$x^2 - 29x + 100 = 0$$

$$(x - 4)(x - 25) = 0$$

$$x = 4 \quad \text{or} \quad x = 25$$

Since $p = D(4) = 4$ and $p = S(4) = 4$, a point of intersection occurs at $(4, 4)$. Since $p = D(25) = 2$ and $p = S(25) = 7$, the graphs *do not* intersect at $x = 25$. The "solution" $x = 25$ is an *extraneous* root introduced by squaring both sides of the equation. Thus, the equilibrium point is $(4, 4)$.

Step 3. *Express the surpluses as integrals.*

$$\text{Consumers' surplus} = \int_0^4 [D(x) - 4]\, dx = \int_0^4 \left[\frac{12}{1 + \sqrt{x}} - 4 \right] dx$$

$$\text{Producers' surplus} = \int_0^4 [4 - S(x)]\, dx = \int_0^4 [2 - \sqrt{x}]\, dx$$

Step 4. *Find the length of each subinterval.*

$$\Delta x = \frac{4 - 0}{4} = 1$$

Step 5. *Find the intervals and the midpoints.*

Intervals: $[0, 1], [1, 2], [2, 3], [3, 4]$
Midpoints: 0.5, 1.5, 2.5, 3.5

Step 6. *Approximate the consumers' surplus.*

$$f(x) = \frac{12}{1 + \sqrt{x}} - 4$$

$$S = [f(0) + 4f(0.5) + 2f(1) + 4f(1.5) + 2f(2) + 4f(2.5)$$

$$+ 2f(3) + 4f(3.5) + f(4)]\left(\frac{1}{6}\right)$$

$$= [8 + 12.117749 + 4 + 5.5755077 + 1.9411255 + 2.5964426$$

$$+ 0.78460969 + 0.71991092 + 0]\left(\frac{1}{6}\right)$$

$$= [35.735345](0.16666667) = 5.9558909$$

Thus, the consumers' surplus is approximately 5.9558909.

Step 7. *Approximate the producers' surplus.*

$$f(x) = 2 + \sqrt{x}$$

$$S = [f(0) + 4f(0.5) + 2f(1) + 4f(1.5) + 2f(2) + 4f(2.5)$$

$$+ 2f(3) + 4f(3.5) + f(4)]\left(\frac{1}{6}\right)$$

$$= [2 + 5.1715729 + 2 + 3.1010205 + 1.1715729 + 1.6754447$$

$$+ 0.53589839 + 0.51668523 + 0]\left(\frac{1}{6}\right)$$

$$= [16.172195](0.16666667) = 2.6953658$$

Thus, the producers' surplus is approximately 2.6953658.

Problem 14   Use Simpson's rule with $n = 6$ to approximate the consumers' surplus and the producers' surplus for

$$p = D(x) = \frac{80}{x + 2} \quad \text{and} \quad p = S(x) = \frac{40}{10 - x}$$

Answer   Consumers' surplus $= 50.91268$, Producers' surplus ≈ 23.348065

Example 15   A solid of revolution (see Section 4-5) is formed by revolving the region bounded by the graphs of an unspecified function $y = f(x)$, $y = 0$, $x = 0$, and $x = 2$ about the x axis. Eleven values of the function f are given in the table.* Use the trapezoid rule to approximate the volume of the solid.

x	0	0.2	0.4	0.6	0.8	1	1.2	1.4	1.6	1.8	2
$f(x)$	1.1	1.4	1.5	1.3	1.1	1	0.5	0.4	0.6	0.8	1

Solution   The volume of the solid is given by

$$V = \pi \int_0^2 [f(x)]^2 \, dx$$

We want to use the trapezoid rule with $n = 10$ to approximate this integral.

$$\Delta x = \frac{2 - 0}{10} = 0.2$$

$$\begin{aligned}
T = \{&[f(0)]^2 + 2[f(0.2)]^2 + 2[f(0.4)]^2 + 2[f(0.6)]^2 + 2[f(0.8)]^2 \\
&+ 2[f(1)]^2 + 2[f(1.2)]^2 + 2[f(1.4)]^2 + 2[f(1.6)]^2 + 2[f(1.8)]^2 \\
&+ [f(2)]^2\} \left(\frac{0.2}{2}\right) \\
= \{&1.21 + 3.92 + 4.5 + 3.38 + 2.42 + 2 + 0.5 + 0.32 + 0.72 \\
&+ 1.28 + 1\}(0.1) \\
= \{&21.25\}(0.1) = 2.125
\end{aligned}$$

Thus, $V \approx \pi T = \pi(2.125) \approx 6.676$.

Problem 15   Repeat Example 15 for the following table of values:

x	1	1.3	1.6	1.9	2.2	2.5	2.8	3.1	3.4	3.7	4
$f(x)$	2.4	2	1.8	1.8	2.3	2.7	3.5	3.5	3.1	2.7	2.2

Answer   65.747

*In actual practice, it is often the case that functions are not known in terms of explicit equations, but only in terms of a table of values obtained from measurements.

EXERCISE 10-2

A   *Construct a table similar to the one in Example 11 to find all five approximating sums. Find the absolute error for each sum.*

1. $\int_0^1 e^x \, dx; \quad n = 10$ 2. $\int_1^2 \ln x \, dx; \quad n = 10$

In Problems 3–12 find the indicated approximating sum.

3. $\int_0^1 \dfrac{1}{1 + x^2} \, dx;$ left-hand rule with $n = 4$

4. $\int_0^1 \sqrt{1 - x^2} \, dx;$ left-hand rule with $n = 6$

5. $\int_0^2 \sqrt{1 + x^2} \, dx;$ right-hand rule with $n = 4$

6. $\int_2^3 \dfrac{1}{\sqrt{x^2 - 1}} \, dx;$ right-hand rule with $n = 6$

7. $\int_1^3 \dfrac{1}{\sqrt{1 + x^2}} \, dx;$ midpoint rule with $n = 5$

8. $\int_{-1}^0 \dfrac{x}{2 + x^3} \, dx;$ midpoint rule with $n = 5$

9. $\int_0^1 e^{-x^2} \, dx;$ trapezoid rule with $n = 4$

10. $\int_1^4 e^{\sqrt{x}} \, dx;$ trapezoid rule with $n = 4$

11. $\int_1^2 \ln(1 + x^2) \, dx;$ Simpson's rule with $n = 5$

12. $\int_1^3 (\ln x)[\ln(1 + x)] \, dx;$ Simpson's rule with $n = 5$

B   *In Problems 13–18 use the midpoint rule with $n = 10$ to approximate the average value of each function over the indicated interval.*

13. $f(x) = \sqrt{1 + \sqrt{x}}; \quad [0, 4]$ 14. $f(x) = \dfrac{1}{1 + \sqrt{x}}; \quad [4, 9]$

15. $f(x) = \dfrac{x^2}{4 + x^2}; \quad [0, 2]$ 16. $f(x) = \dfrac{x}{1 + x^4}; \quad [-5, 5]$

17. $f(x) = x^2 e^{x^2}; \quad [-1, 0]$ 18. $f(x) = \dfrac{\ln(1 + x)}{1 + x^2}; \quad [0, 3]$

In Problems 19–24 use Simpson's rule with $n = 10$ to approximate the area bounded by the graphs of the given equations.

19. $y = \sqrt{4 - x^2}$; $y = 0$, $0 \leq x \leq 2$

20. $y = \sqrt{4 + x^2}$; $y = 0$, $0 \leq x \leq 2$

21. $y = \dfrac{1}{1 + x^2}$; $y = 0$, $0 \leq x \leq 4$

22. $y = \dfrac{1}{1 + \sqrt{x}}$; $y = 0$, $0 \leq x \leq 4$

23. $y = \dfrac{1}{1 + \ln x}$; $y = 0$, $1 \leq x \leq 2$

24. $y = e^{x^2}$; $y = \ln(1 + x^2)$, $0 \leq x \leq 1$

In Problems 25–30 use the trapezoid rule and the values in each table to approximate the volume of the solid formed by revolving the region bounded by the graph of the function about the x axis.

25.

x	0	0.2	0.4	0.6	0.8	1
f(x)	0	1	2	1	1	0

26.

x	0	0.2	0.4	0.6	0.8	1
f(x)	2	2.5	3	3	2.4	1.5

27.

x	2	2.1	2.2	2.3	2.4	2.5
f(x)	1	1.1	1.2	1.6	1.7	2

28.

x	2	2.3	2.6	2.9	3.2	3.5
f(x)	10	9	5	7	8	8

29.

x	2	2.5	3	3.5	4
f(x)	1	1.4	1.6	1.4	1

30.

x	−2	−1.8	−1.6	−1.4	−1.2	−1
f(x)	1	1	2	2	1	1

APPLICATIONS

In Problems 31–37 use Simpson's rule with $n = 10$ to approximate the required definite integrals.

BUSINESS & ECONOMICS

31. *Consumers' and producers' surplus.* Find the consumers' surplus and the producers' surplus for

$$p = D(x) = \frac{9}{1 + \sqrt{x}} \quad \text{and} \quad p = S(x) = 1 + \sqrt{x}$$

32. *Consumers' and producers' surplus.* Find the consumers' surplus and the producers' surplus for

$$p = D(x) = \frac{15}{1 + \sqrt{x}} \quad \text{and} \quad p = S(x) = 3 + \sqrt{x}$$

33. *Marginal analysis.* The marginal cost and revenue equations (in thousands of dollars) for an electronic game are

$$R'(t) = \frac{18}{9 + t^2} \quad \text{and} \quad C'(t) = \frac{2t^2}{9 + t^2}$$

where $R(t)$ and $C(t)$ represent total accumulated revenues and costs, respectively, t years after the game is put into use. What is the useful life of the machine? What is the total profit?

34. *Marginal analysis.* Repeat Problem 33 given

$$R'(t) = \frac{60}{5 + t^2} \quad \text{and} \quad C'(t) = \frac{12t}{5 + t^2}$$

LIFE SCIENCES

35. *Medicine.* The body assimilates a drug at a rate given by

$$R'(t) = 7 - 3\ln(t^2 - 6t + 10) \qquad 0 \le t \le 3$$

where t is the time (in hours) since the drug was administered. Find the total amount of the drug assimilated in the first 3 hours.

36. *Medicine.* The level of concentration of a certain drug in the bloodstream t minutes after it is administered is given by

$$L(t) = 10 - 6\ln(t^2 - 4t + 5) \qquad 0 \le t \le 4$$

Find the average level of concentration for t in the interval $[0, 4]$.

SOCIAL SCIENCES

37. *Learning.* A person learns N items at a rate given by

$$N'(t) = \sqrt{24 + \frac{1}{t^2}} \qquad 1 \le t \le 6$$

where t is the number of hours of continuous study. Find the total number of items N learned from $t = 1$ to $t = 6$ hours of study.

10-3 EULER'S METHOD

EULER'S METHOD:
AN INFORMAL
DEVELOPMENT

In Chapter 8, we developed techniques for finding the solutions to certain types of differential equations. But it is just a fortunate accident when a differential equation has a simple, neat solution expressed in terms of a finite combination of familiar functions—most do not. When a differential equation does not have such a nice solution, it is necessary to find a numerical approximation to the solution. Many different methods can be used to obtain the approximate solution (for example, see Section 9-4). One of the oldest and easiest to understand is **Euler's method.** This method can be used to approximate the solution to any differential equation of the form

$$y' = f(x, y) \qquad y(x_0) = y_0$$

We begin our discussion by considering a specific example. Suppose we are given the differential equation

$$y' = x^2 + y^2 \qquad y(0) = 1 \tag{1}$$

Then,

$$f(x, y) = x^2 + y^2 \qquad x_0 = 0 \qquad y_0 = 1$$

Since this equation cannot be solved by any of the methods presented in Chapter 8, we will approximate the solution. Specifically, we want to approximate the value of the solution $y(x)$ at a sequence of equally spaced points,

$$x_1 = 0.1, x_2 = 0.2, x_3 = 0.3, \ldots$$

First, we find the first-degree Taylor polynomial for $y(x)$ at 0:

$$\begin{aligned} p_1(x) &= y(x_0) + y'(x_0)(x - x_0) \\ &= y(0) + y'(0)x \end{aligned}$$

Now, $y(0) = y_0 = 1$ is the given initial condition and, since y satisfies equation (1), we have

$$y'(0) = f(0, y(0)) = f(0, 1) = 0^2 + 1^2 = 1$$

Thus, we have determined that

$$p_1(x) = 1 + x$$

From our earlier experiences with Taylor polynomials, we know that p_1 can be used to approximate y near $x_0 = 0$. That is,

$$y(x) \approx p_1(x) = 1 + x$$

In particular, if we define

$$y_1 = p_1(x_1) = p_1(0.1) = 1 + 0.1 = 1.1$$

then

$$y(x_1) \approx y_1$$

or

$$y(0.1) \approx 1.1$$

We could use p_1 to approximate y at $x_1, x_2, x_3, \ldots$, but experience has shown us that the accuracy of the approximation provided by a Taylor polynomial decreases as x moves away from the point of expansion. In Chapter 9, we increased the accuracy of the approximation by increasing the degree of the Taylor polynomial. We will use a different approach here. Since first-degree Taylor polynomials provide a good approximation near the point of expansion, we will change the point of expansion. To approximate y at x_2, we will try to find the first-degree Taylor polynomial at x_1. For x near x_1,

$$y(x) \approx y(x_1) + y'(x_1)(x - x_1) \qquad \text{First-degree Taylor polynomial at } x_1$$
$$= y(0.1) + y'(0.1)(x - 0.1)$$

Unfortunately, we do not know the value of either y or y' at $x_1 = 0.1$. But we just concluded that y_1 is an approximation to the value of y at x_1. Furthermore, it is reasonable to assume that $y_1 \approx y(x_1)$ implies

$$y'(x_1) = f(x_1, y(x_1)) \approx f(x_1, y_1)$$

or

$$y'(0.1) \approx f(0.1, 1.1) = (0.1)^2 + (1.1)^2 = 1.22$$

Using these approximate values, we can approximate the first-degree Taylor polynomial at x_1:

$$y(x) \approx y(x_1) + y'(x_1)(x - x_1) \qquad \text{Actual first-degree Taylor}$$
$$\text{polynomial at } x_1$$
$$\approx y_1 + f(x_1, y_1)(x - x_1) \qquad \text{Approximate first-degree}$$
$$\text{Taylor polynomial at } x_1$$
$$= 1.1 + 1.22(x - 0.1)$$

Since this approximation is good near $x_1 = 0.1$, we can use it to approximate y at $x_2 = 0.2$. If

$$y_2 = y_1 + f(x_1, y_1)(x_2 - x_1)$$
$$= 1.1 + 1.22(0.2 - 0.1) = 1.222$$

then

$$y(x_2) \approx y_2$$

or

$$y(0.2) \approx 1.222$$

Since the calculation of y_2 depended only on the known values of x_1 and x_2 and the approximate value of y_1, which was calculated in the preceding step, we can repeat this process indefinitely. Thus,

$$y(x_3) \approx y_3 = y_2 + f(x_2, y_2)(x_3 - x_2)$$
$$y(0.3) \approx 1.222 + [(0.2)^2 + (1.222)^2](0.3 - 0.2) = 1.3753284$$

$$y(x_4) \approx y_4 = y_3 + f(x_3, y_3)(x_4 - x_3)$$
$$y(0.4) \approx 1.3753284 + [(0.3)^2 + (1.3753284)^2](0.4 - 0.3) \approx 1.5734812$$

and, in general,

$$y(x_n) \approx y_n = y_{n-1} + f(x_{n-1}, y_{n-1})(x_n - x_{n-1})$$

Each calculation in this process is called a **step** and the quantity

$$\Delta x = x_n - x_{n-1} = 0.1$$

is called the **step size.** (It is convenient, but not necessary, to use equal step sizes, as we have done here.) The values generated by the first ten steps are listed in Table 2.

TABLE 2

x_n	y_n
0	1
0.1	1.1
0.2	1.222
0.3	1.3753284
0.4	1.5734812
0.5	1.8370655
0.6	2.1995465
0.7	2.719347
0.8	3.5078318
0.9	4.8023202
1.0	7.1895481

We summarize Euler's method in the box for convenient reference.

EULER'S METHOD

If $y(x)$ is the solution to the differential equation

$$y' = f(x, y) \qquad y(x_0) = y_0$$

$x_0, x_1, x_2, \ldots$ is a sequence of values defined by

$$x_n = x_{n-1} + \Delta x \qquad n = 1, 2, \ldots$$

and $y_0, y_1, y_2, \ldots$ is the sequence of values defined by

$$y_n = y_{n-1} + f(x_{n-1}, y_{n-1})\Delta x \qquad n = 1, 2, \ldots$$

then

$$y(x_n) \approx y_n$$

In our next example, we will approximate the solution to a differential equation that has a known solution. This will enable us to examine the accuracy of our approximation.

Example 16   Use Euler's method with $\Delta x = 0.1$ to approximate the solution to the differential equation

$$y' = y - x \qquad y(0) = 2$$

on the interval $[0, 1]$.

Solution   Since $x_0 = 0$ and $\Delta x = 0.1$, the values of x are 0, 0.1, 0.2, 0.3, 0.4, 0.5, 0.6, 0.7, 0.8, 0.9, 1. Since $f(x, y) = y - x$, the formula for generating the sequence of y values is

$$y_n = y_{n-1} + (y_{n-1} - x_{n-1})(0.1)$$

Using this formula, we obtain the following sequence of values:

$$y_0 = 2$$
$$y_1 = 2 + (2 - 0)(0.1) = 2$$
$$y_2 = 2.2 + (2.2 - 0.1)(0.1) = 2.41$$
$$y_3 = 2.41 + (2.41 - 0.2)(0.1) = 2.631$$
$$y_4 = 2.631 + (2.631 - 0.3)(0.1) = 2.8641$$
$$y_5 = 2.8641 + (2.8641 - 0.4)(0.1) = 3.11051$$
$$y_6 = 3.11051 + (3.11051 - 0.5)(0.1) = 3.371561$$
$$y_7 = 3.371561 + (3.371561 - 0.6)(0.1) = 3.6487171$$
$$y_8 = 3.6487171 + (3.6487171 - 0.7)(0.1) = 3.9435888$$
$$y_9 = 3.9435888 + (3.9435888 - 0.8)(0.1) = 4.2579477$$
$$y_{10} = 4.2579477 + (4.2579477 - 0.9)(0.1) = 4.5937425$$

COMPARISON OF APPROXIMATE AND EXACT SOLUTIONS

The solution to this equation is

$$y(x) = 1 + x + e^x$$

This was found by using the integrating factor technique (see Section 8-3). Table 3 uses a step size of 0.1 and illustrates the typical behavior of approximate solutions. The difference between the exact value and the approximate value usually increases with each step. The only way to obtain more accurate approximations is to decrease the step size. Table 4 contains the approximate values obtained by using a step size of 0.00001. This step size produces approximate values accurate to four or more decimal places, but requires 100,000 steps. Needless to say, we did not perform these calculations by hand—we used a digital computer.

TABLE 3
Approximate solution using $\Delta x = 0.1$ and 10 steps

x_n	EULER'S SOLUTION y_n	EXACT SOLUTION $y(x_n)$	DIFFERENCE $y(x_n) - y_n$
0	2	2	0
0.1	2.2	2.2051709	0.0051709
0.2	2.41	2.4214028	0.0114028
0.3	2.631	2.6498588	0.0188588
0.4	2.8641	2.8918247	0.0277247
0.5	3.11051	3.1487213	0.0382113
0.6	3.371561	3.4221188	0.0505578
0.7	3.6487171	3.7137527	0.0650356
0.8	3.9435888	4.0255409	0.0819521
0.9	4.2579477	4.3596031	0.1016554
1	4.5937425	4.7182818	0.1245393

TABLE 4
Approximate solution using $\Delta x = 0.00001$ and 100,000 steps

x_n	EULER'S SOLUTION y_n	EXACT SOLUTION $y(x_n)$	DIFFERENCE $y(x_n) - y_n$
0	2	2	0
0.1	2.2051704	2.2051709	0.0000005
0.2	2.4214015	2.4214028	0.0000013
0.3	2.6498568	2.6498588	0.0000020
0.4	2.8918217	2.8918247	0.0000030
0.5	3.1487172	3.1487213	0.0000041
0.6	3.4221133	3.4221188	0.0000055
0.7	3.7137457	3.7137527	0.0000070
0.8	4.025532	4.0255409	0.0000089
0.9	4.359592	4.3596031	0.0000111
1	4.7182682	4.7182818	0.0000136

Problem 16   Use Euler's method with $\Delta x = 0.1$ to approximate the solution to the differential equation

$$y' = x + y \qquad y(0) = 3$$

on the interval [0, 1]. Find the exact solution and construct a table similar to the one in Example 16 comparing the exact and approximate solutions.

Answer   The exact solution is $y(x) = 4e^x - x - 1$.

	EULER'S SOLUTION	EXACT SOLUTION	DIFFERENCE
x_n	y_n	$y(x_n)$	$y(x_n) - y_n$
0	3	3	0
0.1	3.3	3.3206837	0.0206837
0.2	3.64	3.6856110	0.0456110
0.3	4.024	4.0994352	0.0754352
0.4	4.4564	4.5672988	0.1108988
0.5	4.94204	5.0948851	0.1528451
0.6	5.486244	5.6884752	0.2022312
0.7	6.0948684	6.3550108	0.2601424
0.8	6.7743552	7.1021637	0.3278085
0.9	7.5317908	7.9384124	0.4066216
1	8.3749698	8.8731273	0.4981575

Example 17   Use Euler's method with a step size of 0.2 to approximate the solution of the differential equation

$$y' = 2\sqrt{y} - 8x \qquad y(0) = 1$$

on the interverval [0, 1]. Graph the approximate solution.

Solution

n	x_n	$y_n = y_{n-1} + f(x_{n-1}, y_{n-1})\Delta x$	$f(x_n, y_n) = 2\sqrt{y_n} - 8x_n$
0	0	1	2
1	0.2	$1 + 2(0.2) = 1.4$	0.76643191
2	0.4	$1.4 + (0.76643191)(0.2) = 1.5532864$	-0.70738179
3	0.6	$1.5532864 + (-0.70738179)(0.2) = 1.41181$	-2.4236078
4	0.8	$1.41181 + (-2.4236078)(0.2) = 0.92708847$	-4.4742913
5	1	$0.92708844 + (-4.4742913)(0.2) = 0.03223021$	

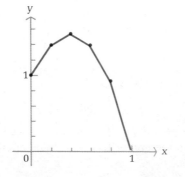

Problem 17   Repeat Example 17 for $y' = x - 2y^2$, $y(0) = 1$.

Answer

n	x_n	y_n
0	0	1
1	0.2	0.6
2	0.4	0.496
3	0.6	0.4775936
4	0.8	0.50635534
5	1	0.56379705

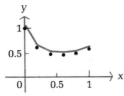

EULER'S METHOD FOR SYSTEMS

Euler's method is easily modified to approximate the solution to a system of differential equations. We simply compute approximate values for each variable in the system.

EULER'S METHOD FOR A SYSTEM OF DIFFERENTIAL EQUATIONS

If $x = x(t)$ and $y = y(t)$ is the solution to the system of differential equations

$$\frac{dx}{dt} = f(x, y) \qquad \frac{dy}{dt} = g(x, y)$$

$$x(t_0) = x_0 \qquad y(t_0) = y_0$$

$t_0, t_1, t_2, \ldots$ is the sequence of values defined by

$$t_n = t_{n-1} + \Delta t \qquad n = 1, 2, \ldots$$

$x_0, x_1, x_2, \ldots$ and $y_0, y_1, y_2, \ldots$ are defined by

$$x_n = x_{n-1} + f(x_{n-1}, y_{n-1})\Delta t$$
$$y_n = y_{n-1} + g(x_{n-1}, y_{n-1})\Delta t \qquad n = 1, 2, \ldots$$

then

$$x(t_n) \approx x_n \qquad \text{and} \qquad y(t_n) \approx y_n$$

Example 18 The prices of two interrelated commodities are p and q. The rate of change of each price is related to the prices by

$$\frac{dp}{dt} = -0.15p + 0.1q + 10$$

$$\frac{dq}{dt} = 0.1p - 0.2q + 15$$

Initially, $p = 150$ and $q = 200$. Use Euler's method with a step size of 1 and five steps to approximate the solution to this system.

Solution $f(p_n, q_n) = -0.15p_n + 0.1q_n + 10$

$g(p_n, q_n) = \quad 0.1p_n - 0.2q_n + 15$

$p_0 = 150$ $\qquad\qquad\qquad\qquad\qquad\qquad$ $q_0 = 200$

$p_1 = 150 + f(150, 200)(1)$ $\qquad\qquad\qquad$ $q_1 = 200 + g(150, 200)(1)$
$\quad = 150 + (7.5)$ $\qquad\qquad\qquad\qquad\quad$ $\quad = 200 - 10$
$\quad = 157.5$ $\qquad\qquad\qquad\qquad\qquad\qquad$ $\quad = 190$

$p_2 = 157.5 + f(157.5, 190)(1)$ $\qquad\qquad\quad$ $q_2 = 190 + g(157.5, 190)(1)$
$\quad = 157.5 + 5.375$ $\qquad\qquad\qquad\qquad$ $\quad = 190 - 7.25$
$\quad = 162.875$ $\qquad\qquad\qquad\qquad\qquad\quad$ $\quad = 182.75$

$p_3 = 162.875 + f(162.875, 182.75)(1)$ $\qquad$ $q_3 = 182.75 + g(162.875, 182.75)(1)$
$\quad = 162.875 + 3.84375$ $\qquad\qquad\qquad$ $\quad = 182.75 - 5.2625$
$\quad = 166.71875$ $\qquad\qquad\qquad\qquad\quad$ $\quad = 177.4875$

$p_4 = 166.71875 + f(166.71875, 177.4875)(1)$ $\quad$ $q_4 = 177.4875 + g(166.71875, 177.4875)(1)$
$\quad = 166.71875 + 2.7409375$ $\qquad\qquad$ $\quad = 177.4875 - 3.825625$
$\quad = 169.45969$ $\qquad\qquad\qquad\qquad\quad$ $\quad = 173.66188$

$p_5 = 169.45969 + f(169.45969, 173.66188)(1)$ $\quad$ $q_5 = 173.66188 + g(169.45969, 173.66188)(1)$
$\quad = 169.45969 + 1.9472345$ $\qquad\qquad$ $\quad = 173.66188 - 2.786407$
$\quad = 171.40692$ $\qquad\qquad\qquad\qquad\quad$ $\quad = 170.87547$

Problem 18 Repeat Example 18 if the initial values are $p = 200$ and $q = 150$.

Answer

p	200	195	191.25	188.4125	186.24313	184.56479
q	150	155	158.5	160.925	162.58125	163.68931

EXERCICSE 10-3

A   In Problems 1–8 use Euler's method with $\Delta x = 0.1$ to approximate the solution to the indicated differential equation on the interval $[0, 1]$. Find the exact solution and construct a table similar to the one in Example 16, comparing the exact and approximate solutions.

1. $y' = 1$; $y(0) = 2$
2. $y' = 2x$; $y(0) = 0$
3. $y' = 3x^2$; $y(0) = 1$
4. $y' = y$; $y(0) = 1$

5. $y' = \dfrac{1}{x + 1}$; $y(0) = 0$
6. $y' = \dfrac{1}{2y}$; $y(0) = 1$

7. $y' = \left(\dfrac{1}{x + 1}\right)y + 2(x + 1)^2$; $y(0) = 1$
8. $y' = y + e^x$; $y(0) = 0$

B   In Problems 9–16 approximate the indicated value of y by using Euler's method with the given step size.

9. $y' = 1 + y^2$; $y(0) = 0$; use $\Delta x = 0.2$ to approximate $y(1)$
10. $y' = \sqrt{1 + y^2}$; $y(0) = 0$; use $\Delta x = 0.2$ to approximate $y(1)$
11. $y' = y^2 - x^2$; $y(1) = -1$; use $\Delta x = 0.4$ to approximate $y(3)$
12. $y' = \sqrt{x^2 + y^2}$; $y(0) = 0$; use $\Delta x = 0.1$ to approximate $y(0.5)$
13. $y' = x + e^{-y}$; $y(0) = 0$; use $\Delta x = 0.1$ to approximate $y(1)$
14. $y' = x + \ln y$; $y(0) = 1$; use $\Delta x = 0.3$ to approximate $y(1.2)$
15. $y' = \ln(x + y)$; $y(1) = 0$; use $\Delta x = 0.2$ to approximate $y(3)$
16. $y' = e^{-xy}$; $y(0) = 0$; use $\Delta x = 0.5$ to approximate $y(3)$

In Problems 17–20 use Euler's method with $\Delta x = 0.2$ to approximate the solution of the indicated equation on the interval $[0, 1]$. Find the exact solution and graph both the approximate solution and the exact solution on the same axes.

17. $y' = -y$; $y(0) = 1$
18. $y' = -y^2$; $y(0) = 1$
19. $y' = 1 + y$; $y(0) = 0$
20. $y' = 1 - y$; $y(0) = 0$

In Problems 21–24 use Euler's method with $\Delta t = 0.2$ to approximate the solution to the given system of differential equations on the interval $[0, 1]$. Compare these values with the exact values from the solutions found in Exercise 8-5.

21. $x' = 2x - y$, $y' = 3x - 2y$; $x(0) = 1$, $y(0) = -1$ (See Problem 3 in Exercise 8-5.)

22. $x' = 5x + 2y$, $y' = 2x + 2y$; $x(0) = 3$, $y(0) = -1$ (See Problem 4 in Exercise 8-5.)

23. $x' = 2x + y$, $y' = 2x + y$; $x(0) = 2$, $y(0) = -1$ (See Problem 5 in Exercise 8-5.)

24. $x' = x - y$, $y' = x - y$; $x(0) = 1$, $y(0) = 2$ (See Problem 6 in Exercise 8-5)

Use Euler's method with $\Delta x = 0.1$ to approximate the solution to each system of differential equations on the interval $[0, 1]$.

25. $x' = x - y + e^{-t}$, $y' = x + y - e^t$; $x(0) = 0$, $y(0) = 0$
26. $x' = tx - t^2 y + t$, $y' = 2tx + t^2 y - 3t$; $x(0) = 0$, $y(0) = 0$

C

27. Construct a table comparing the approximate values of $y(0.5)$ and $y(1)$ obtained by using $\Delta x = 0.5, 0.25$, and 0.1, respectively, given the equation

$$y' = x - y^2 \qquad y(0) = 1$$

28. Repeat Problem 27 for the equation

$$y' = y^2 - x \qquad y(0) = 1$$

29. Compare the approximate solutions obtained by using $\Delta x = 0.2$ on the interval $[1, 2]$ of

$$y' = 2x \qquad y(1) = 1 \qquad \text{and} \qquad y' = \frac{2y}{x} \qquad y(1) = 1$$

30. Compare the approximate solutions obtained by using $\Delta x = 0.2$ on the interval $[0, 1]$ of

$$y' = y \qquad y(0) = 1 \qquad \text{and} \qquad y' = e^x \qquad y(0) = 1$$

31. Let $F(x)$ be defined by

$$F(x) = \int_0^x e^{-t^2} dt$$

(A) Use Simpson's rule with $n = 10$ to approximate $F(1)$.
(B) Show that $F(x)$ is the solution of the differential equation

$$y' = e^{-x^2} \qquad y(0) = 0$$

and use Euler's method with $\Delta x = 0.1$ to approximate $F(1)$.

32. Let $F(x)$ be defined by

$$F(x) = \int_0^x \ln(1 + t^2) dt$$

(A) Use Simpson's rule with $n = 10$ to approximate $F(1)$.
(B) Show that $F(x)$ is the solution of the differential equation

$$y' = \ln(1 + x^2) \qquad y(0) = 0$$

and use Euler's method with $\Delta x = 0.1$ to approximate $F(1)$.

APPLICATIONS

BUSINESS & ECONOMICS

33. *Advertising.* A company is using an extensive newspaper advertising campaign to introduce a new product. The number of people N (in thousands) who have heard of the product satisfies

$$\frac{dN}{dt} = 0.1N(10 - \sqrt{N})$$

If $N(0) = 1$, use Euler's method with $\Delta t = 1$ to approximate $N(7)$.

34. *Sales growth.* The annual sales s (in millions of dollars) of a company satisfy

$$\frac{ds}{dt} = 0.01s(50 - s^{1.5})$$

If the annual sales now ($t = 0$) are \$2 million, use Euler's method with a step size of 1 to approximate the annual sales 5 years from now.

35. *Interrelated markets.* The prices p and q of two commodities in an interrelated market satisfy the following system of differential equations:

$$\frac{dp}{dt} = 2p - q - e^{-t}$$

$$\frac{dq}{dt} = -p + q - e^{-t}$$

If $p(0) = 100$ and $q(0) = 160$, use Euler's method with a step size of 0.5 to approximate $p(3)$ and $q(3)$.

36. *Interrelated markets.* Repeat Problem 35 for the system

$$\frac{dp}{dt} = p - q - \ln(1 + t)$$

$$\frac{dq}{dt} = p - 2q - 2\ln(1 + t)$$

$$p(0) = 150 \qquad q(0) = 100$$

LIFE SCIENCES

37. *Epidemic.* The rate at which a disease spreads through a community is given by

$$\frac{dN}{dt} = 0.1\sqrt{N}(10 - \sqrt{N})$$

where N is the total number (in thousands) of infected individuals at time t. If $N(0) = 1$, use Euler's method with a step size of 1 to approximate $N(8)$.

38. *Population growth.* The population y (in tens of thousands) of a certain species of birds satisfies the growth equation

$$\frac{dy}{dt} = -y + y^2 - 0.1y^3$$

where t is measured in years. If the initial population is 5,000, use Euler's method with a step size of 0.5 to approximate the population after 2 years.

39. *Population growth.* If an ecological system contains two species (say, hares and foxes) that interact with each other in a predator–prey relationship, then the rate of growth of each species is related to the current population of both species. Suppose that the number of hares, y (in thousands), and the number of foxes, x (in thousands), satisfy the system of equations

$$\frac{dy}{dt} = y(0.8 - 0.04x)$$

$$\frac{dx}{dt} = x(0.006y - 0.3)$$

If $y(0) = 55$ and $x(0) = 15$, use Euler's method with a step size of 1 to approximate $y(8)$ and $x(8)$.

40. *Population growth.* Repeat Problem 39 for the system

$$\frac{dy}{dt} = y(0.5 - 0.02x)$$

$$\frac{dx}{dt} = x(0.01y - 0.5)$$

$$y(0) = 60 \qquad x(0) = 20$$

SOCIAL SCIENCES

41. *Rumor spread.* A rumor spreads through a community at a rate given by

$$\frac{dN}{dt} = 0.1N^{2/3}(100 - N^{1/3})$$

where N is the total number of individuals who have heard the rumor at time t. If $N(0) = 1,000$, use Euler's method with a step size of 1 to approximate $N(4)$.

10-4   CHAPTER REVIEW

**IMPORTANT TERMS
AND SYMBOLS**

10-1 Root approximation. root, zero, x intercept, Newton's method, points of intersection, equilibrium point, $x_n = x_{n-1} - \dfrac{f(x_{n-1})}{f'(x_{n-1})}$

10-2 Numerical integration. left-hand end point rule, right-hand end point rule, midpoint rule, trapezoid rule, Simpson's rule, absolute error, area, average value, volume, consumers' and producers' surplus

$x_0 = a, x_1 = a + \Delta x, \ldots, x_n = b,$ $L = [f(x_0) + \cdots + f(x_{n-1})]\Delta x,$

$R = [f(x_1) + \cdots + f(x_n)]\Delta x,$

$M = \left[f\left(\dfrac{x_0 + x_1}{2}\right) + \cdots + f\left(\dfrac{x_{n-1} + x_n}{2}\right) \right]\Delta x,$

$T = [f(x_0) + 2f(x_1) + \cdots + 2f(x_{n-1}) + f(x_n)]\dfrac{\Delta x}{2} = \dfrac{1}{2}(L + R),$

$S = \left[f(x_0) + 4f\left(\dfrac{x_0 + x_1}{2}\right) + 2f(x_1) + 4f\left(\dfrac{x_1 + x_2}{2}\right) + \cdots + 2f(x_{n-1}) \right.$

$\left. + 4f\left(\dfrac{x_{n-1} + x_n}{1}\right) + f(x_n) \right]\dfrac{\Delta x}{6} = \dfrac{1}{3}T + \dfrac{2}{3}M$

10-3 Euler's method. approximate solution, step, step size, Euler's method, $y' = f(x, y),$ $y(x_0) = y_0,$ $y_n = y_{n-1} + f(x_{n-1}, y_{n-1})\Delta x,$ Euler's method for a system of differential equations, $x' = f(x, y),$ $y' = g(x, y),$ $x(t_0) = x_0,$ $y(t_0) = y_0,$ $x_n = x_{n-1} + f(x_{n-1}, y_{n-1})\Delta t,$ $y_n = y_{n-1} + g(x_{n-1}, y_{n-1})\Delta t,$ $x(t_n) \approx x_n,$ $y(t_n) \approx y_n$

EXERCISE 10-4   CHAPTER REVIEW

Work through all the problems in this chapter review and check your answers in the back of the book. (Answers to all review problems are there.) Where weaknesses show up, review appropriate sections in the text. When you are satisfied that you know the material, take the practice test following this review.

A   *In Problems 1–4 find and simplify Newton's formula for x_n. Use this expression and the given value of x_1 to compute x_n for the indicated value of n.*

1. $f(x) = x^2 - 10;$ $x_1 = 3,$ $n = 4$
2. $f(x) = x^2 + x - 1;$ $x_1 = 1,$ $n = 5$

3. $f(x) = \ln x + x + 1$; $x_1 = 0.5$, $n = 5$
4. $f(x) = e^{-x} - x$; $x_1 = 0.5$, $n = 4$

In Problems 5 and 6 construct a table similar to the one in Example 11 (Section 10-2) to find all five approximating sums. Find the absolute error for each sum.

5. $\displaystyle\int_1^3 (2x + 1)\, dx$; $n = 5$

6. $\displaystyle\int_0^2 4x^3\, dx$; $n = 10$

In Problems 7–10 use Euler's method with $\Delta x = 0.1$ to approximate the solution to the given differential equation on the interval $[0, 1]$. Find the solution and construct a table comparing the exact and approximate solutions.

7. $y' = 2xy$; $y(0) = 1$

8. $y' = \dfrac{-4y}{x + 1}$; $y(0) = 1$

9. $y' = y + e^{2x}$; $y(0) = 2$

10. $y' = -3x^2y^2$; $y(0) = 1$

In Problems 11 and 12 sketch the graph of the functions and use Newton's method to find the x intercepts.

11. $f(x) = x^3 - 12x + 3$

12. $f(x) = x^3 - 9x^2 + 15x + 30$

B   In Problems 13 and 14 graph both functions on the same set of axes and find their points of intersection.

13. $f(x) = e^{-x}$; $g(x) = x^3$

14. $f(x) = \dfrac{1}{x^3}$; $g(x) = x + 1$

15. Use the midpoint rule with $n = 10$ to approximate

$$\int_0^1 \frac{x}{\sqrt{1 + x^4}}\, dx$$

16. Use the trapezoid rule with $n = 10$ to approximate

$$\int_0^2 \frac{1}{e^x + e^{-x}}\, dx$$

17. Use Simpson's rule with $n = 10$ to approximate

$$\int_1^3 [\ln(1 + x^2)]^2\, dx$$

18. Use the midpoint rule with $n = 5$ to approximate the average value of $1/(1 + x^3)$ on the interval $[0, 2]$.

19. Use the trapezoid rule with $n = 5$ to approximate the area of the region bounded by the graphs of $y = e^{-x}$, $y = x + 1$, and $x = 1$.

20. Use Simpson's rule with $n = 5$ to approximate the volume of the solid of revolution obtained by revolving the region bounded by $y = 1/(1 + x^2)$, $y = 0$, $x = 0$, and $x = 1$ about the x axis.

In Problems 21–24 use Euler's method and the given step size to approximate the indicated value of the solution to the differential equation.

21. $y' = x + \sqrt{y}$; $y(0) = 0$; use $\Delta x = 0.2$ to approximate $y(1)$

22. $y' = \dfrac{1}{y} + \dfrac{1}{x}$; $y(1) = 1$; use $\Delta x = 0.4$ to approximate $y(3)$

23. $y' = e^{x-y}$; $y(0) = 0$; use $\Delta x = 0.1$ to approximate $y(0.5)$

24. Use $\Delta t = 0.1$ to approximate $x(1)$ and $y(1)$ for the system

$$\frac{dx}{dt} = y - t \qquad\qquad \frac{dy}{dt} = t - x^2$$

$$x(0) = 1 \qquad\qquad y(0) = -1$$

C 25. Apply Newton's method to $f(x) = ax^2 + bx + c$ and derive a formula that approximates the roots of a quadratic polynomial.

26. Approximate the solution to the differential equation

$$y' = \sqrt{x^2 + y^2} \qquad\qquad y(0) = 1$$

on the interval $[0, 2]$ using $\Delta x = 1, 0.5$, and 0.25, respectively. Graph all three approximate solutions on the same coordinate system.

27. The function $f(x) = 3x - x^2 - x^4$ obviously has one intercept at $x = 0$. There is another intercept r near $x = 1$. Use Newton's method to approximate r correct to one decimal place. Then use the midpoint rule to approximate the area under the graph of f from $x = 0$ to $x = r$ with n chosen so that $\Delta x = 0.1$.

28. Use Euler's method with $\Delta x = 0.1$ to approximate the solution to

$$y' = \sqrt{x + y} \qquad\qquad y(1) = 1$$

on the interval $[1, 2]$. Then use Simpson's rule with $n = 5$ to approximate the area under the graph of the solution from $x = 1$ to $x = 2$.

APPLICATIONS

BUSINESS & ECONOMICS 29. *Break-even point.* The cost and revenue functions for a product are

$$C(x) = 10 + 0.5x \qquad \text{and} \qquad R(x) = 2x + \frac{10x}{1 + x^2} \qquad x \geq 1$$

Use Newton's method to approximate the break-even point.

30. *Consumers' and producers' surplus.* Find the consumers' surplus and the producers' surplus for

$$p = D(x) = \frac{56}{2 + \sqrt{x}} \qquad \text{and} \qquad p = S(x) = 3 + \sqrt{x}$$

Use the trapezoid rule with $n = 10$ to approximate the integrals.

31. *Marginal analysis.* The marginal cost and revenue functions for a machine are

$$C'(t) = \frac{2t^2}{4 + t^2} \quad \text{and} \quad R'(t) = \frac{8}{4 + t^2}$$

where $C(t)$ and $R(t)$ are the total cost and revenue functions (in thousands of dollars) t years after the machine has been put into use. Find the useful life of the machine, and use the midpoint rule with $n = 10$ to approximate the total profit earned by the machine during its useful life.

32. *Sales growth.* The annual sales s (in thousands of dollars) of a company satisfy the equation

$$\frac{ds}{dt} = 0.1\sqrt{s}(100 - \sqrt{s})$$

If $s(0) = 50$, use Euler's method with $\Delta t = 1$ to approximate $s(5)$.

LIFE SCIENCES

33. *Medicine.* The level of a particular hormone in the bloodstream at time t is given by

$$L(t) = 10t^3 e^{-t^3}$$

Use the midpoint rule with $n = 10$ to approximate the average hormone level for $0 \le t \le 5$.

34. *Population growth.* Suppose the number of foxes, x (in thousands), and the number of hares, y (in thousands), in an ecological system satisfy the system

$$\frac{dy}{dt} = y(0.9 - 0.05x)$$

$$\frac{dx}{dt} = x(0.005y - 0.2)$$

$$y(0) = 50 \qquad x(0) = 15$$

Use Newton's method with $\Delta t = 1$ to approximate the population of each species after 5 years.

SOCIAL SCIENCES

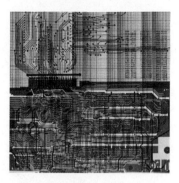

35. *Psychology.* The rate at which workers can assemble electronic components t hours after they begin work is given by

$$N'(t) = \frac{4t^2 + 4t + 15}{4t^2 + 4t + 5}$$

Use the trapezoid rule with $n = 8$ to approximate the total number of components a worker will assemble during the first 4 hours at work.

PRACTICE TEST: **CHAPTER 10**

---

1. Apply Newton's method to the function $f(x) = x^2 - 3$ in order to approximate $\sqrt{3}$.

2. Graph $f(x) = e^{-x}$ and $g(x) = x - 2$ on the same set of axes. Use Newton's method to approximate the x coordinate of the point of intersection.

3. Sketch the graph of $f(x) = x^3 + 6x^2 - 15x + 20$ and use Newton's method to find all the x intercepts.

4. Use Simpson's rule with $n = 5$ to approximate the average value of the function

$$f(x) = \frac{3 + x^3}{1 + x}$$

on the interval $[0, 2]$.

5. Use the trapezoid rule with $n = 4$ to approximate the area bounded by the graphs of $y = \ln(10 + x^2)$, $y = 0$, $x = 0$, and $x = 2$.

6. Use Euler's method with $\Delta x = 0.2$ to approximate the solution to

$$y' = \frac{1}{x^2} + \frac{1}{y^2} \qquad y(1) = 1$$

on the interval $[1, 2]$.

7. Use Euler's method with $\Delta t = 1$ to approximate the solution to the system

$$\frac{dx}{dt} = \frac{ty}{1 + x^2}$$

$$\frac{dy}{dt} = \frac{txy}{1 + y^2}$$

$$x(0) = 1 \qquad y(0) = -1$$

on the interval $[0, 4]$.

8. *Break-even point.* Find the break-even point for

$$R(x) = x\left(1 + \frac{10,000}{x^4}\right) \quad \text{and} \quad C(x) = 20 + 0.8x \qquad x \geq 1$$

9. *Consumers' and producers' surplus.* Find the consumers' surplus and the producers' surplus for

$$p = D(x) = \frac{20}{1 + \sqrt{x}} \quad \text{and} \quad p = S(x) = 2 + \sqrt{x}$$

Use the midpoint rule with n chosen so that $\Delta x = 1$ to approximate the integrals.

10. *Epidemic spread.* The total number of people $N(t)$ who have been infected by a disease at time t satisfies the equation

$$\frac{dN}{dt} = 0.01\sqrt{N}(1,000 - \sqrt{N})$$

If $N(0) = 500$, use Euler's method with $\Delta t = 1$ to approximate $N(4)$.

CHAPTER 11
PROBABILITY AND CALCULUS

CONTENTS

11 PROBABILITY AND CALCULUS

11-1 FINITE PROBABILITY MODELS

SAMPLE SPACE

Before we begin to study the applications of calculus to probability, we will review some of the basic concepts of probability experiments with a finite number of outcomes. The set S of possible outcomes is called a **sample space** for the experiment. Each element in the sample space is called a **simple event.** For example, if three coins are tossed, then a sample space for this experiment is

$$S = \{HHH, HHT, HTH, HTT, THH, THT, TTH, TTT\}$$

where H is heads and T is tails, and each element in the set is a possible outcome of the experiment. If each coin is fair (a head is as likely to occur as a tail), then each simple event in this sample space is **equally likely** to occur. Since there are eight different outcomes in S, the probability of any one of these occurring is

$$P(e) = \tfrac{1}{8} \qquad e \in S$$

RANDOM VARIABLE X

In many situations we may not be interested in each simple event, but in some numerical value associated with the event. For example, in the above coin tossing experiment, we may be interested in the number of heads that turn up rather than in the particular pattern that turns up. Or, in selecting a random sample of students, we may be interested in the proportion that are women rather than which particular students are women. And, in the same way, a craps player is usually interested in the sum of the dots showing on the faces of a pair of dice rather than the pattern of dots on each face.

In each of these examples, we have a rule that assigns to each simple event in a sample space S a single real number. Mathematically speaking, we are dealing with a function. Historically, this particular type of function has been called a *random variable.*

RANDOM VARIABLE

A **random variable** is a function that assigns a numerical value to each simple event in a sample space S.

The term *random variable* is an unfortunate choice, since it is neither random nor a variable—it is a function with a numerical value and it is defined on a sample space. But the terminology has stuck and is now standard, so we will have to live with it. Capital letters, such as X, are used to represent random variables.

Let us return to the experiment of tossing three coins. A sample space S of equally likely simple events is indicated in the first column of Table 1. The second column indicates the number of heads corresponding to a simple event. And the last column indicates the probability of each simple event occurring.

TABLE 1 Tossing three coins

S		NUMBER OF HEADS $X(e_i)$	PROBABILITY $P(e_i)$
e_1:	TTT	0	$\frac{1}{8}$
e_2:	TTH	1	$\frac{1}{8}$
e_3:	THT	1	$\frac{1}{8}$
e_4:	HTT	1	$\frac{1}{8}$
e_5:	THH	2	$\frac{1}{8}$
e_6:	HTH	2	$\frac{1}{8}$
e_7:	HHT	2	$\frac{1}{8}$
e_8:	HHH	3	$\frac{1}{8}$

The random variable X (a function) associates exactly one of the numbers 0, 1, 2, or 3 with each simple event. For example, $X(e_1) = 0$, $X(e_2) = 1$, $X(e_3) = 1$, and so on.

PROBABILITY DISTRIBUTION OF A RANDOM VARIABLE X

We are interested in the probability of the occurrence of each image value of X; that is, in the probability of the occurrence of zero heads, one head, two heads, or three heads in the single toss of three coins. We indicate this probability by

$$p(x) \qquad \text{where} \qquad x \in \{0, 1, 2, 3\}$$

The function p is called the **probability function* of the random vari-**

---

*Formally, the probability function p of the random variable X is defined by $p(x) = P(\{e_i \in S | X(e_i) = x\})$, which, because of its cumbersome nature, is usually simplified to $p(x) = P(X = x)$ or, simply, $p(x)$. We will use the simplified notation.

able X. What is $p(2)$, the probability of getting exactly two heads on the single toss of three coins? Exactly two heads occur if any of the simple events

$$e_5: \quad \text{THH} \qquad e_6: \quad \text{HTH} \qquad e_7: \quad \text{HHT}$$

occurs. Adding the probabilities for these simple events (see Table 1), we obtain $p(2) = \frac{1}{8} + \frac{1}{8} + \frac{1}{8} = \frac{3}{8}$.

Proceeding similarly for $p(0)$, $p(1)$, and $p(3)$, we obtain the results in Table 2. This table is called a **probability distribution for the random variable X.** Probability distributions are also represented graphically, as in Figure 1.

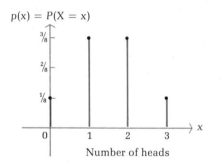

$p(x) = P(X = x)$

FIGURE 1   Probability
distribution

TABLE 2   Probability distribution

NUMBER OF HEADS x	0	1	2	3
PROBABILITY $p(x)$	$\frac{1}{8}$	$\frac{3}{8}$	$\frac{3}{8}$	$\frac{1}{8}$

Note from Table 2 or Figure 1 that

1. $0 \le p(x) \le 1, \quad x \in \{0, 1, 2, 3\}$
2. $p(0) + p(1) + p(3) = \frac{1}{8} + \frac{3}{8} + \frac{3}{8} + \frac{1}{8} = 1$

These are **general properties of any probability distribution** of a random variable X associated with a finite sample space. That is, if we let $\{x_1, x_2, \ldots, x_m\}$ be the image set for the random variable X and let p be the probability function of X, then

1. $0 \le p(x) \le 1, \quad x \in \{x_1, x_2, \ldots, x_m\}$
2. $p(x_1) + p(x_2) + \cdots + p(x_m) = 1$

These properties follow from the way p is defined.

**EXPECTED VALUE
OF A RANDOM
VARIABLE X**

Suppose the experiment of tossing three coins were repeated a large number of times. What would be the average number of heads per toss (the total number of heads in all tosses divided by the total number of tosses)? Consulting the probability distribution in Table 2 or Figure 1, we see that we would expect to toss zero heads one-eighth of the time, one head three-eighths of the time, two heads three-eighths of the time, and three heads one-eighth of the time. Thus, in the long run, we would expect the average number of heads per toss of the three coins, or the *expected value* $E(X)$, to be given by

$$E(X) = 0(\tfrac{1}{8}) + 1(\tfrac{3}{8}) + 2(\tfrac{3}{8}) + 3(\tfrac{1}{8}) = \tfrac{12}{8} = 1.5$$

It is important to note that the expected value is not a value that will necessarily occur in a single experiment (1.5 heads cannot occur in the toss of three coins), but it is average of what occurs over a large number of experiments. Sometimes, we will toss more than 1.5 heads and sometimes less, but if the experiment is repeated many times, the average number of heads per experiment should approach 1.5.

We now make the above discussion precise with the following definition of expected value:

EXPECTED VALUE OF A RANDOM VARIABLE X

Given the probability distribution for the random variable X:

x_i	x_1	x_2	$\cdots$	x_m
p_i	p_1	p_2	$\cdots$	p_m

where $p_i = p(x_i)$, we define the **expected value of X,** denoted by $E(X)$, by the formula

$$E(X) = x_1p_1 + x_2p_2 + \cdots + x_mp_m$$

Example 1   What is the expected value (long-run average) of the number of dots facing up for the roll of a single die?

Solution   If we choose

$$S = \{1, 2, 3, 4, 5, 6\}$$

as our sample space, then each simple event is a numerical outcome reflecting our interest, and each is equally likely. The random variable X in this case is just the identity function (each number is associated with itself). Thus, the probability distribution for X is

x_i	1	2	3	4	5	6
p_i	$\tfrac{1}{6}$	$\tfrac{1}{6}$	$\tfrac{1}{6}$	$\tfrac{1}{6}$	$\tfrac{1}{6}$	$\tfrac{1}{6}$

Hence,

$$\begin{aligned} E(X) &= 1(\tfrac{1}{6}) + 2(\tfrac{1}{6}) + 3(\tfrac{1}{6}) + 4(\tfrac{1}{6}) + 5(\tfrac{1}{6}) + 6(\tfrac{1}{6}) \\ &= \tfrac{21}{6} = 3.5 \end{aligned}$$

Problem 1   Suppose the die in Example 1 is not fair and we obtain (empirically) the probability distribution for X:

x_i	1	2	3	4	5	6
p_i	.14	.13	.18	.20	.11	.24

[Note: Sum = 1]

What is the expected value of X?

Answer $E(X) = 3.73$

MEAN

Since the expected value of a random variable represents the long-run average of repeated experiments, it is often referred to as the *mean.* Traditionally, the greek letter μ is used to denote the mean. Thus,

$$\mu = E(X) = x_1 p_2 + x_2 p_2 + \cdots + x_m p_m$$

is the **mean of the random variable X.** Geometrically, the mean, in some sense, is the center of the values of X and is often referred to as a **measure of central tendency.**

STANDARD
DEVIATION

Another numerical quantity that is used to describe the properties of a random variable is the **standard deviation.** This quantity gives a **measure of the dispersion or spread,** of the random variable X about the mean μ.

STANDARD DEVIATION OF A RANDOM VARIABLE X

Given the probability distribution for the random variable X:

x_i	x_1	x_2	$\cdots$	x_m
p_i	p_1	p_2	$\cdots$	p_m

and the mean

$$\mu = x_1 p_1 + x_2 p_2 + \cdots + x_m p_m$$

we define the **variance of X,** denoted by $V(X)$, by the formula

$$V(X) = (x_1 - \mu)^2 p_1 + (x_2 - \mu)^2 p_2 + \cdots + (x_m - \mu)^2 p_m$$

and the **standard deviation of X,** denoted by σ, by the formula

$$\sigma = \sqrt{V(X)}$$

In other words, the variance is the average of the squares of the distances from each value of X to the mean. Standard deviation is defined using the square root so that it will be expressed in the same units as the values of X.

Returning to the coin tossing experiment, we have already shown that $\mu = E(X) = 1.5$. Thus,

$$V(X) = (0 - 1.5)^2(\tfrac{1}{8}) + (1 - 1.5)^2(\tfrac{3}{8}) + (2 - 1.5)^2(\tfrac{3}{8}) + (3 - 1.5)^2(\tfrac{1}{8})$$
$$= .75$$

and

$$\sigma = \sqrt{V(X)} = \sqrt{.75} \approx .866$$

Example 2   Find the variance and standard deviation for the random variable in Example 1.

Solution $\mu = E(X) = 3.5$
$$V(X) = (1 - 3.5)^2(\tfrac{1}{6}) + (2 - 3.5)^2(\tfrac{1}{6}) + (3 - 3.5)^2(\tfrac{1}{6})$$
$$+ (4 - 3.5)^2(\tfrac{1}{6}) + (5 - 3.5)^2(\tfrac{1}{6}) + (6 - 3.5)^2(\tfrac{1}{6})$$
$$\approx 2.917$$
$$\sigma = \sqrt{V(X)} = \sqrt{2.917} \approx 1.708$$

Problem 2   Find the variance and standard deviation for the random variable in Problem 1.

Answer $V(X) = 2.9571, \quad \sigma \approx 1.720$

The standard deviation is often used to compare different probability distributions. Figure 2 (parts A–D) gives four different probability distributions and their graphs. Notice the relationship between the standard deviation σ and the dispersion of the probability distribution about the mean. The tighter the cluster of the probability distribution about the mean, the smaller the standard deviation.

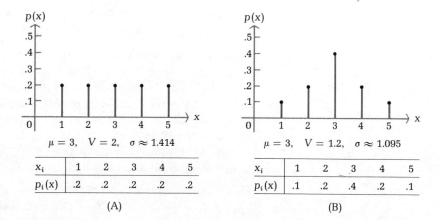

$\mu = 3, \quad V = 2, \quad \sigma \approx 1.414$

x_i	1	2	3	4	5
$p_i(x)$	.2	.2	.2	.2	.2

(A)

$\mu = 3, \quad V = 1.2, \quad \sigma \approx 1.095$

x_i	1	2	3	4	5
$p_i(x)$	.1	.2	.4	.2	.1

(B)

FIGURE 2

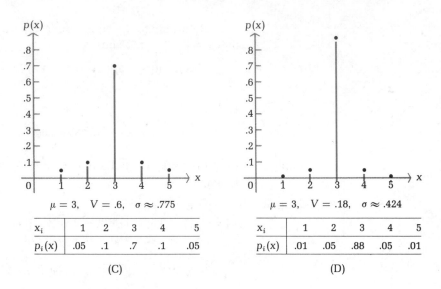

x_i	1	2	3	4	5
$p_i(x)$	.05	.1	.7	.1	.05

(C)

x_i	1	2	3	4	5
$p_i(x)$	.01	.05	.88	.05	.01

(D)

FIGURE 2   *Continued*

EXERCISE 11-1

A   *In Problems 1–4 graph the probability distribution and find the mean and standard deviation of the random variable X.*

1.

x_i	-2	-1	0	1	2
p_i	.1	.2	.4	.2	.1

2.

x_i	-2	-1	0	1	2
p_i	.1	.1	.2	.2	.4

3.

x_i	-2	-1	0	1	2
p_i	.5	.2	.1	.1	.1

4.

x_i	-2	-1	0	1	2
p_i	.3	.1	.1	.1	.4

A spinner is marked from 1 to 10 and each number is as likely to turn up as any other. Problems 5–10 refer to this experiment.

5.   Find a sample space S consisting of equally likely simple events.

6.   The random variable X represents the number that turns up on the spinner. Find the probability distribution for X.

7.   What is the probability of obtaining an even number?

8.   What is the probability of obtaining a number that is exactly divisible by 3?

9.   What is the expected value of X?

10.   What is the standard deviation of X?

B   11.   In tossing two fair coins, what is the expected number of heads?

12.   In a family with two children, excluding multiple births and assuming a boy is as likely as a girl at each birth, what is the expected number of boys?

An experiment consists of tossing a coin four times in succession. Answer the questions in Problems 13–18 regarding this experiment.

13. The random variable X represents the number of heads that occur in four tosses. Find and graph the probability distribution of X.
14. What is the probability of getting two or more heads?
15. What is the probability of getting an even number of heads?
16. What is the probability of getting more heads than tails?
17. Find the expected number of heads.
18. Find the standard deviation of X.

An experiment consists of rolling two fair dice. Answer the questions in Problems 19–24 regarding this experiment.

19. The random variable X represents the sum of the dots on the two up faces of the dice. Find and graph the probability distribution of X.
20. What is the probability that the sum is 7 or 11?
21. What is the probability that the sum is less than 6?
22. What is the probability that the sum is an even number?
23. Find the expected value of the sum of the dots.
24. Find the standard deviation of X.

C 25. After you pay $4 to play a game, a single fair die is rolled and you are paid back the number of dollars equal to the number of dots facing up. For example, if 5 dots turn up, $5 is returned to you for a net gain of $1. If 1 dot turns up, $1 is returned to you for a net gain, or **payoff,** of −$3; and so on. If X is the random variable that represents net gain, or payoff, what is the expected value of X?
26. Repeat Problem 25 with the same game costing $3.50 for each play.
27. A player tosses two coins and wins $3 if two heads appear and $1 if one head appears, but loses $6 if two tails appear. If X is the random variable representing the player's net gain, what is the expected value of X?
28. Repeat Problem 27 if the player wins $5 if two heads appear and $2 if one head appears, but losses $7 if two tails appear.
29. Roulette wheels in the United States generally have thirty-eight equally spaced slots numbered, 00, 0, 1, 2, . . . , 36. A player who bets $1 on any given number wins $35 (and gets the bet back) if the ball comes to rest on the chosen number; otherwise, the $1 bet is lost. If X is the random variable that represents the player's net gain, what is the expected value of X?
30. In roulette (see Problem 29) the numbers from 1 to 36 are evenly divided between red and black. A player who bets $1 on black, wins $1 (and gets the bet back) if the ball comes to rest on black; otherwise (if the ball lands on red, 0, or 00), the $1 bet is lost. If X is the random variable that represents the player's net gain, what is the expected value of X?

APPLICATIONS

BUSINESS & ECONOMICS

31. *Insurance.* The annual premium for a $5,000 insurance policy against the theft of a painting is $150. If the probability that the painting will be stolen during the year is .01, what is your expected gain or loss if you take out this insurance?

32. *Insurance.* Repeat Problem 31 from the point of view of the insurance company.

33. *Decision analysis.* After careful testing and analysis, an oil company is considering drilling in one of two different sites. It is estimated that site *A* will net $30 million if successful (probability .2) and lose $3 million if not (probability .8); site *B* will net $70 million if successful (probability .1) and lose $4 million if not (probability .9). Based on the expected return from each site, which site should the company choose?

34. *Decision analysis.* Repeat Problem 33, assuming additional analysis caused the estimated probability of success in field *B* to be changed from .1 to .11.

LIFE SCIENCES

35. *Genetics.* Suppose that at each birth having a girl is not as likely as having a boy, and that the probability assignments for the number of boys in a family with three children is approximated from past records to be:

NUMBER OF BOYS

x_i	p_i
0	.12
1	.36
2	.38
3	.14

What is the expected number of boys in a three-child family?

36. *Genetics.* A pink-flowering plant is of genotype RW. If two such plants are crossed, we obtain a red plant (RR) with probability .25, a pink plant (RW) with probability .50, and a white plant (WW) with probability .25:

NUMBER OF W GENES PRESENT

x_i	p_i
0	.25
1	.50
2	.25

What is the expected number of W genes present in a crossing of this type?

SOCIAL SCIENCES   37. *Politics.*   A money drive is organized by a campaign committee for a candidate running for public office. Two approaches are considered:

A_1—A general mailing with a followup mailing
A_2—Door-to-door solicitation with followup telephone calls

From campaign records of previous committees, average donations and their corresponding probabilities are estimated to be:

A_1			A_2	
x_i (return per person)	p_i		x_i (return per person)	p_i
$10	.3		$15	.3
$ 5	.2		$ 3	.1
$ 0	.5		$ 0	.6
	1.0			1.0

Which course of action should be taken based on expected return?

11-2   CONTINUOUS RANDOM VARIABLES

All the random variables we considered in the preceding section assumed one of a finite number of possible values. But in many experiments we use random variables that can assume any one of an infinite number of possible values. For example, we may be interested in the life expectancy of a circuit in a computer, the time it takes a rat to find its way through a maze, or the amount of a certain drug present in an individual's bloodstream. In experiments of this type, the set of all possible outcomes forms an interval on the real line. Thus, the life expectancy of a computer's circuit could be any value in the interval $[0, \infty)$, and the transit time of a rat in a maze might always lie in the interval $[5, 60]$. A random variable associated with this kind of experiment is usually called *continuous*.

CONTINUOUS RANDOM VARIABLE

A **continuous random variable** is a random variable with image values that form an interval on the real line.

The term *continuous* is not used in the same sense here as it was used in Section 2-2. In this case, it refers to the fact that the values of the random variable form a continuous set of numbers, such as $[0, \infty)$, rather than a discrete set, such as $\{0, 1, 2, 3\}$. In fact, random variables of the type we considered in the preceding section are often called **discrete random variables** to emphasize the difference between them and continuous random variables.

PROBABILITY DENSITY FUNCTION

In order to work with continuous random variables, we must have some way of defining the probability of an event. Since there are an infinite number of possible outcomes, we cannot define the probability of each outcome by means of a table. Instead, we introduce a new function that is used to compute probabilities. For convenience in stating definitions and formulas, we will assume that the outcomes can be any real number.

PROBABILITY DENSITY FUNCTION

The function $f(x)$ is a **probability density function** for a continuous random variable X if:

1. $f(x) \geq 0$ for all $x \in (-\infty, \infty)$

2. $\displaystyle\int_{-\infty}^{\infty} f(x)\, dx = 1$

3. The probability that x lies in the interval $[c, d]$ is given by

$$P(c \leq x \leq d) = \int_{c}^{d} f(x)\, dx$$

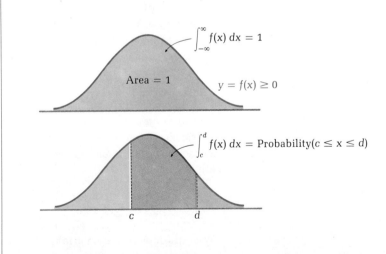

$$\int_{-\infty}^{\infty} f(x)\, dx = 1$$

Area = 1 $y = f(x) \geq 0$

$$\int_{c}^{d} f(x)\, dx = \text{Probability}(c \leq x \leq d)$$

c d

Example 3   Let: $f(x) = \begin{cases} 12x^2 - 12x^3 & 0 \le x \le 1 \\ 0 & \text{otherwise} \end{cases}$

(A)   Verify that f satisfies the first two conditions for a probability density function.

(B)   Compute $P(\frac{1}{4} \le x \le \frac{3}{4})$, $P(x \le \frac{1}{2})$, $P(x \ge \frac{2}{3})$, and $P(x = \frac{1}{3})$.

Solutions   (A)   For $0 \le x \le 1$, we have $f(x) = 12x^2 - 12x^3 = 12x^2(1-x) \ge 0$. Since $f(x) = 0$ for all other values of x, it follows that $f(x) \ge 0$ for all x. Also,

$$\int_{-\infty}^{\infty} f(x)\, dx = \int_0^1 (12x^2 - 12x^3)\, dx = (4x^3 - 3x^4)\Big|_0^1 = (4-3) - (0) = 1$$

(B) $P(\frac{1}{4} \le x \le \frac{3}{4}) = \int_{1/4}^{3/4} f(x)\, dx = \int_{1/4}^{3/4} (12x^2 - 12x^3)\, dx$

$$= (4x^3 - 3x^4)\Big|_{1/4}^{3/4} = {}^{189}\!/_{256} - {}^{13}\!/_{256} = {}^{11}\!/_{16}$$

$P(x \le \frac{1}{2}) = \int_{-\infty}^{1/2} f(x)\, dx = \int_0^{1/2} (12x^2 - 12x^3)\, dx$

$$= (4x^3 - 3x^4)\Big|_0^{1/2} = {}^{5}\!/_{16}$$

$P(x \ge \frac{2}{3}) = \int_{2/3}^{\infty} f(x)\, dx = \int_{2/3}^{1} (12x^2 - 12x^3)\, dx$

$$= (4x^3 - 3x^4)\Big|_{2/3}^{1} = 1 - {}^{16}\!/_{27} = {}^{11}\!/_{27}$$

$P(x = \frac{1}{3}) = \int_{1/3}^{1/3} f(x)\, dx = 0$ Property 1, page 190

Problem 3   Let: $f(x) = \begin{cases} 6x - 6x^2 & 0 \le x \le 1 \\ 0 & \text{otherwise} \end{cases}$

(A)   Verify that f satisfies the first two conditions for a probability density function.

(B)   Compute $P(\frac{1}{3} \le x \le \frac{2}{3})$, $P(x \le \frac{1}{5})$, $P(x \ge \frac{1}{2})$, and $P(x = \frac{1}{4})$.

Answers   (B) ${}^{13}\!/_{27}$, ${}^{13}\!/_{125}$, $\frac{1}{2}$, 0

COMPARING PROBABILITY DISTRIBUTION FUNCTIONS AND PROBABILITY DENSITY FUNCTIONS

The last probability in Example 3 illustrates a fundamental difference between discrete and continuous random variables. In the discrete case, there is a *probability distribution* $p(x)$ that gives the probability of each possible value of the random variable. Thus, if c is one of the values of the random variable, then $P(x = c) = p(c)$. In the continuous case, the *integral of the probability density function* $f(x)$ gives the probability that the outcome lies in a certain interval. If c is any real number, then the probability that the outcome is *exactly* c is

$$P(x = c) = P(c \leq x \leq c) = \int_c^c f(x)\, dx = 0$$

Thus, $P(x = c) = 0$ for *any* number c and, since $f(c)$ is certainly not 0 for all values of c, we see that $f(x)$ does not play the same role for a continuous random variable as $p(x)$ does for a discrete random variable.

The fact that $P(x = c) = 0$ also implies that excluding either end point from an interval does not change the probability that the random variable lies in that interval; that is,

$$P(a < x < b) = P(a < x \leq b) = P(a \leq x < b) = P(a \leq x \leq b)$$

$$= \int_a^b f(x)\, dx$$

Example 4   Use the probability density function in Example 3 to compute $P(.1 < x \leq .2)$ and $P(x > .9)$.

Solution
$$P(.1 < x \leq .2) = \int_{.1}^{.2} f(x)\, dx = \int_{.1}^{.2} (12x^2 - 12x^3)\, dx = (4x^3 - 3x^4)\Big|_{.1}^{.2}$$

$$= .0272 - .0037 = .0235$$

$$P(x > .9) = \int_{.9}^{\infty} f(x)\, dx = \int_{.9}^{1} (12x^2 - 12x^3)\, dx = (4x^3 - 3x^4)\Big|_{.9}^{1}$$

$$= 1 - .9477 = .0523$$

Problem 4   Use the probability density function in Problem 3 to compute $P(.2 \leq x < .4)$ and $P(x < .8)$.

Answer   .248, .896

If $f(x)$ is the probability density function in Example 3, notice that $f(\tfrac{2}{3}) = \tfrac{16}{9} > 1$. Thus, a probability density function can assume values larger than 1. This illustrates another difference between probability density functions and probability distribution functions. In terms of inequalities, a probability distribution function must always satisfy $0 \leq p(x) \leq 1$, while a probability density function need only satisfy $f(x) \geq 0$. Despite these differences, we shall see that there are many similarities in the application of probability distribution functions and probability density functions.

Example 5   The shelf-life (in months) of a certain drug is a continuous random variable with probability density function

$$f(x) = \begin{cases} 50/(x + 50)^2 & x \geq 0 \\ 0 & \text{otherwise} \end{cases}$$

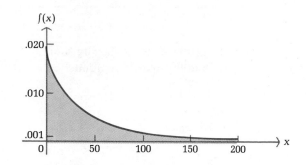

Find the probability that the drug has a shelf-life of:
(A) Between 10 and 20 months
(B) At most 30 months
(C) Over 25 months

Solutions

(A) $P(10 \leq x \leq 20) = \int_{10}^{20} f(x)\, dx = \int_{10}^{20} \frac{50}{(x + 50)^2}\, dx = \left. \frac{-50}{x + 50} \right|_{10}^{20}$

$$= \left(-\frac{50}{70} \right) - \left(-\frac{50}{60} \right) = \frac{5}{42}$$

(B) $P(x \leq 30) = \int_{-\infty}^{30} f(x)\, dx = \int_{0}^{30} \frac{50}{(x + 50)^2}\, dx = \left. \frac{-50}{x + 50} \right|_{0}^{30}$

$$= \left(-\frac{50}{80} \right) - (-1) = \frac{3}{8}$$

(C) $P(x > 25) = \int_{25}^{\infty} f(x)\, dx = \int_{25}^{\infty} \frac{50}{(x + 50)^2}\, dx$

$$= \lim_{R \to \infty} \int_{25}^{R} \frac{50}{(x + 50)^2}\, dx = \lim_{R \to \infty} \left. \frac{-50}{x + 50} \right|_{25}^{R}$$

$$= \lim_{R \to \infty} \left(-\frac{50}{R + 50} + \frac{50}{75} \right) = \frac{2}{3}$$

Problem 5

In Example 5 find the probability that the drug has a shelf-life of:
(A) Between 50 and 100 months
(B) At most 20 months
(C) Over 10 months

Answers   (A) $\frac{1}{6}$ (B) $\frac{2}{7}$ (C) $\frac{5}{6}$

**CUMULATIVE
PROBABILITY
DISTRIBUTION FUNCTION**

Each time we compute the probability for a continuous random variable, we must find the antiderivative of the probability density function. This antiderivative is used so often that it is convenient to give it a name.

CUMULATIVE PROBABILITY DISTRIBUTION FUNCTION

If f is a probability density function, then the associated **cumulative probability distribution function** F is defined by

$$F(x) = P(t \leq x) = \int_{-\infty}^{x} f(t) \, dt$$

Furthermore,

$$P(c \leq x \leq d) = F(d) - F(c)$$

Figure 3 gives a geometric interpretation of these ideas.

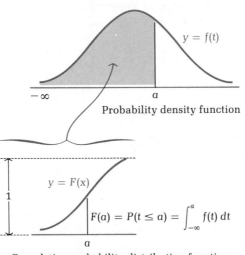

$y = f(t)$

$-\infty$ a

Probability density function

$y = F(x)$

$F(a) = P(t \leq a) = \int_{-\infty}^{a} f(t) \, dt$

a

Cumulative probability distribution function

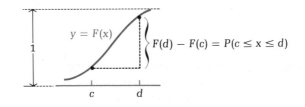

$y = F(x)$

$F(d) - F(c) = P(c \leq x \leq d)$

c d

FIGURE 3

Example 6   Find the cumulative probability distribution function for the probability density function in Example 3, and use it to compute $P(.1 \leq x \leq .9)$.

Solution   If $x < 0$, then

$$F(x) = \int_{-\infty}^{x} f(t) \, dt = \int_{-\infty}^{x} 0 \, dt = 0$$

If $0 \le x \le 1$, then

$$F(x) = \int_{-\infty}^{x} f(t)\, dt = \int_{-\infty}^{0} f(t)\, dt + \int_{0}^{x} f(t)\, dt$$

$$= 0 + \int_{0}^{x} (12t^2 - 12t^3)\, dt = (4t^3 - 3t^4)\Big|_{0}^{x}$$

$$= 4x^3 - 3x^4$$

If $x > 1$, then

$$F(x) = \int_{-\infty}^{x} f(t)\, dt = \int_{-\infty}^{0} f(t)\, dt + \int_{0}^{1} f(t)\, dt + \int_{1}^{x} f(t)\, dt$$

$$= 0 + 1 + 0 = 1$$

Thus,

$$F(x) = \begin{cases} 0 & x < 0 \\ 4x^3 - 3x^4 & 0 \le x \le 1 \\ 1 & x > 1 \end{cases}$$

And

$$P(.1 \le x \le .9) = F(.9) - F(.1) = .9477 - .0037 = .944$$

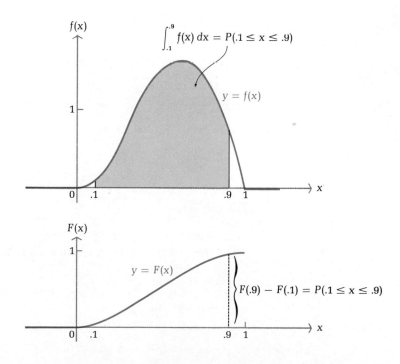

Problem 6

Find the cumulative probability distribution function for the probability density function in Problem 3, and use it to compute $P(.3 \le x \le .7)$.

Answer

$$F(x) = \begin{cases} 0 & x < 0 \\ 3x^2 - 2x^3 & 0 \le x \le 1 \\ 1 & x > 1 \end{cases} \qquad P(.3 \le x \le .7) = .568$$

Example 7

Returning to the shelf-life of a drug in Example 5, suppose a pharmacist wants to be 95% certain that the drug is still good when it is sold. How long is it safe to leave the drug on the shelf?

Solution

Let x be the number of months the drug has been on the shelf when it is sold. The probability that the shelf-life of the drug is less than the number of months it has been sitting on the shelf is $P(0 \le t \le x)$. The pharmacist wants this probability to be .05. Thus, we must solve the equation $P(0 \le t \le x) = .05$ for x. First, we will find the cumulative probability distribution function F. For $x < 0$, we see that $F(x) = 0$. For $x \ge 0$,

$$F(x) = \int_0^x \frac{50}{(50 + t)^2} \, dt = \frac{-50}{50 + t}\Big|_0^x = \frac{-50}{50 + x} - (-1) = 1 - \frac{50}{50 + x}$$

$$= \frac{x}{50 + x}$$

Thus,

$$F(x) = \begin{cases} 0 & x < 0 \\ x/(50 + x) & x \ge 0 \end{cases}$$

Now, to solve the equation $P(0 \le t \le x) = .5$, we solve

$$F(x) - F(0) = .05$$

$$\frac{x}{50 + x} = .05$$

$$x = 2.5 + .05x$$

$$.95x = 2.5$$

$$x \approx 2.6$$

If the drug is sold during the first 2.6 months that it is on the shelf, then the probability that it is still good is .95.

Problem 7

Repeat Example 7 if the pharmacist wants the probability that the drug is still good to be .99.

Answer

Approximately ½ month or 15 days

EXERCISE 11-2

A *Problems 1–12 refer to the continuous random variable X with probability density function*

$$f(x) = \begin{cases} \frac{1}{8}x & 0 \le x \le 4 \\ 0 & \text{otherwise} \end{cases}$$

1. Graph f and verify that f satisfies the first two conditions for a probability density function.
2. Find $P(1 \le x \le 3)$ and illustrate with a graph.
3. Find $P(2 < x < 3)$.   4. Find $P(x \le 2)$.
5. Find $P(x > 3)$.   6. Find $P(x = 1)$.
7. Find $P(x > 5)$.   8. Find $P(x < 5)$.
9. Find and graph the associated cumulative probability distribution function.
10. Use the associated cumulative probability distribution function to find $P(2 \le x \le 4)$ and illustrate with a graph.
11. Use the associated cumulative probability distribution function to find $P(0 < x < 2)$ and illustrate with a graph.
12. Use the associated cumulative probability distribution function to find the value of x that satisfies $P(0 \le t \le x) = \frac{1}{2}$.

B *Problems 13–20 refer to the continuous random variable X with probability density function*

$$f(x) = \begin{cases} 2/(1 + x)^3 & x \ge 0 \\ 0 & \text{otherwise} \end{cases}$$

13. Graph f and verify that f satisfies the first two conditions for a probability density function.
14. Find $P(1 \le x \le 4)$ and illustrate with a graph.
15. Find $P(x > 3)$.   16. Find $P(x \le 2)$.
17. Find $P(x = 1)$.   18. Find $P(x > -1)$.
19. Find and graph the associated cumulative probability distribution function.
20. Use the associated cumulative probability distribution function to find the value of x that satisfies $P(0 \le t \le x) = \frac{3}{4}$.
21. Find the associated cumulative probability distribution function for

$$f(x) = \begin{cases} \frac{3}{2}x - \frac{3}{4}x^2 & 0 \le x \le 2 \\ 0 & \text{otherwise} \end{cases}$$

Graph both functions (on separate sets of axes).
22. Repeat Problem 21 for

$$f(x) = \begin{cases} e^{-x} & x \ge 0 \\ 0 & \text{otherwise} \end{cases}$$

In Problems 23–26 find the associated cumulative probability function, and use it to find the indicated probability.

23. Find $P(1 \le x \le 2)$ for

$$f(x) = \begin{cases} \ln x & 1 \le x \le e \\ 0 & \text{otherwise} \end{cases}$$

24. Find $P(1 \le x \le 2)$ for

$$f(x) = \begin{cases} 3x/(8\sqrt{1+x}) & 0 \le x \le 3 \\ 0 & \text{otherwise} \end{cases}$$

25. Find $P(x \ge 1)$ for

$$f(x) = \begin{cases} xe^{-x} & x \ge 0 \\ 0 & \text{otherwise} \end{cases}$$

26. Find $P(x \ge e)$ for

$$f(x) = \begin{cases} (\ln x)/x^2 & x \ge 1 \\ 0 & \text{otherwise} \end{cases}$$

In Problems 27–30 F(x) is the cumulative probability distribution function for a continuous random variable X. Find the probability density function f(x) associated with each F(x).

27. $F(x) = \begin{cases} 0 & x < 0 \\ x^2 & 0 \le x \le 1 \\ 1 & x > 1 \end{cases}$

28. $F(x) = \begin{cases} 0 & x < 1 \\ \frac{1}{2}x - \frac{1}{2} & 1 \le x \le 3 \\ 1 & x > 3 \end{cases}$

29. $F(x) = \begin{cases} 0 & x < 0 \\ 6x^2 - 8x^3 + 3x^4 & 0 \le x \le 1 \\ 0 & \text{otherwise} \end{cases}$

30. $F(x) = \begin{cases} 1 - (1/x^3) & x \ge 1 \\ 0 & \text{otherwise} \end{cases}$

31. Find the associated cumulative distribution function for

$$f(x) = \begin{cases} x & 0 \le x \le 1 \\ 2 - x & 1 < x \le 2 \\ 0 & \text{otherwise} \end{cases}$$

32. Find the associated cumulative distribution function for

$$f(x) = \begin{cases} \frac{1}{4} & 0 \le x \le 1 \\ \frac{1}{2} & 1 < x \le 2 \\ \frac{1}{4} & 2 < x \le 3 \\ 0 & \text{otherwise} \end{cases}$$

33. If

$$f(x) = \begin{cases} 12x^2 - 12x^3 & 0 \le x \le 1 \\ 0 & \text{otherwise} \end{cases}$$

then

$$F(x) = \begin{cases} 0 & x < 0 \\ 4x^3 - 3x^4 & 0 \le x \le 1 \\ 1 & x > 1 \end{cases}$$

(see Example 6). Use Newton's method (see Section 10-1) to approximate the value of x that satisfies $P(0 \le t \le x) = .4$.

34. Repeat Problem 33 for

$$f(x) = \begin{cases} 6x - 6x^2 & 0 \le x \le 1 \\ 0 & \text{otherwise} \end{cases}$$

and

$$F(x) = \begin{cases} 0 & x < 0 \\ 3x^2 - 2x^3 & 0 \le x \le 1 \\ 1 & x > 1 \end{cases}$$

APPLICATIONS

BUSINESS & ECONOMICS

35. *Time-sharing.* In a computer time-sharing network, the time it takes (in seconds) to respond to a user's request is a continuous random variable with probability density given by

$$f(x) = \begin{cases} \frac{1}{10}e^{-x/10} & x \ge 0 \\ 0 & \text{otherwise} \end{cases}$$

(A) What is the probability that the computer responds within 1 second?

(B) What is the probability that a user must wait over 4 seconds for a response?

36. *Gasoline consumption.* The daily demand for gasoline (in millions of gallons) in a certain area is a continuous random variable with probability density given by

$$f(x) = \begin{cases} \frac{1}{4}xe^{-x/2} & x \ge 0 \\ 0 & \text{otherwise} \end{cases}$$

(A) What is the probability that no more than 1 million gallons are demanded?

(B) What is the probability that 2 million gallons will not be sufficient to meet the daily demand?

37. *Demand.* The weekly demand for hamburger (in thousands of pounds) for a chain of supermarkets is a continuous random variable with probability density given by

$$f(x) = \begin{cases} 0.003x\sqrt{100 - x^2} & 0 \le x \le 10 \\ 0 & \text{otherwise} \end{cases}$$

(A) What is the probability that more than 4,000 pounds of hamburger are demanded?

(B) The manager of the meat department orders 8,000 pounds of hamburger. What is the probability that the demand will not exceed this amount?

(C)   The manager wants the probability that the demand does not exceed the amount ordered to be .9. How much hamburger meat should he order?

38. *Demand.* Repeat Problem 37 if

$$f(x) = \begin{cases} \frac{1}{2,500}x^2(1,000 - x^3)^{1/3} & 0 \le x \le 10 \\ 0 & \text{otherwise} \end{cases}$$

LIFE SCIENCES

39. *Life expectancy.* The life expectancy of a certain microscopic organism (in minutes) is a continuous random variable with probability density function given by

$$f(x) = \begin{cases} \frac{1}{5,000}(10x^3 - x^4) & 0 \le x \le 10 \\ 0 & \text{otherwise} \end{cases}$$

(A)   What is the probability that an organism lives for at least 7 minutes?

(B)   What is the probability that an organism lives for at most 5 minutes?

40. *Shelf-life.* The shelf-life (in days) of a perishable drug is a continuous random variable with probability density function given by

$$f(x) = \begin{cases} (200x)/(100 + x^2)^2 & x \ge 0 \\ 0 & \text{otherwise} \end{cases}$$

(A)   What is the probability that the drug has a shelf-life of at most 10 days?

(B)   What is the probability that the shelf-life exceeds 15 days?

(C)   If the user wants the probability that the drug is still good to be .8, when is the last time it should be used?

SOCIAL SCIENCES

41. *Learning.* The number of words per minute a beginner can type after 1 week of practice is a continuous random variable with probability density given by

$$f(x) = \begin{cases} \frac{1}{400}xe^{-x/20} & x \ge 0 \\ 0 & \text{otherwise} \end{cases}$$

(A)   What is the probability that a beginner can type at least 30 words per minute after 1 week of practice?

(B)   What is the probability that a beginner can type at least 80 words per minute after 1 week of practice?

42. *Learning.* The number of hours it takes a chimpanzee to learn a new task is a continuous random variable with probability density given by

$$f(x) = \begin{cases} \frac{4}{9}x^2 - \frac{4}{27}x^3 & 0 \le x \le 3 \\ 0 & \text{otherwise} \end{cases}$$

(A) What is the probability that the chimpanzee learns the task in the first hour?

(B) What is the probability that the chimpanzee does not learn the task in the first 2 hours?

11-3 EXPECTED VALUE, STANDARD DEVIATION, AND MEDIAN

EXPECTED VALUE AND STANDARD DEVIATION

In Section 11-1 we discussed the expected value, variance, and standard deviation for discrete random variables. The formulas for these quantities can be generalized to the continuous case by replacing the finite summation operation with integration. Compare the formulas below with those in Section 11-1.

EXPECTED VALUE AND STANDARD DEVIATION

FOR A CONTINUOUS RANDOM VARIABLE

Let $f(x)$ be the probability density function for a continuous random variable X. The **expected value, or mean, of X** is

$$\mu = E(X) = \int_{-\infty}^{\infty} x f(x)\, dx$$

The **variance** is

$$V(X) = \int_{-\infty}^{\infty} (x - \mu)^2 f(x)\, dx$$

and the **standard deviation** is

$$\sigma = \sqrt{V(X)}$$

REMARK

The standard deviation of a continuous random variable measures the dispersion of the probability density function about the mean, just as in the discrete case. This is illustrated in Figure 4 (next page). The probability density function in Figure 4A has a standard deviation of 1. Most of the area under the curve is near the mean. In Figure 4B the standard deviation is four times as large, and the area under the graph is much more spread out.

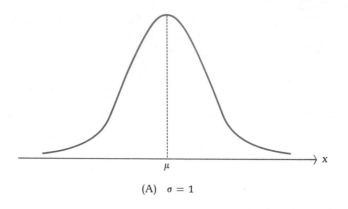

(A) $\sigma = 1$

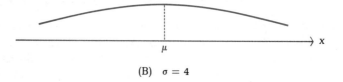

FIGURE 4

(B) $\sigma = 4$

Example 8   Find the mean, variance, and standard deviation for

$$f(x) = \begin{cases} 12x^2 - 12x^3 & 0 \le x \le 1 \\ 0 & \text{otherwise} \end{cases}$$

Solution $\mu = E(x) = \displaystyle\int_{-\infty}^{\infty} x f(x)\, dx = \int_0^1 x(12x^2 - 12x^3)\, dx = \int_0^1 (12x^3 - 12x^4)\, dx$

$$= \left(3x^4 - \frac{12}{5}x^5\right)\Big|_0^1 = \frac{3}{5}$$

$$V(X) = \int_{-\infty}^{\infty} (x - \mu)^2 f(x)\, dx = \int_0^1 \left(x - \frac{3}{5}\right)^2 (12x^2 - 12x^3)\, dx$$

$$= \int_0^1 \left(x^2 - \frac{6}{5}x + \frac{9}{25}\right)(12x^2 - 12x^3)\, dx$$

$$= \int_0^1 \left(\frac{108}{25}x^2 - \frac{468}{25}x^3 + \frac{132}{5}x^4 - 12x^5\right) dx$$

$$= \left(\frac{36}{25}x^3 - \frac{117}{25}x^4 + \frac{132}{25}x^5 - 2x^6\right)\Big|_0^1 = \frac{1}{25}$$

$$\sigma = \sqrt{V(X)} = \sqrt{\frac{1}{25}} = \frac{1}{5}$$

Problem 8   Find the mean, variance, and standard deviation for

$$f(x) = \begin{cases} 6x - 6x^2 & 0 \le x \le 1 \\ 0 & \text{otherwise} \end{cases}$$

Answer $\mu = \frac{1}{2}$, $V(X) = \frac{1}{20}$, $\sigma \approx .2236$

ALTERNATE FORMULA
FOR VARIANCE
The term $(x - \mu)^2$ in the formula for $V(X)$ introduces some complicated algebraic manipulations in the evaluation of the integral. We can use the properties of the definite integral to simplify this formula. Thus,

$$V(X) = \int_{-\infty}^{\infty} (x - \mu)^2 f(x) \, dx$$

$$= \int_{-\infty}^{\infty} (x^2 - 2x\mu + \mu^2) f(x) \, dx \qquad \text{Expand } (x - \mu)^2$$

$$= \int_{-\infty}^{\infty} [x^2 f(x) - 2x\mu f(x) + \mu^2 f(x)] \, dx \qquad \text{Multiply by } f(x)$$

$$= \int_{\infty}^{\infty} x^2 f(x) \, dx - \int_{-\infty}^{\infty} 2x\mu f(x) \, dx + \int_{-\infty}^{\infty} \mu^2 f(x) \, dx \qquad \text{Property 4, page 190}$$

$$= \int_{-\infty}^{\infty} x^2 f(x) \, dx - 2\mu \int_{-\infty}^{\infty} x f(x) \, dx + \mu^2 \int_{-\infty}^{\infty} f(x) \, dx \qquad \text{Property 3, page 190}$$

$$= \int_{-\infty}^{\infty} x^2 f(x) \, dx - 2\mu(\mu) + \mu^2(1) \qquad \int_{-\infty}^{\infty} x f(x) \, dx = \mu, \quad \int_{-\infty}^{\infty} f(x) \, dx = 1$$

$$= \int_{-\infty}^{\infty} x^2 f(x) \, dx - \mu^2$$

In general, it will be easier to evaluate $\int_{-\infty}^{\infty} x^2 f(x) \, dx$ than to evaluate $\int_{-\infty}^{\infty} (x - \mu)^2 f(x) \, dx$.

ALTERNATE FORMULA FOR VARIANCE

$$V(X) = \int_{-\infty}^{\infty} x^2 f(x) \, dx - \mu^2$$

Example 9   Use the alternate formula for variance to compute the variance in Example 8.

Solution   From Example 8, we have $\mu = \int_{-\infty}^{\infty} x f(x) \, dx = \frac{3}{5}$.

$$\int_{-\infty}^{\infty} x^2 f(x)\, dx = \int_0^1 x^2(12x^2 - 12x^3)\, dx = \int_0^1 (12x^4 - 12x^5)\, dx$$

$$= \left(\frac{12}{5}x^5 - \frac{12}{6}x^6\right)\Big|_0^1 = \frac{2}{5}$$

$$V(X) = \int_{-\infty}^{\infty} x^2 f(x)\, dx - \mu^2 = \frac{2}{5} - \left(\frac{3}{5}\right)^2 = \frac{1}{25}$$

Problem 9   Use the alternate formula for variance to compute the variance in Problem 8.

Answer $\frac{1}{20}$

Example 10   Find the mean, variance, and standard deviation for

$$f(x) = \begin{cases} 3/x^4 & x \ge 1 \\ 0 & \text{otherwise} \end{cases}$$

Solution

$$\mu = \int_{-\infty}^{\infty} xf(x)\, dx = \int_1^{\infty} x\frac{3}{x^4}\, dx = \lim_{R \to \infty} \int_1^R \frac{3}{x^3}\, dx$$

$$= \lim_{R \to \infty} \left[-\frac{3}{2}\left(\frac{1}{x^2}\right)\right]\Big|_1^R = \lim_{R \to \infty} \left[-\frac{3}{2}\left(\frac{1}{R^2}\right) + \frac{3}{2}\right] = \frac{3}{2}$$

$$\int_{-\infty}^{\infty} x^2 f(x)\, dx = \int_1^{\infty} x^2\frac{3}{x^4}\, dx = \lim_{R \to \infty} \int_1^R \frac{3}{x^2}\, dx = \lim_{R \to \infty} \left(-\frac{3}{x}\right)\Big|_1^R$$

$$= \lim_{R \to \infty} \left(-\frac{3}{R} + 3\right) = 3$$

$$V(X) = \int_{-\infty}^{\infty} x^2 f(x)\, dx - \mu^2 = 3 - \left(\frac{3}{2}\right)^2 = \frac{3}{4}$$

$$\sigma = \sqrt{V(X)} = \sqrt{\tfrac{3}{4}} = \frac{\sqrt{3}}{2} \approx .8660$$

Problem 10   Find the mean, variance, and standard deviation for

$$f(x) = \begin{cases} 24/x^4 & x \ge 2 \\ 0 & \text{otherwise} \end{cases}$$

Answer $\mu = 3, \quad V(X) = 3, \quad \sigma = \sqrt{3} \approx 1.732$

Example 11   The life expectancy (in hours) for a particular brand of light bulbs is a continuous random variable with probability density function

$$f(x) = \begin{cases} \frac{1}{100} - \frac{1}{20,000}x & 0 \le x \le 200 \\ 0 & \text{otherwise} \end{cases}$$

(A)   What is the average life expectancy of one of these light bulbs?
(B)   What is the probability that a bulb will last longer than this average?

Solutions   (A)   Since the value of this random variable is the number of hours a bulb lasts, the average life expectancy is just the expected value of of the random variable. Thus,

$$E(X) = \int_{-\infty}^{\infty} xf(x)\,dx = \int_{0}^{200} x\left(\frac{1}{100} - \frac{1}{20,000}x\right)dx$$

$$= \int_{0}^{200}\left(\frac{1}{100}x - \frac{1}{20,000}x^2\right)dx = \left(\frac{1}{200}x^2 - \frac{1}{60,000}x^3\right)\Big|_{0}^{200}$$

$$= \frac{200}{3} \quad \text{or} \quad 66\frac{2}{3} \text{ hours}$$

(B)   The probability that a bulb lasts longer than $66\frac{2}{3}$ hours is

$$P\left(x > \frac{200}{3}\right) = \int_{200/3}^{\infty} f(x)\,dx = \int_{200/3}^{200}\left(\frac{1}{100} - \frac{1}{20,000}x\right)$$

$$= \left(\frac{1}{100}x - \frac{1}{40,000}x^2\right)\Big|_{200/3}^{200} = 1 - \frac{5}{9} = \frac{4}{9}$$

Problem 11   Repeat Example 11 if the probability density function is

$$f(x) = \begin{cases} \frac{1}{200} - \frac{1}{90,000}x & 0 \le x \le 300 \\ 0 & \text{otherwise} \end{cases}$$

Answers   (A)   125 hours   (B) $\frac{133}{288}$

MEDIAN   Another measurement often used to describe the properties of a random variable is the median. The **median** is the value of the random variable that divides the area under the graph of the probability density function into two equal parts (see Fig. 5). If x_m is the median, then x_m must satisfy

$$P(x \le x_m) = \frac{1}{2}$$

Generally, this equation is solved by first finding the cumulative probability distribution function.

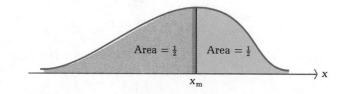

FIGURE 5

Example 12   Find the median of the continuous random variable with probability density function

$$f(x) = \begin{cases} 3/x^4 & x \ge 1 \\ 0 & \text{otherwise} \end{cases}$$

Solution   **Step 1.** *Find the cumulative probability distribution function.* For $x < 1$, we have $F(x) = 0$. If $x \geq 1$, then

$$F(x) = \int_{-\infty}^{x} f(t)\, dt = \int_{1}^{x} \frac{3}{t^4}\, dt = -\frac{1}{t^3}\Big|_{1}^{x} = -\frac{1}{x^3} + 1 = 1 - \frac{1}{x^3}$$

Step 2. *Solve the equation $P(x \leq x_m) = \frac{1}{2}$ for x_m.*

$$F(x_m) = P(x \leq x_m)$$

$$1 - \frac{1}{x_m^3} = \frac{1}{2}$$

$$\frac{1}{2} = \frac{1}{x_m^3}$$

$$x_m^3 = 2$$

$$x_m = \sqrt[3]{2}$$

Thus, the median is $\sqrt[3]{2} \approx 1.26$.

Problem 12   Find the median of the continuous random variable with probability density function

$$f(x) = \begin{cases} 24/x^4 & x \geq 2 \\ 0 & \text{otherwise} \end{cases}$$

Answer $x_m = \sqrt[3]{16} = 2\sqrt[3]{2} \approx 2.52$

Example 13   In Example 11, find the median life expectancy of a light bulb.

Solution   **Step 1.** *Find the cumulative probability distribution function.* If $x < 0$, we have $F(x) = 0$. If $0 \leq x \leq 200$, then

$$F(x) = \int_{-\infty}^{x} f(t)\, dt = \int_{0}^{x} \left(\frac{1}{100} - \frac{1}{20{,}000}t \right) dt$$

$$= \left(\frac{1}{100}t - \frac{1}{40{,}000}t^2 \right)\Big|_{0}^{x} = \frac{1}{100}x - \frac{1}{40{,}000}x^2$$

If $x > 200$, then

$$F(x) = \int_{-\infty}^{x} f(t)\, dt = \int_{-\infty}^{0} f(t)\, dt + \int_{0}^{200} f(t)\, dt + \int_{200}^{x} f(t)\, dt$$

$$= 0 + 1 + 0 = 1$$

Thus,

$$F(x) = \begin{cases} 0 & x < 0 \\ \frac{1}{100}x - \frac{1}{40{,}000}x^2 & 0 \leq x \leq 200 \\ 1 & x > 200 \end{cases}$$

Step 2. *Solve the equation $P(x \le x_m) = \frac{1}{2}$ for x_m.*

$$F(x_m) = P(x \le x_m) = \frac{1}{2}$$

The solution must occur for $0 \le x_m \le 200$

$$\frac{1}{100}x_m - \frac{1}{40,000}x_m^2 = \frac{1}{2}$$

$$x_m^2 - 400x_m + 20,000 = 0$$

This quadratic equation has two solutions, $200 + 100\sqrt{2}$ and $200 - 100\sqrt{2}$. Since x_m must lie in the interval [0, 200], the second root is the correct answer. Thus, the median life expectancy is $200 - 100\sqrt{2} \approx 58.58$ hours.

Problem 13   In Problem 11, find the median life expectancy of a light bulb.

Answer $x_m = 450 - 150\sqrt{5} \approx 114.59$ hours

If you compare Examples 11 and 13, and Examples 10 and 12, you will see that the mean and the median generally are not equal (see Fig. 6).

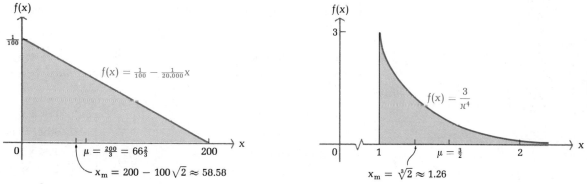

FIGURE 6

EXERCISE 11-3

A   *Problems 1–8 refer to the random variable X with probability density function*

$$f(x) = \begin{cases} \frac{1}{2}x & 0 \le x \le 2 \\ 0 & \text{otherwise} \end{cases}$$

1. Find the mean.
2. Find the variance.
3. Find the standard deviation.
4. Find the probability that the random variable is less than the mean.
5. Find the probability that the random variable is within 1 standard deviation of the mean.

6. Find the associated cumulative probability distribution function.
7. Find the median.
8. Find the probability that the random variable lies between the median and the mean.

B   Problems 9–16 refer to the continuous random variable X with probability density function

$$f(x) = \begin{cases} 4/x^5 & x \geq 1 \\ 0 & \text{otherwise} \end{cases}$$

9. Find the mean.
10. Find the variance.
11. Find the standard deviation.
12. Find the probability that the random variable is greater than the mean.
13. Find the probability that the random variable is within 2 standard deviations of the mean.
14. Find the associated cumulative probability density function.
15. Find the median.
16. Find the probability that the random variable is between the mean and the median.

In Problems 17–20 find the mean, variance, and standard deviation of the continuous random variable with the indicated probability density function.

17. $f(x) = \begin{cases} \frac{1}{3} & 2 \leq x \leq 5 \\ 0 & \text{otherwise} \end{cases}$

18. $f(x) = \begin{cases} 3x^2 & 0 \leq x \leq 1 \\ 0 & \text{otherwise} \end{cases}$

19. $f(x) = \begin{cases} 1/(2\sqrt{1+x}) & 0 \leq x \leq 3 \\ 0 & \text{otherwise} \end{cases}$

20. $f(x) = \begin{cases} \ln x & 1 \leq x \leq e \\ 0 & \text{otherwise} \end{cases}$

In Problems 21–24 find the median of the continuous random variable with the indicated probability density function.

21. $f(x) = \begin{cases} 1/x & 1 \leq x \leq e \\ 0 & \text{otherwise} \end{cases}$

22. $f(x) = \begin{cases} 2/(1+x)^2 & 0 \leq x \leq 1 \\ 0 & \text{otherwise} \end{cases}$

23. $f(x) = \begin{cases} 1/(1+x)^2 & x \geq 0 \\ 0 & \text{otherwise} \end{cases}$

24. $f(x) = \begin{cases} e^{-x} & x \geq 0 \\ 0 & \text{otherwise} \end{cases}$

C   In Problems 25–28 use Newton's method from Section 10-1 to approximate the median of the random variable with the indicated probability density function.

25. $f(x) = \begin{cases} 12x^2 - 12x^3 & 0 \leq x \leq 1 \\ 0 & \text{otherwise} \end{cases}$

26. $f(x) = \begin{cases} 6x - 6x^2 & 0 \leq x \leq 1 \\ 0 & \text{otherwise} \end{cases}$

27. $f(x) = \begin{cases} \ln x & 1 \le x \le e \\ 0 & \text{otherwise} \end{cases}$

28. $f(x) = \begin{cases} xe^{-x} & x \ge 0 \\ 0 & \text{otherwise} \end{cases}$

The **quartile points** for a probability density function are the values x_1, x_2, x_3 that divide the area under the graph of the function into four equal parts. Find the quartile points for the probability density functions in Problems 29 and 30.

29. $f(x) = \begin{cases} \frac{1}{2}x & 0 \le x \le 2 \\ 0 & \text{otherwise} \end{cases}$

30. $f(x) = \begin{cases} 1/(1 + x)^2 & x \ge 0 \\ 0 & \text{otherwise} \end{cases}$

APPLICATIONS

BUSINESS & ECONOMICS

31. *Profit.* A building contractor's profit (in thousands of dollars) on each unit in a subdivision is a continuous random variable with probability density given by

$$f(x) = \begin{cases} \frac{1}{8}(10 - x) & 6 \le x \le 10 \\ 0 & \text{otherwise} \end{cases}$$

(A) What is the contractor's expected profit?
(B) What is the median profit?

32. *Product life.* The life expectancy (in years) of an automobile battery is a continuous random variable with probability density given by

$$f(x) = \begin{cases} \frac{1}{2}e^{-x/2} & x \ge 0 \\ 0 & \text{otherwise} \end{cases}$$

Find the median life expectancy.

33. *Water consumption.* The daily consumption of water (in millions of gallons) in a small city is a continuous random variable with probability density given by

$$f(x) = \begin{cases} 1/(1 + x^2)^{3/2} & x \ge 0 \\ 0 & \text{otherwise} \end{cases}$$

Find the expected daily consumption.

34. *Demand.* The weekly demand for hamburger (in thousands of pounds) for a chain of supermarkets is a continuous random variable with probability density given by

$$f(x) = \begin{cases} 0.003x\sqrt{100 - x^2} & 0 \le x \le 10 \\ 0 & \text{otherwise} \end{cases}$$

(See Problem 37 in Exercise 11-2.) Find the median demand for hamburger meat.

LIFE SCIENCES   35. *Life expectancy.* The life expectancy of a certain microscopic organism (in minutes) is a continuous random variable with probability density function given by

$$f(x) = \begin{cases} \frac{1}{5,000}(10x^3 - x^4) & 0 \le x \le 10 \\ 0 & \text{otherwise} \end{cases}$$

(See Problem 39 in Exercise 11-2.) Find the mean life expectancy of one of these organisms.

36. *Shelf-life.* The shelf-life (in days) of a perishable drug is a continuous random variable with probability density given by

$$f(x) = \begin{cases} (200x)/(100 + x^2)^2 & x \ge 0 \\ 0 & \text{otherwise} \end{cases}$$

(See Problem 40 in Exercise 11-2.) Find the median shelf-life of this drug.

SOCIAL SCIENCES   37. *Learning.* The number of hours it takes a chimpanzee to learn a new task is a continuous random variable with probability density given by

$$f(x) = \begin{cases} \frac{4}{9}x^2 - \frac{4}{27}x^3 & 0 \le x \le 3 \\ 0 & \text{otherwise} \end{cases}$$

(See Problem 42 in Exercise 11-2.) What is the expected number of hours it will take a chimpanzee to learn the task?

11-4   SPECIAL PROBABILITY DENSITY FUNCTIONS

In this section we will examine several important probability density functions. In actual practice, we do not usually construct a probability density function for each experiment. Instead, we try to select a known probability density function that seems to give a reasonable description of the experiment. Thus, it is important to be familiar with the properties and applications of a variety of probability density functions.

UNIFORM
DISTRIBUTION   To begin, suppose the outcome of an experiment can be any number in a certain finite interval [a, b]. If we believe that the probability of the

outcome lying in a small interval of fixed length is independent of the location of this small interval within $[a, b]$, then we say that the continuous random variable for this experiment is **uniformly distributed** on the interval $[a, b]$. The **uniform probability density function** is

$$f(x) = \begin{cases} \dfrac{1}{b-a} & a \leq x \leq b \\ 0 & \text{otherwise} \end{cases}$$

Since $f(x) \geq 0$ and

$$\int_{-\infty}^{\infty} f(x)\, dx = \int_{a}^{b} \frac{1}{b-a}\, dx = \frac{x}{b-a}\Big|_{a}^{b} = \frac{b}{b-a} - \frac{a}{b-a} = 1$$

f satisfies the necessary conditions for a probability density function.

If F is the associated cumulative probability distribution function, then for $x < a$, $F(x) = 0$. For $a \leq x \leq b$, we have

$$F(x) = \int_{-\infty}^{x} f(t)\, dt = \int_{a}^{x} \frac{1}{b-a}\, dt = \frac{t}{b-a}\Big|_{a}^{x}$$

$$= \frac{x}{b-a} - \frac{a}{b-a} = \frac{x-a}{b-a}$$

For $x > b$, $F(x) = 1$.

Now we calculate the mean, median, and standard deviation for the uniform probability density function:

MEAN

$$\mu = \int_{-\infty}^{\infty} x f(x)\, dx = \int_{a}^{b} \frac{x}{b-a}\, dx = \frac{x^2}{2(b-a)}\Big|_{a}^{b} = \frac{b^2}{2(b-a)} - \frac{a^2}{2(b-a)}$$

$$= \frac{b^2 - a^2}{2(b-a)} = \frac{(b-a)(b+a)}{2(b-a)} = \frac{1}{2}(a+b)$$

MEDIAN

$$F(x_m) = P(x \leq x_m)$$

$$\frac{x_m - a}{b - a} = \frac{1}{2}$$

$$x_m - a = \frac{1}{2}(b - a)$$

$$x_m = a + \frac{1}{2}(b - a) = \frac{1}{2}(a + b)$$

STANDARD DEVIATION

$$\int_{-\infty}^{\infty} x^2 f(x)\, dx = \int_{a}^{b} \frac{x^2}{b-a}\, dx = \frac{x^3}{3(b-a)}\Big|_{a}^{b} = \frac{b^3}{3(b-a)} - \frac{a^3}{3(b-a)}$$

$$= \frac{b^3 - a^3}{3(b-a)} = \frac{(b-a)(b^2 + ab + a^2)}{3(b-a)} = \frac{1}{3}(b^2 + ab + a^2)$$

$$V(X) = \int_{-\infty}^{\infty} x^2 f(x)\, dx - \mu^2 = \frac{1}{3}(b^2 + ab + a^2) - \left[\frac{1}{2}(a+b)\right]^2$$

$$= \frac{1}{3}(b^2 + ab + a^2) - \frac{1}{4}(a^2 + 2ab + b^2)$$

$$= \frac{4(b^2 + ab + a^2) - 3(a^2 + 2ab + b^2)}{12}$$

$$= \frac{b^2 - 2ab + a^2}{12} = \frac{1}{12}(b-a)^2$$

$$\sigma = \sqrt{V(X)} = \sqrt{\tfrac{1}{12}(b-a)^2} = \frac{1}{\sqrt{12}}(b-a)$$

These properties are summarized in the box.

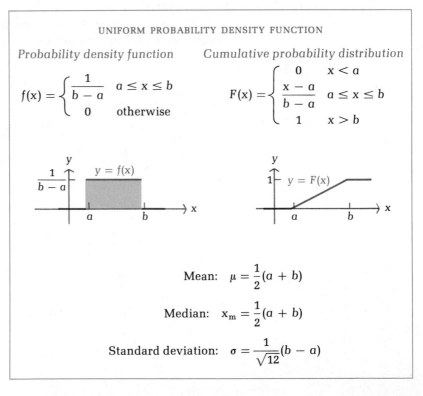

UNIFORM PROBABILITY DENSITY FUNCTION

Probability density function   *Cumulative probability distribution*

$$f(x) = \begin{cases} \dfrac{1}{b-a} & a \le x \le b \\ 0 & \text{otherwise} \end{cases}$$

$$F(x) = \begin{cases} 0 & x < a \\ \dfrac{x-a}{b-a} & a \le x \le b \\ 1 & x > b \end{cases}$$

Mean: $\mu = \dfrac{1}{2}(a+b)$

Median: $x_m = \dfrac{1}{2}(a+b)$

Standard deviation: $\sigma = \dfrac{1}{\sqrt{12}}(b-a)$

Example 14   Standard electrical current is uniformly distributed between 110 and 120 volts. What is the probability that the current is between 113 and 118 volts?

Solution   Since we are told that the current is uniformly distributed on the interval [110, 120], we choose the uniform probability density function

$$f(x) = \begin{cases} \frac{1}{10} & 110 \le x \le 120 \\ 0 & \text{otherwise} \end{cases}$$

Then

$$P(113 \le x \le 118) = \int_{113}^{118} \frac{1}{10} \, dx = \frac{x}{10} \Big|_{113}^{118} = \frac{118}{10} - \frac{113}{10} = \frac{1}{2}$$

Problem 14   In Example 14, what is the probability that the current is at least 116 volts?

Answer $\frac{2}{5}$

BETA DISTRIBUTION   A continuous random variable has a **beta distribution*** and is referred to as a **beta random variable** if its probability density function is

$$f(x) = \begin{cases} (\beta + 1)(\beta + 2)x^{\beta}(1 - x) & 0 \le x \le 1 \\ 0 & \text{otherwise} \end{cases}$$

where β is a constant, $\beta \ge 0$. Then

$$f(x) = (\beta + 1)(\beta + 2)x^{\beta}(1 - x) \ge 0 \qquad 0 \le x \le 1$$

$$\int_{-\infty}^{\infty} f(x) \, dx = \int_{0}^{1} (\beta + 1)(\beta + 2)x^{\beta}(1 - x) \, dx$$

$$= \int_{0}^{1} (\beta + 1)(\beta + 2)(x^{\beta} - x^{\beta+1}) \, dx$$

$$= (\beta + 1)(\beta + 2)\left(\frac{x^{\beta+1}}{\beta + 1} - \frac{x^{\beta+2}}{\beta + 2}\right)\Big|_{0}^{1}$$

$$= (\beta + 1)(\beta + 2)\left(\frac{1}{\beta + 1} - \frac{1}{\beta + 2}\right)$$

$$= (\beta + 2) - (\beta + 1) = 1$$

Thus, f is a probability density function.

   If $F(x)$ is the associated cumulative probability distribution function, then for $x < 0$, $F(x) = 0$. For $0 \le x \le 1$, we have

---

*There is a more general definition of a beta distribution, but we will not consider it here.

$$F(x) = \int_{-\infty}^{x} f(t)\, dt = \int_{0}^{x} (\beta + 1)(\beta + 2)t^{\beta}(1 - t)\, dt$$

$$= (\beta + 1)(\beta + 2)\left(\frac{t^{\beta+1}}{\beta + 1} - \frac{t^{\beta+2}}{\beta + 2}\right)\Big|_{0}^{x}$$

$$= (\beta + 1)(\beta + 2)\left(\frac{x^{\beta+1}}{\beta + 1} - \frac{x^{\beta+2}}{\beta + 2}\right)$$

$$= (\beta + 2)x^{\beta+1} - (\beta + 1)x^{\beta+2}$$

And for $x > 1$,

$$F(x) = 1$$

It is not possible to solve the equation $F(x_m) = \frac{1}{2}$ for an arbitrary value of β. Given a specific value of β, the median can be approximated using the numerical methods discussed in Chapter 10. By straightforward (but tedious) integration we can show that

$$\mu = \frac{\beta + 1}{\beta + 3} \quad \text{and} \quad \sigma = \sqrt{\frac{2(\beta + 1)}{(\beta + 4)(\beta + 3)^2}}$$

The calculations are not included here.

BETA PROBABILITY DENSITY FUNCTION

$$f(x) = \begin{cases} (\beta + 1)(\beta + 2)x^{\beta}(1 - x) & 0 \le x \le 1 \\ 0 & \text{otherwise} \end{cases} \quad \text{where} \quad \beta \ge 0$$

$$F(x) = \begin{cases} 0 & x < 0 \\ (\beta + 2)x^{\beta+1} - (\beta + 1)x^{\beta+2} & 0 \le x \le 1 \\ 1 & x > 1 \end{cases}$$

Mean: $\mu = \dfrac{\beta + 1}{\beta + 3}$

Standard deviation: $\sigma = \sqrt{\dfrac{2(\beta + 1)}{(\beta + 4)(\beta + 3)^2}}$

Figure 7 shows the graphs of $f(x)$ for some typical values of β.

Example 15   The annual percentage of correct income tax forms filed with the Internal Revenue Service is a beta random variable with $\beta = 8$.
(A)   What is the probability that at least half the returns filed are correct?
(B)   What is the expected percentage of correct returns?

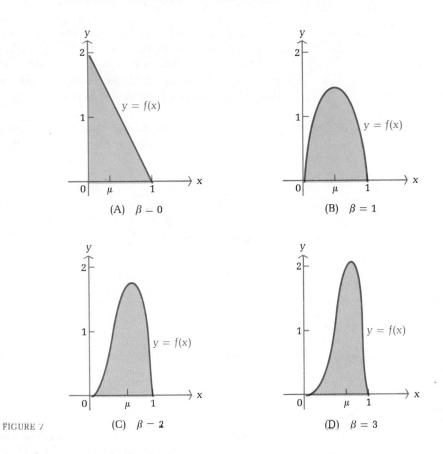

FIGURE 7

Solutions

$$f(x) = \begin{cases} 90x^8(1 - x) & 0 \le x \le 1 \\ 0 & \text{otherwise} \end{cases}$$

(A) $P\left(x \ge \dfrac{1}{2}\right) = \displaystyle\int_{1/2}^{1} 90x^8(1 - x)\,dx = \int_{1/2}^{1} (90x^8 - 90x^9)\,dx$

$$= (10x^9 - 9x^{10})\Big|_{1/2}^{1} = 1 - \frac{11}{2^{10}} \approx .989$$

(B) $\mu = E(X) = \dfrac{\beta + 1}{\beta + 3} = \dfrac{8 + 1}{8 + 3} = \dfrac{9}{11} \approx .818$

Thus, we expect 82% of the returns to be correct.

Problem 15 In Example 15, what is the probability that at least 90% of the returns are correct?

Answer   .264

Example 16   A psychologist is studying the learning abilities of children in a certain age group. She has determined that on the average 75% of the children can learn to perform a particular task in 5 minutes. She believes that the percentage of children that can learn the task in 5 minutes is a continuous beta random variable. What is an appropriate value of β?

Solution   Since the average percentage of children that learned the task is .75 and the mean for any beta distribution is $(\beta + 1)/(\beta + 3)$, the value of β must satisfy

$$\frac{\beta + 1}{\beta + 3} = .75$$

Solving, we obtain

$$\beta + 1 = .75(\beta + 3)$$
$$\beta + 1 = .75\beta + 2.25$$
$$.25\beta = 1.25$$
$$\beta = 5$$

Problem 16   In Example 16, what is the probability that at least 75% of the children will learn the task in 5 minutes?

Answer   .555

EXPONENTIAL DISTRIBUTION   A continuous random variable has an **exponential distribution** and is referred to as an **exponential random variable if** its probability density function is

$$f(x) = \begin{cases} \frac{1}{\lambda}e^{-x/\lambda} & x \geq 0 \\ 0 & \text{otherwise} \end{cases}$$

where λ is a positive constant. Since $f(x) \geq 0$ and

$$\int_{-\infty}^{\infty} f(x)\,dx = \int_{0}^{\infty} \frac{1}{\lambda}e^{-x/\lambda}\,dx = \lim_{R \to \infty}\int_{0}^{R} \frac{1}{\lambda}e^{-x/\lambda}\,dx = \lim_{R \to \infty}(-e^{-x/\lambda})\Big|_{0}^{R}$$
$$= \lim_{R \to \infty}(-e^{-R/\lambda} + 1) = 1$$

f satisfies the conditions for a probability density function. If F is the cumulative distribution function, we see that $F(x) = 0$ for $x < 0$. For $x \geq 0$, we have

$$F(x) = \int_{-\infty}^{x} f(t)\,dt = \int_{0}^{x} \frac{1}{\lambda}e^{-t/\lambda}\,dt = -e^{-t/\lambda}\Big|_{0}^{x} = 1 - e^{-x/\lambda}$$

MEDIAN

$$F(x_m) = P(x \le x_m) = \frac{1}{2}$$

$$1 - e^{-x_m/\lambda} = \frac{1}{2}$$

$$\frac{1}{2} = e^{-x_m/\lambda}$$

$$\ln\frac{1}{2} = -\frac{x_m}{\lambda}$$

$$x_m = -\lambda \ln\frac{1}{2} = \lambda \ln 2 \qquad Note: \ \ \ln\frac{1}{2} = -\ln 2$$

Integration by parts can be used to show that $\mu = \lambda$ and $\sigma = \lambda$. The calculations are not included here.

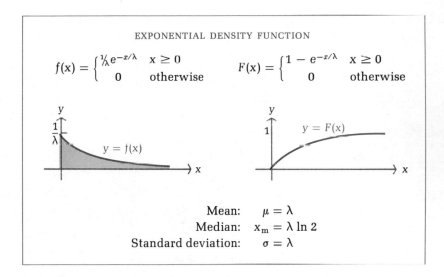

EXPONENTIAL DENSITY FUNCTION

$$f(x) = \begin{cases} \frac{1}{\lambda}e^{-x/\lambda} & x \ge 0 \\ 0 & \text{otherwise} \end{cases} \qquad F(x) = \begin{cases} 1 - e^{-x/\lambda} & x \ge 0 \\ 0 & \text{otherwise} \end{cases}$$

$y = f(x)$

$y = F(x)$

Mean: $\mu = \lambda$
Median: $x_m = \lambda \ln 2$
Standard deviation: $\sigma = \lambda$

Example 17   The number of units of a certain item sold each week in a chain of department stores is an exponential random variable. The average number of items sold each week is 10,000. What is the probability that 15,000 or more units will be sold in one week?

Solution   If we let x represent the number of units sold (in thousands), then the appropriate probability density function is

$$f(x) = \begin{cases} \frac{1}{10}e^{-x/10} & x \ge 0 \\ 0 & \text{otherwise} \end{cases}$$

Thus,

$$P(x \geq 15) = \int_{15}^{\infty} \frac{1}{10} e^{-x/10} \, dx = \lim_{R \to \infty} (-e^{-x/10}) \Big|_{15}^{R} = \lim_{R \to \infty} (-e^{-R/10} + e^{-15/10})$$

$$= e^{-1.5} \approx .223$$

Problem 17   In Example 17, what is the probability that at most 5,000 units will be sold?

Answer $1 - e^{-0.5} \approx .393$

NORMAL DISTRIBUTION   A continuous random variable has a **normal distribution** and is said to be a **normal random variable** with mean μ and standard deviation σ if its probability density function is

$$f(x) = \frac{1}{\sigma\sqrt{2\pi}} e^{-1/2(x-\mu/\sigma)^2}$$

(see Fig. 8). The normal probability density function is the most important of all the probability density functions. Unfortunately, its antiderivative cannot be expressed in terms of elementary functions. A table of values or numerical integration must be used to evaluate the integrals that arise in problems dealing with normal random variables. Since our goal is to illustrate the applications of calculus to probability, we will leave the treatment of normal variables to a statistics text.

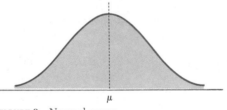

μ

FIGURE 8   Normal curve.

EXERCISE 11-4

A   *Problems 1–4 refer to a continuous random variable X that is uniformly distributed on the interval* [0, 2].

1. Find the probability density function for X.
2. Find the associated cumulative probability distribution function for X.
3. Find the mean, median, and standard deviation.
4. Find the probability that the random variable is within 1 standard deviation of the mean.

Problems 5–8 refer to a beta random variable X with $\beta = 3$.

5. Find the probability density function for X.
6. Find the cumulative probability distribution function for X.
7. Find the mean and standard deviation.
8. Find the probability that the random variable is within 1 standard deviation of the mean.

Problems 9–12 refer to an exponential random variable X with $\lambda = \frac{1}{2}$.

9. Find the probability density function for X.
10. Find the cumulative probability distribution function for X.
11. Find the mean, median, and standard deviation.
12. Find the probability that the random variable is within 1 standard deviation of the mean.

B   *Problems 13–16 refer to a beta random variable with $\beta = \frac{1}{2}$.*

13. Find the probability density function for X.
14. Find the cumulative probability distribution function for X.
15. Find the mean and standard deviation.
16. Find the probability that the random variable is within 1 standard deviation of the mean.

Problems 17–20 refer to a beta random variable X with mean $\mu = .4$.

17. Find β.
18. Find the probability density function.
19. Find the cumulative probability distribution function.
20. Find the standard deviation.

Problems 21–24 refer to an exponential random variable X with median $x_m = 2$.

21. Find λ.
22. Find the probability density function.
23. Find the cumulative probability distribution function.
24. Find the mean and standard deviation.

25. Compute the following probabilities for an exponential random variable with mean λ:
    (A) $P(0 \leq x \leq \lambda)$ (B) $P(0 \leq x \leq 2\lambda)$ (C) $P(0 \leq x \leq 3\lambda)$
26. The point where a probability density function assumes its maximum value is sometimes referred to as the **mode.** Find the mode of the probability density function

$$f(x) = \begin{cases} (\beta + 1)(\beta + 2)x^{\beta}(1 - x) & 0 \leq x \leq 1 \\ 0 & \text{otherwise} \end{cases}$$

27. If the random variable $\overset{.}{X}$ is the time at which a device malfunctions, then the failure rate is given by

$$\frac{f(x)}{1 - F(x)}$$

where $f(x)$ is the probability density function and $F(x)$ is the cumulative probability distribution function for X. Find the failure rate for:
(A) A uniform random variable
(B) An exponential random variable

28. A continuous random variable X is said to have a *Pareto distribution* if its probability density function is given by

$$f(x) = \begin{cases} p/x^{p+1} & x \geq 1 \\ 0 & \text{otherwise} \end{cases}$$

where p is a constant, $p > 0$.
(A) Find the mean. What restrictions must you place on p?
(B) Find the variance and standard deviation. What restrictions must you place on p?
(C) Find the median.

APPLICATIONS

BUSINESS & ECONOMICS

29. *Waiting time.* The time (in minutes) applicants must wait for an officer to give them a driver's examination is uniformly distributed on the interval [0, 40]. What is the probability that an applicant must wait more than 25 minutes?

30. *Business failures.* The percentage of computer hobby stores that fail during the first year of operation is a beta random variable with $\beta = 4$.
(A) What is the expected percentage of failures?
(B) What is the probability that over 50% of the stores fail during the first year?

31. *Absenteeism.* The percentage of assembly line workers that are absent one Monday each month is a beta random variable. The mean percentage is 50%.
(A) What is the appropriate value of β?
(B) What is the probability that no more than 75% of the workers will be absent on one Monday each month?

32. *Waiting time.* The waiting time (in minutes) for customers at a drive-in bank is an exponential random variable. The average (mean) time a customer waits is 4 minutes. What is the probability that a customer waits more than 5 minutes?

33. *Communication.* The length of time for telephone conversations (in minutes) is exponentially distributed. The average (mean) length of a conversation is 3 minutes. What is the probability that a conversation lasts less than 2 minutes?

34. *Component failure.* The life expectancy (in years) of a component in a microcomputer is an exponential random variable. Half of the components fail in the first 3 years. The company that manufactures the component offers a 1 year warranty. What is the probability that a component will fail during the warranty period?

LIFE SCIENCES

35. *Nutrition.* The percentage of the daily requirement of vitamin D present in an 8 ounce serving of milk is a beta random variable with $\beta = .2$.
    (A) What is the expected percentage of vitamin D per serving?
    (B) What is the probability that a serving contains at least 50% of the daily requirement?

36. *Medicine.* A scientist is measuring the percentage of a drug present in the bloodstream 10 minutes after an injection. The results indicate that the percentage of the drug present is a beta random variable with mean $\mu = .75$.
    (A) What is the value of β?
    (B) What is the probability that no more than 25% of the drug is present 10 minutes after an injection?

37. *Survival time.* The time of death (in years) after patients have contracted a certain disease is exponentially distributed. The probability that a patient dies within 1 year is .3.
    (A) What is the expected time of death?
    (B) What is the probability that a patient survives longer than the expected time of death?

38. *Survival time.* Repeat Problem 37 if the probability that a patient dies within 1 year is .5.

SOCIAL SCIENCES

39. *Education.* The percentage of entering freshmen that complete the first year of college is a beta random variable with $\beta = 17$.
    (A) What is the expected percentage of students that complete the first year?
    (B) What is the probability that more than 95% of the students complete the first year?

40. *Psychology.* The time (in seconds) it takes rats to find their way through a maze is exponentially distributed. The average (mean) time is 30 seconds. What is the probability that it takes a rat over 1 minute to find a path through the maze?

11-5   JOINT PROBABILITY DENSITY FUNCTIONS

Sometimes an experiment has several outcomes. For example, students who take aptitude tests for admission to college usually receive two scores—one measuring verbal ability and another measuring quantitative ability. In this type of experiment, we must use a joint probability density function.

JOINT PROBABILITY DENSITY FUNCTIONS

A function $f(x, y)$ is a **joint probability density function** for the continuous random variables X and Y if:

1. $f(x, y) \geq 0$ for $-\infty < x < \infty$, $-\infty < y < \infty$

2. $\displaystyle\int_{-\infty}^{\infty} \int_{-\infty}^{\infty} f(x, y)\, dy\, dx = 1$

3. $\displaystyle P((x, y) \in R) = \iint_R f(x, y)\, dy\, dx$ for any region R in the xy plane

---

Example 18

Let: $f(x, y) = \begin{cases} x + y & 0 \leq x \leq 1, 0 \leq y \leq 1 \\ 0 & \text{otherwise} \end{cases}$

(A) Verify that f satisfies the first two conditions for a joint probability density function.
(B) Find the probability that $0 \leq x \leq \tfrac{1}{2}$ and $0 \leq y \leq \tfrac{1}{2}$.
(C) Find the probability that $0 \leq y \leq x$, where $0 \leq x \leq 1$.

Solutions

(A) Clearly, $f(x, y) \geq 0$ when $0 \leq x \leq 1$ and $0 \leq y \leq 1$.

$$\int_{-\infty}^{\infty} \int_{-\infty}^{\infty} f(x, y)\, dy\, dx = \int_0^1 \left[\int_0^1 (x + y)\, dy \right] dx$$

$$= \int_0^1 \left[\left(xy + \frac{1}{2}y^2 \right)\Big|_{y=0}^{y=1} \right] dx$$

$$= \int_0^1 \left(x + \frac{1}{2} \right) dx$$

$$= \left(\frac{1}{2}x^2 + \frac{1}{2}x \right)\Big|_{x=0}^{x=1} = 1$$

(B) Let $R = \{(x, y)|0 \le x \le \frac{1}{2}, \quad 0 \le y \le \frac{1}{2}\}$. Then

$$P((x, y) \in R) = \iint\limits_{R} f(x, y) \, dy \, dx$$

$$= \int_{0}^{1/2} \left[\int_{0}^{1/2} (x + y) \, dy \right] dx$$

$$= \int_{0}^{1/2} \left[\left(xy + \frac{1}{2}y^2 \right) \Big|_{y=0}^{y=1/2} \right] dx$$

$$= \int_{0}^{1/2} \left(\frac{1}{2}x + \frac{1}{8} \right) dx = \left(\frac{1}{4}x^2 + \frac{1}{8}x \right) \Big|_{x=0}^{x=1/2}$$

$$= \frac{1}{16} + \frac{1}{16} = \frac{1}{8}$$

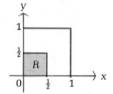

(C) Let $R_1 = \{(x, y)|0 \le y \le x, \quad 0 \le x \le 1\}$. Then

$$P((x, y) \in R_1) = \iint\limits_{R_1} f(x, y) \, dy \, dx$$

$$= \int_{0}^{1} \left[\int_{0}^{x} (x + y) \, dy \right] dx$$

$$= \int_{0}^{1} \left[\left(xy + \frac{1}{2}y^2 \right) \Big|_{y=0}^{y=x} \right] dx$$

$$= \int_{0}^{1} \frac{3}{2}x^2 \, dx = \frac{1}{2}x^3 \Big|_{x=0}^{x=1} = \frac{1}{2}$$

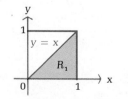

Problem 18   Repeat Example 18 for $f(x, y) = \begin{cases} 4xy & 0 \leq x \leq 1, \quad 0 \leq y \leq 1 \\ 0 & \text{otherwise} \end{cases}$

Answers   (B) $\frac{1}{16}$ (C) $\frac{1}{2}$

Example 19   The price X (in dollars) and the sales Y (in thousands of units) of a certain item are continuous random variables with joint probability density function

$$f(x, y) = \begin{cases} xe^{-xy} & 0.50 \leq x \leq 1.50, \, y \geq 0 \\ 0 & \text{otherwise} \end{cases}$$

(A)   What is the probability that the price will be between \$1.00 and \$1.25 and the sales will not exceed 8,000 units?

(B)   What is the probability that the price will be less than \$0.75 and the sales will exceed 1,000 units?

(C)   What is the probability that the total revenue exceeds \$1,000?

Solutions   (A)   Let $R_1 = \{(x, y) | 1 \leq x \leq 1.25, \quad 0 \leq y \leq 8\}$. Then

$$P((x, y) \in R_1) = \iint\limits_{R_1} f(x, y) \, dy \, dx$$

$$= \int_1^{1.25} \left[\int_0^8 xe^{-xy} \, dy \right] dx$$

$$= \int_1^{1.25} \left[(-e^{-xy}) \Big|_{y=0}^{y=8} \right] dx$$

$$= \int_1^{1.25} (1 - e^{-8x}) \, dx$$

$$= \left(x + \frac{1}{8}e^{-8x} \right) \Big|_{x=1}^{x=1.25}$$

$$\approx 1.250006 - 1.000042$$

$$= .249964$$

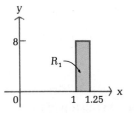

(B)   Let $R_2 = \{(x, y)|0.5 \le x \le 0.75, \quad y \ge 1\}$. Then

$$P((x, y) \in R_2) = \iint\limits_{R_2} f(x, y)\, dy\, dx$$

$$= \int_{0.5}^{0.75} \left[\int_1^\infty xe^{-xy}\, dy \right] dx$$

$$= \int_{0.5}^{0.75} \left[\lim_{R \to \infty}(-e^{-xy}) \Big|_{y=1}^{y=R} \right] dx$$

$$= \int_{0.5}^{0.75} \left[\lim_{R \to \infty}(-e^{-Rx} + e^{-x}) \right] dx = \int_{0.5}^{0.75} e^{-x}\, dx$$

$$= -e^{-x} \Big|_{x=0.5}^{x=0.75}$$

$$= -e^{-0.75} + e^{-0.5} \approx .134$$

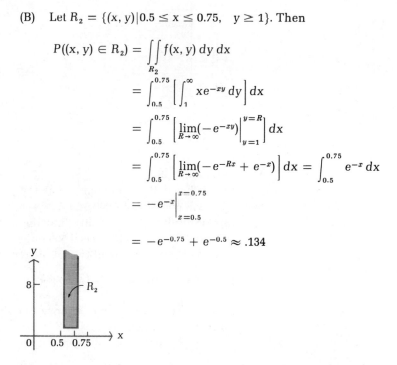

(C)   Since x is the price per unit and y is the number of units (in thousands), the total revenue in thousands of dollars is given by $R = xy$. If we define R_3 by

$$R_3 = \{(x, y)|0.50 \le x \le 1.50, \quad y \ge 0, \quad xy \ge 1\}$$

then the probability that the revenue exceeds \$1,000 is

$$P((x, y) \in R_3) = \iint\limits_{R_3} f(x, y)\, dy\, dx$$

$$= \int_{0.5}^{1.5} \left[\int_{1/x}^\infty xe^{-xy}\, dy \right] dx$$

$$= \int_{0.5}^{1.5} \left[\lim_{R \to \infty}(-e^{-xy}) \Big|_{y=1/x}^{y=R} \right] dx$$

$$= \int_{0.5}^{1.5} \left[\lim_{R \to \infty}(-e^{-Rx} + e^{-1}) \right] dx$$

$$= \int_{0.5}^{1.5} e^{-1}\, dx$$

$$= e^{-1}x \Big|_{x=0.5}^{x=1.5} = e^{-1} \approx .368$$

See the figure at the top of the next page.

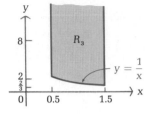

Problem 19

In Example 19:

(A)   What is the probability that the price is at least $0.80 and sales will not exceed 2,000 units?

(B)   What is the probability that the price is between $0.60 and $1.20 and the sales will exceed 500 units?

(C)   What is the probability that the revenue exceeds $2,000?

Answers

(A)   .624   (B)   .384   (C)   .135

Expected value and variance can also be extended to joint probability density functions.

EXPECTED VALUE AND VARIANCE FOR
JOINT PROBABILITY DENSITY FUNCTIONS

If $f(x, y)$ is the joint probability density function for two continuous random variables X and Y, then the **expected value of X** is

$$\mu_X = E(X) = \int_{-\infty}^{\infty} \int_{-\infty}^{\infty} xf(x, y) \, dy \, dx$$

the **expected value of Y** is

$$\mu_Y = E(Y) = \int_{-\infty}^{\infty} \int_{-\infty}^{\infty} yf(x, y) \, dy \, dx$$

and the **covariance of X and Y** is

$$C(X, Y) = \int_{-\infty}^{\infty} \int_{-\infty}^{\infty} (x - \mu_X)(y - \mu_Y)f(x, y) \, dy \, dx$$

Furthermore, the covariance can be computed by using the formula

$$C(X, Y) = \int_{-\infty}^{\infty} \int_{-\infty}^{\infty} xyf(x, y) \, dy \, dx - \mu_X\mu_Y$$

Example 20 Find μ_X, μ_Y, and $C(X, Y)$ for two continuous random variables X and Y with joint probability density function

$$f(x, y) = \begin{cases} \tfrac{2}{3}x + \tfrac{4}{3}y & 0 \le x \le 1, 0 \le y \le 1 \\ 0 & \text{otherwise} \end{cases}$$

Solution

$$\mu_X = \int_{-\infty}^{\infty} \int_{-\infty}^{\infty} xf(x, y)\, dy\, dx = \int_0^1 \left[\int_0^1 (\tfrac{2}{3}x^2 + \tfrac{4}{3}xy)\, dy \right] dx$$

$$= \int_0^1 \left[(\tfrac{2}{3}x^2y + \tfrac{2}{3}xy^2) \Big|_{y=0}^{y=1} \right] dx = \int_0^1 (\tfrac{2}{3}x^2 + \tfrac{2}{3}x)\, dx$$

$$= (\tfrac{2}{9}x^3 + \tfrac{1}{3}x^2) \Big|_{x=0}^{x=1} = \tfrac{2}{9} + \tfrac{1}{3} = \tfrac{5}{9}$$

$$\mu_Y = \int_{-\infty}^{\infty} \int_{-\infty}^{\infty} yf(x, y)\, dy\, dx = \int_0^1 \left[\int_0^1 (\tfrac{2}{3}xy + \tfrac{4}{3}y^2)\, dy \right] dx$$

$$= \int_0^1 \left[(\tfrac{1}{3}xy^2 + \tfrac{4}{9}y^3) \Big|_{y=0}^{y=1} \right] dx = \int_0^1 (\tfrac{1}{3}x + \tfrac{4}{9})\, dx$$

$$= (\tfrac{1}{6}x^2 + \tfrac{4}{9}x) \Big|_{x=0}^{x=1} = \tfrac{1}{6} + \tfrac{4}{9} = \tfrac{11}{18}$$

$$\int_{-\infty}^{\infty} \int_{-\infty}^{\infty} xyf(x, y)\, dy\, dx = \int_0^1 \left[\int_0^1 (\tfrac{2}{3}x^2y + \tfrac{4}{3}xy^2)\, dy \right] dx$$

$$= \int_0^1 \left[(\tfrac{1}{3}x^2y^2 + \tfrac{4}{9}xy^3) \Big|_{y=0}^{y=1} \right] dx = \int_0^1 (\tfrac{1}{3}x^2 + \tfrac{4}{9}x)\, dx$$

$$= (\tfrac{1}{9}x^3 + \tfrac{2}{9}x^2) \Big|_{x=0}^{x=1} = \tfrac{1}{9} + \tfrac{2}{9} = \tfrac{1}{3}$$

$$C(X, Y) = \int_{-\infty}^{\infty} \int_{-\infty}^{\infty} xyf(x, y)\, dy\, dx - \mu_X\mu_Y = \tfrac{1}{3} - \tfrac{5}{9}(\tfrac{11}{18}) = -\tfrac{1}{162}$$

Problem 20 Repeat Example 20 for: $f(x, y) = \begin{cases} 6xy^2 & 0 \le x \le 1, 0 \le y \le 1 \\ 0 & \text{otherwise} \end{cases}$

Answer $\mu_X = \tfrac{2}{3}, \mu_Y = \tfrac{3}{4}, C(X, Y) = 0$

Example 21 The time X (in microseconds) it takes a computer to compile a user's source program and the time Y (in microseconds) it takes to execute the program are continuous random variables with joint probability density function

$$f(x, y) = \begin{cases} 15/x^6y^4 & x \ge 1, y \ge 1 \\ 0 & \text{otherwise} \end{cases}$$

What is the expected compilation time?

Solution $\quad \mu_X = \int_{-\infty}^{\infty} \int_{-\infty}^{\infty} xf(x, y)\, dy\, dx = \int_{1}^{\infty} \int_{1}^{\infty} \frac{15}{x^5 y^4}\, dy\, dx$

$$= \int_{1}^{\infty} \left[\lim_{R \to \infty} \left(-\frac{5}{x^5 y^3} \right) \Big|_{y=1}^{y=R} \right] dx$$

$$= \int_{1}^{\infty} \lim_{R \to \infty} \left(-\frac{5}{x^5 R^3} + \frac{5}{x^5} \right) dx$$

$$= \int_{1}^{\infty} \frac{5}{x^5}\, dx = \lim_{R \to \infty} \left(-\frac{5}{4x^4} \right) \Big|_{x=1}^{x=R}$$

$$= \lim_{R \to \infty} \left(-\frac{5}{4R^4} + \frac{5}{4} \right)$$

$$= 1.25 \text{ microseconds}$$

Problem 21   In Example 21, what is the expected execution time?

Answer   1.5 microseconds

EXERCISE 11-5

A   Problems 1–10 refer to the continuous random variables X and Y with joint probability density function

$$f(x, y) = \begin{cases} \tfrac{1}{10}(x^2 + 2y) & 0 \le x \le 1, 0 \le y \le 3 \\ 0 & \text{otherwise} \end{cases}$$

1. Find $P((x, y) \in R)$ for $R = \{(x, y)\,|\,0 \le x \le \tfrac{1}{2},\ \ 0 \le y \le \tfrac{1}{2}\}$.
2. Find $P((x, y) \in R)$ for $R = \{(x, y)\,|\,\tfrac{1}{2} \le x \le 1,\ \ 2 \le y \le 3\}$.
3. Find $P((x, y) \in R)$ for $R = \{(x, y)\,|\,x \le \tfrac{1}{4},\ \ y \ge 1\}$.
4. Find $P((x, y) \in R)$ for $R = \{(x, y)\,|\,y \le 2\}$.
5. Find $P((x, y) \in R)$ for $R = \{(x, y)\,|\,x \ge \tfrac{3}{4}\}$.
6. Find $P((x, y) \in R)$ for $R = \{(x, y)\,|\,0 \le x \le 1,\ \ y = 2\}$.
7. Find $P((x, y) \in R)$ for $R = \{(x, y)\,|\,0 \le x \le 1,\ \ x \le y \le 3x\}$.
8. Find $P((x, y) \in R)$ for $R = \{(x, y)\,|\,3x + y \le 3\}$.
9. Find μ_X and μ_Y.
10. Find $C(X, Y)$.

B   Problems 11–20 refer to the continuous random variables X and Y with joint probability density function

$$f(x, y) = \begin{cases} 6/x^4 y^3 & x \ge 1, y \ge 1 \\ 0 & \text{otherwise} \end{cases}$$

11. Find $P((x, y) \in R)$ where $R = \{(x, y)\,|\,1 \le x \le 2,\ \ 1 \le y \le 3\}$.
12. Find $P((x, y) \in R)$ where $R = \{(x, y)\,|\,x \ge 3,\ \ 1 \le y \le 2\}$.
13. Find $P((x, y) \in R)$ where $R = \{(x, y)\,|\,2 \le x \le 3,\ \ y \ge 4\}$.
14. Find $P((x, y) \in R)$ where $R = \{(x, y)\,|\,x \ge 2,\ \ y \ge 3\}$.

15. Find $P((x, y) \in R)$ where $R = \{(x, y) | 1 \le x \le 2\}$.
16. Find $P((x, y) \in R)$ where $R = \{(x, y) | y \ge 2\}$.
17. Find $P((x, y) \in R)$ where $R = \{(x, y) | y \le x\}$.
18. Find $P((x, y) \in R)$ where $R = \{(x, y) | y^2 \le x \le y^3\}$.
19. Find μ_X and μ_Y.
20. Find $C(X, Y)$.

Problems 21–26 refer to the continuous random variables X and Y with joint probability density function

$$f(x, y) = \begin{cases} \frac{1}{4}(x + y) & 0 \le y \le x, 0 \le x \le 2 \\ 0 & \text{otherwise} \end{cases}$$

21. Find $P((x, y) \in R)$ where $R = \{(x, y) | 0 \le y \le x^2, \quad 0 < x \le 1\}$.
22. Find $P((x, y) \in R)$ where $R = \{(x, y) | x \ge 1\}$.
23. Find $P((x, y) \in R)$ where $R = \{(x, y) | x + y \le 2\}$.
24. Find $P((x, y) \in R)$ where $R = \{(x, y) | x + 2y \ge 2\}$.
25. Find μ_X and μ_Y.
26. Find $C(X, Y)$.

C   27. If X and Y are continuous random variables with joint probability density function

$$f(x, y) = \begin{cases} \dfrac{1}{(b - a)(d - c)} & a \le x \le b, c \le y \le d \\ 0 & \text{otherwise} \end{cases}$$

find μ_X, μ_Y, and $C(X, Y)$.

28. If X and Y are continuous random variables with joint probability density function

$$f(x, y) = \begin{cases} (\alpha + 1)(\beta + 1)x^\alpha y^\beta & 0 \le x \le 1, 0 \le y \le 1 \\ 0 & \text{otherwise} \end{cases}$$

where α and β are constants, $\alpha > 0$, $\beta > 0$, find μ_X, μ_Y, and $C(X, Y)$.

APPLICATIONS

BUSINESS & ECONOMICS   29. *Sales.* The price X (in dollars) and the sales Y (in thousands of units) of a certain item are continuous random variables with joint probability density function

$$f(x, y) = \begin{cases} xe^{-xy} & 1 \le x \le 2, y \ge 0 \\ 0 & \text{otherwise} \end{cases}$$

(A) What is the probability that the sales exceed 5,000 units?
(B) What is the probability that the revenue exceeds $2,000?
(C) What is the expected price?

30. *Sales.* Repeat Problem 29 for

$$f(x, y) = \begin{cases} xe^{-2xy} & 1 \le x \le 3, y \ge 0 \\ 0 & \text{otherwise} \end{cases}$$

31. *Waiting time.* The time X (in minutes) that customers spend waiting in line for service and the time Y (in minutes) it takes to serve a customer are continuous random variables with joint probability density function

$$f(x, y) = \begin{cases} \frac{1}{4}e^{-[(x/4)+y]} & x \ge 0, y \ge 0 \\ 0 & \text{otherwise} \end{cases}$$

(A) What is the probability that the time spent waiting in line does not exceed 3 minutes?

(B) What is the probability that the service time does not exceed 2 minutes?

(C) What is the probability that the total of the waiting time and the service time does not exceed 5 minutes?

32. *Waiting time.* Repeat Problem 31 for

$$f(x, y) = \begin{cases} \frac{1}{6}e^{-[(x/3)+(y/2)]} & x \ge 0, y \ge 0 \\ 0 & \text{otherwise} \end{cases}$$

LIFE SCIENCES

33. *Medicine.* A patient is injected with a drug that has two active ingredients. The percentage of the first ingredient, X, and the percentage of the second ingredient, Y, that will be present in the patient's bloodstream 5 minutes after the injection are continuous random variables with joint probability density function

$$f(x, y) = \begin{cases} 30x^2y^3(1 - xy) & 0 \le x \le 1, 0 \le y \le 1 \\ 0 & \text{otherwise} \end{cases}$$

(A) What is the probability that more than 50% of each ingredient will be present 5 minutes after the injection?

(B) What is the expected percentage of each ingredient present 5 minutes after the injection?

34. *Nutrition.* The percentage, X, of the daily requirement of vitamin A and the percentage, Y, of the daily requirement of vitamin D present in an 8 ounce serving of milk are continuous random variables with joint probability density function

$$f(x, y) = \begin{cases} 8xy^2(1 - x^2y^3) & 0 \leq x \leq 1, 0 \leq y \leq 1 \\ 0 & \text{otherwise} \end{cases}$$

(A) What is the probability that a serving contains more than 50% of each vitamin?

(B) What is the expected percentage of each vitamin present in a serving?

SOCIAL SCIENCES   **35.** *Testing.* Students entering a university are given a mathematics placement examination and a foreign language placement examination. The percentage, X, of correct answers on the math exam and the percentage, Y, of correct answers on the language exam are continuous random variables with joint probability density function

$$f(x, y) = \begin{cases} x^2 + 2y^2 & 0 \leq x \leq 1, 0 \leq y \leq 1 \\ 0 & \text{otherwise} \end{cases}$$

(A) What is the probability that a student scores over 70% on both exams?

(B) What is the expected score on each exam?

11-6   CHAPTER REVIEW

IMPORTANT TERMS AND SYMBOLS

11-1 Finite probability models. sample space, simple event, random variable, probability function, probability distribution of a random variable, expected value, mean, variance, standard deviation,

$$\mu = E(X) = x_1p_1 + x_2p_2 + \cdots + x_mp_m,$$
$$V(X) = (x_1 - \mu)^2p_1 + (x_2 - \mu)^2p_2 + \cdots + (x_m - \mu)^2p_m,$$
$$\sigma = \sqrt{V(X)}$$

11-2 Continuous random variables. continuous random variable, probability density function, $P(c \leq x \leq d) = \int_c^d f(x)\,dx$, cumulative probability distribution function, $F(x) = P(t \leq x) = \int_{-\infty}^x f(t)\,dt$

11-3 Expected value, standard deviation, and median. mean, expected value, variance, standard deviation, median,

$$\mu = E(X) = \int_{-\infty}^{\infty} xf(x)\,dx, \; V(X) = \int_{-\infty}^{\infty} (x - \mu)^2f(x)\,dx = \int_{-\infty}^{\infty} x^2f(x)\,dx - \mu^2,$$
$$\sigma = \sqrt{V(X)}, \quad P(x \leq x_m) = \frac{1}{2}$$

11-4 Special probability density functions.

Uniform probability
density function: $f(x) = \begin{cases} \dfrac{1}{b-a} & a \leq x \leq b \\ 0 & \text{otherwise} \end{cases}$

$$F(x) = \begin{cases} 0 & x < a \\ \dfrac{x-a}{b-a} & a \leq x \leq b \\ 1 & x > b \end{cases}$$

$\mu = \dfrac{1}{2}(a+b)$ $\qquad x_m = \dfrac{1}{2}(a+b)$ $\qquad \sigma = \dfrac{1}{\sqrt{12}}(b-a)$

Beta probability
density function: $f(x) = \begin{cases} (\beta+1)(\beta+2)x^{\beta}(1-x) & 0 \leq x \leq 1 \\ 0 & \text{otherwise} \end{cases}$

$$F(x) = \begin{cases} 0 & x < 0 \\ (\beta+2)x^{\beta+1} - (\beta+1)x^{\beta+2} & 0 \leq x \leq 1 \\ 1 & x > 1 \end{cases}$$

$\mu = \dfrac{\beta+1}{\beta+3}$ $\qquad \sigma = \sqrt{\dfrac{2(\beta+1)}{(\beta+4)(\beta+3)^2}}$

Exponential
density function: $f(x) = \begin{cases} \frac{1}{\lambda}e^{-x/\lambda} & x \geq 0, \lambda > 0 \\ 0 & \text{otherwise} \end{cases}$

$$F(x) = \begin{cases} 1 - e^{-x/\lambda} & x \geq 0 \\ 0 & \text{otherwise} \end{cases}$$

$\mu = \lambda$ $\qquad x_m = \lambda \ln 2$ $\qquad \sigma = \lambda$

Normal distribution

11-5 Joint probability density functions. joint probability density functions, expected value, covariance, $P((x, y) \in R) = \displaystyle\iint_R f(x, y)\, dy\, dx$,

$\mu_X = \displaystyle\int_{-\infty}^{\infty} \int_{-\infty}^{\infty} xf(x, y)\, dy\, dx$, $\quad \mu_Y = \displaystyle\int_{-\infty}^{\infty} \int_{-\infty}^{\infty} yf(x, y)\, dy\, dx$,

$C(X, Y) = \displaystyle\int_{-\infty}^{\infty} \int_{-\infty}^{\infty} (x - \mu_X)(y - \mu_Y)f(x, y)\, dy\, dx$

$\qquad = \displaystyle\int_{-\infty}^{\infty} \int_{-\infty}^{\infty} xyf(x, y)\, dy\, dx - \mu_X\mu_Y$

EXERCISE 11-6   CHAPTER REVIEW

A   *A spinner can land on any one of eight different sectors, and each sector is as likely to turn up as any other. The sectors are numbered in the figure. An experiment consists of spinning the dial once and recording the number in the sector that the spinner lands on. Problems 1–4 refer to this experiment.*

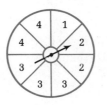

1. Find a sample space and probability distribution for this experiment.
2. What is the probability that the spinner stops on an even-numbered sector?
3. Find the expected value, variance, and standard deviation for the probability distribution found in Problem 1.
4. After paying $3 to play, you spin the dial and are paid back the number of dollars corresponding to the number on the sector where the spinner stopped. What is the expected value of this game?

Problems 5–8 refer to the continuous random variable X with probability density function

$$f(x) = \begin{cases} 1 - \frac{1}{2}x & 0 \le x \le 2 \\ 0 & \text{otherwise} \end{cases}$$

5. Find $P(0 \le x \le 1)$ and illustrate with a graph.
6. Find the mean, variance, and standard deviation.
7. Find and graph the associated cumulative probability distribution function.
8. Find the median.

Problems 9–12 refer to the continuous random variable X with probability density function

$$f(x) = \begin{cases} \frac{5}{2}x^{-7/2} & x \ge 1 \\ 0 & \text{otherwise} \end{cases}$$

9. Find $P(1 \le x \le 4)$ and illustrate with a graph.
10. Find the mean, variance, and standard deviation.

11. Find and graph the associated cumulative probability distribution function.
12. Find the median.

B   Problems 13–16 refer to the continuous random variables X and Y with joint probability density function

$$f(x, y) = \begin{cases} \frac{1}{40}(2x + y) & 0 \le x \le 4, 0 \le y \le 2 \\ 0 & \text{otherwise} \end{cases}$$

13. Find $P((x, y) \in R)$ where $R = \{(x, y)|0 \le x \le 3, \ 0 \le y \le 1\}$.
14. Find $P((x, y) \in R)$ where $R = \{(x, y)|x \le 2\}$.
15. Find $P((x, y) \in R)$ where $R = \{(x, y)|x + y \le 4\}$.
16. Find μ_X, μ_Y, and $C(X, Y)$.

Problems 17–20 refer to a beta random variable X with $\beta = 5$.

17. Find and graph the probability density function.
18. Find $P(\frac{1}{4} \le x \le \frac{3}{4})$.
19. Find and graph the associated cumulative probability distribution function.
20. Find the mean and standard deviation.

Problems 21–24 refer to an exponentially distributed random variable X.

21. If $P(4 \le x) = e^{-2}$, find the probability density function.
22. Find $P(0 \le x \le 2)$.
23. Find the associated cumulative probability distribution function.
24. Find the mean, standard deviation, and median.

C   25. If X is a beta random variable with mean $\mu = .8$, what is the value of β?
26. Find the mean and the median of the continuous random variable with probability density function

$$f(x) = \begin{cases} 50/(x + 5)^3 & x \ge 0 \\ 0 & \text{otherwise} \end{cases}$$

27. If X and Y are continuous random variables with joint probability density function

$$f(x, y) = \begin{cases} \frac{2}{7}(x + y) & 0 \le y \le x, 1 \le x \le 2 \\ 0 & \text{otherwise} \end{cases}$$

what is the probability that the product of X and Y is greater than 1?

28. Suppose $f(x)$ and $g(x)$ are both probability density functions and a and b are constants. If $h(x)$ is defined by

$$h(x) = af(x) + bg(x)$$

determine conditions on a and b that ensure that h is also a probability density function. If μ_f, μ_g, and μ_h are the respective means of these three functions, how are μ_f, μ_g, and μ_h related?

APPLICATIONS

BUSINESS & ECONOMICS

29. *Demand.* The manager of a movie theater has determined that the weekly demand for popcorn (in pounds) is a continuous random variable with probability density function

$$f(x) = \begin{cases} \frac{1}{50}(1 - 0.01x) & 0 \le x \le 100 \\ 0 & \text{otherwise} \end{cases}$$

(A) If the manager has 50 pounds of popcorn on hand at the beginning of the week, what is the probability that this will be enough to meet the weekly demand?

(B) If the manager wants the probability that the supply on hand exceeds the weekly demand to be .96, how much popcorn must be on hand at the beginning of the week?

30. *Credit applications.* The percentage of applications for a national credit card that are processed on the same day they are received is a beta random variable with $\beta = 1$.

(A) What is the probability that at least 20% of the applications received are processed the same day they arrive?

(B) What is the expected percentage of applications processed the same day they arrive?

31. *Computer failure.* A computer manufacturer has determined that the time between failures for its computers is an exponentially distributed random variable with a mean failure time of 4,000 hours. Suppose a particular computer has just been repaired.

(A) What is the probability that the computer operates for the next 4,000 hours without a failure?

(B) What is the probability that the computer fails in the next 1,000 hours?

32. *Sales.* The price X (in dollars) and the sales Y (in thousands of units) for a certain item are continuous random variables with joint probability density function

$$f(x, y) = \begin{cases} \tfrac{1}{4}xe^{-xy/2} & 2 \le x \le 4, y \ge 0 \\ 0 & \text{otherwise} \end{cases}$$

(A) What is the probability that the sales exceed 2,000 units?
(B) What is the probability that the revenue exceeds $1,000?

LIFE SCIENCES

33. *Medicine.* The shelf-life (in months) of a certain drug is a continuous random variable with probability density function

$$f(x) = \begin{cases} 10/(x + 10)^2 & x \ge 0 \\ 0 & \text{otherwise} \end{cases}$$

(A) What is the probability that the drug is still usable after 5 months?
(B) What is the median shelf-life?

34. *Life expectancy.* The life expectancy (in months) after dogs have contracted a certain disease is an exponentially distributed random variable. The probability of surviving more than 1 month is e^{-2}. After contracting this disease:
(A) What is the probability of surviving more than 2 months?
(B) What is the mean life expectancy?

SOCIAL SCIENCES

35. *Testing.* The percentage of correct answers on a college entrance examination is a beta random variable. The mean score is 75%. What is the probability that a student answers over 50% of the questions correctly?

PRACTICE TEST: **CHAPTER 11**

1. Find the mean, variance, and standard deviation for the discrete random variable X with probability distribution

x_i	1	2	3	4	5
p_i	.1	.3	.2	.1	.3

2. Find the mean, variance, and standard deviation for the continuous random variable X with probability density function

$$f(x) = \begin{cases} \tfrac{1}{4}(1 + x) & 0 \le x \le 2 \\ 0 & \text{otherwise} \end{cases}$$

3. If X is a continuous random variable with probability density function

$$f(x) = \begin{cases} 10/9x^2 & 1 \le x \le 10 \\ 0 & \text{otherwise} \end{cases}$$

find $P(1 \le x \le 5)$ and illustrate this with a graph.

4. Find the associated cumulative probability distribution function $F(x)$ and the median for the random variable in Problem 3. Graph $F(x)$ and locate the median on your graph.

5. If X and Y are continuous random variables with joint probability density function

$$f(x, y) = \begin{cases} \tfrac{1}{8}(x + 4y) & 0 \le y \le x, 0 \le x \le 2 \\ 0 & \text{otherwise} \end{cases}$$

find $P((x, y) \in R)$ for:

(A) $R = \{(x, y) | 1 \le x \le 2, \quad 0 \le y \le 1\}$

(B) $R = \{(x, y) | x + y \le 2\}$

6. The life expectancy of a certain brand of light bulbs (in hundreds of hours) is an exponential random variable with a mean life expectancy of 500 hours. What is the probability that a bulb lasts over 500 hours?

7. The percentage of completed Social Security application forms that contain errors is a beta random variable with mean $\mu = .4$. What is the appropriate value of β?

8. The daily demand for doughnuts in a chain of bakeries (in hundreds of dozens) is a continuous random variable with probability density function

$$f(x) = \begin{cases} \tfrac{1}{8}(6 - x) & 2 \le x \le 6 \\ 0 & \text{otherwise} \end{cases}$$

What is the expected demand and the median demand?

9. The price X (in dollars) and the sales Y (in thousands of units) for a certain item are continuous random variables with joint probability density function

$$f(x, y) = \begin{cases} xe^{-4xy} & 0 \le x \le 4, y \ge 0 \\ 0 & \text{otherwise} \end{cases}$$

Find the probability that the revenue exceeds $2,000.

APPENDIX A
REVIEW: SETS AND ALGEBRA

CONTENTS

A REVIEW: SETS AND ALGEBRA

This appendix is a concise review of topics from elementary and intermediate algebra that are important to an understanding of much of the subject matter in this book. No attempt is made to give complete developments for all topics. In most cases summaries of key ideas and results are presented, followed by examples, matched problems, and exercises. It is hoped that this review material will serve to sharpen rusty algebraic skills, remind you of forgotten concepts, and serve as a convenient reference when needed.

A-1   SETS

SET PROPERTIES AND SET NOTATION

In this section we will review a few key ideas from set theory. Set concepts and notation not only help us talk about certain mathematical ideas with greater clarity and precision, but are indispensable to a clear understanding of probability.

We can think of a **set** as any collection of objects specified in such a way that we can tell whether any given object is or is not in the collection. Capital letters, such as A, B, and C, are often used to designate particular sets. Each object in a set is called a **member** or **element** of the set. Symbolically,

$$a \in A \quad \text{means} \quad \text{``}a \text{ is an element of set } A\text{''}$$
$$a \notin A \quad \text{means} \quad \text{``}a \text{ is not an element of set } A\text{''}$$

A set without any elements is called the **empty** or **null set.** For example, the set of all people over 10 feet tall is an empty set. Symbolically,

> Ø   represents   "the empty or null set"

A set is usually described either by listing all its elements between braces { }, or by enclosing a rule within braces that determines the elements of the set. Thus, if $P(x)$ is a statement about x, then

> $S = \{x \mid P(x)\}$ means   "S is the set of all x such that $P(x)$ is true"

Recall that the vertical bar in the symbolic form is read "such that." The following example illustrates the rule and listing methods of representing sets.

Example 1

| RULE | LISTING |

$$\{x \mid x \text{ is a weekend day}\} = \{\text{Saturday, Sunday}\}$$
$$\{x \mid x^2 = 4\} = \{-2, 2\}$$
$$\{x \mid x \text{ is an odd positive counting number}\} = \{1, 3, 5, \ldots\}$$

The three dots . . . in the last set in Example 1 indicate that the pattern established by the first three entries continues indefinitely. The first two sets in Example 1 are **finite sets** (intuitively, the elements can be counted); the last set is an **infinite set** (if we try to count the elements, there is no end). When listing the elements in a set we do not list an element more than once.

Problem 1   Let G be the set of all numbers such that $x^2 = 9$.
(A)   Denote G by the rule method.
(B)   Denote G by the listing method.
(C)   Indicate whether the following are true or false: $3 \in G$, $9 \notin G$.

Answers   (A) $\{x \mid x^2 = 9\}$ (B) $\{-3, 3\}$ (C)   True, True

If each element of a set A is also an element of set B, we say that A is a **subset** of B. For example, the set of all women students in a class is a subset of the whole class. Note that the definition allows a set to be a subset of itself. If set A and set B have exactly the same elements, then the two sets are said to be **equal**. Symbolically,

> $A \subset B$ means   "A is a subset of B"
> $A = B$ means   "A and B have exactly the same elements"

It can be proved that

$\varnothing$ is a subset of every set

Example 2   List all the subsets of the set $\{a, b, c\}$.

Solution $\{a, b, c\}, \{a, b\}, \{a, c\}, \{b, c\}, \{a\}, \{b\}, \{c\}, \varnothing$

Problem 2   List all the subsets of the set $\{1, 2\}$.

Answer $\{1, 2\}, \{1\}, \{2\}, \varnothing$

SET OPERATIONS   The **union** of sets A and B, denoted by $A \cup B$, is the set of all elements formed by combining all the elements of A and all the elements of B into one set. Symbolically,

---

UNION

$$A \cup B = \{x \mid x \in A \quad \textbf{or} \quad x \in B\}$$

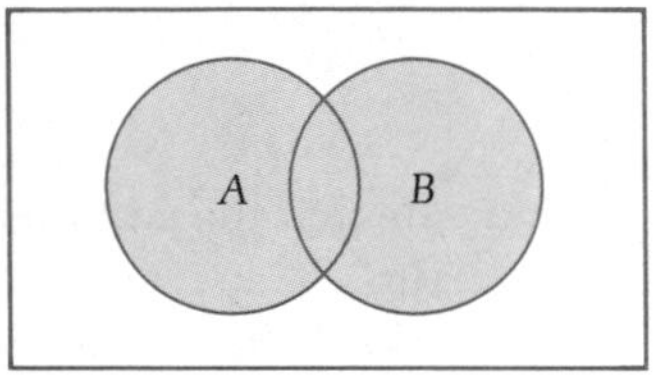

Here we use the word "or" in the way it is always used in mathematics; that is, x may be an element of set A or set B or both.

Venn diagrams are useful aids in visualizing set relationships. The union of two sets can be illustrated as shown in Figure 1.

The **intersection** of sets A and B, denoted by $A \cap B$, is the set of elements in set A that are also in set B. Symbolically,

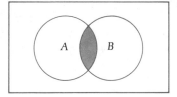

FIGURE 1 $A \cup B$ is the shaded region.

INTERSECTION

$$A \cap B = \{x \mid x \in A \quad \textbf{and} \quad x \in B\}$$

This relationship is easily visualized in the Venn diagram shown in Figure 2.

If $A \cap B = \varnothing$, then the sets A and B are said to be **disjoint;** this is illustrated in Figure 3.

The set of all elements under consideration is called the **universal set** U. Once the universal set is determined for a particular discussion, all other sets in that discussion must be subsets of U. We now define one more operation on sets called the *complement*. The **complement** of A (relative to U), denoted by A', is the set of elements in U that are not in A (see Fig. 4). Symbolically,

FIGURE 2 $A \cap B$ is the shaded region.

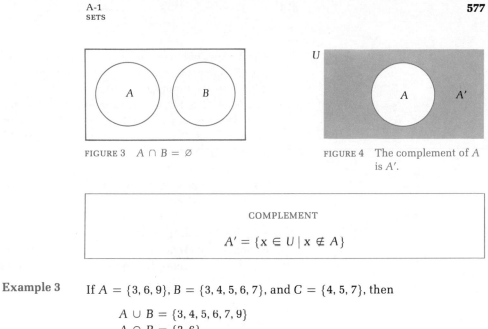

FIGURE 3 $A \cap B = \emptyset$

FIGURE 4   The complement of A is A'.

COMPLEMENT

$$A' = \{x \in U \mid x \notin A\}$$

Example 3   If $A = \{3, 6, 9\}$, $B = \{3, 4, 5, 6, 7\}$, and $C = \{4, 5, 7\}$, then

$A \cup B = \{3, 4, 5, 6, 7, 9\}$

$A \cap B = \{3, 6\}$

$A \cap C = \emptyset$ A and C are disjoint

C' relative to B is $\{3, 6\}$

Problem 3   If $R = \{1, 2, 3, 4\}$, $S = \{1, 3, 5, 7\}$, and $T = \{2, 4\}$, find:
(A) $R \cup S$ (B) $R \cap S$ (C) $S \cap T$ (D) T' relative to R

Answers   (A) $\{1, 2, 3, 4, 5, 7\}$ (B) $\{1, 3\}$ (C) $\emptyset$ (D) $\{1, 3\}$

Example 4   From a survey involving 100 college students, a marketing research company found that seventy-five students owned stereos, forty-five owned cars, and thirty-five owned cars and stereos.
(A)   How many students owned either a car or a stereo?
(B)   How many students did not own either a car or a stereo?

Solutions   Venn diagrams are very useful for this type of problem. If we let

$U =$ Set of students in sample (100)

$S =$ Set of students who own stereos (75)

$C =$ Set of students who own cars (45)

$S \cap C =$ Set of students who own cars and stereo (35)

then

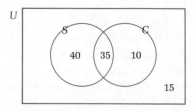

Place the number in the intersection first, then work outward.

(A)   The number of students who own either a car or a stereo is the number of students in the set $S \cup C$. You might be tempted to say that this is just the number of students in S plus the number of students in C, $75 + 45 = 120$, but this sum is larger than the sample we started with! What is wrong? We have actually counted the number in the intersection (35) twice. The correct answer, as seen in the Venn diagram, is

$$40 + 35 + 10 = 85$$

(B)   The number of students who do not own either a car or a stereo is the number of students in the set $(S \cup C)'$; that is, 15.

Problem 4   Referring to Example 4:
(A)   How many students owned a car but not a stereo?
(B)   How many students did not own both a car and a stereo?

Answers   (A)   10 [the number in $S' \cap C$]   (B)   65 [the number in $(S \cap C)'$]

Note in Example 4 and Problem 4 that the word "and" is associated with "intersection" and the word "or" is associated with "union."

EXERCISE A-1

A   *Indicate true (T) or false (F).*

1. $4 \in \{2, 3, 4\}$ 2. $6 \notin \{2, 3, 4\}$
3. $\{2, 3\} \subset \{2, 3, 4\}$ 4. $\{3, 2, 4\} = \{2, 3, 4\}$
5. $\{3, 2, 4\} \subset \{2, 3, 4\}$ 6. $\{3, 2, 4\} \in \{2, 3, 4\}$
7. $\varnothing \subset \{2, 3, 4\}$ 8. $\varnothing = \{0\}$

In Problems 9–14 write the resulting set using the listing method.

9. $\{1, 3, 5\} \cup \{2, 3, 4\}$ 10. $\{3, 4, 6, 7\} \cup \{3, 4, 5\}$
11. $\{1, 3, 4\} \cap \{2, 3, 4\}$ 12. $\{3, 4, 6, 7\} \cap \{3, 4, 5\}$
13. $\{1, 5, 9\} \cap \{3, 4, 6, 8\}$ 14. $\{6, 8, 9, 11\} \cap \{3, 4, 5, 7\}$

B   *In Problems 15–20 write the resulting set using the listing method.*

15. $\{x \mid x - 2 = 0\}$ 16. $\{x \mid x + 7 = 0\}$
17. $\{x \mid x^2 = 49\}$ 18. $\{x \mid x^2 = 100\}$
19. $\{x \mid x$ is an odd number between 1 and 9, inclusive$\}$
20. $\{x \mid x$ is a month starting with $M\}$
21.   For $U = \{1, 2, 3, 4, 5\}$ and $A = \{2, 3, 4\}$, find A'.
22.   For $U = \{7, 8, 9, 10, 11\}$ and $A = \{7, 11\}$, find A'.

Problems 23–34 refer to the Venn diagram:

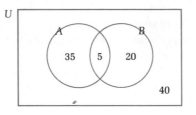

How many elements are in each of the indicated sets?

23. A 24. U 25. A' 26. B'
27. $A \cup B$ 28. $A \cap B$ 29. $A' \cap B$ 30. $A \cap B'$
31. $(A \cap B)'$ 32. $(A \cup B)'$ 33. $A' \cap B'$ 34. U'

35. If $R = \{1, 2, 3, 4\}$ and $T = \{2, 4, 6\}$, find:
    (A) $\{x \mid x \in R \ \text{or} \ x \in T\}$ (B) $R \cup T$
36. If $R = \{1, 3, 4\}$ and $T = \{2, 4, 6\}$, find:
    (A) $\{x \mid x \in R \ \text{and} \ x \in T\}$ (B) $R \cap T$
37. For $P = \{1, 2, 3, 4\}, Q = \{2, 4, 6\}$, and $R = \{3, 4, 5, 6\}$, find $P \cup (Q \cap R)$.
38. For P, Q, and R in Problem 37, find $P \cap (Q \cup R)$.

C   *Venn diagrams may be of help in Problems 39–44.*

39. If $A \cup B = B$, can we always conclude that $A \subset B$?
40. If $A \cap B = B$, can we always conclude that $B \subset A$?
41. If A and B are arbitrary sets, can we always conclude that $A \cap B \subseteq B$?
42. If $A \cap B = \varnothing$, can we always conclude that $B = \varnothing$?
43. If $A \subset B$ and $x \in A$, can we always conclude that $x \in B$?
44. If $A \subset B$ and $x \in B$, can we always conclude that $x \in A$?

45. How many subsets does each of the following sets have? Also, try to discover a formula in terms of n for a set with n elements.
    (A) $\{a\}$ (B) $\{a, b\}$ (C) $\{a, b, c\}$
46. How do the sets $\varnothing, \{\varnothing\}$, and $\{0\}$ differ from each other?

APPLICATIONS

BUSINESS & ECONOMICS

Problems 47–58 refer to the following survey: From a survey involving 1,000 business administration majors, it was found that 600 had taken finite math, 500 had taken calculus, and 300 had taken both. Let

$F = $ Set of students in sample who had taken finite math
$C = $ Set of students in sample who had taken calculus

Following the procedures in Example 4, find the number of students in each set described below.

47. $F \cup C$ **48.** $F \cap C$ **49.** $(F \cup C)'$

50. $(F \cap C)'$ **51.** $F' \cap C$ **52.** $F \cap C'$

53. Set of students who had taken either finite math or calculus

54. Set of students who had taken both finite math and calculus

55. Set of students who had not taken either finite math or calculus

56. Set of students who had not taken both courses

57. Set of students who had taken calculus, but not finite math

58. Set of students who had taken finite math, but not calculus

59. The management of a company, a president and three vice presidents denoted by the set $\{P, V_1, V_2, V_3\}$, wish to select a committee of two people from among themselves. How many ways can this committee be formed; that is, how many two-person subsets can be formed from a set of four people?

60. The management of the company in Problem 59 decide for or against certain measures as follows: The president has two votes and each vice president has one vote. Three favorable votes are needed to pass a measure. List all minimal winning coalitions; that is, list all subsets of $\{P, V_1, V_2, V_3\}$ that represent exactly three votes.

LIFE SCIENCES

Blood types. When receiving a blood transfusion, a recipient must have all the antigens of the donor. A person may have one or more of the three antigens A, B, and Rh, or none at all. Eight blood types are possible, as indicated in the Venn diagram, where U is the set of all people under consideration:

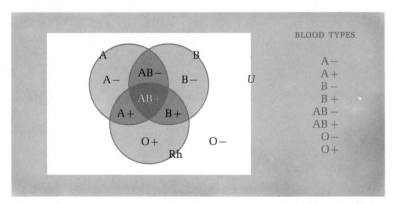

An A− person has A antigens but no B or Rh; an O+ person has Rh but neither A nor B; an AB− person has A and B antigens but no Rh; and so on.

Using the Venn diagram, indicate which of the eight blood types are included in each set.

61. A ∩ Rh   **62.** A ∩ B   **63.** A ∪ Rh

64. A ∪ B   **65.** (A ∪ B)′   **66.** (A ∪ B ∪ Rh)′

67. A′ ∩ B   **68.** Rh′ ∩ A

SOCIAL SCIENCES   *Group structures.* R. D. Luce and A. D. Perry, in a study on group structure (*Psychometrika*, 1949, 14: 95–116), used the idea of sets to formally define the notion of "clique" within a group. Let G be the set of all persons in the group and let $C \subset G$. Then C is a clique provided that:

1. C contains at least three elements
2. For every $a, b \in C$, a R b and b R a
3. For every $a \notin C$, there is at least one $b \in C$ such that a Ꞧ b or b Ꞧ a or both

[*Note:* Interpret "a R b" to mean "a relates to b," "a likes b," "a is as wealthy as b," and so on. Of course, "a Ꞧ b" would mean "a does not like b," and so on.]

69. Translate statement 2 into ordinary English.

70. Translate statement 3 into ordinary English.

A-2   THE REAL NUMBER SYSTEM

Before we start reviewing algebra it is appropriate to review briefly the structure of the **real number** system, since it is actually the algebra of real numbers that concerns us. Table 1 and Figure 5 break down the real number system into its important subsets and show how these sets are related to each other. Note that the set of natural numbers is a subset of the integers; the set of integers is a subset of the rational numbers; and the

TABLE 1   The set of real numbers

SYMBOL	NUMBER SET	DESCRIPTION	EXAMPLES
N	Natural numbers	Counting numbers (also called positive integers)	1, 2, 3, …
I	Integers	Set of natural numbers, their negatives, and 0	…, −2, −1, 0, 1, 2, …
Q	Rationals	Any number that can be represented as the quotient of two integers a/b, where $b \neq 0$	$-4; -\frac{3}{5}; 0; 7; \frac{2}{3}; 1.62$
R	Reals	Set of all rational and irrational numbers	$-\sqrt{7}; -4; -\frac{2}{3}; 0; 5; \frac{3}{4};$ $57.35; \sqrt{13}; \pi$

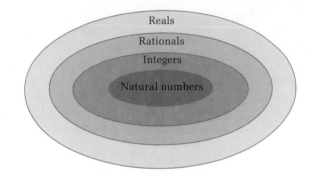

FIGURE 5 The set of real numbers and its important subsets.

set of rational numbers is a subset of the real numbers. An irrational number is any real number that cannot be expressed as the quotient of two integers. The properties of the real numbers provide us with the basic manipulative rules that are used in algebra.

A one-to-one correspondence exists between the set of real numbers and points on a line; that is, each real number corresponds to exactly one point, and each point to exactly one real number. A line with a real number associated with each point, as in Figure 6, is called a **real number line** or simply a real line.

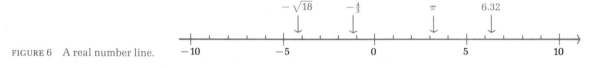

FIGURE 6 A real number line.

A-3   EXPONENTS AND RADICALS

Exponent and radical forms are encountered with great frequency wherever algebra is used. Manipulative skills can be sharpened by reviewing the basic definitions, properties, and exercises that follow.

INTEGER EXPONENTS

Table 2 lists definitions for integer exponents and the five basic properties associated with these definitions.

Example 5

Simplify and express the answers using positive exponents only.*

(A) $(2x^3)(3x^5) \boxed{= 2 \cdot 3x^{3+5}} = 6x^8$ (B) $x^5 x^{-9} = x^{-4} = \dfrac{1}{x^4}$

*The dashed boxes indicate steps that are usually done mentally.

TABLE 2

DEFINITION OF a^n	PROPERTIES OF EXPONENTS
Given: n is an integer a is a real number	Given: n and m are integers a and b are real numbers

DEFINITION OF a^n

1. For n a positive integer
$$a^n = a \cdot a \cdot \cdots \cdot a, \quad n \text{ factors of } a$$
Example: $5^4 = 5 \cdot 5 \cdot 5 \cdot 5$

2. For $n = 0$
$$a^0 = 1, \quad a \neq 0$$
0^0 is not defined
Example: $12^0 = 1$

3. For n a negative integer
$$a^n = \frac{1}{a^{-n}}, \quad a \neq 0$$

Example: $5^{-2} = \frac{1}{5^2}$

Note: It follows from the above that
$$a^{-n} = \frac{1}{a^n}, \quad a \neq 0$$
for all integers n

PROPERTIES OF EXPONENTS

1. $a^m a^n = a^{m+n}$

2. $(a^n)^m = a^{mn}$

3. $(ab)^m = a^m b^m$

4. $\left(\dfrac{a}{b}\right)^m = \dfrac{a^m}{b^m}, \quad b \neq 0$

5. $\dfrac{a^m}{a^n} = a^{m-n} = \dfrac{1}{a^{n-m}}, \quad a \neq 0$

(C) $\dfrac{x^5}{x^7} \; \boxed{= x^{5-7}} = x^{-2} = \dfrac{1}{x^2}$

or $\dfrac{x^5}{x^7} \; \boxed{= \dfrac{1}{x^{7-5}}} = \dfrac{1}{x^2}$

(D) $\dfrac{x^{-3}}{y^{-4}} = \dfrac{y^4}{x^3}$

(E) $(u^{-3}v^2)^{-2} \; \boxed{= (u^{-3})^{-2}(v^2)^{-2}} = u^6 v^{-4} = \dfrac{u^6}{v^4}$

(F) $\left(\dfrac{y^{-5}}{y^{-2}}\right)^{-2} \; \boxed{= \left(\dfrac{y^{-5}}{y^{-2}}\right)^{-2}} = \dfrac{y^{10}}{y^4} = y^6$

(G) $\dfrac{4m^{-3}n^{-5}}{6m^{-4}n^3} \; \boxed{= \dfrac{2m^{-3-(-4)}}{3n^{3-(-5)}}} = \dfrac{2m}{3n^8}$

(H) $\left(\dfrac{x^{-3}x^3}{n^{-2}}\right)^{-3} = \left(\dfrac{x^0}{n^{-2}}\right)^{-3} = \left(\dfrac{1}{n^{-2}}\right)^{-3} = \dfrac{1}{n^6}$

Problem 5 Simplify and express the answers using positive exponents only.

(A) $(3y^4)(2y^3)$ (B) $m^2 m^{-6}$ (C) $(u^3 v^{-2})^{-2}$

(D) $\left(\dfrac{y^{-6}}{y^{-2}}\right)^{-1}$ (E) $\dfrac{8x^{-2}y^{-4}}{6x^{-5}y^2}$ (F) $\left(\dfrac{m^{-3}}{x^2 x^{-2}}\right)^{-2}$

Answers   (A) $6y^7$ (B) $1/m^4$ (C) v^4/u^6

(D) y^4 (E) $4x^3/3y^6$ (F) m^6

SCIENTIFIC NOTATION

Writing and working with very large or very small numbers in standard decimal notation is often awkward, even with electronic hand calculators. It is often convenient to represent numbers of this type in **scientific notation;** that is, as the product of a number between 1 and 10 and a power of 10.

Example 6

DECIMAL FRACTIONS AND SCIENTIFIC NOTATION

$$7 = 7 \times 10^0 \qquad\qquad 0.5 = 5 \times 10^{-1}$$
$$67 = 6.7 \times 10 \qquad\qquad 0.45 = 4.5 \times 10^{-1}$$
$$580 = 5.8 \times 10^2 \qquad\qquad 0.0032 = 3.2 \times 10^{-3}$$
$$43,000 = 4.3 \times 10^4 \qquad\qquad 0.000045 = 4.5 \times 10^{-5}$$
$$73,400,000 = 7.34 \times 10^7 \qquad 0.000000391 = 3.91 \times 10^{-7}$$

Note that the power of 10 used corresponds to the number of places we move the decimal to form a number between 1 and 10. The power is positive if the decimal is moved to the left and negative if it is moved to the right. Positive exponents are associated with numbers greater than or equal to 10; negative exponents are associated with numbers less than 1.

Problem 6

Write each number in scientific notation.
(A) 370   (B) 47,300,000,000   (C) 0.047   (D) 0.000000089

Answers

(A) 3.7×10^2 (B) 4.73×10^{10}
(C) 4.7×10^{-2} (D) 8.9×10^{-8}

RATIONAL EXPONENTS AND RADICALS

Recall that

r	is a **square root** of b if	$r^2 = b$
r	is a **cube root** of b if	$r^3 = b$

and, in general,

r	is an **nth root** of b if	$r^n = b$

1. If b is positive and n is even, then there are two real number nth roots of b. One root is the negative of the other. For example, -2 and 2 are both square roots of 4 since $(-2)^2 = 4$ and $2^2 = 4$.
2. If b is negative and n is even, then there are no real number nth roots of b, since no real number raised to an even power can be negative. For example, -1 has no real number square roots, since no real number squared can be -1.
3. If b is any real number and n is odd, then there is exactly one real number nth root of b. This root is negative if b is negative and positive if b is positive. For example, -3 is the real cube root of -27 since $(-3)^3 = -27$.

How do we symbolize these nth roots? For n a natural number greater than 1 we use

$$b^{1/n} \quad \text{or} \quad \sqrt[n]{b}$$

to represent one of the **real nth roots of b.** Which one? The symbols represent the real nth root of b if n is odd (see case 3 above) and the positive real nth root of b if b is positive and n is even (see case 1 above). Note that we write $\sqrt{b}$ to indicate $\sqrt[2]{b}$.

Example 7

(A) $4^{1/2} = \sqrt{4} = 2 \quad (\sqrt{4} \neq \pm 2)$ (B) $-4^{1/2} = -\sqrt{4} = -2$

(C) $(-4)^{1/2}$ and $\sqrt{-4}$ are not real numbers

(D) $8^{1/3} = \sqrt[3]{8} = 2$ (E) $(-8)^{1/3} = \sqrt[3]{-8} = -2$

Problem 7

Evaluate each of the following:

(A) $16^{1/2}$ (B) $-\sqrt{16}$ (C) $\sqrt[3]{-27}$

(D) $(-9)^{1/2}$ (E) $(\sqrt[4]{81})^3$

Answers

(A) 4   (B) -4 (C) -3

(D) Not a real number   (E) 27

For m and n natural numbers without common prime factors, b a real number, and b nonnegative when n is even, we define $b^{m/n}$ as follows (both definitions are equivalent under the indicated restrictions):

$$b^{m/n} = \begin{cases} (b^{1/n})^m = (\sqrt[n]{b})^m \\ (b^m)^{1/n} = (\sqrt[n]{b^m}) \end{cases}$$

Thus,

$$8^{2/3} = (8^{1/3})^2 = 2^2 = 4 \quad \text{or} \quad 8^{2/3} = (8^2)^{1/3} = 64^{1/3} = 4 .$$

We complete the definition of rational exponents with

$$b^{-m/n} = \frac{1}{b^{m/n}} \quad b \neq 0$$

All the properties of exponents listed in Table 2 also hold for the rational exponent forms just defined, provided b is not negative when n is even.

Example 8   (A) $(3x^{1/3})(2x^{1/2}) = 6x^{1/3 + 1/2} = 6x^{5/6} = 6\sqrt[6]{x^5}$

(B) $(-8)^{5/3} = [(-8)^{1/3}]^5 = (-2)^5 = -32$

(C) $(2x^{1/3}y^{-2/3})^3 = 8xy^{-2}$ or $\dfrac{8x}{y^2}$

(D) $\left(\dfrac{4x^{1/3}}{x^{1/2}}\right)^{1/2} = \dfrac{4^{1/2}x^{1/6}}{x^{1/4}} = \dfrac{2}{x^{1/4-1/6}} = \dfrac{2}{x^{1/12}} = \dfrac{2}{\sqrt[12]{x}}$

Problem 8   Simplify each and express answers using positive exponents only. If rational exponents appear in final answers, convert to radical forms.

(A) $9^{3/2}$ (B) $(-27)^{4/3}$ (C) $(5y^{1/4})(2y^{1/3})$

(D) $(2x^{-3/4}y^{1/4})^4$ (E) $\left(\dfrac{8x^{1/2}}{x^{2/3}}\right)^{1/3}$

Answers   (A) 27   (B) 81   (C) $10y^{7/12} = 10\sqrt[12]{y^7}$
(D) $16y/x^3$ (E) $2/x^{1/18} = 2/\sqrt[18]{x}$

RADICAL PROPERTIES   We conclude this rather rapid review of exponents and radicals with three useful properties of radicals that follow directly from the exponent properties. If n is a natural number greater than 1, and x and y are positive real numbers, then

1. $\sqrt[n]{x^n} = (x^n)^{1/n} = x^{n/n} = x$

2. $\sqrt[n]{x}\,\sqrt[n]{y} = x^{1/n}\,y^{1/n} = (xy)^{1/n} = \sqrt[n]{xy}$

3. $\dfrac{\sqrt[n]{x}}{\sqrt[n]{y}} = \dfrac{x^{1/n}}{y^{1/n}} = \left(\dfrac{x}{y}\right)^{1/n} = \sqrt[n]{\dfrac{x}{y}}$

We will use these special relationships to simplify several radical forms.

Example 9   All variables are restricted to positive real numbers.

(A) $\sqrt[3]{x^3y^6} = \sqrt[3]{(xy^2)^3} = xy^2$
   or $\sqrt[3]{x^3y^6} = (x^3y^6)^{1/3} = x^{3/3}y^{6/3} = xy^2$

(B) $\sqrt{8x^3y^8} = \sqrt{(4x^2y^8)(2x)} = \sqrt{4x^2y^8}\,\sqrt{2x} = 2xy^4\sqrt{2x}$
   or $\sqrt{8x^3y^8} = (8x^3y^8)^{1/2} = [(4x^2y^8)(2x)]^{1/2}$
   $= (2^2x^2y^8)^{1/2}(2x)^{1/2} = 2xy^4\sqrt{2x}$

(C) $\sqrt[3]{-8x^5y^7} = \sqrt[3]{(-8x^3y^6)(x^2y)} = \sqrt[3]{-8x^3y^6}\,\sqrt[3]{x^2y} = -2xy^2\sqrt[3]{x^2y}$
   or $\sqrt[3]{-8x^5y^7} = (-8x^5y^7)^{1/3} = [(-2)^3x^3y^6)(x^2y)]^{1/3}$
   $= [(-2)^3x^3y^6]^{1/3}(x^2y)^{1/3} = -2xy^2\sqrt[3]{x^2y}$

(D) $\dfrac{6x^2}{\sqrt{3x}} = \dfrac{6x^2}{\sqrt{3x}} \cdot \dfrac{\sqrt{3x}}{\sqrt{3x}} = \dfrac{6x^2\sqrt{3x}}{\sqrt{9x^2}} = \dfrac{6x^2\sqrt{3x}}{3x} = 2x\sqrt{3x}$

(E) $6\sqrt{\dfrac{x}{3}} = 6\sqrt{\dfrac{x}{3} \cdot \dfrac{3}{3}} = 6\sqrt{\dfrac{3x}{9}} = 6\dfrac{\sqrt{3x}}{\sqrt{9}} = 6\dfrac{\sqrt{3x}}{3} = 2\sqrt{3x}$

Note that in the last two parts of Example 9, we **rationalized denominators;** that is, we performed operations to remove radicals from the denominators. This is a useful operation in some problems.

Problem 9   Simplify as in Example 9.

    (A) $\sqrt{12x^5y^6}$ (B) $\sqrt[3]{-27x^7y^5}$ (C) $\dfrac{8y^3}{\sqrt{2y}}$ (D) $4\sqrt{\dfrac{y}{2}}$

Answers   (A) $2x^2y^3\sqrt{3x}$ (B) $-3x^2y\sqrt[3]{xy^2}$ (C) $4y^2\sqrt{2y}$ (D) $2\sqrt{2y}$

EXERCISE A-3

Simplify each expression and write answers using positive exponents only.

A   1. $(2x^3)(3x^7)$ 2. $\dfrac{4u^3}{2u^7}$ 3. $\dfrac{2x^3y^8}{6x^7y^2}$

    4. $\dfrac{(2xy^3)^2}{(4x^2y)^3}$ 5. $(2 \times 10^{-3})(4 \times 10^8)$

    6. $(3 \times 10^{-4})(2 \times 10^{12})$ 7. $\dfrac{y^{-3}}{y^5}$

    8. $\dfrac{x^8}{x^{-4}}$ 9. $(u^{-5}v^3)^{-2}$ 10. $(x^3y^{-2})^{-3}$

    11. $\dfrac{6 \times 10^{12}}{2 \times 10^7}$ 12. $\dfrac{9 \times 10^8}{3 \times 10^9}$ 13. $25^{3/2}$

    14. $(-8)^{2/3}$ 15. $(\sqrt{25})^3$ 16. $(\sqrt[3]{-8})^2$

    17. $(x^4y^2)^{1/2}$ 18. $\left(\dfrac{x^9}{y^{12}}\right)^{1/3}$

B   19. $\dfrac{2a^6b^{-2}}{16a^{-3}b^2}$ 20. $\dfrac{8x^{-3}y^{-1}}{6x^2y^{-4}}$ 21. $(3x^0y^{-2})^{-2}$

    22. $(2x^{-3}y^0)^{-3}$ 23. $\left(\dfrac{6mn^{-2}}{3m^{-1}n^2}\right)^{-3}$ 24. $\left(\dfrac{2x^{-3}y^2}{4xy^{-1}}\right)^{-2}$

    25. $x^{4/5}x^{-2/5}$ 26. $\dfrac{m^{2/3}}{m^{-1/3}}$ 27. $x^{1/4}x^{-3/4}$

    28. $y^{-2/3}y^{1/3}$ 29. $\left(\dfrac{8x^{-4}y^3}{27x^2y^{-3}}\right)^{1/3}$ 30. $\left(\dfrac{25x^5y^{-1}}{16x^{-3}y^{-5}}\right)^{1/2}$

Simplify as in Example 9.

31. $\sqrt{8m^3n^4}$

32. $\sqrt[3]{8u^6v^8}$

33. $\dfrac{6x^2}{\sqrt{2x}}$

34. $\dfrac{12m^3}{\sqrt{4m}}$

35. $8\sqrt{\dfrac{x}{2}}$

36. $9\sqrt{\dfrac{y}{3}}$

Convert each numeral to scientific notation and simplify. Express the answer in scientific notation and as a decimal fraction.

37. $\dfrac{(900,000)(0.00002)}{0.0006}$

38. $\dfrac{(60,000)(0.000003)}{(0.0004)(1,500,000)}$

39. $\dfrac{(1,250,000)(0.00038)}{0.0423}$

40. $\dfrac{(0.00000082)(230,000)}{(430,000)(0.0082)}$

C   *Simplify each expression and write answers using positive exponents only.*

41. $\dfrac{x^{-1}+y^{-1}}{x+y}$.

42. $(x^{-1}+y^{-1})^{-1}$

43. $\left(\dfrac{9x^{1/3}x^{1/2}}{x^{-1/6}}\right)^{1/2}$

44. $\left(\dfrac{8y^{1/3}y^{-1/4}}{y^{-1/12}}\right)^2$

45. $\left(\dfrac{x^m}{x^{m-2}}\right)^{1/2}$,   m positive

46. $\left(\dfrac{y^{m+2}}{y^m}\right)^{1/2}$,   m positive

A-4   POLYNOMIALS—BASIC OPERATIONS

The rules behind manipulating algebraic expressions have their basis in the operational properties of real numbers. We list a few of these properties for ready reference. Let R be the set of real numbers and let x, y, z be arbitrary elements of R; then

COMMUTATIVE PROPERTIES
$$x + y = y + x$$
$$xy = yx$$

DISTRIBUTIVE PROPERTIES
$$x(y + z) = xy + xz$$
$$(y + z)x = yx + zx$$

ASSOCIATIVE PROPERTIES
$$(x + y) + z = x + (y + z)$$
$$(xy)z = x(yz)$$

SUBTRACTION RULE
$$x - y = x + (-y)$$

These properties are either used or assumed almost any time you work with algebraic expressions.

POLYNOMIALS

A **polynomial in one variable** is an algebraic expression of the form

$$a_n x^n + a_{n-1} x^{n-1} + \cdots + a_1 x + a_0$$

where n is a nonnegative integer and the **coefficients** $a_0, a_1, \ldots, a_n$ are real numbers. If $a_n \neq 0$, then the **degree of the polynomial** is n, the highest power of x present without a zero coefficient.

There are **polynomials with more than one variable.** These are algebraic expressions having terms such as $3x^2 y^3 z$—the exponents are nonnegative integers and the coefficients are real numbers. The sum of the powers of the variables in a term is the **degree** of the term. A nonzero constant has degree zero.

In general, the degree of a polynomial is the degree of the term of the highest degree without a zero coefficient.

POLYNOMIALS

$x^3 - 3x^2 + 5x - 1$	$3x^2 - 2xy + y^2$	x	7
Degree 3	Degree 2	Degree 1	Degree 0

NONPOLYNOMIALS

$$\frac{1}{x} + x \qquad 1 + \sqrt{x} \qquad 5x^{-3} + 2x^2 - 5 \qquad \sqrt{x^2 - 2x + 1} \qquad \frac{2x - 1}{3x + 1}$$

**ADDITION,
SUBTRACTION,
AND SIMPLIFICATION**

Addition and subtraction of polynomials can often be thought of in terms of removing parentheses, or other symbols of grouping, and then combining like terms. The process of removing parentheses is based primarily on three of the operational properties of the real numbers: (1) Any expression involving subtraction can be converted to addition; in other words, $x - y = x + (-y)$. (2) Multiplication distributes over addition; that is, $x(y + z) = xy + xz$. (3) Parentheses may be inserted or removed at will relative to addition; that is, $(x + y) + z = x + (y + z)$. Out of these properties evolves a *mechanical rule:* To remove a pair of parentheses (or other grouping symbols), each term within the pair is multiplied by the coefficient of the parentheses.

Example 10

(A) $2(x^2 - 3) - (x^2 + 2x - 3) \; \boxed{= 2(x^2 - 3) - 1(x^2 + 2x - 3)}$
$$= 2x^2 - 6 - x^2 - 2x + 3$$
$$= x^2 - 2x - 3$$

(B) $(2x - 3y) - [x - 2(3x - y)] = (2x - 3y) - [x - 6x + 2y]$
$$= 2x - 3y - x + 6x - 2y = 7x - 5y$$

(C) $3x - \{5 - 3[x - x(3 - x)]\} = 3x - \{5 - 3[x - 3x + x^2]\}$
$$= 3x - \{5 - 3x + 9x - 3x^2\}$$
$$= 3x - 5 + 3x - 9x + 3x^2$$
$$= 3x^2 - 3x - 5$$

Problem 10   Remove symbols of grouping and simplify.
(A) $3(x^2 + 2x) - (x^2 - 2x + 1)$
(B) $(4x + 2y) - [3x - 5(x - 3y)]$
(C) $2m - \{7 - 2[m - m(4 + m)]\}$

Answers   (A) $2x^2 + 8x - 1$ (B) $6x - 13y$ (C) $-2m^2 - 4m - 7$

MULTIPLICATION   The distributive property is an important tool in multiplying polynomials. It leads to the following simple *mechanical procedure:* To multiply
two polynomials, multiply each term of one by each term of the other and
combine like terms. The commutative and associative properties are also
used in simplifying the result.

Example 11   Multiply $(2x - 3)(2x^2 + 3x - 2)$.

Solution
$$(2x - 3)(2x^2 + 3x - 2)$$
$$= (2x - 3)2x^2 + (2x - 3)3x + (2x - 3)(-2)$$
$$= 4x^3 - 6x^2 + 6x^2 - 9x - 4x + 6$$
$$= 4x^3 - 13x + 6$$

or
$$
\begin{array}{r}
2x^2 + 3x - 2 \\
2x - 3 \\
\hline
4x^3 + 6x^2 - 4x \\
-6x^2 - 9x + 6 \\
\hline
4x^3 \qquad -13x + 6
\end{array}
$$

Problem 11   Multiply $(2y - 3)(3y^2 - 2y + 3)$.

Answer $6y^3 - 13y^2 + 12y - 9$

Example 12   Multiply $(2x - 3y)(3x + 2y)$ mentally.

Solution

$$(2x - 3y)(3x + 2y) = 6x^2 - 5xy - 6y^2$$

Perform the operations as indicated in the diagram. The two products in
step 2 are like terms and can be combined mentally.

Problem 12   Multiply mentally.
(A) $(x - 2)(x - 3)$ (B) $(2a + 3b)(2a - 3b)$
(C) $(3m - 5n)(4m + 3n)$

Answers   (A) $x^2 - 5x + 6$ (B) $4a^2 - 9b^2$ (C) $12m^2 - 11mn - 15n^2$

FACTORING

Just as the distributive property allows us to distribute multiplication over an algebraic sum,

$$a(b + c + d + e) = ab + ac + ad + ae$$

it also allows us to take out a factor common to all terms and write the sum as a product:

$$ab + ac + ad + ae = a(b + c + d + e)$$

The first process is referred to as **multiplication** (or **expansion**) and the second process is called **factoring.**

Example 13

Factor out factors common to all terms.
(A) $6x^2 + 2xy - 4x = 2x(3x + y - 2)$
(B) $x^2 - x = x(x - 1)$
(C) $2x^3y - 8x^2y^2 - 6xy^3 = 2xy(x^2 - 4xy - 3y^2)$
(D) $3x(2x - y) + 5y(2x - y) = (2x - y)(3x + 5y)$

Problem 13

Factor out factors common to all terms.
(A) $8m^2 - 4mn + 6m$ (B) $3m^2 + m$
(C) $3u^3v - 6u^2v^2 - 3uv^3$ (D) $2x(3x - 2) - 7(3x - 2)$

Answers

(A) $2m(4m - 2n + 3)$ (B) $m(3m + 1)$ (C) $3uv(u^2 - 2uv - v^2)$
(D) $(3x - 2)(2x - 7)$

You should now be able to perform the following multiplications without too much effort:

$$(x + 5)(x - 3) \quad \text{and} \quad (2x - y)(x + 3y)$$

But can you reverse the process? Can you find integers $a, b, c,$ and d so that

$$2x^2 - 5x - 3 = (ax + b)(cx + d)$$

Being able to carry out this type of factoring—when it is possible—is very useful. Though it is not as easy as multiplying, there is a systematic approach to the problem that can be readily mastered.

We are interested in factoring second-degree polynomials of the types

$$ax^2 + bx + c \quad \text{and} \quad ax^2 + bxy + cy^2 \qquad (1)$$

with integer coefficients, into products of first-degree factors with integer coefficients. If the product ac has integer factors that add up to b, that is, if

$$ac = pq \quad \text{and} \quad p + q = b \qquad (2)$$

then it can be shown that equations (1) can be factored into first-degree factors with integer coefficients. If no integers p and q exist so that conditions (2) are satisfied, then equations (1) will not have first-degree factors with integer coefficients. Once we find p and q—if they exist—our work is almost finished. We can then write equations (1) in the form

$$ax^2 + px + qx + c \quad \text{and} \quad ax^2 + pxy + qxy + cy^2 \qquad (3)$$

and the factoring can be completed in a couple of easy steps. An example will help clarify the process.

Consider the polynomial

$$x^2 - 4x - 12$$

Comparing it with the standard form in (1) we see that

$$a = 1 \quad b = -4 \quad c = -12$$

Now we seek factors of

$$ac = (1)(-12) = -12$$

that add up to $b = -4$. After a little trial and error we find that if $p = -6$ and $q = 2$, then

$$(-6)(2) = -12 \qquad pq = ac$$

and

$$(-6) + (2) = -4 \qquad p + q = b$$

Thus, $x^2 - 4x - 12$ has first-degree factors with integer coefficients. Now we write $x^2 - 4x - 12$ in the form (3) and factor out common factors from the first two terms and the last two terms:

$$
\begin{aligned}
x^2 - 4x - 12 &= x^2 - 6x + 2x - 12 \\
&= x(x - 6) + 2(x - 6) \qquad \text{Note that } (x - 6) \text{ is} \\
&= (x - 6)(x + 2) \qquad\qquad \text{now a common factor} \\
& \text{and can be taken out}
\end{aligned}
$$

Note that it does not matter whether we let $p = -6$ and $q = 2$ or vice versa. If we let $p = 2$ and $q = -6$, we obtain the same result:

$$
\begin{aligned}
x^2 - 4x - 12 &= x^2 + 2x - 6x - 12 \\
&= x(x + 2) - 6(x + 2) \\
&= (x - 6)(x + 2)
\end{aligned}
$$

Example 14 Factor, if possible, using integer coefficients.
(A) $x^2 - 10xy - 24y^2$ (B) $x^2 - 3x + 4$
(C) $6x^2 + 5xy - 4y^2$

Solutions (A) $x^2 - 10xy - 24y^2$

$$ac = (1)(-24) = -24$$
$$pq = (-12)(2) = -24$$

and

$$p + q = (-12) + (2) = -10 = b$$

Thus,

$$x^2 - 10xy - 24y^2 = x^2 - 12xy + 2xy - 24y^2$$
$$= x(x - 12y) + 2y(x - 12y)$$
$$= (x - 12y)(x + 2y)$$

(B) $x^2 - 3x + 4$

$$ac = (1)(4) = 4$$

Possible integer factors of 4 are

p	q
1	4
-1	-4
2	2
-2	2

None of these pairs add up to $b = -3$. Therefore, $x^2 - 3x + 4$ cannot be factored using integer coefficients.

(C) $6x^2 + 5xy - 4y^2$

$$ac = (6)(-4) = -24$$
$$pq = (-3)(8) = -24$$

and

$$p + q = (-3) + 8 = 5 = b$$

Therefore,

$$6x^2 + 5xy - 4y^2 = 6x^2 - 3xy + 8xy - 4y^2$$
$$= 3x(2x - y) + 4y(2x - y)$$
$$= (3x + 4y)(2x - y)$$

Problem 14 Factor, if possible, using integer coefficients.
(A) $x^2 + 3x - 18$ (B) $x^2 - 2xy + 12y^2$ (C) $8x^2 + 6xy - 9y^2$

Answers (A) $(x + 6)(x - 3)$ (B) Not factorable   (C) $(4x - 3y)(2x + 3y)$

If we multiply $(A - B)$ and $(A + B)$, we obtain

$$(A + B)(A - B) = A^2 - B^2$$

a difference of two squares. Writing the result from right to left, we obtain the very useful factoring formula

$$A^2 - B^2 = (A + B)(A - B)$$

This finds wider use than you might expect and it should be memorized. The sum of two squares,

$$A^2 + B^2$$

does not factor using integer coefficients. Try it to see why.

Example 15   Factor, if possible, using integer coefficients.
(A) $x^2 - y^2 = (x + y)(x - y)$
(B) $9m^2 - 25n^2 = (3m)^2 - (5n)^2 = (3m + 5n)(3m - 5n)$
(C) $4u^2 + v^2$ does not factor with integer coefficients

Problem 15   Factor, if possible, using integer coefficients.
(A) $R^2 - T^2$ (B) $x^2 + 4y^2$ (C) $16A^2 - 9B^2$

Answers   (A) $(R - T)(R + T)$ (B)   Not factorable
(C) $(4A + 3B)(4A - 3B)$

Now let us look at a couple of problems that combine some of the techniques discussed above. In general, in starting a factoring process we first see if we can factor out a factor common to all terms, and then we proceed as in Example 14 or 15.

Example 16   Factor as far as possible using integers as coefficients.
(A) $4x^3 - 14x^2 + 6x = 2x(2x^2 - 7x + 3)$
Now we see if we can factor $2x^2 - 7x + 3$ further:

$$ac = (2)(3) = 6$$
$$pq = (-1)(-6) = 6$$
$$p + q = (-1) + (-6) = -7 = b$$

Therefore,

$$2x^2 - 7x + 3 = 2x^2 - x - 6x + 3$$
$$= x(2x - 1) - 3(2x - 1)$$
$$= (2x - 1)(x - 3)$$

and

$$4x^3 - 14x^2 + 6x = 2x(2x - 1)(x - 3)$$

(B) $18x^3 - 8x = 2x(9x^2 - 4)$
$$= 2x(3x + 2)(3x - 2)$$

Problem 16   Factor as far as possible using integers as coefficients.
(A) $8x^3y + 20x^2y^2 - 12xy^3$ (B) $3x^3 - 48x$

Answers   (A) $4xy(2x - y)(x + 3y)$ (B) $3x(x - 4)(x + 4)$

EXERCISE A-4

A   *Perform the indicated operations and simplify.*

1. $2u^2(u^2 + 3u)$ 2. $3x^2(x^3 + 2x^2)$
3. $2(m + 3n) + 4(m - 2n)$ 4. $3(x - 2y) + 2(3x + y)$
5. $(x + 3y) - (2x - 5y)$ 6. $(2x - y) - (3x - 5y)$
7. $(m - 7)(m + 3)$ 8. $(x + 9)(x - 2)$
9. $(3x - 5)(2x + 1)$ 10. $(4x - 3)(x - 2)$
11. $(x - 3y)(x + 3y)$ 12. $(2x + 3)(2x - 3)$

Factor out factors common to all terms.

13. $8m + 16m^2$ 14. $5x^2 - x$
15. $12x^3 - 6x^2 - 3x$ 16. $6y^3 - 4y^2 + 2y$
17. $y(y - 4) + 2(y - 4)$ 18. $x(x + 4) - 2(x + 4)$
19. $x(x + 3y) + 5y(x + 3y)$ 20. $m(m - 4n) - 5n(m - 4n)$

Factor as far as possible using integers as coefficients.

21. $y^2 - 2y - 8$ 22. $x^2 + 2x - 8$
23. $x^2 + 8xy + 15y^2$ 24. $m^2 - 9mn + 20n^2$
25. $x^2 - 25$ 26. $u^2 - 100$

B   *Perform the indicated operations and simplify.*

27. $-2(y - 7) - 3(2y + 1) - (-5y + 7)$
28. $2(x - 1) - 3(2x - 3) - (4x - 5)$
29. $4m - 3[4 - 2(m - 1)]$ 30. $3y - 2[2y - (y - 7)]$
31. $(3x + 7y)(2x - 5y)$ 32. $(6x - 4y)(5x + 3y)$
33. $(2x - 5y)^2$ 34. $(3x + 4y)^2$
35. $(8x - 3y)(8x + 3y)$ 36. $(5u + 7v)(5u - 7v)$
37. $(x - 3y)(x^2 - 3xy + y^2)$ 38. $(m + 2n)(m^2 - 4mn - n^2)$

Factor as far as possible using integers as coefficients.

39. $2x^2 - 7xy + 6y^2$ 40. $3x^2 - 11xy + 6y^2$
41. $6x^2 + 7xy - 3y^2$ 42. $6x^2 - xy - 12y^2$
43. $3x^2 - 2x - 4$ 44. $4u^2 - 3u - 4$
45. $25x^2 - 4y^2$ 46. $u^2v^2 - x^2$

47. $2x^4 - 24x^3 + 40x^2$

48. $x^3 - 11x^2 + 24x$

49. $2x^3 + 8xy^2$

50. $12u^3 + 27u$

51. $4u^3v + 14u^2v^2 + 6uv^3$

52. $3m^3n - 15m^2n^2 + 18mn^3$

C   *Perform the indicated operations and simplify.*

53. $2x - 3\{x + 2[x - (x + 5)] + 1\}$

54. $u - \{u - [u - (u - 1)]\}$

55. $(2x^2 + 2x - 1)(3x^2 - 2x + 1)$

56. $(x^2 - 3x + 5)(2x^2 + x - 2)$

57. $(2x + 3)(x - 5) - (3x - 1)^2$

58. $(3x - 1)(x + 2) - (2x - 3)^2$

59. $2\{(x - 3)(x^2 - 2x + 1) - x[3 - x(x - 2)]\}$

60. $-3x\{x[x - x(2 - x)] - (x + 2)(x^2 - 3)\}$

A-5   ALGEBRAIC FRACTIONS

Quotients of polynomials are called **rational expressions.** For example,

$$\frac{1}{x} \qquad \frac{1}{y - 3} \qquad \frac{u + 7}{3u^2 - 5u + 6} \qquad \frac{x^2 - 3xy + y^2}{2x^3y^4}$$

are all rational expressions.

MULTIPLICATION AND DIVISION

Excluding division by zero, rational expressions become real numbers when we substitute real numbers for the variables. Hence, all properties of the real numbers apply to these expressions. In particular:

MULTIPLICATION AND DIVISION

If P, Q, R, and S represent polynomials, then

1. $\dfrac{P}{Q} \cdot \dfrac{R}{S} = \dfrac{PR}{QS}$

2. $\dfrac{P}{Q} \div \dfrac{R}{S} = \dfrac{P}{Q} \cdot \dfrac{S}{R} = \dfrac{PS}{QR}$

excluding any division by zero.

Example 17

(A) $\dfrac{y}{2x} \cdot \dfrac{y - 3}{3x^2} = \dfrac{y(y - 3)}{2x(3x^2)} = \dfrac{y^2 - 3y}{6x^3}$

(B) $\dfrac{2x}{3y} \div \dfrac{y - 2}{x + 3} = \dfrac{2x}{3y} \cdot \dfrac{x + 3}{y - 2} = \dfrac{2x(x + 3)}{3y(y - 2)} = \dfrac{2x^2 + 6x}{3y^2 - 6y}$

Problem 17   Perform the indicated operations.

(A) $\dfrac{2m}{3n} \cdot \dfrac{n+2}{m-4}$ (B) $\dfrac{2m}{3n} \div \dfrac{m-4}{n+2}$

Answers   (A) $\dfrac{2mn+4m}{3mn-12n}$ (B) $\dfrac{2mn+4m}{3mn-12n}$

Note that if P, Q, and K are polynomials, where Q and K are not zero, then

$$\frac{PK}{QK} = \frac{P}{Q} \cdot \frac{K}{K} = \frac{P}{Q} \cdot 1 = \frac{P}{Q}$$

Thus:

FUNDAMENTAL PRINCIPLE OF FRACTIONS
If P, Q, and K are polynomials,

$$\frac{PK}{QK} = \frac{P}{Q} \qquad Q \text{ and } K \text{ not } 0$$

In words, this principle states that we may multiply the numerator and denominator of a rational form by a nonzero polynomial, or divide the numerator and denominator by a nonzero polynomial. The principle is the basis of all canceling used to reduce fractional forms to lower terms (canceling common factors from numerator and denominator). And, worked from right to left, it is used to raise rational forms to higher terms (multiplying numerator and denominator by common factors). The latter operation is fundamental to the processes of addition and subtraction of rational forms.

Example 18   (A) $\dfrac{9x^2(x-3)}{6x(x+2)(x-3)} = \dfrac{3x}{2(x+2)}$ Lower terms

(B) $\dfrac{2y}{x(x+2)} = \dfrac{2y}{x(x+2)} \cdot \dfrac{2x(y-2)}{2x(y-2)} = \dfrac{4xy(y-2)}{2x^2(x+2)(y-2)}$ Higher terms

Problem 18   (A) Reduce to lowest terms: (B) Raise to higher terms:

$$\frac{8u(u+3)(u-2)}{12u^2(u-2)}$$

$$\frac{3x}{4(x-3)} = \frac{?}{8x(x-3)(x+2)}$$

Answers   (A) $\dfrac{2(u+3)}{3u}$ (B) $6x^2(x+2)$

Example 19

(A) $(x^2 - 4) \cdot \dfrac{2x - 3}{x + 2} = \dfrac{(\cancel{x + 2})^1 (x - 2)}{1} \cdot \dfrac{(2x - 3)}{(\cancel{x + 2})_1} = (x - 2)(2x - 3)$

(B) $(x + 4) \div \dfrac{2x^2 - 32}{6xy} = \dfrac{(x + 4)}{1} \cdot \dfrac{6xy}{2(x - 4)(x + 4)} = \dfrac{3xy}{x - 4}$

(C) $\dfrac{x^2 - 9y^2}{x^2 - 6xy + 9y^2} \div \dfrac{2x^2 + 6xy}{6x^2y} = \dfrac{(x - 3y)(x + 3y)}{(x - 3y)^2} \cdot \dfrac{6x^2y}{2x(x + 3y)}$

$$= \dfrac{3xy}{x - 3y}$$

Problem 19

Perform the indicated operations and reduce to lowest terms.

(A) $\dfrac{x + 5}{x^2 - 9} \cdot (x + 3)$

(B) $\dfrac{2x^2 - 8}{4x} \div (x + 2)$

(C) $\dfrac{x^2 - 4x + 4}{4x^2y - 8xy} \div \dfrac{x^2 + x - 6}{6x^2 + 18x}$

Answers

(A) $\dfrac{x + 5}{x - 3}$ (B) $\dfrac{x - 2}{2x}$ (C) $\dfrac{3}{2y}$

ADDITION AND SUBTRACTION

The operations of addition and subtraction of rational expressions are based on the corresponding properties of real numbers. Thus:

ADDITION AND SUBTRACTION

If D, P, Q, and K represent polynomials, then

1. $\dfrac{P}{D} + \dfrac{Q}{D} = \dfrac{P + Q}{D}$

2. $\dfrac{P}{D} - \dfrac{Q}{D} = \dfrac{P - Q}{D}$

3. $\dfrac{P}{D} = \dfrac{PK}{DK}$

where D and K are not zero.

Verbally, if the denominators of two rational expressions are the same, we may either add or subtract the expressions by adding or subtracting the numerators and then place the result over the common denominator. If the denominators are not the same, then we use property 3 to change the form of each fraction so they have a common denominator. Then we can use either property 1 or property 2.

Even though any common denominator will do, the problem will generally become less involved if the least common denominator (LCD) is used. The LCD is found as follows:

LEAST COMMON DENOMINATOR (LCD)

1. Factor each denominator completely, including the numerical coefficients.
2. Form a product by selecting as factors each different factor that occurs in the denominators.
3. Raise each factor in this product to the highest power it occurs in the denominators. The resulting product is the LCD.

Example 20

(A) $\dfrac{x-4}{x+2} - \dfrac{x-2}{x+2} = \dfrac{x-4-(x-2)}{x+2} = \dfrac{x-4-x+2}{x+2} = \dfrac{-2}{x+2}$

(B) $\dfrac{1}{3y^2} - \dfrac{1}{6y} + 1 = \dfrac{2(1)}{2(3y^2)} - \dfrac{y(1)}{y(6y)} + \dfrac{6y^2}{6y^2}$

$= \dfrac{2 - y + 6y^2}{6y^2}$ $\qquad$ LCD $= 6y^2$

(C) $\dfrac{3}{x^2-1} - \dfrac{2}{x^2+2x+1} = \dfrac{3}{(x-1)(x+1)} - \dfrac{2}{(x+1)^2}$

$= \dfrac{3(x+1)}{(x-1)(x+1)^2} - \dfrac{2(x-1)}{(x-1)(x+1)^2}$

$= \dfrac{3(x+1) - 2(x-1)}{(x-1)(x+1)^2} = \dfrac{3x+3-2x+2}{(x-1)(x+1)^2}$

$= \dfrac{x+5}{(x-1)(x+1)^2}$ $\qquad$ LCD $= (x-1)(x+1)^2$

Problem 20   Combine into single fractions and simplify.

(A) $\dfrac{x-2}{x-3} - \dfrac{x+2}{x-3}$ $\qquad$ (B) $\dfrac{1}{y} + \dfrac{1}{4y^2} - 1$ $\qquad$ (C) $\dfrac{4}{x^2-4} - \dfrac{3}{x^2-x-2}$

Answers   (A) $\dfrac{-4}{x-3}$ or $\dfrac{4}{3-x}$ $\qquad$ (B) $\dfrac{1+4y-4y^2}{4y^2}$ $\qquad$ (C) $\dfrac{1}{(x+2)(x+1)}$

EXERCISE A-5

A *Reduce to lowest terms; that is, remove all common factors from the numerator and denominator.*

1. $\dfrac{21x^2y}{28xy^2}$

2. $\dfrac{12x^3y^2}{9xy^3}$

3. $\dfrac{8x^2(x-y)^2}{6x(x-y)(x+y)}$

4. $\dfrac{12m^2n(m+2)(m-3)}{8mn^2(m-3)^2}$

Perform the indicated operations and reduce to lowest terms.

5. $\dfrac{3x^2}{4}\cdot\dfrac{16y}{12x^3}$

6. $\dfrac{2x^2}{3y^2}\cdot\dfrac{9y}{4x}$

7. $\dfrac{a}{4c}\cdot\dfrac{a^2}{12c^2}$

8. $\dfrac{2x}{3y}\cdot\dfrac{4x}{6y^2}$

9. $\dfrac{3x^2y}{x-y}\div\dfrac{6xy}{x-y}$

10. $\dfrac{x+3}{2x^2}\div\dfrac{x+3}{4x}$

11. $\dfrac{x-8}{x+2}-\dfrac{x-9}{x+2}$

12. $\dfrac{u-3}{u+5}-\dfrac{u-8}{u+5}$

13. $\dfrac{2}{x}-\dfrac{1}{3}$

14. $\dfrac{3x}{y}+\dfrac{1}{4}$

15. $\dfrac{u}{v^2}-\dfrac{1}{v}+\dfrac{u^2}{v^3}$

16. $\dfrac{1}{x}-\dfrac{y}{x^2}+\dfrac{y^2}{x^3}$

17. $\dfrac{2u}{4u^2-9}-\dfrac{3}{4u^2-9}$

18. $\dfrac{x}{x^2-9}-\dfrac{3}{x^2-9}$

19. $\dfrac{3}{2x}-\dfrac{2}{x+2}$

20. $\dfrac{2}{3y}-\dfrac{3}{y+3}$

Which of these equations (if any) are false?

21. $\dfrac{7}{x}+\dfrac{2}{y}=\dfrac{9}{x+y}$

22. $\dfrac{1}{x}-\dfrac{x-3}{x}=\dfrac{1-x-3}{x}$

23. $\dfrac{3+x}{5+x}=\dfrac{3}{5}$

24. $\dfrac{x^2-y^2}{x-y}=x-y$

B *Perform the indicated operations and reduce to lowest terms.*

25. $\dfrac{x-2}{4y}\cdot\dfrac{12y^2}{x^2+x-6}$

26. $\dfrac{4x}{x-4}\cdot\dfrac{x^2-6x+8}{8x^2}$

27. $\dfrac{6x^2}{4x^2y-12xy}\div\dfrac{3x^2+12x}{x^2+x-12}$

28. $\dfrac{2x^2+4x}{12x^2y}\div\dfrac{x^2+6x+8}{6x}$

29. $(x^2 - x - 12) \div \dfrac{x^2 - 9}{x^2 - 3x}$

30. $\dfrac{2x^2 + 7x + 3}{4x^2 - 1} \div (x + 3)$

31. $\dfrac{4x - 3}{18x^3} + \dfrac{3}{4x} - \dfrac{2x - 1}{6x^2}$

32. $\dfrac{3x + 8}{4x^2} - \dfrac{2x - 1}{x^3} - \dfrac{5}{8x}$

33. $5 + \dfrac{x}{x + 1} - \dfrac{x}{x - 1}$

34. $\dfrac{1}{x + 2} + 3 - \dfrac{2}{x - 2}$

35. $\dfrac{3x}{3x^2 - 12} + \dfrac{1}{2x^2 + 4x}$

36. $\dfrac{2m}{3m^2 - 48} + \dfrac{m}{4m + m^2}$

37. $\dfrac{1}{x^2 - y^2} - \dfrac{1}{x^2 - 2xy + y^2}$

38. $\dfrac{3}{m^2 - 1} - \dfrac{2}{m^2 - 2m + 1}$

C   39. $\dfrac{x^2 - xy}{xy^2 + y^2} \div \left(\dfrac{x^2 - y^2}{x^2 + 2xy + y^2} \div \dfrac{x^2 - 2xy + y^2}{x^2y + xy^2} \right)$

40. $\left(\dfrac{x^2 - xy}{xy + y^2} \div \dfrac{x^2 - y^2}{x^2 + 2xy + y^2} \right) \div \dfrac{x^2 - 2xy + y^2}{x^2y + xy^2}$

41. $\dfrac{x^2}{x^2 + 2x + 1} + \dfrac{1}{3x + 3} - \dfrac{1}{6}$

42. $\dfrac{x}{x^2 - x - 2} - \dfrac{1}{x^2 + 5x - 14} - \dfrac{2}{x^2 + 8x + 7}$

A-6   LINEAR EQUATIONS AND INEQUALITIES

The equation

$$3 - 2(x + 3) = \frac{x}{3} - 5$$

and the inequality

$$\frac{x}{2} + 2(3x - 1) \geq 5$$

are both first degree (linear) in one variable. A **solution** of an equation (or inequality) involving a single variable is a number, which when substituted for the variable, makes the equation (or inequality) true. The set of all solutions is called the **solution set.** When we say that we **solve** an equation (or inequality), we mean that we find its solution set.

Knowing what is meant by the solution set is one thing; finding it is another. We start by recalling the idea of equivalent equations and equivalent inequalities. We say that two equations (or two inequalities) are **equivalent** if they both have the same solution set. The basic idea in solving equations and inequalities is to perform operations on these forms that produce simpler equivalent forms, and to continue the process until we reach an equation or inequality with an obvious solution.

LINEAR EQUATIONS

The following properties of equality are fundamental to the equation solving process. For a, b, and c real numbers:

1. If $a = b$, then $a + c = b + c$ Addition property
2. If $a = b$, then $a - c = b - c$ Subtraction property
3. If $a = b$, then $ca = cb$ Multiplication property
4. If $a = b$ and $c \neq 0$, then $\dfrac{a}{c} = \dfrac{b}{c}$ Division property

An equivalent equation will result if the original equation is altered by use of any of the above properties, except for multiplication or division by zero. Several examples should serve to remind you of the process of solving linear equations.

Example 21

Solve $3x - 9 = 7x + 3$ and check.

Solution

$$3x - 9 = 7x + 3$$ Try to isolate x on the left side

$$3x - 9 + 9 = 7x + 3 + 9$$ Addition property

$$3x = 7x + 12$$

$$3x - 7x = 7x + 12 - 7x$$ Subtraction property

$$-4x = 12$$

$$\frac{-4x}{-4} = \frac{12}{-4}$$ Division property

$$x = -3$$

Check: $3(-3) - 9 \overset{?}{=} 7(-3) + 3$

$$-18 \overset{\checkmark}{=} -18$$

Problem 21

Solve $2x - 8 = 5x + 4$ and check.

Answer

$x = -4$

Example 22

Solve $8x - 3(x - 4) = 3(x - 4) + 6$.

Solution

$$8x - 3(x - 4) = 3(x - 4) + 6$$
$$8x - 3x + 12 = 3x - 12 + 6$$
$$5x + 12 = 3x - 6$$
$$2x = -18$$
$$x = -9$$

Problem 22   Solve $3x - 2(2x - 5) = 2(x + 3) - 8$.

Answer $x = 4$

Example 23   What operations can we perform on

$$\frac{x + 2}{2} - \frac{x}{3} = 5$$

to eliminate the denominators? If we can find a number that is exactly divisible by each denominator, then we can use the multiplication property of equality to clear the denominators. The LCD of the fractions, 6, is exactly what we are looking for! Actually, any common denominator will do, but the LCD results in a simpler equivalent equation. Thus, we multiply both sides of the equation by 6:

$$6\left(\frac{x + 2}{2} - \frac{x}{3}\right) = 6 \cdot 5$$

$$\overset{3}{\cancel{6}} \cdot \frac{(x + 2)}{\underset{1}{\cancel{2}}} - \overset{2}{\cancel{6}} \cdot \frac{x}{\underset{1}{\cancel{3}}} = 30$$

$$3(x + 2) - 2x = 30$$

$$3x + 6 - 2x = 30$$

$$x = 24$$

Problem 23   Solve $\dfrac{x + 1}{3} - \dfrac{x}{4} = \dfrac{1}{2}$.

Answer $x = 2$

A frequent application of algebra requires changing formulas or equations to alternate equivalent forms. The following examples are typical.

Example 24   Solve the simple interest formula $A = P + Prt$ for
(A) r in terms of the other variables
(B) P in terms of the other variables

Solutions   (A) $A = P + Prt$ Reverse equation

$\qquad\qquad P + Prt = A$ Now isolate r on the left side

$\qquad\qquad\quad Prt = A - P$ Divide both members by Pt

$\qquad\qquad\qquad r = \dfrac{A - P}{Pt}$

(B) $A = P + Prt$

$$P + Prt = A \qquad \text{Reverse equation}$$

$$P(1 + rt) = A \qquad \text{Factor out } P$$

$$P = \frac{A}{1 + rt} \qquad \text{Divide by } (1 + rt)$$

Problem 24   Solve $M = Nt + Nr$ for
(A) t (B) N

Answers   (A) $t = \dfrac{M - Nr}{N}$ (B) $N = \dfrac{M}{t + r}$

LINEAR INEQUALITIES   Before we start solving linear inequalities let us recall what we mean by $<$ (less than) and $>$ (greater than). If a and b are real numbers, then we write

$$a < b$$

if there exists a positive number p such that $a + p = b$. Certainly, we would expect that if a positive number were added to any real number, the sum would be larger than the original. That is essentially what the definition states.

We write

$$a > b$$

if there exists a positive number such that $a - p = b$.

Example 25   (A) $3 < 5$ Since $3 + 2 = 5$
(B) $-6 < -2$ Since $-6 + 4 = -2$
(C) $0 > -10$ Since $0 - 10 = -10$

Problem 25   Replace each question mark with either $<$ or $>$.
(A) $2 ? 8$ (B) $-20 ? 0$ (C) $-3 ? -30$

Answers   (A) $<$ (B) $<$ (C) $>$

The inequality symbols have a very clear geometric interpretation on the real number line. If $a < b$, then a is to the left of b on the number line; if $c > d$, then c is to the right of d (Fig. 7).

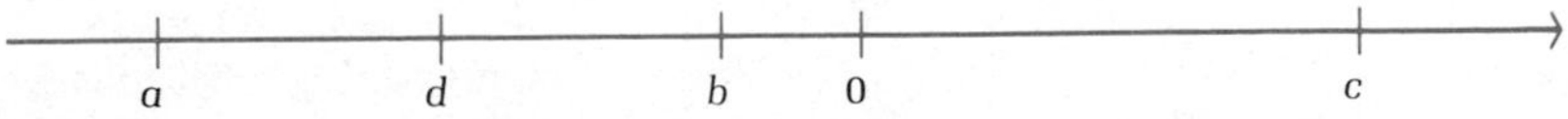

FIGURE 7 $a < b, c > d$

Now let us return to the problem of solving linear inequalities. The procedures used are almost the same as those used to solve linear equations, *but with two important exceptions.* We state the following basic properties of inequalities and compare them with the equality properties stated at the beginning of the section. For a, b, and c real numbers:

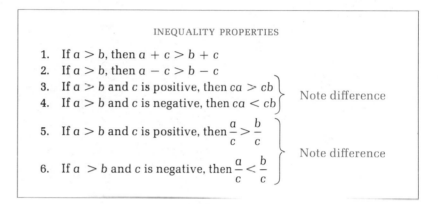

INEQUALITY PROPERTIES

1. If $a > b$, then $a + c > b + c$
2. If $a > b$, then $a - c > b - c$
3. If $a > b$ and c is positive, then $ca > cb$ ⎫
4. If $a > b$ and c is negative, then $ca < cb$ ⎬ Note difference
5. If $a > b$ and c is positive, then $\dfrac{a}{c} > \dfrac{b}{c}$ ⎫
   ⎬ Note difference
6. If $a > b$ and c is negative, then $\dfrac{a}{c} < \dfrac{b}{c}$ ⎭

Similar properties hold if each inequality sign is reversed, or if $>$ is replaced with $\geq$ (greater than or equal) and $<$ is replaced with $\leq$ (less than or equal). Thus, we find that we can perform essentially the same operations on inequalities that we perform on equations, with the exception that **the sense of the inequality reverses if we multiply or divide both sides by a negative number.** Otherwise, the sense of the inequality does not change. For example, if we start with the true statement

$$-3 > -7$$

and multiply both sides by 2, we obtain

$$-6 > -14$$

and the sense of the inequality stays the same. But if we multiply both sides of $-3 > -7$ by -2, then the left side becomes 6 and the right side becomes 14, so we must write

$$6 < 14$$

to have a true statement. That is, the sense of the inequality reverses.

Recall that the double inequality $a \leq x \leq b$ means that $a \leq x$ and $x \leq b$. Other variations, as well as a useful interval notation, are indicated in Table 3. Note that an end point on a line graph is a solid dot if it is included in the inequality; otherwise, it is a hollow dot.

TABLE 3

INTERVAL NOTATION	INEQUALITY NOTATION	LINE GRAPH
$[a, b]$	$a \leq x \leq b$	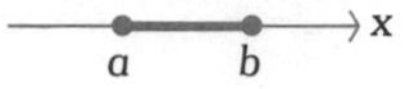
$[a, b)$	$a \leq x < b$	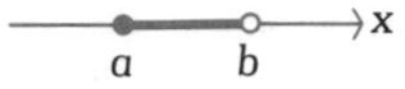
$(a, b]$	$a < x \leq b$	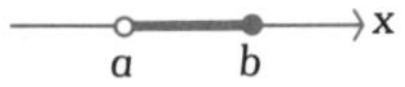
(a, b)	$a < x < b$	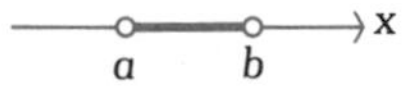
$(-\infty, a]$	$x \leq a$	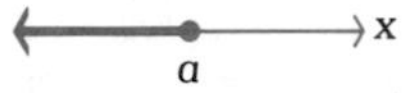
$(-\infty, a)$	$x < a$	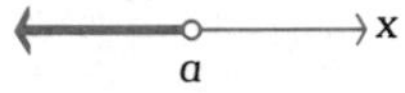
$[b, \infty)$	$x \geq b$	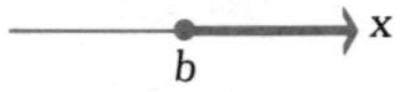
(b, ∞)	$x > b$	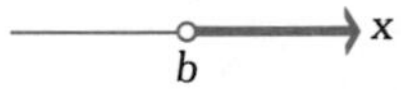

Example 26   Solve and graph $2(2x + 3) < 6(x - 2) + 10$.

Solution

$$2(2x + 3) < 6(x - 2) + 10$$
$$4x + 6 < 6x - 12 + 10$$
$$4x + 6 < 6x - 2$$
$$-2x + 6 < -2$$
$$-2x < -8$$
$$x > 4 \quad \text{or} \quad (4, \infty)$$

Notice that the sense of the inequality reverses when we divide both sides by -2

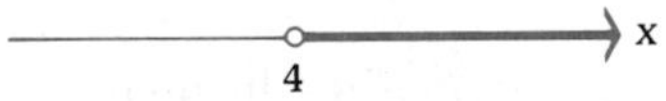

Problem 26   Solve and graph $3(x - 1) \leq 5(x + 2) - 5$.

Answer $x \geq -4 \quad \text{or} \quad [-4, \infty)$

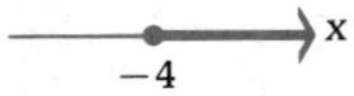

Note that in the graph of $x > 4$ we use a hollow circle for 4, since the point 4 is not included in the graph. For the graph of $x \geq -4$, we use a solid circle for -4, since the point -4 is included in the graph.

Example 27   Solve and graph $-3 < 2x + 3 \leq 9$.

Solution   We are looking for all numbers x such that $2x + 3$ is between -3 and 9, including 9 but not -3. We proceed as above, except we try to isolate x in the middle:

$$-3 < 2x + 3 \leq 9$$

$$-3 - 3 < 2x + 3 - 3 \leq 9 - 3$$

$$-6 < 2x \leq 6$$

$$\frac{-6}{2} < \frac{2x}{2} \leq \frac{6}{2}$$

$$-3 < x \leq 3 \quad \text{or} \quad (-3, 3]$$

Problem 27   Solve and graph $-8 \leq 3x - 5 < 7$.

Answer $-1 \leq x < 4 \quad \text{or} \quad [-1, 4)$

APPLICATIONS   To realize the full potential of algebra, we must be able to translate real-world problems into mathematical forms. In short, we must be able to do *word problems*.

Example 28   It costs a record company $6,000 to prepare a record album—recording costs, album design costs, etc. These costs represent a one-time **fixed cost.** Manufacturing, marketing, and royalty costs (all **variable costs**) are $2.50 per album. If the album is sold to record shops for $4 each, how many must be sold for the company to **break even?**

Solution   Let

x = Number of records sold
C = Cost for producing x records
R = Revenue (return) on sales of x records

The company breaks even if $R = C$, with

C = Fixed costs + Variable costs
  = $6,000 + $2.50x
R = $4x

Find x such that $R = C$; that is, such that

$$4x = 6,000 + 2.50x$$

$$1.50x = 6,000$$

$$x = 4,000$$

Check: For x = 4,000,

$$C = 6{,}000 + 2.50x \qquad \text{and} \qquad R = 4x$$
$$= 6{,}000 + 2.50(4{,}000) \qquad\qquad\qquad = 4(4{,}000)$$
$$= 6{,}000 + 10{,}000 \qquad\qquad\qquad\qquad = \$16{,}000$$
$$= \$16{,}000$$

Thus, the company must sell 4,000 records to break even; any sales over 4,000 will result in a profit; and sales under 4,000 will result in a loss.

Problem 28

What is the break-even point in Example 28 if fixed costs are $9,000 and variable costs are $2.80 per record?

Answer   7,500

There are many different types of applications that use algebra; so many, in fact, that no single approach will apply to all. However, the following suggestions may help you get started:

SUGGESTIONS FOR SOLVING WORD PROBLEMS

1. Read the problem very carefully.
2. Write down important facts and relationships.
3. Identify unknown quantities in terms of a single letter—if possible.
4. Write an equation or inequality relating unknown quantities and facts in the problem.
5. Solve the equation (or inequality).
6. Write all solutions asked for in the original problem.
7. Check the solution(s) in the original problem.

EXERCISE A-6

A   *Solve.*

1. $2m + 9 = 5m - 6$
2. $3y - 4 = 6y - 19$
3. $x + 5 < -4$
4. $x - 3 > -2$
5. $-3x \geq -12$
6. $-4x \leq 8$

Solve and graph.

7. $-4x - 7 > 5$
8. $-2x + 8 < 4$
9. $2 \leq x + 3 \leq 5$
10. $-3 < y - 5 < 8$

Solve.

11. $\dfrac{y}{7} - 1 = \dfrac{1}{7}$
12. $\dfrac{m}{5} - 2 = \dfrac{3}{5}$

13. $\dfrac{x}{3} > -2$

14. $\dfrac{y}{-2} \le -1$

15. $\dfrac{y}{3} = 4 - \dfrac{y}{6}$

16. $\dfrac{x}{4} = 9 - \dfrac{x}{2}$

B 17. $10x + 25(x - 3) = 275$

18. $-3(4 - x) = 5 - (x + 1)$

19. $3 - y \le 4(y - 3)$

20. $x - 2 \ge 2(x - 5)$

21. $\dfrac{x}{5} - \dfrac{x}{6} = \dfrac{6}{5}$

22. $\dfrac{y}{4} - \dfrac{y}{3} = \dfrac{1}{2}$

23. $\dfrac{m}{5} - 3 < \dfrac{3}{5} - m$

24. $u - \dfrac{2}{3} > \dfrac{u}{3} + 2$

25. $0.1(x - 7) + 0.05x = 0.8$

26. $0.4(u + 5) - 0.3u = 17$

Solve and graph.

27. $2 \le 3x - 7 < 14$

28. $-4 \le 5x + 6 < 21$

29. $-4 \le \frac{9}{5}C + 32 \le 68$

30. $-1 \le \frac{2}{3}t + 5 \le 11$

Solve for the indicated variable.

31. $3x - 4y = 12$, for y

32. $y = -\frac{2}{3}x + 8$, for x

33. $Ax + By = C$, for y

34. $y = mx + b$, for m

35. $F = \frac{9}{5}C + 32$, for C

36. $C = \frac{5}{9}(F - 32)$, for F

37. $A = Bm - Bn$, for B

38. $U = 3C - 2CD$, for C

C 39. $-2\{2 - [1 - 2(y + 1)]\} = 2(y + 5) - 4$

40. $-2\{3 + [2x - (x - 4)]\} = 2[(x + 2) - 3]$

41. $\dfrac{2m - 3}{9} - \dfrac{m + 5}{6} = \dfrac{3 - m}{2} - 1$

42. $\dfrac{3n + 4}{3} - \dfrac{2 - n}{5} = \dfrac{n - 2}{15} - 1$

43. $\dfrac{3y}{7} - \dfrac{y - 4}{3} \ge 4 + \dfrac{2y}{7}$

44. $\dfrac{x}{3} - \dfrac{x - 2}{2} \le \dfrac{x}{4} - 4$

APPLICATIONS

BUSINESS & ECONOMICS 45. A jazz concert brought in $60,000 on the sale of 8,000 tickets. If the tickets sold for $6 and $10 each, how many of each type of ticket was sold?

46. An all-day parking meter takes only dimes and quarters. If it contains 100 coins with a total value of $14.50, how many of each type of coin is in the meter?

47. You have $12,000 to invest. If part is invested at 10% and the rest at 15%, how much should be invested at each rate to yield 12% on the total amount?

48. An investor has $20,000 to invest. If part is invested at 8% and the rest at 12%, how much should be invested at each rate to yield 11% on the total amount?

49. *Cost-of-living index.* The cost-of-living index for several years is given in the table.

YEAR	COST-OF-LIVING INDEX
1945	62
1950	82
1955	93
1960	103
1965	110

(A) What would a net monthly salary have to be in 1965 to have the same purchasing power of a net monthly salary of $500 in 1945? [*Hint:* To have the same purchasing power, the ratio of a salary in 1965 to a salary in 1945 would have to be the same as the ratio of the cost-of-living index in 1965 to the cost-of-living index in 1945. Thus, if x is the net monthly salary in 1965, to find x we need to solve the equation $x/500 = 110/62$.]

(B) Find the net monthly salary needed in 1950 to have the same purchasing power of a net salary of $600 in 1960.

50. *Break-even analysis.* For a business to realize a profit, it is clear that revenue R must be greater than costs C; that is, a profit will result only if $R > C$ (the company breaks even when $R = C$). A record manufacturer has a weekly cost equation $C = 300 + 1.5x$ and a revenue equation $R = 2x$, where x is the number of records produced and sold in a week. How many records must be sold for the company to make a profit?

LIFE SCIENCES

51. *Wildlife management.* A naturalist for a fish and game department estimated the total number of rainbow trout in a certain lake using the popular capture–mark–recapture technique. He netted, marked,

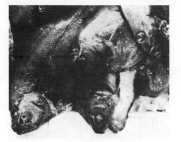

and released 200 rainbow trout. A week later, allowing for thorough mixing, he again netted 200 trout and found eight marked ones among them. Assuming that the proportion of marked fish in the second sample was the same as the proportion of all marked fish in the total population, estimate the number of rainbow trout in the lake.

52. *Ecology.* If the temperature for a 24 hour period in the Antarctica ranged between $-49°F$ and $14°F$ (that is, $-49 \le F \le 14$), what was the range in Celsius degrees? [*Note:* $F = \frac{9}{5} C + 32$.]

SOCIAL SCIENCES

53. *Psychology.* The IQ (intelligence quotient) is found by dividing the mental age (MA), as indicated on standard tests, by the chronological age (CA) and multiplying by 100. For example, if a child has a mental age of twelve and a chronological age of eight, the calculated IQ is 150. If a 9-year-old girl has an IQ of 140, compute her mental age.

54. *Anthropology.* In their study of genetic groupings, anthropologists use a ratio called the *cephalic index*. This is the ratio of the breadth of the head to its length (looking down from above) expressed as a percentage. Symbolically,

$$C = \frac{100B}{L}$$

where C is the cephalic index, B is the breadth, and L is the length. If an Indian tribe in Baja California (Mexico) had an average cephalic index of 66 and the average breadth of their heads was 6.6 inches, what was the average length of their heads?

A-7   QUADRATIC EQUATIONS AND INEQUALITIES

QUADRATIC EQUATIONS

A **quadratic equation** in one variable is any equation that can be written in the form

$$ax^2 + bx + c = 0 \qquad a \ne 0$$

where x is a variable and a, b, and c are constants. We will refer to this form as the **standard form.** The equations

$$5x^2 - 3x + 7 = 0 \qquad \text{and} \qquad 18 = 32t^2 - 12t$$

are both quadratic equations since they are either in the standard form or can be transformed into this form.

We will restrict our review to finding real solutions to quadratic equations.

**SOLUTION BY
SQUARE ROOT**

The easiest type of quadratic equation to solve is the special form where the first-degree term is missing:

$$ax^2 + c = 0 \qquad a \neq 0$$

The method makes use of the definition of square root given in Section A-3.

Example 29

Solve by the square root method.
(A) $x^2 - 7 = 0$ (B) $2x^2 - 10 = 0$ (C) $3x^2 + 27 = 0$

Solutions

(A) $x^2 - 7 = 0$

$\qquad x^2 = 7$ What real number squared is 7?

$\qquad x = \pm \sqrt{7}$ Short for $\sqrt{7}$ and $-\sqrt{7}$

(B) $2x^2 - 10 = 0$

$\qquad 2x^2 = 10$

$\qquad x^2 = 5$ What real number squared is 5?

$\qquad x = \pm \sqrt{5}$

(C) $3x^2 + 27 = 0$

$\qquad 3x^2 = -27$

$\qquad x^2 = -9$ What real number squared is -9?

No real solution. (Why?)

Problem 29

Solve by the square root method.
(A) $x^2 - 6 = 0$ (B) $3x^2 - 12 = 0$ (C) $x^2 + 4 = 0$

Answers

(A) $\pm \sqrt{6}$ (B) ± 2 (C) No real solution

**SOLUTION BY
FACTORING**

If the left side of a quadratic equation when written in standard form can be factored, then the equation can be solved very quickly. The method of solution by factoring rests on the following important property of real numbers: *If a and b are real numbers, then ab = 0 if and only if a = 0 or b = 0 (or both).*

Example 30

Solve by factoring, if possible.
(A) $3x^2 - 6x - 24 = 0$ (B) $3y^2 = 2y$ (C) $x^2 - 2x - 1 = 0$

Solutions

(A) $3x^2 - 6x - 24 = 0$ Divide both sides by 3, since 3 is a factor of each coefficient

$\qquad x^2 - 2x - 8 = 0$ Factor the left side, if possible

$\qquad (x - 4)(x + 2) = 0$

$\qquad x - 4 = 0$ or $x + 2 = 0$

$\qquad \quad x = 4$ or $\qquad x = -2$

(B) $\qquad 3y^2 = 2y$ We lose the solution $y = 0$ if both

$\qquad 3y^2 - 2y = 0$ sides are divided by y ($3y^2 = 2y$ and

$\qquad y(3y - 2) = 0$ $3y = 2$ are not equivalent)

$$y = 0 \quad \text{or} \quad 3y - 2 = 0$$
$$3y = 2$$
$$y = \tfrac{2}{3}$$

(C) $x^2 - 2x - 1 = 0$

This equation cannot be factored using integer coefficients. We will solve this type of equation by another method, considered below.

Problem 30   Solve by factoring, if possible.

(A) $2x^2 + 4x - 30 = 0$ (B) $2x^2 = 3x$ (C) $2x^2 - 8x + 3 = 0$

Answers   (A) $-5, 3$ (B) $0, \tfrac{3}{2}$
(C)   Cannot be factored using integer coefficients

The factoring and square root methods are fast and easy to use when they apply. However, there are quadratic equations that look simple but cannot be solved by either method. For example, as was noted in Example 30C, the polynomial in

$$x^2 - 2x - 1 = 0$$

cannot be factored using integer coefficients. This brings us to the well-known and widely used quadratic formula.

QUADRATIC FORMULA   There is a method called *completing the square* that will work for all quadratic equations. After briefly reviewing this method, we will then use it to develop the famous quadratic formula—a formula that will enable us to solve any quadratic equation quite mechanically.

The method of **completing the square** is based on the process of transforming a quadratic equation in standard form,

$$ax^2 + bx + c = 0$$

into the form

$$(x + A)^2 = B$$

where A and B are constants. Then, this last equation can easily be solved (if it has a real solution) by the square root method discussed above.

Consider the equation

$$x^2 - 2x - 1 = 0 \tag{1}$$

Since the left side does not factor using integer coefficients, we add 1 to each side to remove the constant term from the left side:

$$x^2 - 2x \quad\ \ = 1 \tag{2}$$

Now we try to find a number that we can add to each side to make the left side a square of a first-degree polynomial. Note the following two squares:

$$(x + m)^2 = x^2 + 2mx + m^2 \qquad (x - m)^2 = x^2 - 2mx + m^2$$

We see that the third term on the right is the square of one-half the coefficient of x in the second term on the right. To complete the square in equation (2), we add the square of one-half the coefficient of x, $(-\tfrac{2}{2})^2 = 1$, to each side. (This rule works only when the coefficient of x^2 is 1, that is, $a = 1$.) Thus,

$$x^2 - 2x + 1 = 1 + 1$$

The left side is the square of $x - 1$, and we write

$$(x - 1)^2 = 2$$

What number squared is 2?

$$x - 1 = \pm \sqrt{2}$$
$$x = 1 \pm \sqrt{2}$$

And equation (1) is solved!

Let us try the method on the general quadratic equation

$$ax^2 + bx + c = 0 \qquad a \neq 0 \tag{3}$$

and solve it once and for all for x in terms of the coefficients a, b, and c. We start by multiplying both sides of (3) by $1/a$ to obtain

$$x^2 + \frac{b}{a}x + \frac{c}{a} = 0$$

Add $-c/a$ to both members:

$$x^2 + \frac{b}{a}x \qquad = -\frac{c}{a}$$

Now we complete the square on the left side by adding the square of one-half the coefficient of x, that is, $(b/2a)^2 = b^2/4a^2$, to each side:

$$x^2 + \frac{b}{a}x + \frac{b^2}{4a^2} = \frac{b^2}{4a^2} - \frac{c}{a}$$

Writing the left member as a square and combining the right side into a single fraction, we obtain

$$\left(x + \frac{b}{2a}\right)^2 = \frac{b^2 - 4ac}{4a^2}$$

Now we solve by the square root method:

$$x + \frac{b}{2a} = \pm \sqrt{\frac{b^2 - 4ac}{4a^2}}$$

$$x = -\frac{b}{2a} \pm \frac{\sqrt{b^2 - 4ac}}{2a}$$

When this is written as a single fraction, it becomes the quadratic formula:

QUADRATIC FORMULA

$$x = \frac{-b \pm \sqrt{b^2 - 4ac}}{2a}$$

TABLE 4

$b^2 - 4ac$	$ax^2 + bx + c = 0$
Positive	Two real solutions
Zero	One real solution
Negative	No real solutions

This formula is generally used to solve quadratic equations when the square root or factoring methods do not work. The quantity $b^2 - 4ac$ under the radical is called the **discriminant,** and it gives us the useful information about solutions listed in Table 4.

Example 31   Solve $x^2 - 2x - 1 = 0$ using the quadratic formula.

Solution $x^2 - 2x - 1 = 0$

$$x = \frac{-b \pm \sqrt{b^2 - 4ac}}{2a} \qquad \begin{aligned} a &= 1 \\ b &= -2 \\ c &= -1 \end{aligned}$$

$$= \frac{-(-2) \pm \sqrt{(-2)^2 - 4(1)(-1)}}{2(1)}$$

$$= \frac{2 \pm \sqrt{8}}{2} = \frac{2 \pm 2\sqrt{2}}{2} = 1 \pm \sqrt{2}$$

Check: $x^2 \quad 2x \quad 1 = 0$

When $x = 1 + \sqrt{2}$,
$(1 + \sqrt{2})^2 - 2(1 + \sqrt{2}) - 1 = 1 + 2\sqrt{2} + 2 - 2 - 2\sqrt{2} - 1 = 0$
When $x = 1 - \sqrt{2}$,
$(1 - \sqrt{2})^2 - 2(1 - \sqrt{2}) - 1 = 1 - 2\sqrt{2} + 2 - 2 + 2\sqrt{2} - 1 = 0$

Problem 31   Solve $2x^2 - 4x - 3 = 0$ using the quadratic formula.

Answer $(2 \pm \sqrt{10})/2$

If we try to solve $x^2 - 6x + 11 = 0$ using the quadratic formula, we obtain

$$x = \frac{6 \pm \sqrt{-8}}{2}$$

which is not a real number. (Why?)

QUADRATIC INEQUALITIES   Solving quadratic inequalities of the type

$$x^2 - x - 6 \geq 0 \qquad \text{or} \qquad 2x^2 + 5x - 3 < 0$$

is best illustrated by example.

Example 32   Find the real values of x that make $\sqrt{x^2 - x - 6}$ a real number.

Solution   For this radical form to be real, $x^2 - x - 6$ must not be negative. That is, we must find x so that

$$x^2 - x - 6 \geq 0$$

This is a quadratic inequality. To solve it, we factor the left side to obtain

$$(x - 3)(x + 2) \geq 0$$

The product on the left side will equal zero if either factor is zero; that is, if $x = 3$ or $x = -2$. When will the product be greater than zero (positive)? The product will be positive when both factors are positive or both factors have the same sign):

$(x - 3)$ positive $(x - 3)$ negative
$\quad x - 3 > 0$ $\quad x - 3 < 0$
$\qquad x > 3$ $\qquad x < 3$

$(x + 2)$ positive $(x + 2)$ negative
$\quad x + 2 > 0$ $\quad x + 2 < 0$
$\qquad x > -2$ $\qquad x < -2$

Summarizing these results relative to a real number line, the solution becomes obvious:

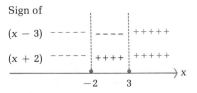

Thus, we see that both factors have the same sign (and the product is positive) when

$$x < -2 \quad \text{or} \quad x > 3$$

Combining the above results, we see that

$$(x - 3)(x + 2) \geq 0$$

when

$$x \leq -2 \quad \text{or} \quad x \geq 3$$

Problem 32   Solve $2x^2 + 5x - 3 < 0$.

Answer $-3 < x < \frac{1}{2}$

EXERCISE A-7

Find only real solutions in the problems below. If there are no real solutions, say so.

A   *Solve by the square root method.*

1. $x^2 - 4 = 0$ 2. $x^2 - 9 = 0$
3. $2x^2 - 22 = 0$ 4. $3m^2 - 21 = 0$

Solve by factoring.

5. $2u^2 - 8u - 24 = 0$ 6. $3x^2 - 18x + 15 = 0$
7. $x^2 = 2x$ 8. $n^2 = 3n$

Solve by using the quadratic formula.

9. $x^2 - 6x - 3 = 0$ 10. $m^2 + 8m + 3 = 0$
11. $3u^2 + 12u + 6 = 0$ 12. $2x^2 - 20x - 6 = 0$

Solve, using any method.

B   13. $2x^2 = 4x$ 14. $2x^2 = -3x$
15. $4u^2 - 9 = 0$ 16. $9y^2 - 25 = 0$
17. $8x^2 + 20x = 12$ 18. $9x^2 - 6 = 15x$
19. $x^2 = 1 - x$ 20. $m^2 = 1 - 3m$
21. $2x^2 - 6x - 3$ 22. $2x^2 = 4x - 1$
23. $y^2 - 4y = -8$ 24. $x^2 - 2x = -3$
25. $(x + 4)^2 = 11$ 26. $(y - 5)^2 = 7$

Solve each quadratic inequality in Problems 27–34.

27. $x^2 - x - 12 \geq 0$ 28. $x^2 - 2x - 8 \geq 0$
29. $x^2 + x - 12 < 0$ 30. $x^2 + 3x - 10 < 0$
31. $2x^2 + x > 1$ 32. $2x^2 + x > 6$
33. $x^2 \geq 4x - 4$ 34. $x^2 < 2x - 1$

C   35. Solve $A = P(1 + r)^2$ for r in terms of A and P; that is, isolate r on the left side of the equation (with coefficient 1) and end up with an algebraic expression on the right side involving A and P but not r. Write the answer using positive square roots only.

36. Solve $x^2 + mx + n = 0$ for x in terms of m and n.

APPLICATIONS

BUSINESS & ECONOMICS   37. *Supply and demand.* The demand equation for a certain brand of popular records is $d = 3{,}000/p$. Notice that as the price (p) goes up, the number of records people are willing to buy (d) goes down, and vice versa. The supply equation is given by $s = 1{,}000p - 500$. Notice again, as the price (p) goes up, the number of records a supplier is willing to sell (s) goes up. At what price will supply equal demand; that is, at what price will $d = s$? In economic theory the price at which supply equals demand is called the **equilibrium point**—the point where the price ceases to change.

LIFE SCIENCES

38. *Ecology.* An important element in the erosive force of moving water is its velocity. To measure the velocity v (in feet per second) of a stream we have only to find a hollow L-shaped tube, place one end under the water pointing upstream and the other end pointing straight up a couple of feet out of the water. The water will then be pushed up the tube a certain distance h (in feet) above the surface of the stream. Physicists have shown that $v^2 = 64h$. Approximately how fast is a stream flowing if $h = 1$ foot? If $h = 0.5$ foot?

SOCIAL SCIENCES

39. *Safety research.* It is of considerable importance to know the least number of feet d in which a car can be stopped, including reaction time of the driver, at various speeds v (in miles/hour). Safety re-

search has produced the formula $d = 0.044v^2 + 1.1v$. If it took a car 550 feet to stop, estimate the car's speed at the moment the stopping process was started. You might find a hand calculator of help in this problem.

A-8   APPENDIX REVIEW

IMPORTANT TERMS AND SYMBOLS

A-1 Sets. element of, member of, null set, empty set, finite set, infinite set, subset, equal sets, union, intersection, disjoint, universal set, complement, $\varnothing$, $A \subset B$, $A \cup B$, $A \cap B$, A'

A-2 The real number system. natural number, integer, rational number, irrational number, real number, real number line

A-3 Exponents and radicals. exponent, exponent properties, scientific notation, square root, n th root, radical, radical properties, rationalizing denominators, b^n, b^0, b^{-n}, $b^{m/n}$, $\sqrt{b}$, $\sqrt[n]{b^m}$

A-4 Polynomials—basic operations. commutative properties, associative properties, distributive properties, polynomial, coefficient of a term, degree of a term, degree of a polynomial, factoring, $a_n x^n + a_{n-1} x^{n-1} + \cdots + a_1 x + a_0$

A-5 Algebraic fractions. rational expression, fundamental principle of fractions, lower terms, higher terms, least common denominator, LCD

A-6 Linear equations and inequalities. linear equation, linear inequality, solution, solution set, equivalent equations, equivalent inequalities, inequality properties, interval notation, $a < b$, $a > b$, $a \le x \le b$, $[a, b]$, $[a, b)$, $(a, b]$, (a, b), $(-\infty, b)$, $(-\infty, b]$, (a, ∞), $[a, \infty)$

A-7 Quadratic equations and inequalities. quadratic equation, standard form, completing the square, quadratic formula, discriminant, quadratic inequality, $ax^2 + bx + c = 0, a \ne 0$, $x = \dfrac{-b \pm \sqrt{b^2 - 4ac}}{2a}$

EXERCISE A-8   APPENDIX REVIEW

Work through all the problems in this appendix review and check your answers in the back of the book. (Answers to all review problems are there.) Where weaknesses show up, review appropriate sections in the

text. When you are satisfied that you know the material, take the practice test following this review.

A   1.   True (T) or false (F)?
(A) $7 \notin \{4, 6, 8\}$ (B) $\{8\} \subset \{4, 6, 8\}$
(C) $\emptyset \in \{4, 6, 8\}$ (D) $\emptyset \subset \{4, 6, 8\}$

2.   Simplify and write the answer using positive exponents only:

$$\frac{8x^{-2}y^3}{6x^{-1}y^{-4}}$$

In Problems 3 and 4 perform the indicated operations and simplify.

3. $(3x - 4y)(2x + 3y)$ 4. $\dfrac{3}{x - 2} - \dfrac{2}{3}$

5.   Factor as far as possible using integer coefficients:
(A) $x^2 + 3x - 18$ (B) $4x^2 - 25$

6.   Solve, and graph on a real number line: $-3x + 8 > -4$

Solve.

7. $x^2 = 5x$ 8. $\dfrac{u}{5} = \dfrac{u}{6} + \dfrac{6}{5}$

B   9.   If $A = \{1, 2, 3\}$ and $B = \{2, 3, 4\}$, find:
(A) $A \cup B$ (B) $\{x \mid x \in A \text{ and } x \in B\}$

10.   Given the Venn diagram shown, with the number of elements indicated in each part, how many elements are in each of the following sets:

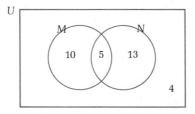

(A) $M \cup N$ (B) $M \cap N$ (C) $(M \cup N)'$ (D) $M \cap N'$

11.   In a freshman class of 100 students, seventy are taking English, forty-five are taking math, and twenty-five are taking both English and math.
(A)   How many students are taking either English or math?
(B)   How many students are taking English and not math?

In Problems 12 and 13, simplify and write answers using positive exponents only.

12. $\left(\dfrac{4x^{-2}y}{6xy^{-3}}\right)^{-2}$

13. $(9x^{-4}y^6)^{3/2}$

14. Rationalize the denominator and simplify: $\dfrac{4x}{\sqrt{2x}}$

15. Write in scientific notation, and then multiply: $(220{,}000{,}000)(0.0003)$

In Problems 16–19 perform the indicated operations and simplify.

16. $(2x + 1)(x^2 - 3x + 5)$

17. $(3x - 1)(x + 2) - (2x - 3)^2$

18. $\dfrac{2x^2 - 4x}{4x^2} \div \dfrac{x^2 - 4x + 4}{2x^3 - 8x}$

19. $\dfrac{3}{x^2 - 4} - \dfrac{2}{x^2 - 4x + 4}$

In Problems 20 and 21 factor as far as possible using integer coefficients.

20. $6x^3 + 22x^2 - 8x$

21. $9x^3 - 6x^2 + 6x$

Solve.

22. $\dfrac{x}{12} - \dfrac{x - 3}{3} = \dfrac{1}{2}$

23. $-5 < 2x - 3 \le 5$

24. $3x^2 - 21 = 0$

25. $2x^2 - 3x - 1 = 0$

26. $3x^2 + x \ge 2$

27. $x^2 - 3x - 10 < 0$

Solve for y in terms of x.

28. $2x - 3y = 6$

29. $xy - y = 3$

30. Simplify: $\sqrt[3]{16x^4y^7}$

C   31. If $A \cap B = A$, then is it always true that $A \subset B$?

32. Solve: $\dfrac{x}{12} - \dfrac{x - 4}{4} \le 2 + \dfrac{x}{6}$

33. Solve $x^2 + jx + k = 0$ for x in terms of j and k.

PRACTICE TEST: **APPENDIX A**

1. If $U = \{2, 4, 5, 6, 8\}$, $M = \{2, 4, 5\}$, and $N = \{5, 6\}$, find:
   (A) $M \cup N$ (B) $M \cap N$ (C) $(M \cup N)'$ (D) $M \cap N'$
2. Indicate true (T) or false (F) for U, M, and N in Problem 1:
   (A) $N \subset M$ (B) $\varnothing \subset U$ (C) $6 \notin M$ (D) $5 \in N$
3. Simplify, and express the answer using positive exponents only:
   $\left(\dfrac{16x^{-2}y^4}{25x^4y^{-4}}\right)^{1/2}$

4. Rationalize the denominator and simplify: $\dfrac{4x^2y^3}{\sqrt{xy}}$

In Problems 5–7 perform the indicated operations and simplify.

5. $3(x - 2)(2x + 3) - (x^2 - x + 2)$

6. $\dfrac{3}{x^2 - 4} - \dfrac{1}{x + 2}$

7. $\dfrac{m^4 - m^2}{8mn^3} \div \dfrac{m^3 - 2m^2 + m}{4m^3n^2}$

Solve.

8. $-2 \le \dfrac{x}{2} - 3 < 3$

9. $\dfrac{1}{3} - \dfrac{x - 1}{2} = \dfrac{1}{6}$

10. $2x^2 = 3x + 1$

11. $2x^2 + 5x > 3$

12. A survey company sampled 1,000 students at a university. Out of the sample, it was found that 550 smoked cigarettes, 820 drank alcoholic beverages, and 470 did both.
    (A) How many smoked or drank?
    (B) How many drank, but did not smoke?

TABLES

TABLE I Exponential Functions (e^x and e^{-x})

x	e^x	e^{-x}	x	e^x	e^{-x}	x	e^x	e^{-x}
0.00	1.0000	1.00 000	0.50	1.6487	0.60 653	1.00	2.7183	0.36 788
0.01	1.0101	0.99 005	0.51	1.6653	0.60 050	1.01	2.7456	0.36 422
0.02	1.0202	0.98 020	0.52	1.6820	0.59 452	1.02	2.7732	0.36 059
0.03	1.0305	0.97 045	0.53	1.6989	0.58 860	1.03	2.8011	0.35 701
0.04	1.0408	0.96 079	0.54	1.7160	0.58 275	1.04	2.8292	0.35 345
0.05	1.0513	0.95 123	0.55	1.7333	0.57 695	1.05	2.8577	0.34 994
0.06	1.0618	0.94 176	0.56	1.7507	0.57 121	1.06	2.8864	0.34 646
0.07	1.0725	0.93 239	0.57	1.7683	0.56 553	1.07	2.9154	0.34 301
0.08	1.0833	0.92 312	0.58	1.7860	0.55 990	1.08	2.9447	0.33 960
0.09	1.0942	0.91 393	0.59	1.8040	0.55 433	1.09	2.9743	0.33 622
0.10	1.1052	0.90 484	0.60	1.8221	0.54 881	1.10	3.0042	0.33 287
0.11	1.1163	0.89 583	0.61	1.8404	0.54 335	1.11	3.0344	0.32 956
0.12	1.1275	0.88 692	0.62	1.8589	0.53 794	1.12	3.0649	0.32 628
0.13	1.1388	0.87 810	0.63	1.8776	0.53 259	1.13	3.0957	0.32 303
0.14	1.1503	0.86 936	0.64	1.8965	0.52 729	1.14	3.1268	0.31 982
0.15	1.1618	0.86 071	0.65	1.9155	0.52 205	1.15	3.1582	0.31 664
0.16	1.1735	0.85 214	0.66	1.9348	0.51 685	1.16	3.1899	0.31 349
0.17	1.1853	0.84 366	0.67	1.9542	0.51 171	1.17	3.2220	0.31 037
0.18	1.1972	0.83 527	0.68	1.9739	0.50 662	1.18	3.2544	0.30 728
0.19	1.2092	0.82 696	0.69	1.9937	0.50 158	1.19	3.2871	0.30 422
0.20	1.2214	0.81 873	0.70	2.0138	0.49 659	1.20	3.3201	0.30 119
0.21	1.2337	0.81 058	0.71	2.0340	0.49 164	1.21	3.3535	0.29 820
0.22	1.2461	0.80 252	0.72	2.0544	0.48 675	1.22	3.3872	0.29 523
0.23	1.2586	0.79 453	0.73	2.0751	0.48 191	1.23	3.4212	0.29 229
0.24	1.2712	0.78 663	0.74	2.0959	0.47 711	1.24	3.4556	0.28 938
0.25	1.2840	0.77 880	0.75	2.1170	0.47 237	1.25	3.4903	0.28 650
0.26	1.2969	0.77 105	0.76	2.1383	0.46 767	1.26	3.5254	0.28 365
0.27	1.3100	0.76 338	0.77	2.1598	0.46 301	1.27	3.5609	0.28 083
0.28	1.3231	0.75 578	0.78	2.1815	0.45 841	1.28	3.5966	0.27 804
0.29	1.3364	0.74 826	0.79	2.2034	0.45 384	1.29	3.6328	0.27 527
0.30	1.3499	0.74 082	0.80	2.2255	0.44 933	1.30	3.6693	0.27 253
0.31	1.3634	0.73 345	0.81	2.2479	0.44 486	1.31	3.7062	0.26 982
0.32	1.3771	0.72 615	0.82	2.2705	0.44 043	1.32	3.7434	0.26 714
0.33	1.3910	0.71 892	0.83	2.2933	0.43 605	1.33	3.7810	0.26 448
0.34	1.4049	0.71 177	0.84	2.3164	0.43 171	1.34	3.8190	0.26 185
0.35	1.4191	0.70 469	0.85	2.3396	0.42 741	1.35	3.8574	0.25 924
0.36	1.4333	0.69 768	0.86	2.3632	0.42 316	1.36	3.8962	0.25 666
0.37	1.4477	0.69 073	0.87	2.3869	0.41 895	1.37	3.9354	0.25 411
0.38	1.4623	0.68 386	0.88	2.4109	0.41 478	1.38	3.9749	0.25 158
0.39	1.4770	0.67 706	0.89	2.4351	0.41 066	1.39	4.0149	0.24 908
0.40	1.4918	0.67 032	0.90	2.4596	0.40 657	1.40	4.0552	0.24 660
0.41	1.5068	0.66 365	0.91	2.4843	0.40 252	1.41	4.0960	0.24 414
0.42	1.5220	0.65 705	0.92	2.5093	0.39 852	1.42	4.1371	0.24 171
0.43	1.5373	0.65 051	0.93	2.5345	0.39 455	1.43	4.1787	0.23 931
0.44	1.5527	0.64 404	0.94	2.5600	0.39 063	1.44	4.2207	0.23 693
0.45	1.5683	0.63 763	0.95	2.5857	0.38 674	1.45	4.2631	0.23 457
0.46	1.5841	0.63 128	0.96	2.6117	0.38 289	1.46	4.3060	0.23 224
0.47	1.6000	0.62 500	0.97	2.6379	0.37 908	1.47	4.3492	0.22 993
0.48	1.6161	0.61 878	0.98	2.6645	0.37 531	1.48	4.3939	0.22 764
0.49	1.6323	0.61 263	0.99	2.6912	0.37 158	1.49	4.4371	0.22 537
0.50	1.6487	0.60 653	1.00	2.7183	0.36 788	1.50	4.4817	0.22 313

x	e^x	e^{-x}	x	e^x	e^{-x}	x	e^x	e^{-x}
1.50	4.4817	0.22 313	2.00	7.3891	0.13 534	2.50	12.182	0.082 085
1.51	4.5267	0.22 091	2.01	7.4633	0.13 399	2.51	12.305	0.081 268
1.52	4.5722	0.21 871	2.02	7.5383	0.13 266	2.52	12.429	0.080 460
1.53	4.6182	0.21 654	2.03	7.6141	0.13 134	2.53	12.554	0.079 659
1.54	4.6646	0.21 438	2.04	7.6906	0.13 003	2.54	12.680	0.078 866
1.55	4.7115	0.21 225	2.05	7.7679	0.12 873	2.55	12.807	0.078 082
1.56	4.7588	0.21 014	2.06	7.8460	0.12 745	2.56	12.936	0.077 305
1.57	4.8066	0.20 805	2.07	7.9248	0.12 619	2.57	13.066	0.076 536
1.58	4.8550	0.20 598	2.08	8.0045	0.12 493	2.58	13.197	0.075 774
1.59	4.9037	0.20 393	2.09	8.0849	0.12 369	2.59	13.330	0.075 020
1.60	4.9530	0.20 190	2.10	8.1662	0.12 246	2.60	13.464	0.074 274
1.61	5.0028	0.19 989	2.11	8.2482	0.12 124	2.61	13.599	0.073 535
1.62	5.0531	0.19 790	2.12	8.3311	0.12 003	2.62	13.736	0.072 803
1.63	5.1039	0.19 593	2.13	8.4149	0.11 884	2.63	13.874	0.072 078
1.64	5.1552	0.19 398	2.14	8.4994	0.11 765	2.64	14.013	0.071 361
1.65	5.2070	0.19 205	2.15	8.5849	0.11 648	2.65	14.154	0.070 651
1.66	5.2593	0.19 014	2.16	8.6711	0.11 533	2.66	14.296	0.069 948
1.67	5.3122	0.18 825	2.17	8.7583	0.11 418	2.67	14.440	0.069 252
1.68	5.3656	0.18 637	2.18	8.8463	0.11 304	2.68	14.585	0.068 563
1.69	5.4195	0.18 452	2.19	8.9352	0.11 192	2.69	14.732	0.067 881
1.70	5.4739	0.18 268	2.20	9.0250	0.11 080	2.70	14.880	0.067 206
1.71	5.5290	0.18 087	2.21	9.1157	0.10 970	2.71	15.029	0.066 537
1.72	5.5845	0.17 907	2.22	9.2073	0.10 861	2.72	15.180	0.065 875
1.73	5.6407	0.17 728	2.23	9.2999	0.10 753	2.73	15.333	0.065 219
1.74	5.6973	0.17 552	2.24	9.3933	0.10 646	2.74	15.487	0.064 570
1.75	5.7546	0.17 377	2.25	9.4877	0.10 540	2.75	15.643	0.063 928
1.76	5.8124	0.17 204	2.26	9.5831	0.10 435	2.76	15.800	0.063 292
1.77	5.8700	0.17 033	2.27	9.6794	0.10 331	2.77	15.959	0.062 662
1.78	5.9299	0.16 864	2.28	9.7767	0.10 228	2.78	16.119	0.062 039
1.79	5.9895	0.16 696	2.29	9.8749	0.10 127	2.79	16.281	0.061 421
1.80	6.0496	0.16 530	2.30	9.9742	0.10 026	2.80	16.445	0.060 810
1.81	6.1104	0.16 365	2.31	10.074	0.099 261	2.81	16.610	0.060 205
1.82	6.1719	0.16 203	2.32	10.176	0.098 274	2.82	16.777	0.059 606
1.83	6.2339	0.16 041	2.33	10.278	0.097 296	2.83	16.945	0.059 013
1.84	6.2965	0.15 882	2.34	10.381	0.096 328	2.84	17.116	0.058 426
1.85	6.3598	0.15 724	2.35	10.486	0.095 369	2.85	17.288	0.057 844
1.86	6.4237	0.15 567	2.36	10.591	0.094 420	2.86	17.462	0.057 269
1.87	6.4883	0.15 412	2.37	10.697	0.093 481	2.87	17.637	0.056 699
1.88	6.5535	0.15 259	2.38	10.805	0.092 551	2.88	17.814	0.056 135
1.89	6.6194	0.15 107	2.39	10.913	0.091 630	2.89	17.993	0.055 576
1.90	6.6859	0.14 957	2.40	11.023	0.090 718	2.90	18.174	0.055 023
1.91	6.7531	0.14 808	2.41	11.134	0.089 815	2.91	18.357	0.054 476
1.92	6.8210	0.14 661	2.42	11.246	0.088 922	2.92	18.541	0.053 934
1.93	6.8895	0.14 515	2.43	11.359	0.088 037	2.93	18.728	0.053 397
1.94	6.9588	0.14 370	2.44	11.473	0.087 161	2.94	18.916	0.052 866
1.95	7.0287	0.14 227	2.45	11.588	0.086 294	2.95	19.106	0.052 340
1.96	7.0993	0.14 086	2.46	11.705	0.085 435	2.96	19.298	0.051 819
1.97	7.1707	0.13 946	2.47	11.822	0.084 585	2.97	19.492	0.051 303
1.98	7.2427	0.13 807	2.48	11.941	0.083 743	2.98	19.688	0.050 793
1.99	7.3155	0.13 670	2.49	12.061	0.082 910	2.99	19.886	0.050 287
2.00	7.3891	0.13 534	2.50	12.182	0.082 085	3.00	20.086	0.049 787

TABLE I   (Continued)

x	e^x	e^{-x}	x	e^x	e^{-x}	x	e^x	e^{-x}
3.00	20.086	0.049 787	3.50	33.115	0.030 197	4.00	54.598	0.018 316
3.01	20.287	0.049 292	3.51	33.448	0.029 897	4.01	55.147	0.018 133
3.02	20.491	0.048 801	3.52	33.784	0.029 599	4.02	55.701	0.017 953
3.03	20.697	0.048 316	3.53	34.124	0.029 305	4.03	56.261	0.017 774
3.04	20.905	0.047 835	3.54	34.467	0.029 013	4.04	56.826	0.017 597
3.05	21.115	0.047 359	3.55	34.813	0.028 725	4.05	57.397	0.017 422
3.05	21.328	0.046 888	3.56	35.163	0.028 439	4.06	57.974	0.017 249
3.07	21.542	0.046 421	3.57	35.517	0.028 156	4.07	58.557	0.017 077
3.08	21.758	0.045 959	3.58	35.874	0.027 876	4.08	59.145	0.016 907
3.09	21.977	0.045 502	3.59	36.234	0.027 598	4.09	59.740	0.016 739
3.10	22.198	0.045 049	3.60	36.598	0.027 324	4.10	60.340	0.016 573
3.11	22.421	0.044 601	3.61	36.966	0.027 052	4.11	60.947	0.016 408
3.12	22.646	0.044 157	3.62	37.338	0.026 783	4.12	61.559	0.016 245
3.13	22.874	0.043 718	3.63	37.713	0.026 516	4.13	62.178	0.016 083
3.14	23.104	0.043 283	3.64	38.092	0.026 252	4.14	62.803	0.015 923
3.15	23.336	0.042 852	3.65	38.475	0.025 991	4.15	63.434	0.015 764
3.16	23.571	0.042 426	3.66	38.861	0.025 733	4.16	64.072	0.015 608
3.17	23.807	0.042 004	3.67	39.252	0.025 476	4.17	64.715	0.015 452
3.18	24.047	0.041 586	3.68	39.646	0.025 223	4.18	65.366	0.015 299
3.19	24.288	0.041 172	3.69	40.045	0.024 972	4.19	66.023	0.015 146
3.20	24.533	0.040 762	3.70	40.447	0.024 724	4.20	66.686	0.014 996
3.21	24.779	0.040 357	3.71	40.854	0.024 478	4.21	67.357	0.014 846
3.22	25.028	0.039 955	3.72	41.264	0.024 234	4.22	68.033	0.014 699
3.23	25.280	0.039 557	3.73	41.679	0.023 993	4.23	68.717	0.014 552
3.24	25.534	0.039 164	3.74	42.098	0.023 754	4.24	69.408	0.014 408
3.25	25.790	0.038 774	3.75	42.521	0.023 518	4.25	70.105	0.014 264
3.26	26.050	0.038 388	3.76	42.948	0.023 284	4.26	70.810	0.014 122
3.27	26.311	0.038 006	3.77	43.380	0.023 052	4.27	71.522	0.013 982
3.28	26.576	0.037 628	3.78	43.816	0.022 823	4.28	72.240	0.013 843
3.29	26.843	0.037 254	3.79	44.256	0.022 596	4.29	72.966	0.013 705
3.30	27.113	0.036 883	3.80	44.701	0.022 371	4.30	73.700	0.013 569
3.31	27.385	0.036 516	3.81	45.150	0.022 148	4.31	74.440	0.013 434
3.32	27.660	0.036 153	3.82	45.604	0.021 928	4.32	75.189	0.013 300
3.33	27.938	0.035 793	3.83	46.063	0.021 710	4.33	75.944	0.013 168
3.34	28.219	0.035 437	3.84	46.525	0.021 494	4.34	76.708	0.013 037
3.35	28.503	0.035 084	3.85	46.993	0.021 280	4.35	77.478	0.012 907
3.36	28.789	0.034 735	3.86	47.465	0.021 068	4.36	78.257	0.012 778
3.37	29.079	0.034 390	3.87	47.942	0.020 858	4.37	79.044	0.012 651
3.38	29.371	0.034 047	3.88	48.424	0.020 651	4.38	79.838	0.012 525
3.39	29.666	0.033 709	3.89	48.911	0.020 445	4.39	80.640	0.012 401
3.40	29.964	0.033 373	3.90	49.402	0.020 242	4.40	81.451	0.012 277
3.41	30.265	0.033 041	3.91	49.899	0.020 041	4.41	82.269	0.012 155
3.42	30.569	0.032 712	3.92	50.400	0.019 841	4.42	83.096	0.012 034
3.43	30.877	0.032 387	3.93	50.907	0.019 644	4.43	83.931	0.011 914
3.44	31.187	0.032 065	3.94	51.419	0.019 448	4.44	84.775	0.011 796
3.45	31.500	0.031 746	3.95	51.935	0.019 255	4.45	85.627	0.011 679
3.46	31.817	0.031 430	3.96	52.457	0.019 063	4.46	86.488	0.011 562
3.47	32.137	0.031 117	3.97	52.985	0.018 873	4.47	87.357	0.011 447
3.48	32.460	0.030 807	3.98	53.517	0.018 686	4.48	88.235	0.011 333
3.49	32.786	0.030 501	3.99	54.055	0.018 500	4.49	89.121	0.011 221
3.50	33.115	0.030 197	4.00	54.598	0.018 316	4.50	90.017	0.011 109

x	e^x	e^{-x}	x	e^x	e^{-x}	x	e^x	e^{-x}
4.50	90.017	0.011 109	5.00	148.41	0.006 7379	7.50	1,808.0	0.000 5531
4.51	90.922	0.010 998	5.05	156.02	0.006 4093	7.55	1,900.7	0.000 5261
4.52	91.836	0.010 889	5.10	164.02	0.006 0967	7.60	1,998.2	0.000 5005
4.53	92.759	0.010 781	5.15	172.43	0.005 7994	7.65	2,100.6	0.000 4760
4.54	93.691	0.010 673	5.20	181.27	0.005 5166	7.70	2,208.3	0.000 4528
4.55	94.632	0.010 567	5.25	190.57	0.005 2475	7.75	2,321.6	0.000 4307
4.56	95.583	0.010 462	5.30	200.34	0.004 9916	7.80	2,440.6	0.000 4097
4.57	96.544	0.010 358	5.35	210.61	0.004 7482	7.85	2,565.7	0.000 3898
4.58	97.514	0.010 255	5.40	221.41	0.004 5166	7.90	2,697.3	0.000 3707
4.59	98.494	0.010 153	5.45	232.76	0.004 2963	7.95	2,835.6	0.000 3527
4.60	99.484	0.010 052	5.50	244.69	0.004 0868	8.00	2,981.0	0.000 3355
4.61	100.48	0.009 9518	5.55	257.24	0.003 8875	8.05	3,133.8	0.000 3191
4.62	101.49	0.009 8528	5.60	270.43	0.003 6979	8.10	3,294.5	0.000 3035
4.63	102.51	0.009 7548	5.65	284.29	0.003 5175	8.15	3,463.4	0.000 2887
4.64	103.54	0.009 6577	5.70	298.87	0.003 3460	8.20	3,641.0	0.000 2747
4.65	104.58	0.009 5616	5.75	314.19	0.003 1828	8.25	3,827.6	0.000 2613
4.66	105.64	0.009 4665	5.80	330.30	0.003 0276	8.30	4,023.9	0.000 2485
4.67	106.70	0.009 3723	5.85	347.23	0.002 8799	8.35	4,230.2	0.000 2364
4.68	107.77	0.009 2790	5.90	365.04	0.002 7394	8.40	4,447.1	0.000 2249
4.69	108.85	0.009 1867	5.95	383.75	0.002 6058	8.45	4,675.1	0.000 2139
4.70	109.95	0.009 0953	6.00	403.43	0.002 4788	8.50	4,914.8	0.000 2035
4.71	111.05	0.009 0048	6.05	424.11	0.002 3579	8.55	5,166.8	0.000 1935
4.72	112.17	0.008 9152	6.10	445.86	0.002 2429	8.60	5,431.7	0.000 1841
4.73	113.30	0.008 8265	6.15	468.72	0.002 1335	8.65	5,710.1	0.000 1751
4.74	114.43	0.008 7386	6.20	492.75	0.002 2094	8.70	6,002.9	0.000 1666
4.75	115.58	0.008 6517	6.25	518.01	0.001 9305	8.75	6,310.7	0.000 1585
4.76	116.75	0.008 5656	6.30	544.57	0.001 8363	8.80	6,634.2	0.000 1507
4.77	117.92	0.008 4804	6.35	573.10	0.001 7467	8.85	6,974.4	0.000 1434
4.78	119.10	0.008 3960	6.40	601.85	0.001 6616	8.90	7,332.0	0.000 1364
4.79	120.30	0.008 3125	6.45	632.70	0.001 5805	8.95	7,707.9	0.000 1297
4.80	121.51	0.008 2297	6.50	665.14	0.001 5034	9.00	8,103.1	0.000 1234
4.81	122.73	0.008 1479	6.55	699.24	0.001 4301	9.05	8,518.5	0.000 1174
4.82	123.97	0.008 0668	6.60	735.10	0.001 3604	9.10	8,955.3	0.000 1117
4.83	125.21	0.007 9865	6.65	772.78	0.001 2940	9.15	9,414.4	0.000 1062
4.84	126.47	0.007 9071	6.70	812.41	0.001 2309	9.20	9,897.1	0.000 1010
4.85	127.74	0.007 8284	6.75	854.06	0.001 1709	9.25	10,405	0.000 0961
4.86	129.02	0.007 7505	6.80	897.85	0.001 1138	9.30	10,938	0.000 0914
4.87	130.32	0.007 6734	6.85	943.88	0.001 0595	9.35	11,499	0.000 0870
4.88	131.63	0.007 5970	6.90	992.27	0.001 0078	9.40	12,088	0.000 0827
4.89	132.95	0.007 5214	6.95	1,043.1	0.000 9586	9.45	12,708	0.000 0787
4.90	134.29	0.007 4466	7.00	1,096.6	0.000 9119	9.50	13,360	0.000 0749
4.91	135.64	0.007 3725	7.05	1,152.9	0.000 8674	9.55	14,045	0.000 0712
4.92	137.00	0.007 2991	7.10	1,212.0	0.000 8251	9.60	14,765	0.000 0677
4.93	138.38	0.007 2265	7.15	1,274.1	0.000 7849	9.65	15,522	0.000 0644
4.94	139.77	0.007 1546	7.20	1,339.4	0.000 7466	9.70	16,318	0.000 0613
4.95	141.17	0.007 0834	7.25	1,408.1	0.000 7102	9.75	17,154	0.000 0583
4.96	142.59	0.007 0129	7.30	1,480.3	0.000 6755	9.80	18,034	0.000 0555
4.97	144.03	0.006 9431	7.35	1,556.2	0.000 6426	9.85	18,958	0.000 0527
4.98	145.47	0.006 8741	7.40	1,636.0	0.000 6113	9.90	19,930	0.000 0502
4.99	146.94	0.006 8057	7.45	1,719.9	0.000 5814	9.95	20,952	0.000 0477
5.00	148.41	0.006 7379	7.50	1,808.0	0.000 5531	10.00	22,026	0.000 0454

N	0	1	2	3	4	5	6	7	8	9
1.0	0.0000	0.004321	0.008600	0.01284	0.01703	0.02119	0.02531	0.02938	0.03342	0.03743
1.1	0.04139	0.04532	0.04922	0.05308	0.05690	0.06070	0.06446	0.06819	0.07188	0.07555
1.2	0.07918	0.08279	0.08636	0.08991	0.09342	0.09691	0.1004	0.1038	0.1072	0.1106
1.3	0.1139	0.1173	0.1206	0.1239	0.1271	0.1303	0.1335	0.1367	0.1399	0.1430
1.4	0.1461	0.1492	0.1523	0.1553	0.1584	0.1614	0.1644	0.1673	0.1703	0.1732
1.5	0.1761	0.1790	0.1818	0.1847	0.1875	0.1903	0.1931	0.1959	0.1987	0.2014
1.6	0.2041	0.2068	0.2095	0.2122	0.2148	0.2175	0.2201	0.2227	0.2253	0.2279
1.7	0.2304	0.2330	0.2355	0.2380	0.2405	0.2430	0.2455	0.2480	0.2504	0.2529
1.8	0.2553	0.2577	0.2601	0.2625	0.2648	0.2673	0.2695	0.2718	0.2742	0.2765
1.9	0.2788	0.2810	0.2833	0.2856	0.2878	0.2900	0.2923	0.2945	0.2967	0.2989
2.0	0.3010	0.3032	0.3054	0.3075	0.3096	0.3118	0.3139	0.3160	0.3181	0.3201
2.1	0.3222	0.3243	0.3263	0.3284	0.3304	0.3324	0.3345	0.3365	0.3385	0.3404
2.2	0.3424	0.3444	0.3464	0.3483	0.3502	0.3522	0.3541	0.3560	0.3579	0.3598
2.3	0.3617	0.3636	0.3655	0.3674	0.3692	0.3711	0.3729	0.3747	0.3766	0.3784
2.4	0.3802	0.3820	0.3838	0.3856	0.3874	0.3892	0.3909	0.3927	0.3945	0.3962
2.5	0.3979	0.3997	0.4014	0.4031	0.4048	0.4065	0.4082	0.4099	0.4116	0.4133
2.6	0.4150	0.4166	0.4183	0.4200	0.4216	0.4232	0.4249	0.4265	0.4281	0.4298
2.7	0.4314	0.4330	0.4346	0.4362	0.4378	0.4393	0.4409	0.4425	0.4440	0.4456
2.8	0.4472	0.4487	0.4502	0.4518	0.4533	0.4548	0.4564	0.4579	0.4594	0.4609
2.9	0.4624	0.4639	0.4654	0.4669	0.4683	0.4698	0.4713	0.4728	0.4742	0.4757
3.0	0.4771	0.4786	0.4800	0.4814	0.4829	0.4843	0.4857	0.4871	0.4886	0.4900
3.1	0.4914	0.4928	0.4942	0.4955	0.4969	0.4983	0.4997	0.5011	0.5024	0.5038
3.2	0.5051	0.5065	0.5079	0.5092	0.5105	0.5119	0.5132	0.5145	0.5159	0.5172
3.3	0.5185	0.5198	0.5211	0.5224	0.5237	0.5250	0.5263	0.5276	0.5289	0.5302
3.4	0.5315	0.5328	0.5340	0.5353	0.5366	0.5378	0.5391	0.5403	0.5416	0.5428
3.5	0.5441	0.5453	0.5465	0.5478	0.5490	0.5502	0.5514	0.5527	0.5539	0.5551
3.6	0.5563	0.5575	0.5587	0.5599	0.5611	0.5623	0.5635	0.5647	0.5658	0.5670
3.7	0.5682	0.5694	0.5705	0.5717	0.5729	0.5740	0.5752	0.5763	0.5775	0.5786
3.8	0.5798	0.5809	0.5821	0.5832	0.5843	0.5855	0.5866	0.5877	0.5888	0.5899
3.9	0.5911	0.5922	0.5933	0.5944	0.5955	0.5966	0.5977	0.5988	0.5999	0.6010
4.0	0.6021	0.6031	0.6042	0.6053	0.6064	0.6075	0.6085	0.6096	0.6107	0.6117
4.1	0.6128	0.6138	0.6149	0.6160	0.6170	0.6180	0.6191	0.6201	0.6212	0.6222
4.2	0.6232	0.6243	0.6253	0.6263	0.6274	0.6284	0.6294	0.6304	0.6314	0.6325
4.3	0.6335	0.6345	0.6355	0.6365	0.6375	0.6385	0.6395	0.6405	0.6415	0.6425
4.4	0.6435	0.6444	0.6454	0.6464	0.6474	0.6484	0.6493	0.6503	0.6513	0.6522
4.5	0.6532	0.6542	0.6551	0.6561	0.6571	0.6580	0.6590	0.6599	0.6609	0.6618
4.6	0.6628	0.6637	0.6646	0.6656	0.6665	0.6675	0.6684	0.6693	0.6702	0.6712
4.7	0.6721	0.6730	0.6739	0.6749	0.6758	0.6767	0.6776	0.6785	0.6794	0.6803
4.8	0.6812	0.6821	0.6830	0.6839	0.6848	0.6857	0.6866	0.6875	0.6884	0.6893
4.9	0.6902	0.6911	0.6920	0.6928	0.6937	0.6946	0.6955	0.6964	0.6972	0.6981
5.0	0.6990	0.6998	0.7007	0.7016	0.7024	0.7033	0.7042	0.7050	0.7059	0.7067
5.1	0.7076	0.7084	0.7093	0.7101	0.7110	0.7118	0.7126	0.7135	0.7143	0.7152
5.2	0.7160	0.7168	0.7177	0.7185	0.7193	0.7202	0.7210	0.7218	0.7226	0.7235
5.3	0.7243	0.7251	0.7259	0.7267	0.7275	0.7284	0.7292	0.7300	0.7308	0.7316
5.4	0.7324	0.7332	0.7340	0.7348	0.7356	0.7364	0.7372	0.7380	0.7388	0.7396

N	0	1	2	3	4	5	6	7	8	9
5.5	0.7404	0.7412	0.7419	0.7427	0.7435	0.7443	0.7451	0.7459	0.7466	0.7474
5.6	0.7482	0.7490	0.7497	0.7505	0.7513	0.7520	0.7528	0.7536	0.7543	0.7551
5.7	0.7559	0.7566	0.7574	0.7582	0.7589	0.7597	0.7604	0.7612	0.7619	0.7627
5.8	0.7634	0.7642	0.7649	0.7657	0.7664	0.7672	0.7679	0.7686	0.7694	0.7701
5.9	0.7709	0.7716	0.7723	0.7731	0.7738	0.7745	0.7752	0.7760	0.7767	0.7774
6.0	0.7782	0.7789	0.7796	0.7803	0.7810	0.7818	0.7825	0.7832	0.7839	0.7846
6.1	0.7853	0.7860	0.7868	0.7875	0.7882	0.7889	0.7896	0.7903	0.7910	0.7917
6.2	0.7924	0.7931	0.7938	0.7945	0.7952	0.7959	0.7966	0.7973	0.7980	0.7987
6.3	0.7993	0.8000	0.8007	0.8014	0.8021	0.8028	0.8035	0.8041	0.8048	0.8055
6.4	0.8062	0.8069	0.8075	0.8082	0.8089	0.8096	0.8102	0.8109	0.8116	0.8122
6.5	0.8129	0.8136	0.8142	0.8149	0.8156	0.8162	0.8169	0.8176	0.8182	0.8189
6.6	0.8195	0.8202	0.8209	0.8215	0.8222	0.8228	0.8235	0.8241	0.8248	0.8254
6.7	0.8261	0.8267	0.8274	0.8280	0.8287	0.8293	0.8299	0.8306	0.8312	0.8319
6.8	0.8325	0.8331	0.8338	0.8344	0.8351	0.8357	0.8363	0.8370	0.8376	0.8382
6.9	0.8388	0.8395	0.8401	0.8407	0.8414	0.8420	0.8426	0.8432	0.8439	0.8445
7.0	0.8451	0.8457	0.8463	0.8470	0.8476	0.8482	0.8488	0.8494	0.8500	0.8506
7.1	0.8513	0.8519	0.8525	0.8531	0.8537	0.8543	0.8549	0.8555	0.8561	0.8567
7.2	0.8573	0.8579	0.8585	0.8591	0.8597	0.8603	0.8609	0.8615	0.8621	0.8627
7.3	0.8633	0.8639	0.8645	0.8651	0.8657	0.8663	0.8669	0.8675	0.8681	0.8686
7.4	0.8692	0.8698	0.8704	0.8710	0.8716	0.8722	0.8727	0.8733	0.8739	0.8745
7.5	0.8751	0.8756	0.8762	0.8768	0.8774	0.8779	0.8785	0.8791	0.8797	0.8802
7.6	0.8808	0.8814	0.8820	0.8825	0.8831	0.8837	0.8842	0.8848	0.8854	0.8859
7.7	0.8865	0.8871	0.8876	0.8882	0.8887	0.8893	0.8899	0.8904	0.8910	0.8915
7.8	0.8921	0.8927	0.8932	0.8938	0.8943	0.8949	0.8954	0.8960	0.8965	0.8971
7.9	0.8976	0.8982	0.8987	0.8993	0.8998	0.9004	0.9009	0.9015	0.9020	0.9025
8.0	0.9031	0.9036	0.9042	0.9047	0.9053	0.9058	0.9063	0.9069	0.9074	0.9079
8.1	0.9085	0.9090	0.9096	0.9101	0.9106	0.9112	0.9117	0.9122	0.9128	0.9133
8.2	0.9138	0.9143	0.9149	0.9154	0.9159	0.9165	0.9170	0.9175	0.9180	0.9186
8.3	0.9191	0.9196	0.9201	0.9206	0.9212	0.9217	0.9222	0.9227	0.9232	0.9238
8.4	0.9243	0.9248	0.9253	0.9258	0.9263	0.9269	0.9274	0.9279	0.9284	0.9289
8.5	0.9294	0.9299	0.9304	0.9309	0.9315	0.9320	0.9325	0.9330	0.9335	0.9340
8.6	0.9345	0.9350	0.9355	0.9360	0.9365	0.9370	0.9375	0.9380	0.9385	0.9390
8.7	0.9395	0.9400	0.9405	0.9410	0.9415	0.9420	0.9425	0.9430	0.9435	0.9440
8.8	0.9445	0.9450	0.9455	0.9460	0.9465	0.9469	0.9474	0.9479	0.9484	0.9489
8.9	0.9494	0.9499	0.9504	0.9509	0.9513	0.9518	0.9523	0.9528	0.9533	0.9538
9.0	0.9542	0.9547	0.9552	0.9557	0.9562	0.9566	0.9571	0.9576	0.9581	0.9586
9.1	0.9590	0.9595	0.9600	0.9605	0.9609	0.9614	0.9619	0.9624	0.9628	0.9633
9.2	0.9638	0.9643	0.9647	0.9652	0.9657	0.9661	0.9666	0.9671	0.9675	0.9680
9.3	0.9685	0.9689	0.9694	0.9699	0.9703	0.9708	0.9713	0.9717	0.9722	0.9727
9.4	0.9731	0.9736	0.9741	0.9745	0.9750	0.9754	0.9759	0.9763	0.9768	0.9773
9.5	0.9777	0.9782	0.9786	0.9791	0.9795	0.9800	0.9805	0.9809	0.9814	0.9818
9.6	0.9823	0.9827	0.9832	0.9836	0.9841	0.9845	0.9850	0.9854	0.9859	0.9863
9.7	0.9868	0.9872	0.9877	0.9881	0.9886	0.9890	0.9894	0.9899	0.9903	0.9908
9.8	0.9912	0.9917	0.9921	0.9926	0.9930	0.9934	0.9939	0.9943	0.9948	0.9952
9.9	0.9956	0.9961	0.9965	0.9969	0.9974	0.9978	0.9983	0.9987	0.9991	0.9996

TABLE III   Natural logarithms (ln N = log$_e$ N)

ln 10 = 2.3026	5 ln 10 = 11.5130	9 ln 10 = 20.7233
2 ln 10 = 4.6052	6 ln 10 = 13.8155	10 ln 10 = 23.0259
3 ln 10 = 6.9078	7 ln 10 = 16.1181	
4 ln 10 = 9.2103	8 ln 10 = 18.4207	

N	.00	.01	.02	.03	.04	.05	.06	.07	.08	.09
1.0	0.0000	0.0100	0.0198	0.0296	0.0392	0.0488	0.0583	0.0677	0.0770	0.0862
1.1	0.0953	0.1044	0.1133	0.1222	0.1310	0.1398	0.1484	0.1570	0.1655	0.1740
1.2	0.1823	0.1906	0.1989	0.2070	0.2151	0.2231	0.2311	0.2390	0.2469	0.2546
1.3	0.2624	0.2700	0.2776	0.2852	0.2927	0.3001	0.3075	0.3148	0.3221	0.3293
1.4	0.3365	0.3436	0.3507	0.3577	0.3646	0.3716	0.3784	0.3853	0.3920	0.3988
1.5	0.4055	0.4121	0.4187	0.4253	0.4318	0.4383	0.4447	0.4511	0.4574	0.4637
1.6	0.4700	0.4762	0.4824	0.4886	0.4947	0.5008	0.5068	0.5128	0.5188	0.5247
1.7	0.5306	0.5365	0.5423	0.5481	0.5539	0.5596	0.5653	0.5710	0.5766	0.5822
1.8	0.5878	0.5933	0.5988	0.6043	0.6098	0.6152	0.6206	0.6259	0.6313	0.6366
1.9	0.6419	0.6471	0.6523	0.6575	0.6627	0.6678	0.6729	0.6780	0.6831	0.6881
2.0	0.6931	0.6981	0.7031	0.7080	0.7129	0.7178	0.7227	0.7275	0.7324	0.7372
2.1	0.7419	0.7467	0.7514	0.7561	0.7608	0.7655	0.7701	0.7747	0.7793	0.7839
2.2	0.7885	0.7930	0.7975	0.8020	0.8065	0.8109	0.8154	0.8198	0.8242	0.8286
2.3	0.8329	0.8372	0.8416	0.8459	0.8502	0.8544	0.8587	0.8629	0.8671	0.8713
2.4	0.8755	0.8796	0.8838	0.8879	0.8920	0.8961	0.9002	0.9042	0.9083	0.9123
2.5	0.9163	0.9203	0.9243	0.9282	0.9322	0.9361	0.9400	0.9439	0.9478	0.9517
2.6	0.9555	0.9594	0.9632	0.9670	0.9708	0.9746	0.9783	0.9821	0.9858	0.9895
2.7	0.9933	0.9969	1.0006	1.0043	1.0080	1.0116	1.0152	1.0188	1.0225	1.0260
2.8	1.0296	1.0332	1.0367	1.0403	1.0438	1.0473	1.0508	1.0543	1.0578	1.0613
2.9	1.0647	1.0682	1.0716	1.0750	1.0784	1.0818	1.0852	1.0886	1.0919	1.0953
3.0	1.0986	1.1019	1.1053	1.1086	1.1119	1.1151	1.1184	1.1217	1.1249	1.1282
3.1	1.1314	1.1346	1.1378	1.1410	1.1442	1.1474	1.1506	1.1537	1.1569	1.1600
3.2	1.1632	1.1663	1.1694	1.1725	1.1756	1.1787	1.1817	1.1848	1.1878	1.1909
3.3	1.1939	1.1969	1.2000	1.2030	1.2060	1.2090	1.2119	1.2149	1.2179	1.2208
3.4	1.2238	1.2267	1.2296	1.2326	1.2355	1.2384	1.2413	1.2442	1.2470	1.2499
3.5	1.2528	1.2556	1.2585	1.2613	1.2641	1.2669	1.2698	1.2726	1.2754	1.2782
3.6	1.2809	1.2837	1.2865	1.2892	1.2920	1.2947	1.2975	1.3002	1.3029	1.3056
3.7	1.3083	1.3110	1.3137	1.3164	1.3191	1.3218	1.3244	1.3271	1.3297	1.3324
3.8	1.3350	1.3376	1.3403	1.3429	1.3455	1.3481	1.3507	1.3533	1.3558	1.3584
3.9	1.3610	1.3635	1.3661	1.3686	1.3712	1.3737	1.3762	1.3788	1.3813	1.3838
4.0	1.3863	1.3888	1.3913	1.3938	1.3962	1.3987	1.4012	1.4036	1.4061	1.4085
4.1	1.4110	1.4134	1.4159	1.4183	1.4207	1.4231	1.4255	1.4279	1.4303	1.4327
4.2	1.4351	1.4375	1.4398	1.4422	1.4446	1.4469	1.4493	1.4516	1.4540	1.4563
4.3	1.4586	1.4609	1.4633	1.4656	1.4679	1.4702	1.4725	1.4748	1.4770	1.4793
4.4	1.4816	1.4839	1.4861	1.4884	1.4907	1.4929	1.4951	1.4974	1.4996	1.5019
4.5	1.5041	1.5063	1.5085	1.5107	1.5129	1.5151	1.5173	1.5195	1.5217	1.5239
4.6	1.5261	1.5282	1.5304	1.5326	1.5347	1.5369	1.5390	1.5412	1.5433	1.5454
4.7	1.5476	1.5497	1.5518	1.5539	1.5560	1.5581	1.5602	1.5623	1.5644	1.5665
4.8	1.5686	1.5707	1.5728	1.5748	1.5769	1.5790	1.5810	1.5831	1.5851	1.5872
4.9	1.5892	1.5913	1.5933	1.5953	1.5974	1.5994	1.6014	1.6034	1.6054	1.6074
5.0	1.6094	1.6114	1.6134	1.6154	1.6174	1.6194	1.6214	1.6233	1.6253	1.6273
5.1	1.6292	1.6312	1.6332	1.6351	1.6371	1.6390	1.6409	1.6429	1.6448	1.6467
5.2	1.6487	1.6506	1.6525	1.6544	1.6563	1.6582	1.6601	1.6620	1.6639	1.6658
5.3	1.6677	1.6696	1.6715	1.6734	1.6752	1.6771	1.6790	1.6808	1.6827	1.6845
5.4	1.6864	1.6882	1.6901	1.6919	1.6938	1.6956	1.6974	1.6993	1.7011	1.7029

Note: ln 35, 200 = ln (3.52 × 10⁴) = ln 3.52 + 4 ln 10
       ln 0.00864 = ln (8.64 × 10⁻³) = ln 8.64 − 3 ln 10

N	.00	.01	.02	.03	.04	.05	.06	.07	.08	.09
5.5	1.7047	1.7066	1.7084	1.7102	1.7120	1.7138	1.7156	1.7174	1.7192	1.7210
5.6	1.7228	1.7246	1.7263	1.7281	1.7299	1.7317	1.7334	1.7352	1.7370	1.7387
5.7	1.7405	1.7422	1.7440	1.7457	1.7475	1.7492	1.7509	1.7527	1.7544	1.7561
5.8	1.7579	1.7596	1.7613	1.7630	1.7647	1.7664	1.7681	1.7699	1.7716	1.7733
5.9	1.7750	1.7766	1.7783	1.7800	1.7817	1.7834	1.7851	1.7867	1.7884	1.7901
6.0	1.7918	1.7934	1.7951	1.7967	1.7984	1.8001	1.8017	1.8034	1.8050	1.8066
6.1	1.8083	1.8099	1.8116	1.8132	1.8148	1.8165	1.8181	1.8197	1.8213	1.8229
6.2	1.8245	1.8262	1.8278	1.8294	1.8310	1.8326	1.8342	1.8358	1.8374	1.8390
6.3	1.8405	1.8421	1.8437	1.8453	1.8469	1.8485	1.8500	1.8516	1.8532	1.8547
6.4	1.8563	1.8579	1.8594	1.8610	1.8625	1.8641	1.8656	1.8672	1.8687	1.8703
6.5	1.8718	1.8733	1.8749	1.8764	1.8779	1.8795	1.8810	1.8825	1.8840	1.8856
6.6	1.8871	1.8886	1.8901	1.8916	1.8931	1.8946	1.8961	1.8976	1.8991	1.9006
6.7	1.9021	1.9036	1.9051	1.9066	1.9081	1.9095	1.9110	1.9125	1.9140	1.9155
6.8	1.9169	1.9184	1.9199	1.9213	1.9228	1.9242	1.9257	1.9272	1.9286	1.9301
6.9	1.9315	1.9330	1.9344	1.9359	1.9373	1.9387	1.9402	1.9416	1.9430	1.9445
7.0	1.9459	1.9473	1.9488	1.9502	1.9516	1.9530	1.9544	1.9559	1.9573	1.9587
7.1	1.9601	1.9615	1.9629	1.9643	1.9657	1.9671	1.9685	1.9699	1.9713	1.9727
7.2	1.9741	1.9755	1.9769	1.9782	1.9796	1.9810	1.9824	1.9838	1.9851	1.9865
7.3	1.9879	1.9892	1.9906	1.9920	1.9933	1.9947	1.9961	1.9974	1.9988	2.0001
7.4	2.0015	2.0028	2.0042	2.0055	2.0069	2.0082	2.0096	2.0109	2.0122	2.0136
7.5	2.0149	2.0162	2.0176	2.0189	2.0202	2.0215	2.0229	2.0242	2.0255	2.0268
7.6	2.0281	2.0295	2.0308	2.0321	2.0334	2.0347	2.0360	2.0373	2.0386	2.0399
7.7	2.0412	2.0425	2.0438	2.0451	2.0464	2.0477	2.0490	2.0503	2.0510	2.0520
7.8	2.0541	2.0554	2.0567	2.0580	2.0592	2.0605	2.0618	2.0631	2.0643	2.0656
7.9	2.0669	2.0681	2.0694	2.0707	2.0719	2.0732	2.0744	2.0757	2.0769	2.0782
8.0	2.0794	2.0807	2.0819	2.0832	2.0844	2.0857	2.0869	2.0882	2.0894	2.0906
8.1	2.0919	2.0931	2.0943	2.0956	2.0968	2.0980	2.0992	2.1005	2.1017	2.1029
8.2	2.1041	2.1054	2.1066	2.1078	2.1090	2.1102	2.1114	2.1126	2.1138	2.1150
8.3	2.1163	2.1175	2.1187	2.1199	2.1211	2.1223	2.1235	2.1247	2.1258	2.1270
8.4	2.1282	2.1294	2.1306	2.1318	2.1330	2.1342	2.1353	2.1365	2.1377	2.1389
8.5	2.1401	2.1412	2.1424	2.1436	2.1448	2.1459	2.1471	2.1483	2.1494	2.1506
8.6	2.1518	2.1529	2.1541	2.1552	2.1564	2.1576	2.1587	2.1599	2.1610	2.1622
8.7	2.1633	2.1645	2.1656	2.1668	2.1679	2.1691	2.1702	2.1713	2.1725	2.1736
8.8	2.1748	2.1759	2.1770	2.1782	2.1793	2.1804	2.1815	2.1827	2.1838	2.1849
8.9	2.1861	2.1872	2.1883	2.1894	2.1905	2.1917	2.1928	2.1939	2.1950	2.1961
9.0	2.1972	2.1983	2.1994	2.2006	2.2017	2.2028	2.2039	2.2050	2.2061	2.2072
9.1	2.2083	2.2094	2.2105	2.2116	2.2127	2.2138	2.2148	2.2159	2.2170	2.2181
9.2	2.2192	2.2203	2.2214	2.2225	2.2235	2.2246	2.2257	2.2268	2.2279	2.2289
9.3	2.2300	2.2311	2.2322	2.2332	2.2343	2.2354	2.2364	2.2375	2.2386	2.2396
9.4	2.2407	2.2418	2.2428	2.2439	2.2450	2.2460	2.2471	2.2481	2.2492	2.2502
9.5	2.2513	2.2523	2.2534	2.2544	2.2555	2.2565	2.2576	2.2586	2.2597	2.2607
9.6	2.2618	2.2628	2.2638	2.2649	2.2659	2.2670	2.2680	2.2690	2.2701	2.2711
9.7	2.2721	2.2732	2.2742	2.2752	2.2762	2.2773	2.2783	2.2793	2.2803	2.2814
9.8	2.2824	2.2834	2.2844	2.2854	2.2865	2.2875	2.2885	2.2895	2.2905	2.2915
9.9	2.2925	2.2935	2.2946	2.2956	2.2966	2.2976	2.2986	2.2996	2.3006	2.3016

TABLE IV Trigonometric functions

ANGLE		SINE	COSINE	TANGENT	ANGLE		SINE	COSINE	TANGENT
Degree	*Radian*				*Degree*	*Radian*			
0	0.000	0.000	1.000	0.000					
1	0.017	0.017	1.000	0.017	46	0.803	0.719	0.695	1.036
2	0.035	0.035	0.999	0.035	47	0.820	0.731	0.682	1.072
3	0.052	0.052	0.999	0.052	48	0.838	0.743	0.669	1.111
4	0.070	0.070	0.998	0.070	49	0.855	0.755	0.656	1.150
5	0.087	0.087	0.996	0.087	50	0.873	0.766	0.643	1.192
6	0.105	0.105	0.995	0.105	51	0.890	0.777	0.629	1.235
7	0.122	0.122	0.993	0.123	52	0.908	0.788	0.616	1.280
8	0.140	0.139	0.990	0.141	53	0.925	0.799	0.602	1.327
9	0.157	0.156	0.988	0.158	54	0.942	0.809	0.588	1.376
10	0.175	0.174	0.985	0.176	55	0.960	0.819	0.574	1.428
11	0.192	0.191	0.982	0.194	56	0.977	0.829	0.559	1.483
12	0.209	0.208	0.978	0.213	57	0.995	0.839	0.545	1.540
13	0.227	0.225	0.974	0.231	58	1.012	0.848	0.530	1.600
14	0.244	0.242	0.970	0.249	59	1.030	0.857	0.515	1.664
15	0.262	0.259	0.966	0.268	60	1.047	0.866	0.500	1.732
16	0.279	0.276	0.961	0.287	61	1.065	0.875	0.485	1.804
17	0.297	0.292	0.956	0.306	62	1.082	0.883	0.469	1.881
18	0.314	0.309	0.951	0.325	63	1.100	0.891	0.454	1.963
19	0.332	0.326	0.946	0.344	64	1.117	0.899	0.438	2.050
20	0.349	0.342	0.940	0.364	65	1.134	0.906	0.423	2.145
21	0.367	0.358	0.934	0.384	66	1.152	0.914	0.407	2.246
22	0.384	0.375	0.927	0.404	67	1.169	0.921	0.391	2.356
23	0.401	0.391	0.921	0.424	68	1.187	0.927	0.375	2.475
24	0.419	0.407	0.914	0.445	69	1.204	0.934	0.358	2.605
25	0.436	0.423	0.906	0.466	70	1.222	0.940	0.342	2.748
26	0.454	0.438	0.899	0.488	71	1.239	0.946	0.326	2.904
27	0.471	0.454	0.891	0.510	72	1.257	0.951	0.309	3.078
28	0.489	0.469	0.883	0.532	73	1.274	0.956	0.292	3.271
29	0.506	0.485	0.875	0.554	74	1.292	0.961	0.276	3.487
30	0.524	0.500	0.866	0.577	75	1.309	0.966	0.259	3.732
31	0.541	0.515	0.857	0.601	76	1.326	0.970	0.242	4.011
32	0.559	0.530	0.848	0.625	77	1.344	0.974	0.225	4.332
33	0.576	0.545	0.839	0.649	78	1.361	0.978	0.208	4.705
34	0.593	0.559	0.829	0.675	79	1.379	0.982	0.191	5.145
35	0.611	0.574	0.819	0.700	80	1.396	0.985	0.174	5.671
36	0.628	0.588	0.809	0.727	81	1.414	0.988	0.156	6.314
37	0.646	0.602	0.799	0.754	82	1.431	0.990	0.139	7.115
38	0.663	0.616	0.788	0.781	83	1.449	0.993	0.122	8.144
39	0.681	0.629	0.777	0.810	84	1.466	0.995	0.105	9.514
40	0.698	0.643	0.766	0.839	85	1.484	0.996	0.087	11.43
41	0.716	0.656	0.755	0.869	86	1.501	0.998	0.070	14.30
42	0.733	0.669	0.743	0.900	87	1.518	0.999	0.052	19.08
43	0.750	0.682	0.731	0.933	88	1.536	0.999	0.035	28.64
44	0.768	0.695	0.719	0.966	89	1.553	1.000	0.017	57.29
45	0.785	0.707	0.707	1.000	90	1.571	1.000	0.000	

ANSWERS

CHAPTER 1 EXERCISE 1-2

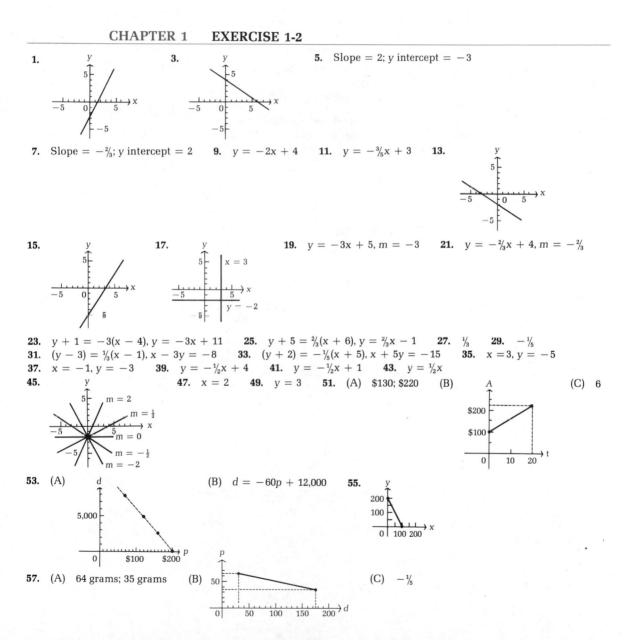

1.

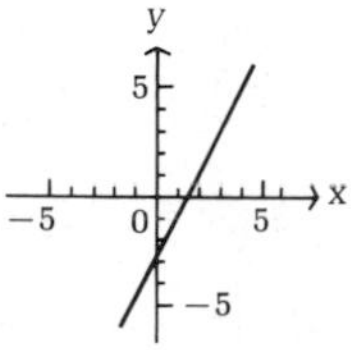

3.

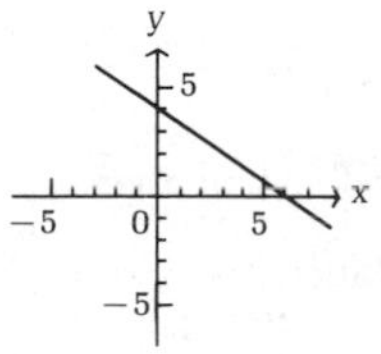

5. Slope = 2; y intercept = −3

7. Slope = $-\frac{2}{3}$; y intercept = 2 **9.** $y = -2x + 4$ **11.** $y = -\frac{3}{5}x + 3$ **13.**

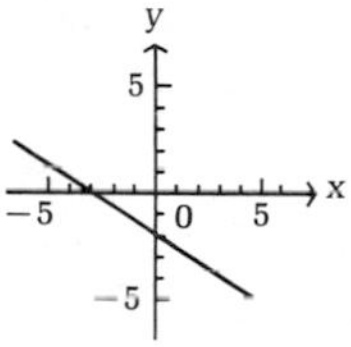

15.

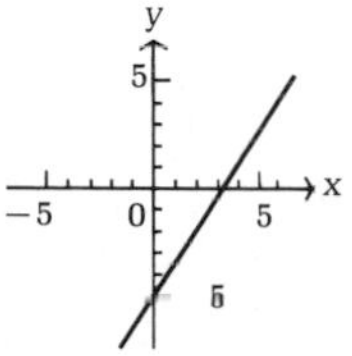

17.

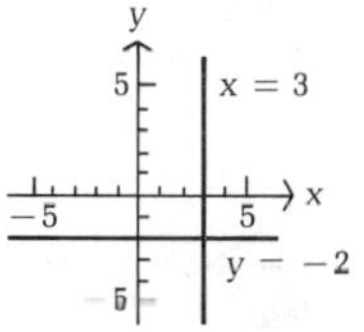

19. $y = -3x + 5, m = -3$ **21.** $y = -\frac{2}{3}x + 4, m = -\frac{2}{3}$

23. $y + 1 = -3(x - 4), y = -3x + 11$ **25.** $y + 5 = \frac{2}{3}(x + 6), y = \frac{2}{3}x - 1$ **27.** $\frac{1}{3}$ **29.** $-\frac{1}{5}$
31. $(y - 3) = \frac{1}{3}(x - 1), x - 3y = -8$ **33.** $(y + 2) = -\frac{1}{5}(x + 5), x + 5y = -15$ **35.** $x = 3, y = -5$
37. $x = -1, y = -3$ **39.** $y = -\frac{1}{2}x + 4$ **41.** $y = -\frac{1}{2}x + 1$ **43.** $y = \frac{1}{2}x$
45.

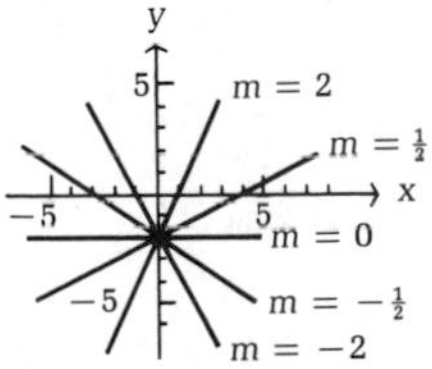

47. $x = 2$ **49.** $y = 3$ **51.** (A) \$130; \$220 (B)

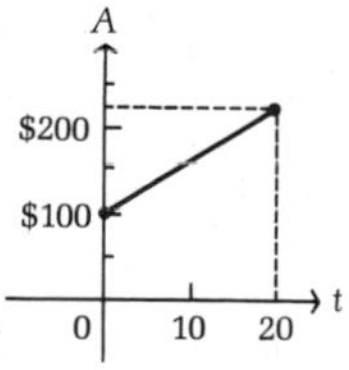

(C) 6

53. (A)

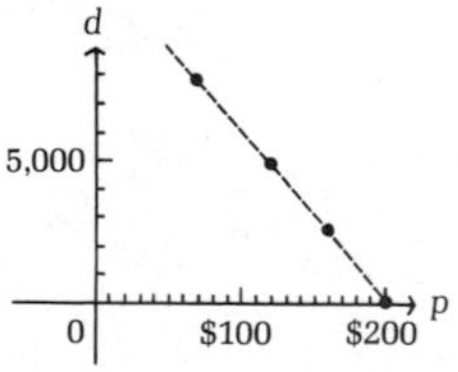

(B) $d = -60p + 12{,}000$ **55.**

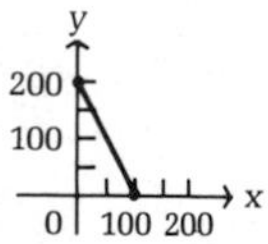

57. (A) 64 grams; 35 grams (B)

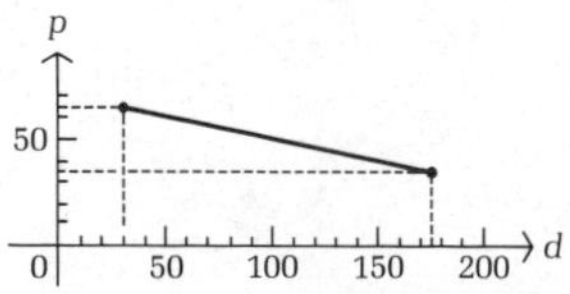

(C) $-\frac{1}{5}$

EXERCISE 1-3

1. Function   **3.** Not a function   **5.** Function   **7.** Function   **9.** Not a function   **11.** Function

13. Function $(x \neq 1)$ **15.** Function   **17.** Not a function; when $x = 4$, $y = \pm 2$

19. Not a function; when $x = 0$, $y = 0, 1$ **21.** Function   **23.** Function   **25.** 4   **27.** -5 **29.** -6

31. -2 **33.** -12 **35.** -1 **37.** -6 **39.** 12   **41.** $\frac{3}{4}$

43. Domain $= \{1, 2, 3\}$; Range $= \{1, 2, 3\}$; not a function

45. Domain $= \{-1, 0, 1, 2, 3, 4\}$; Range $= \{-2, -1, 0, 1, 2\}$; a function

47. Domain $= \{0, 1, 2, 3\}$; Range $= \{0, 2, 4, 6\}$; a function

49. Domain $= \{0, 1, 4\}$; Range $= \{-2, -1, 0, 1, 2\}$; not a function   **51.** 13   **53.** -3 **55.** 5   **57.** $\sqrt{2}$

59. $e^2 - e$ **61.** $\sqrt{u}$ **63.** $(2 + h)^2 - (2 + h) = h^2 + 3h + 2$ **65.** $2(a + h) + 1 = 2a + 2h + 1$

67. $\sqrt{2a + 1}$ **69.** $\sqrt{2x + 1}$ **71.** $\sqrt{a^2 - a}$ **73.** $\sqrt{x^2 - x}$ **75.** $\dfrac{[2(2 + h) + 1] - [2(2) + 1]}{h} = 2$

77. $\dfrac{[(2 + h)^2 - (2 + h)] - [2^2 - 2]}{h} = 3 + h$ **79.** All nonnegative real numbers

81. All real numbers, except $x = -3, 5$ **83.** All real numbers x such that $x \geq 1$

85. All real numbers x, except $x = -2, 3$ **87.** (A) 1   (B) 0   (C) 2   (D) 6   **89.** $C(x) = 4x$

91. $C(F) = \frac{5}{9}(F - 32)$ **93.** $IQ = 100(MA/12)$

EXERCISE 1-4

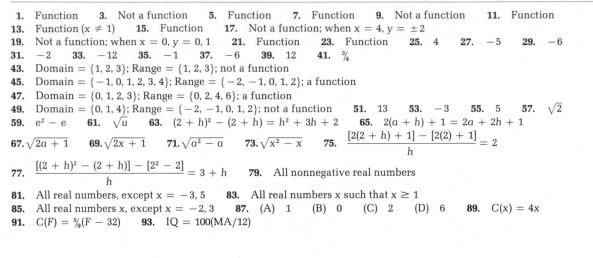

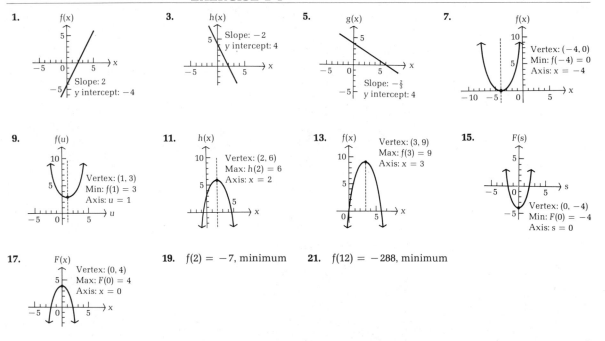

19. $f(2) = -7$, minimum   **21.** $f(12) = -288$, minimum

23. $A(25) = 625$, maximum

25.

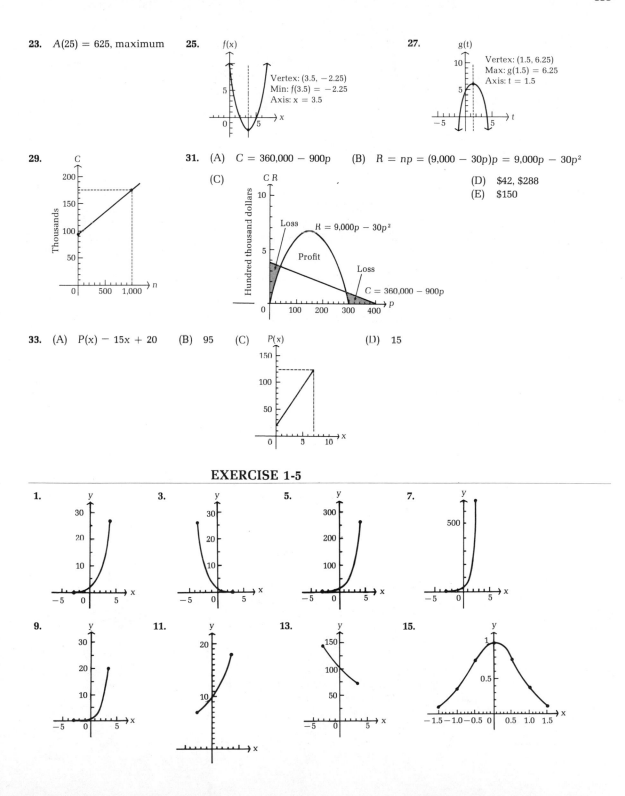

f(x)

Vertex: $(3.5, -2.25)$
Min: $f(3.5) = -2.25$
Axis: $x = 3.5$

27.

g(t)

Vertex: $(1.5, 6.25)$
Max: $g(1.5) = 6.25$
Axis: $t = 1.5$

29.

C

Thousands

n

31. (A) $C = 360{,}000 - 900p$ (B) $R = np = (9{,}000 - 30p)p = 9{,}000p - 30p^2$

(C)

C R

Hundred thousand dollars

Loss

$R = 9{,}000p - 30p^2$

Profit

Loss

$C = 360{,}000 - 900p$

p

(D) $42, $288

(E) $150

33. (A) $P(x) - 15x + 20$ (B) 95 (C)

P(x)

(D) 15

EXERCISE 1-5

1. y

3. y

5. y

7. y

9. y

11. y

13. y

15. y

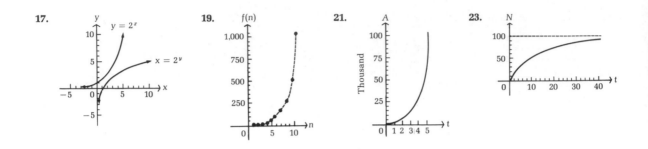

17. $y = 2^x$, $x = 2^y$

19. $f(n)$

21. A (Thousand)

23. N

EXERCISE 1-6

1. $27 = 3^3$ **3.** $10^0 = 1$ **5.** $8 = 4^{3/2}$ **7.** $\log_7 49 = 2$ **9.** $\log_4 8 = \frac{3}{2}$ **11.** $\log_b A = u$ **13.** 3
15. -3 **17.** 3 **19.** $\log_b P - \log_b Q$ **21.** $5 \log_b L$ **23.** $\log_b p - \log_b q - \log_b r - \log_b s$ **25.** $x = 9$
27. $y = 2$ **29.** $b = 10$ **31.** $x = 2$ **33.** $y = -2$ **35.** $b = 100$ **37.** $5 \log_b x - 3 \log_b y$
39. $\frac{1}{3} \log_b N$ **41.** $2 \log_b x + \frac{1}{3} \log_b y$ **43.** $\log_b 50 - 0.2t \log_b 2$ **45.** $\log_b P + t \log_b(1 + r)$
47. $\log_e 100 - 0.01t$ **49.** $x = 2$ **51.** $x = 8$ **53.** $\log_b 1 = 0, b > 0, b \neq 1$ **55.** $y = c10^{0.8x}$

EXERCISE 1-7 CHAPTER REVIEW

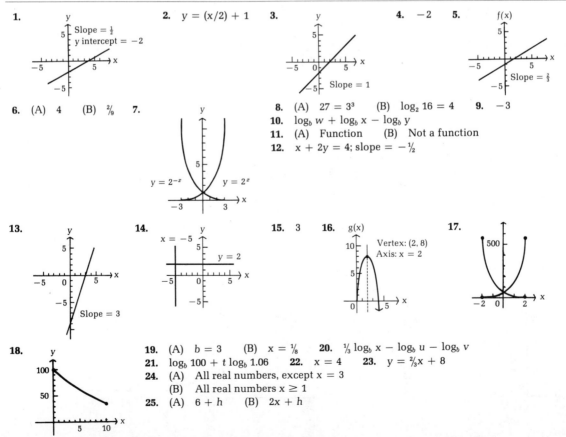

1. Slope $= \frac{1}{2}$, y intercept $= -2$

2. $y = (x/2) + 1$

3. Slope $= 1$

4. -2

5. $f(x)$, Slope $= \frac{2}{3}$

6. (A) 4 (B) $\frac{2}{9}$

7. $y = 2^{-x}$, $y = 2^x$

8. (A) $27 = 3^3$ (B) $\log_2 16 = 4$ **9.** -3
10. $\log_b w + \log_b x - \log_b y$
11. (A) Function (B) Not a function
12. $x + 2y = 4$; slope $= -\frac{1}{2}$

13. Slope $= 3$

14. $x = -5$, $y = 2$

15. 3 **16.** $g(x)$, Vertex: (2, 8), Axis: $x = 2$

17.

18.

19. (A) $b = 3$ (B) $x = \frac{1}{8}$ **20.** $\frac{1}{3} \log_b x - \log_b u - \log_b v$
21. $\log_b 100 + t \log_b 1.06$ **22.** $x = 4$ **23.** $y = \frac{2}{3}x + 8$
24. (A) All real numbers, except $x = 3$
(B) All real numbers $x \geq 1$
25. (A) $6 + h$ (B) $2x + h$

26. 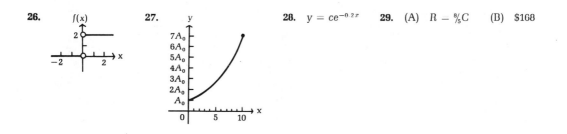 **27.** **28.** $y = ce^{-0.2x}$ **29.** (A) $R - \frac{8}{5}C$ (B) \$168

PRACTICE TEST: CHAPTER 1

1.

Slope: $-\frac{1}{2}$
x intercept: 6
y intercept: 3

2. $y = -\frac{3}{2}x + 2$ **3.** $x - 2y = 8$ **4.**

$x = 2$

$y = -3$

5. (A) -11 (B) $1 - a^2$ **6.** (A) $\sqrt{8} = 2\sqrt{2}$ (B) $a + 2\sqrt{a}$ **7.** $\log_b 5 + \frac{1}{2}\log_b w - 3\log_b x \quad \log_b y$
8. (A) $y = \frac{1}{2}$ (B) $x = 4$ (C) $b = 2$ **9.** **10.** (A) $V = -1,800t + 20,000$
(B) \$9,200

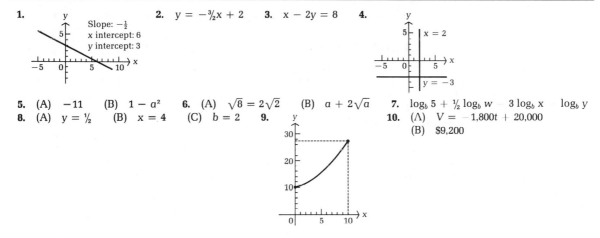

CHAPTER 2   EXERCISE 2-2

1. $x = 3, x = 4$ **3.** (A) Yes (B) 1 (C) 1 **5.** (A) Yes (B) -1 (C) -1 **7.** None
9. $x = 3$ **11.** $x = -3, x = 3$ **13.** $x = -2, x = 3$ **15.** $x = -3, x = 3$ **17.** 47 **19.** 4 **21.** $\frac{5}{3}$
23. -3 **25.** $\frac{5}{6}$ **27.** 3 **29.** 6 **31.** $\sqrt{8} = 2\sqrt{2}$ **33.** Does not exist
35. (A) 5, 6, 7, 8, 9 (B) 6; 6 (C) Does not exist; 6 (D) 12; 12 (E) Yes
37. (A) $t_2; t_4$ (B) 10%; 10% (C) 30%; 10% (D) Does not exist; 80%

EXERCISE 2-3

1. $\Delta x = 3; \Delta y = 45; \Delta y/\Delta x = 15$ **3.** 12 **5.** 12 **7.** 12 **9.** 15 **11.** (A) $12 + 3\Delta x$ (B) 12
13. (A) $24 + 3\Delta x$ (B) 24
15. (A) 5 meters per second (B) $3 + \Delta x$ meters per second (C) 3 meters per second
17. (A) 5 (B) $3 + \Delta x$ (C) 3 (D) $y = 3x - 1$ **19.** 3 meters per second
21. (A) \$200 per year (B) \$450 per year
23. (A) -110 square millimeters per day (B) -15 square millimeters per day
25. (A) 0.6 birth per year (B) 8 births per year

EXERCISE 2-4

1. $f'(x) = 2; f'(1) = 2; f'(2) = 2; f'(3) = 2$ **3.** $f'(x) = 6 - 2x; f'(1) = 4; f'(2) = 2; f'(3) = 0$
5. $v = f'(x) = 8x - 2; f'(1) = 6$ feet per second; $f'(3) = 22$ feet per second; $f'(5) = 38$ feet per second

7. (A) $m = f'(x) = 2x$ (B) $m_1 = f'(-2) = -4; m_2 = f'(0) = 0; m_3 = f'(2) = 4$
  (C) $y = -4x - 4; y = 0; y = 4x - 4$ (D)

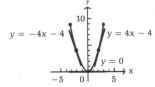

9. (A) $f'(x) = 3x^2 + 2$ (B) $f'(1) = 5; f'(3) = 29$

11. (A)   Marginal cost $= C'(x) = 10 - 2x$ (B) $C'(1) = \$800$ per 100 unit increase in production; $C'(3) = \$400$ per 100 unit increase in production; $C'(4) = \$200$ per 100 unit increase in production (Notice that as production goes up, cost per unit increase in production goes down.)

13. (A) $N'(t) = 14 - 2t$ (B) $N'(1) = 12$ phrases per hour; $N'(3) = 8$ phrases per hour; $N'(6) = 2$ phrases per hour

EXERCISE 2-5

1. 0   **3.** 0   **5.** $12x^{11}$ **7.** 1   **9.** $8x^3$ **11.** $2x^5$ **13.** $-10x^{-6}$ **15.** $-x^{-2/3}$ **17.** $15x^4 - 6x^2$

19. $-12x^{-5} - 4x^{-3}$ **21.** $-6x^{-3}$ **23.** $-\frac{1}{3}x^{-4/3}$ **25.** $-6x^{-3/2} + 6x^{-3} + 1$

27. (A) $m = 6 - 2x$ (B) $2; -2$ (C) $x = 3$ **29.** (A) $m = x^2 - 6x$ (B) $-8; -8$ (C) $x = 0, 6$

31. (A) $v = 176 - 32x$ (B)   176 feet per second; 80 feet per second; -16 feet per second   (C) $x = 5.5$ seconds

33. (A) $v = 40 - 10x$ (B)   40 feet per second; 10 feet per second; -20 feet per second   (C) $x = 4$ seconds

35. $2x - 3 - 10x^{-3}$ **37.** $-x^{-2} + 9x^{-5/2}$

39. (A) $N'(x) = 60 - 2x$ (B) $N'(10) = 40$ (at the \$10,000 level of advertising, there would be an increase of 40 units of sales per \$1,000 increase in advertising); $N'(20) = 20$ (at the \$20,000 level of advertising, there would be an increase of only 20 units of sales per \$1,000 increase in advertising); the effect of advertising levels off as the amount spent increases.

41. (A) -1.37 beats per minute   (B) -0.58 beat per minute

43. (A)   25 items per hour   (B)   8.33 items per hour

EXERCISE 2-6

1. $2x^3(2x) + (x^2 - 2)(6x^2) = 10x^4 - 12x^2$ **3.** $(x - 3)(2) + (2x - 1)(1) = 4x - 7$

5. $\dfrac{(x - 3)(1) - x(1)}{(x - 3)^2} = \dfrac{-3}{(x - 3)^2}$ **7.** $\dfrac{(x - 2)(2) - (2x + 3)(1)}{(x - 2)^2} = \dfrac{-7}{(x - 2)^2}$

9. $(x^2 + 1)(2) + (2x - 3)(2x) = 6x^2 - 6x + 2$ **11.** $\dfrac{(2x - 3)(2x) - (x^2 + 1)(2)}{(2x - 3)^2} = \dfrac{2x^2 - 6x - 2}{(2x - 3)^2}$

13. $(2x + 1)(2x - 3) + (x^2 - 3x)(2) = 6x^2 - 10x - 3$ **15.** $(2x - x^2)(5) + (5x + 2)(2 - 2x) = -15x^2 + 16x + 4$

17. $\dfrac{(x^2 + 2x)(5) - (5x - 3)(2x + 2)}{(x^2 + 2x)^2} = \dfrac{-5x^2 + 6x + 6}{(x^2 + 2x)^2}$ **19.** $\dfrac{(x^2 - 1)(2x - 3) - (x^2 - 3x + 1)(2x)}{(x^2 - 1)^2} = \dfrac{3x^2 - 4x + 3}{(x^2 - 1)^2}$

21. $(2x^4 - 3x^3 + x)(2x - 1) + (x^2 - x + 5)(8x^3 - 9x^2 + 1)$ **23.** $\dfrac{(4x^2 + 5x - 1)(6x - 2) - (3x^2 - 2x + 3)(8x + 5)}{(4x^2 + 5x - 1)^2}$

25. $9x^{1/3}(3x^2) + (x^3 + 5)(3x^{-2/3})$ **27.** $\dfrac{(x^2 - 3)(2x^{-2/3}) - 6x^{1/3}(2x)}{(x^2 - 3)^2}$

29. $x^{-2/3}(3x^2 - 4x) + (x^3 - 2x^2)(-\frac{2}{3}x^{-5/3})$ **31.** $\dfrac{(x^2 + 1)[(2x^2 - 1)(2x) + (x^2 + 3)(4x)] - (2x^2 - 1)(x^2 + 3)(2x)}{(x^2 + 1)^2}$

33. (A) $d'(x) = \dfrac{-50,000(2x + 10)}{(x^2 + 10x + 25)^2} = \dfrac{-100,000}{(x + 5)^3}$ (B) $d'(5) = -100$ radios per \$1 increase in price; $d'(10) = -30$ radios per \$1 increase in price

35. (A) $N'(x) = \dfrac{(x + 32)(100) - (100x + 200)}{(x + 32)^2} = \dfrac{3,000}{(x + 32)^2}$ (B) $N'(4) = 2.31; N'(68) = 0.30$

EXERCISE 2-7

1. $y = u^3, u = 2x + 5$ **3.** $y = u^8, u = x^3 - x^2$ **5.** $y = u^{1/3}, u = x^3 + 3x$ **7.** $6(2x + 5)^2$

9. $8(x^3 - x^2)^7(3x^2 - 2x)$ **11.** $(x^3 + 3x)^{-2/3}(x^2 + 1)$ **13.** $24x(x^2 - 2)^3$ **15.** $-6(x^2 + 3x)^{-4}(2x + 3)$

17. $x(x^2 + 8)^{-1/2}$ **19.** $1/\sqrt[3]{(3x + 4)^2}$ **21.** $\frac{1}{2}(x^2 - 4x + 2)^{-1/2}(2x - 4) = (x - 2)/(x^2 - 4x + 2)^{1/2}$

23. $(-1)(2x + 4)^{-2}(2) = -2/(2x + 4)^2$ **25.** $(-1)(4x^2 - 4x + 1)^{-2}(8x - 4) = -4/(2x - 1)^3$

27. $18x^2(x^2 + 1)^2 + 3(x^2 + 1)^3 = 3(x^2 + 1)^2(7x^2 + 1)$ **29.** $\frac{2x^3 4(x^3 - 7)^3 3x^2 - (x^3 - 7)^4 6x^2}{4x^6} = \frac{3(x^3 - 7)^3(3x^3 + 7)}{2x^4}$

31. $(2x - 3)^2[3(2x^2 + 1)^2(4x)] + (2x^2 + 1)^3[2(2x - 3)(2)] = 4(2x^2 + 1)^2(2x - 3)(8x^2 - 9x + 1)$

33. $4x^2[(\frac{1}{2})(x^2 - 1)^{-1/2}(2x)] + (x^2 - 1)^{1/2}(8x) = (12x^3 - 8x)/\sqrt{x^2 - 1}$

35. $\frac{(x - 3)^{1/2}(2) - 2x[(\frac{1}{2})(x - 3)^{-1/2}]}{x - 3} = \frac{x - 6}{(x - 3)^{3/2}}$ **37.** $y = -x + 3$

39. (A) $\overline{C}'(x) = 2(2x - 8)2 = 8x - 32$ (B) $\overline{C}'(2) = -16; \overline{C}'(4) = 0; \overline{C}'(6) = 16$. An increase in production at the 2,000 level will reduce costs; at the 4,000 level, no increase or decrease will occur; and at the 6,000 level, an increase in production will increase the costs.

41. (A) $f'(n) = n(n - 2)^{-1/2} + 2(n - 2)^{1/2} = \frac{3n - 4}{(n - 2)^{1/2}}$ (B) $f'(11) = 9\frac{2}{3}$ (rate of learning is $9\frac{2}{3}$ units per minute at the $n = 11$ level); $f'(27) = 15\frac{2}{5}$ (rate of learning is $15\frac{2}{5}$ units per minute at the $n = 27$ level)

EXERCISE 2-8

1. (A) $C'(x) = 60$ (B) $R(x) = xp(x) = 200x - (x^2/30)$ (C) $R'(x) = 200 - (x/15)$

(D) $R'(1,500) = 100$ (revenue is increasing at $100 per unit increase in production at the 1,500 output level); $R'(4,500) = -100$ (revenue is decreasing $100 per unit increase at the 4,500 output level)

(E)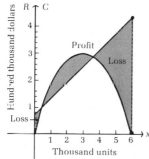

(F) $P(x) = R(x) - C(x) = -(x^2/30) + 140x - 72,000$ (G) $P'(x) = -(x/15) + 140$

(H) $P'(1,500) = 40$ (profit is increasing at $40 per unit increase in production at the 1,500 output level); $P'(3,000) = -60$ (profit is decreasing at $60 per unit increase in production at the 3,000 output level)

3. (A) $\overline{C}(x) = (72,000/x) + 60; \overline{R}(x) = xp/x = 200 - (x/30); \overline{P}(x) = \overline{R}(x) - \overline{C}(x) = 140 - (x/30) - (72,000/x)$

(B) $\overline{C}'(x) = -72,000/x^2; \overline{R}'(x) = -\frac{1}{30}; \overline{P}'(x) = -\frac{1}{30} + 72,000/x^2$

(C) $\overline{P}'(1,000) = \$0.039$ (profit is increasing at a rate of 3.9¢ per unit at an output level of 1,000 units per week); $\overline{P}'(6,000) = -\$0.031$ (profit is decreasing at a rate of 3.1¢ per unit at an output level of 6,000 units per week)

EXERCISE 2-9   CHAPTER REVIEW

1. $12x^3 - 4x$ **2.** $x^{-1/2} - 3 = (1/x^{1/2}) - 3$ **3.** 0 **4.** 0 **5.** $(2x - 1)(3) + (3x + 2)(2) = 12x + 1$

6. $(x^2 - 1)(3x^2) + (x^3 - 3)(2x) = 5x^4 - 3x^2 - 6x$ **7.** $\frac{(x^2 + 2)2 - 2x(2x)}{(x^2 + 2)^2} = \frac{4 - 2x^2}{(x^2 + 2)^2}$

8. $(-1)(3x + 2)^{-2}3 = -3/(3x + 2)^2$ **9.** $3(2x - 3)^2 2 = 6(2x - 3)^2$ **10.** $-2(x^2 + 2)^{-3}2x = -4x/(x^2 + 2)^3$

11. $12x^3 + 6x^{-4}$ **12.** $(2x^2 - 3x + 2)(2x + 2) + (x^2 + 2x - 1)(4x - 3) = 8x^3 + 3x^2 - 12x + 7$

13. $\frac{(x - 1)^2 2 - (2x - 3)2(x - 1)}{(x - 1)^4} = \frac{4 - 2x}{(x - 1)^3}$ **14.** $x^{-1/2} - 2x^{-3/2} = \frac{1}{\sqrt{x}} - \frac{2}{\sqrt{x^3}}$

15. $(x^2 - 1)[2(2x + 1)2] + (2x + 1)^2(2x) = 2(2x + 1)(4x^2 + x - 2)$ **16.** $\frac{1}{3}(x^3 - 5)^{-2/3}3x = \dfrac{x}{\sqrt[3]{(x^3 - 5)^2}}$

17. $-\frac{1}{3}(3x^2 - 2)^{-4/3}6x = \dfrac{-2x}{\sqrt[3]{(3x^2 - 2)^4}}$ **18.** $\dfrac{(2x - 3)4(x^2 + 2)^3 2x - (x^2 + 2)^4 2}{(2x - 3)^2} = \dfrac{2(x^2 + 2)^3(7x^2 - 12x - 2)}{(2x - 3)^2}$

19. (A) $m = f'(1) = 2$ (B) $y = 2x + 3$ **20.** (A) $m = f'(x) = 10 - 2x$ (B) $x = 5$

21. (A) $v = f'(x) = 32x - 4$ (B) $f'(3) = 92$ feet per second

22. (A) $v = f'(x) = 96 - 32x$ (B) $x = 3$ seconds   **23.** (A) 3   (B) 3   (C) Yes

24. (A) 6   (B) Does not exist   (C) No   **25.** None   **26.** $x = -5$ **27.** $x = 3, x = -2$ **28.** None

29. $\dfrac{2(3) - 3}{3 + 5} = \dfrac{3}{8}$ **30.** $2(3^2) - 3 + 1 = 16$ **31.** -1 **32.** 4   **33.** $f'(x) = 2x$ **34.** $x = -\frac{3}{2}, x = 2$

35. $\frac{5}{8}$

36. (A) $\overline{C}'(x) = 2x - 10$ (B) $\overline{C}'(3) = -4$ (average cost per unit is decreasing at \$4 per unit as production increases); $\overline{C}'(5) = 0$ (average cost per unit does not change for a small change in production); $\overline{C}'(7) = 4$ (average cost per unit is increasing at \$4 per unit as production increases at a 700 output level)

37. $C'(9) = -1$ part per million per meter; $C'(99) = -0.001$ part per million per meter

38. (A) 10 items per hour   (B) 5 items per hour

PRACTICE TEST: CHAPTER 2

1. $6x - x^{-1/2} = 6x - (1/x^{1/2})$ **2.** $(x^2 + 2)(2) + (2x - 3)(2x) = 6x^2 - 6x + 4$

3. $\dfrac{(x^2 + 1)(6x) - (3x^2 - 5)(2x)}{(x^2 + 1)^2} = \dfrac{16x}{(x^2 + 1)^2}$ **4.** $3(2x^3 - 3x + 1)^2(6x^2 - 3)$

5. $(x^2 - 1)^3(2) + (2x + 1)[3(x^2 - 1)^2 2x] = 2(x^2 - 1)^2(7x^2 + 3x - 1)$ **6.** $-\frac{1}{4}(2x^2 - 3)^{-5/4}4x = -x/\sqrt[4]{(2x^2 - 3)^5}$

7. (A) $m = f'(x) = 8 - 2x$ (B) 4   (C) $y = 4x + 4$ (D) $x = 4$

8. (A) $v = f'(x) = 80 - 20x$ (B) 20 feet per second   (C) $x = 4$ seconds   **9.** -1 **10.** (A) 9   (B) 0

11. $x = 0, x = 2$ (B) -2 **12.** $\overline{P}'(x) = -\frac{1}{20} + 4{,}000/x^2$

CHAPTER 3   EXERCISE 3-1

1. $y' = 6x; 6$ **3.** $y' = 3x/y; 3$ **5.** $y' = 1/(2y + 1); \frac{1}{3}$ **7.** $y' = -y/x; -\frac{3}{2}$ **9.** $y' = -2y/(2x + 1); 4$

11. $y' = (6 - 2y)/x; -1$ **13.** $y' = (2xy - 3x^2)/(2y - x^2); 8$

15. $y = -x + 5$ **17.** $y = \frac{2}{5}x - \frac{12}{5}; y = \frac{3}{5}x + \frac{12}{5}$

19. (A) $y' = -x/y$ (B) $3x + 4y = 25; 3x - 4y = 25$ (C)

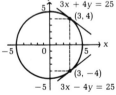

21. $y' = 1/[3(1 + y)^2 + 1]; \frac{1}{13}$ **23.** $y' = 3(x - 2y)^2/[6(x - 2y)^2 + 4y]; \frac{3}{10}$ **25.** $p' = 1/(2p - 2)$

27. $p' = -\sqrt{10{,}000 - p^2}/p$ **29.** $dL/dv = -(L + m)/(V + n)$

EXERCISE 3-2

1. 240   **3.** $\frac{9}{4}$ **5.** $\frac{1}{2}$ **7.** $dA/dt \approx 126$ square feet per second   **9.** 3,768 cubic centimeters per minute

11. $-\frac{9}{4}$ feet per second   **13.** $\frac{20}{3}$ feet per second

15. (A) $dC/dt = \$15{,}000$ per week   (B) $dR/dt = -\$50{,}000$ per week   (C) $dP/dt = -\$65{,}000$ per week

17. Approximately 100 cubic feet per minute

EXERCISE 3-3

1. $6x - 4$ **3.** 0 **5.** $40x^3$ **7.** 0 **9.** $-6x^{-4}$ **11.** $6x$ **13.** $6x^{-3} + 12x^{-4}$ **15.** $-15(2x - 1)^{-7/2}$
17. $24(1 - 2x)$ **19.** $24x^2(x^2 - 1) + 6(x^2 - 1)^2 = 6(x^2 - 1)(5x^2 - 1)$ **21.** $-12/y^3$ **23.** $-(6y^3 + 8x^2)/9y^5$
25. $-24(2x - 1)^{-4}$ **27.** $-12x/y^5$

EXERCISE 3-4

1. $(a, b), (c, e)$ **3.** $(a, b), (b, d), (f, g)$ **5.** d, f
7. (A) Positive $(0, \infty)$, negative $(-\infty, 0)$ (B) Positive $(-\infty, \infty)$
9. (A) Positive $(2, \infty)$, negative $(-\infty, 2)$ (B) Positive $(-\infty, \infty)$
11. Falling $(-\infty, 0)$; rising $(0, \infty)$; concave upward $(-\infty, \infty)$; horizontal tangent at $x = 0$

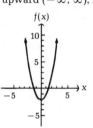

13. Rising $(-\infty, 4)$; falling $(4, \infty)$; concave downward $(-\infty, \infty)$; horizontal tangent at $x = 4$

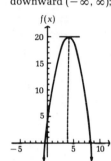

15. Rising $(-\infty, 0),(0, \infty)$; concave downward $(-\infty, 0)$; concave upward $(0, \infty)$; horizontal tangent at $x = 0$; inflection point at $x = 0$

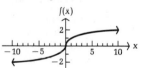

17. Rising $(-\infty, 0), (0, \infty)$ [$f'(0)$ is not defined]; concave upward $(-\infty, 0)$; concave downward $(0, \infty)$; inflection point at $x = 0$

19. Rising $(-\infty, -1)$, $(1, \infty)$; falling $(-1, 1)$; concave downward $(-\infty, 0)$; concave upward $(0, \infty)$; horizontal tangents at $x = -1, 1$; inflection point at $x = 0$

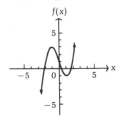

21. Rising $(-\infty, -1)$, $(2, \infty)$; falling $(-1, 2)$; concave downward $(-\infty, \frac{1}{2})$; concave upward $(\frac{1}{2}, \infty)$; horizontal tangents at $x = -1, 2$; inflection point at $x = \frac{1}{2}$

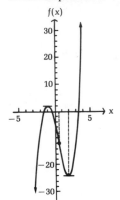

23. (A) Decreasing $(0, 3)$; increasing $(3, 6)$ (B) $\overline{C}(x)$

25. Decreasing   **27.** Increases for all t in $(0, 9)$

EXERCISE 3-5

1. c_2, c_5 **3.** c_1, c_2, c_4, c_8 **5.** (A) Local minimum   (B) Neither
7. (A) Local maximum   (B) Neither   (C) Neither   **9.** (A) Local minimum   (B) Neither
11. $f(-1) = 1$ is a local maximum; $f(0) = 0$ is a local minimum; $f(1) = 1$ is a local maximum
13. $f(-2) = 35$ is a local maximum; $f(4) = -73$ is a local minimum   **15.** $f(1) = 2$ is a local minimum
17. $f(2) = 4$ is a local minimum; $f(-2) = -4$ is a local maximum   **19.** $f(-2) = \frac{3}{4}$ is a local minimum
21. $f(1) = 0$ is a local minimum   **23.** $f(0) = 0$ is a local maximum; $f(4) = 8$ is a local minimum
25. Local minimum at $x = 3$ **27.** Local maximum at $x = 2$ **29.** Local maximum at $x = 2$

31. Local maximum at $x = -3$; local minimum at $x = 3$; inflection point at $x = 0$

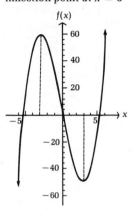

33. Local minimum at $x = 0$

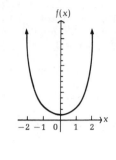

35. Local maximum at $x = -2$; local minimum at $x = 3$; inflection point at $(0.5, -12.5)$

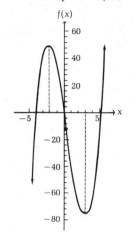

37. Local maximum at $x = -1$; local minimum at $x = 1$

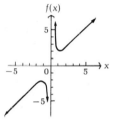

39. Local maximum at $x = -8$; local minimum at $x = 0$

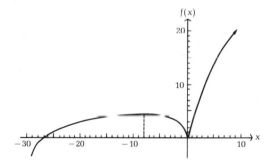

EXERCISE 3-6

1. Min $f(x) = f(2) = 1$; no maximum　　**3.** Max $f(x) = f(4) = 26$　**5.** Min $f(x) = f(2) = 2$; Max $f(x) = f(5) = 11$

7. Max $f(x) = f(1) = f(4) = 10$; Min $f(x) = f(0) = f(3) = 6$　**9.** Max $f(x) = f(0) = 126$; Min $f(x) = f(2) = -26$

11. Max $f(x) = f(1) = 0$; Min $f(x) = f(-1) = -2$　　**13.** Exactly in half　　**15.** 15 and -15

17. A square of side 25 centimeters; Max area = 625 square centimeters　　**19.** 3,000; \$3

21. \$15 per day; \$1,125 per day　　**23.** 40 trees; 1,600 pounds　　**25.** $(10 - 2\sqrt{7})/3 = 1.57$ inch squares

27. 20 feet by 40 feet (with one of the short sides the expensive side)　　**29.** $x = 5.1$ miles

31. 4 days; 20 bacteria per cubic centimeter　　**33.** 2 milligrams　　**35.** 1 month; 2 feet　　**37.** 4 years from now

EXERCISE 3-7

1. $dy = (24x - 3x^2)\, dx$　　**3.** $dy = \left(2x - \dfrac{x^2}{3}\right) dx$　　**5.** $dy = -\dfrac{295}{x^{3/2}}\, dx$　　**7.** $dy = \dfrac{150}{x^2}\, dx$

9. $dy = 1.4$, $\Delta y = 1.44$　　**11.** $dy = 3$, $\Delta y = 2.73$　　**13.** 2.03　　**15.** 3.04　　**17.** 120 cubic inches

19. $dy = \dfrac{6x - 2}{3(3x^2 - 2x + 1)^{2/3}}\, dx$　　**21.** $dy = 3.9$, $\Delta y = 3.83$　　**23.** 40 unit increase; 20 unit increase　　**25.** $-\$6$, \$4

27. -1.37 per minute; -0.58 per minute　　**29.** 1.26 square millimeters　　**31.** 3 words per minute

33. (A) 2,100 increase　　(B) 4,800 increase　　(C) 2,100 increase

EXERCISE 3-8   CHAPTER REVIEW

1. $dy/dx = 9x^2/4y$; $dy/dx|_{(1,2)} = \frac{9}{8}$ **2.** $dy/dt = 216$ **3.** $f''(x) = -6 - 60x^3$ **4.** $(a, c_1), (c_3, c_5), (c_5, c_6)$

5. $(c_1, c_3), (c_6, b)$ **6.** $(a, c_2), (c_4, c_5), (c_7, b)$ **7.** c_3 **8.** c_6 **9.** c_1, c_3, c_5 **10.** c_6 **11.** c_2, c_4, c_5, c_7

12. Decreasing $(0, 3)$; increasing $(3, 5)$ **13.** Upward $(0, 5)$; no downward   **14.** $x = 3$ only

15. Relative minimum at $x = 3$ **16.** Max $f(x) = f(0) = 10$, Min $f(x) = f(3) = 1$ **17.**

18. $dy = (-1 + 6x)\,dx$ **19.** $dy/dx = (6x - 2y)/x$; $dy/dx|_{(2,3)} = 3$

20. $dy/dt = -3/2$ **21.** $d^2y/dx^2 = 6x + \frac{1}{4}x^{-3/2}$

22. $d^2y/dx^2 = -2(2x^2 - y^2)/y^3 = -6/y^3$ **23.** Decreasing $(-\infty, 0)$; increasing $(0, \infty)$

24. Concave downward for all x, except $x = 0$ **25.** $x = 0$ only

26. Relative minimum at $x = 0$ **27.** Min $f(x) = f(0) = -1$; no absolute maximum

28.   **29.** Relative maximum at $x = 0$; relative minimum at $x = 2$

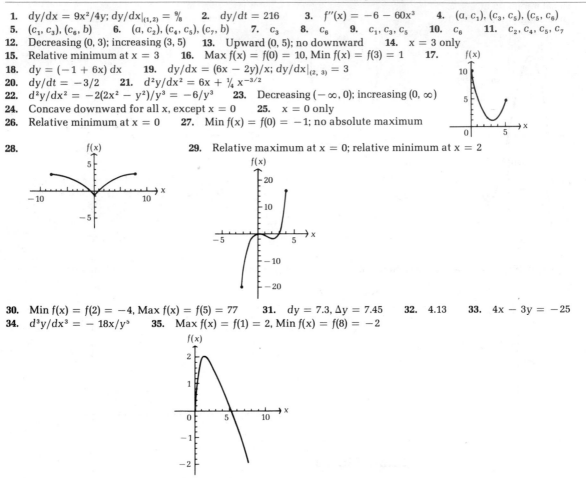

30. Min $f(x) = f(2) = -4$, Max $f(x) = f(5) = 77$ **31.** $dy = 7.3$, $\Delta y = 7.45$ **32.** 4.13   **33.** $4x - 3y = -25$

34. $d^3y/dx^3 = -18x/y^5$ **35.** Max $f(x) = f(1) = 2$, Min $f(x) = f(8) = -2$

36. $dy = -0.0031$, $\Delta y = -0.0031$ **37.** \$6,000 per week   **38.** $x = 3,000$; \$175,000   **39.** \$37 per night

40. -0.01 **41.** 3 days   **42.** 2.4 items   **43.** In 2 years

PRACTICE TEST: CHAPTER 3

1. $y' = (3y - 2x)/(8y - 3x)$ **2.** $-3(1 - 2x)^{-5/2} = -3/\sqrt{(1 - 2x)^5}$

3. $f(0) = 7$ is a local maximum; $f(3) = -20$ is a local minimum

4. Concave downward $(-\infty, \frac{3}{2})$; concave upward $(\frac{3}{2}, \infty)$; inflection point at $x = \frac{3}{2}$

5.

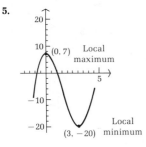

6. $f(4) = -1$ is a local minimum   **7.** Max $f(-4) = 3$; Min $f(4) = -1$

8. $-\frac{3}{2}$ feet per second   **9.** Each number is 6; maximum product $= 36$

10. $\Delta y = 0.61$, $dy = 0.6$

11. Volume of ivory shell ≈ 1.2 cubic inches

12. Let $R(x) = (3,600 + 300x)(10 - 0.5x)$ where x is the number of 50¢ decreases. Max $R(x) = R(4) = \$38,400$ per month with 4,800 subscribers at a rate of $8 per month

CHAPTER 4   EXERCISE 4-1

1. $7x + C$ **3.** $(x^7/7) + C$ **5.** $2t^4 + C$ **7.** $u^2 + u + C$ **9.** $x^3 + x^2 - 5x + C$

11. $(s^5/5) - \frac{4}{3}s^0 + C$ **13.** $y = 40x^5 + C$ **15.** $P = 24x - 3x^2 + C$ **17.** $y = \frac{1}{3}u^6 - u^3 - u + C$

19. $4x^{3/2} + C$ **21.** $-4x^{-2} + C$ **23.** $2\sqrt{u} + C$ **25.** $-(x^{-2}/8) + C$ **27.** $-(u^{-4}/8) + C$

29. $2x^{3/2} - 2x^{1/2} + C$ **31.** $6x^{5/3} - 6x^{4/3} - 2x + C$ **33.** $2x^{3/2} + 4x^{1/2} + C = 2\sqrt{x^3} + 4\sqrt{x} + C$

35. $\frac{3}{5}x^{5/3} + 2x^{-2} + C$ **37.** $y - x^2 - 3x + 5$ **39.** $C(x) = 2x^3 - 2x^2 + 3,000$ **41.** $x = 40\sqrt{t}$

43. $y = 2x^2 - 3x + 1$ **45.** $x^2 + x^{-1} + C$ **47.** $\frac{1}{2}x^2 + x^{-2} + C$ **49.** $M = t - 2\sqrt{t} + 5$

51. $y = 3x^{5/3} + 3x^{2/3} - 6$ **53.** $P(x) = 50x - 0.02x^2$; $P(100) = \$4,800$ **55.** $v(t) - 300 - 50\sqrt{t}$; $v(15) = \$106,000$

57. $W(h) = 0.0005h^3$; $W(70) = 171.5$ pounds   **59.** 19,400   **61.** $N(t) = 20 - 2\sqrt{t}$; $N(36) - 8$ facts

EXERCISE 4-2

1. $\dfrac{(x^2 - 4)^6}{6} + C$ **3.** $\frac{2}{3}(2x^2 - 1)^{3/2} + C$ **5.** $\dfrac{(3x - 2)^8}{24} + C$ **7.** $\dfrac{(x^2 + 3)^8}{16} + C$ **9.** $\dfrac{(3x^2 + 7)^{3/2}}{9} + C$

11. $\dfrac{(2x^4 + 3)^{1/2}}{4} + C$ **13.** $\frac{1}{3}(x^2 - 2x - 3)^{3/2} + C$ **15.** $-\frac{1}{18}(3t^2 + 1)^{-3} + C$ **17.** $-\frac{2}{3}(4 - x^3)^{1/2} + C$

19. $\dfrac{-1}{12(x^4 + 2x^2 + 1)^3} + C$ **21.** $R(x) = \sqrt{x^2 + 9} - 3$; $R(4) - \$2,000$

23. $E(t) = 12,000 - \dfrac{10,000}{\sqrt{t + 1}}$; $E(15) = 9,500$ students

EXERCISE 4-3

1. 5   **3.** 5   **5.** 2   **7.** 48   **9.** $-\frac{7}{3}$ **11.** 2   **13.** -2 **15.** 14   **17.** $5^6 = 15,625$ **19.** 12

21. $\frac{1}{6}(15^{3/2} - 5^{3/2})$ **23.** $\frac{3}{4}(2^{2/3} - 3^{2/3})$

25. (A) $C(4) - C(2) = \displaystyle\int_2^4 1\, dx = \$2$ thousand per day   (B) $R(4) - R(2) = \displaystyle\int_2^4 (10 - 2x)\, dx = \$8$ thousand per day

     (C) $P(4) - P(2) = \displaystyle\int_2^4 [R'(x) - C'(x)]\, dx = \$6$ thousand per day

27. $\displaystyle\int_0^5 500(t - 12)\, dt = -\$23,750$; $\displaystyle\int_5^{10} 500(t - 12)\, dt = -\$11,250$ **29.** $\displaystyle\int_{49}^{64} -295x^{-3/2}\, dx \approx -10.5$ beats per minute

31. $\displaystyle\int_{10}^{20} (12 + 0.006t^2)\, dt = 134$ billion cubic feet   **33.** $\displaystyle\int_1^9 \dfrac{25}{\sqrt{t}}\, dt = 100$ items

EXERCISE 4-4

1. $\frac{1}{6}$ **3.** 7   **5.** $\frac{7}{3}$ **7.** 9   **9.** 15   **11.** 32   **13.** 36   **15.** 9   **17.** $2\frac{1}{2}$ **19.** $\frac{23}{3}$ **21.** $1\frac{1}{3}$

23. Consumers' surplus $= 1$; producers' surplus $= \frac{4}{3}$ **25.** 5 years; $25,000

EXERCISE 4-5

1. (A) 120 (B) 124 **3.** (A) 123 (B) 124 **5.** (A) −4 (B) −5.33 **7.** (A) −5 (B) −5.33
9. (A) 1.63 (B) Not possible at this time **11.** (A) 1.59 (B) Not possible at this time **13.** 250
15. 2 **17.** $^{45}\!/_{28} \approx 1.61$ **19.** 26π **21.** 64π **23.** $32\pi/3$ **25.** 0.261 **27.** 6.54

29. (A) $I = -200t + 600$ (B) $\frac{1}{3}\int_0^3 (-200t + 600)\,dt = 300$ **31.** $16,000 **33.** 10°C

EXERCISE 4-6 CHAPTER REVIEW

1. $t^3 - t^2 + C$ **2.** 12 **3.** $-3t^{-1} - 3t + C$ **4.** $7\frac{1}{2}$ **5.** $y = f(x) = x^3 - 2x + 4$ **6.** 12
7. $2[5 + 17] = 44$ **8.** $\frac{1}{8}(6x - 5)^{4/3} + C$ **9.** 2 **10.** $-2x^{-1} - \frac{3}{5}x^{5/3} + C$ **11.** $(20^{3/2} - 8)/3$
12. $y = f(x) = 2x^{3/2} + x^{-1} + C$ **13.** $y = 3x^2 + x - 4$ **14.** −2 **15.** $6\frac{1}{2}$ **16.** $\pi/2$
17. $\left|\int_{-2}^{2} (x^2 - 4)\,dx\right| + \int_{2}^{4} (x^2 - 4)\,dx = {}^{64}\!/_3 = 21\frac{1}{3}$ **18.** $\frac{3}{8}(15)^{4/3}$ **19.** $y = 2x^{1/2} + x^{-2} + C$
20. $y = \frac{2}{9}(x^3 + 4)^{3/2} + \frac{2}{9}$ **21.** $^{46}\!/_3$ **22.** $\frac{1}{4}[f(0.5) + f(1.5) + f(2.5) + f(3.5)] = 0.394$
23. (A) $43\pi \approx 135$ (B) $128\pi/3 \approx 134$ **24.** $P(x) = 100x - 0.01x^2$; $P(10) = $999
25. $\int_0^{15} (60 - 4t)\,dt = 450$ thousand barrels **26.** $\int_{10}^{40}\left(150 - \frac{x}{10}\right)dx = $4,425$ **27.** 109 items

28. 1 square centimeter **29.** $\int_{50}^{60} 0.0015h^2\,dh = 45.5$ pounds **30.** 5,200 **31.** $\int_0^4 (12t - 3t^2)\,dt = 32$ thousand

PRACTICE TEST: CHAPTER 4

1. $^{3}\!/_8$ **2.** $\frac{2}{9}(x^3 + 9)^{3/2} + C$ **3.** $-2x^{-2} + (x^2/2) + C$ **4.** $8\frac{2}{3}$ **5.** $f(x) = 1 + 6x - x^2$ **6.** $\frac{1}{3}$ **7.** 2
8. 46; 48 **9.** 6π **10.** 27 **11.** $\int_0^2 (40 - 4t)\,dt = 72$ thousand ounces; $\int_2^4 (40 - 4t)\,dt = 56$ thousand ounces

CHAPTER 5 EXERCISE 5-3

1. $1/t$ **3.** $1/(x - 3)$ **5.** $3/(x - 1)$ **7.** $2z + (3/z)$ **9.** $\ln|t| + C$ **11.** $\ln|x - 3| + C$
13. $3\ln|x| + C$ **15.** $x + \ln|x| + C$ **17.** $\ln e - \ln 1 = 1$ **19.** $\ln 4 - \ln 1 \approx 1.3863$ **21.** $7/x$ **23.** $\frac{1}{2}x$
25. $\dfrac{1}{2x\sqrt{\ln x}} + \dfrac{1}{2x}$ **27.** $\dfrac{4}{x + 1}$ **29.** $\dfrac{2t + 3}{t^2 + 3t}$ **31.** $6x^2 + \dfrac{2x}{x^2 + 1}$ **33.** $\dfrac{1 - 2\ln x}{x^3}$ **35.** $\ln x$
37. $(2x + 1) + (2x + 1)\ln(x^2 + x) = (2x + 1)[1 + \ln(x^2 + 1)]$ **39.** 0 **41.** $\frac{1}{5}\ln|5t - 3| + C$
43. $\frac{1}{4}\ln|2x^2 + 1| + C$ **45.** e^3 **47.** $\ln 10 \approx 2.3026$ **49.** $\ln|-1| - \ln|-2| = -\ln 2 \approx -0.6931$ **51.** $\dfrac{x}{x^2 + 1}$

53. $\dfrac{6x - 2}{3x^2 - 2x}(\log_{10} e)$ **55.** $\dfrac{2}{x - 1} - \dfrac{3}{x + 1}$ **57.** $\dfrac{2}{x - 1} + \dfrac{1}{2x}$
59. $\frac{1}{2}\ln|x^2 - 4x + 3| + C$ **61.** $\frac{1}{2}\ln|2x^2 + 5| + C$ **63.** $\frac{1}{2}(\ln 3 - \ln 8) \approx \ln\sqrt{^3\!/_8}$ **65.** $\frac{1}{2}(\ln 13 - \ln 7) = \ln\sqrt{^{13}\!/_7}$
67. $\ln 2$; $\ln 3$; $\ln x$

EXERCISE 5-4

1. e^t **3.** $8e^{8x}$ **5.** $6e^{2x}$ **7.** $-8e^{-4t}$ **9.** $e^t + C$ **11.** $e^{8x} + C$ **13.** $\frac{1}{3}e^{3x} + C$ **15.** $-\frac{1}{2}e^{-2t} + C$
17. $e - 1 \approx 1.7183$ **19.** $e^2 - 1 \approx 6.3891$ **21.** $(6x - 2)e^{3x^2 - 2x}$ **23.** $(e^x + e^{-x})/2$ **25.** $2e^{2x} - 6x$
27. $xe^x + e^x$ **29.** $-3e^{-0.03x}$ **31.** $4(e^{2x} - 1)^3(2e^{2x}) = 8e^{2x}(e^{2x} - 1)^3$ **33.** $7^x \ln 7$ **35.** $4e^{4x} + 4x^3e^{x^4}$
37. $\dfrac{2e^{2x}(x^2 - x + 1)}{(x^2 + 1)^2}$ **39.** $(x^2 + 1)(-e^{-x}) + e^{-x}(2x) = e^{-x}(2x - x^2 - 1)$ **41.** $(e^{-x}/x) - e^{-x}\ln x$
43. $\frac{1}{2}e^{x^2 + 1} + C$ **45.** $[(e^x + e^{-x})/2] + C$ **47.** $(4^x/\ln 4) + C$ **49.** $-\frac{1}{2}(e^{-1} - 1) \approx 0.3161$
51. $e^x + xe^x \ln x + e^x \ln x$ **53.** $(2x + 1)(10^{x^2 + x})(\ln 10)$ **55.** $-\frac{1}{2}e^{1 - x^2}$ or $-1/2e^{x^2 - 1}$ **57.** $6^{x^2}/(2\ln 6)$
59. $\ln(e^x + e^{-x}) + C$ **61.** e^x **63.** x/e^{x^2}

EXERCISE 5-5

1. $4,953   **3.** $9,931.71   **5.** $R'(x) = (100 - 5x)e^{-0.05x}$ **7.** $P'(x) = 0.005e^{0.0005x}$
9. $-$27,145$ per year; $-$18,196$ per year; $-$11,036$ per year   **11.** Solve $2 = e^{0.02t}$ to obtain $t \approx 35$ years
13. $I = I_0 e^{-0.00942x}$ $x \approx 74$ feet   **15.** $Q = 3e^{-0.04t}$; $Q(10) = 2.01$ millimeters   **17.** 24,200 years (approximately)
19. 104 times; 67 times   **21.** (A) 7 people; 353 people   (B) 400

EXERCISE 5-6

1. $xe^x - e^x + C$ **3.** $\dfrac{x^3}{3} \ln x - \dfrac{x^3}{9} + C$ **5.** $-xe^{-x} - e^{-x} + C$ **7.** $\frac{1}{2}e^{x^2} + C$

9. $(xe^x - 4e^x)\Big|_0^1 = -3e + 4 \approx -4.1548$ **11.** $(x \ln 2x - x)\Big|_1^3 = (3 \ln 6 - 3) - (\ln 2 - 1) \approx 2.6821$

13. $\ln(x^2 + 1) + C$ **15.** $\dfrac{(\ln x)^2}{2} + C$ **17.** $\frac{2}{3}x(x + 1)^{3/2} - \frac{4}{15}(x + 1)^{5/2} + C$ **19.** $(x^2 - 2x + 2)e^x + C$

21. $\dfrac{xe^{ax}}{a} - \dfrac{e^{ax}}{a^2} + C$ **23.** $\left(-\dfrac{\ln x}{x} - \dfrac{1}{x}\right)\Big|_1^e = -\dfrac{2}{e} + 1 \approx 0.2642$ **25.** $x(\ln x)^2 - 2x \ln x + 2x + C$
27. $P(t) = t^2 + te^{-t} + e^{-t} - 1$

EXERCISE 5-7

1. $\frac{1}{3}$ **3.** 2   **5.** Diverges   **7.** 1   **9.** Diverges   **11.** Diverges   **13.** 1   **15.** $\displaystyle\int_2^{2.5}\left(-\dfrac{x}{2} + 2\right)dx \approx .94$

17. $\frac{1}{4}\displaystyle\int_1^{\infty} e^{-t/4}\, dt \approx .78$ **19.** 1   **21.** Diverges   **23.** $.05\displaystyle\int_3^{\infty} e^{-.05x}\, dx \approx .86$ **25.** $.2\displaystyle\int_0^5 e^{-.2t}\, dt \approx .63$

27. $\displaystyle\int_9^{\infty} \dfrac{dx}{(x + 1)^2} = .1$

EXERCISE 5-8   CHAPTER REVIEW

1. $2e^{2x-3}$ **2.** $1/(x - 2)$ **3.** $\ln|x + 1| + C$ **4.** $e^{x-1} + C$ **5.** $(e^4 - 1) \approx 53.5982$ **6.** $\frac{1}{2}$
7. $\frac{1}{2}e^{x^2} + C$ **8.** $\dfrac{e^{-2x}}{5x} - 2e^{-2x} \ln 5x$ **9.** $-2e^{-1} + .1 \approx 0.2642$ **10.** $\dfrac{x^2}{2} \ln x - \dfrac{x^2}{4} + C$

11. $-\ln(e^{-x} + 3) + C$ **12.** $-(e^x + 2)^{-1} + C$ **13.** $\dfrac{1}{\sqrt{x}} + \dfrac{3x^2}{x^3 + 1}$ **14.** $\frac{1}{3}$ **15.** 1   **16.** $\dfrac{(\ln x)^3}{3} + C$

17. 2   **18.** $\dfrac{2x - 1}{x^2 - x} \ln_5 e$ **19.** $5^{x^2-1}2x(\ln 5)$ **20.** $\dfrac{10^{2x}}{2 \ln 10} + C$ **21.** 13.9 years (approximately)

22. $R'(x) = (1,000 - 20x)e^{-0.02x}$ **23.** $p'(x) = -20e^{-0.02x}$ **24.** $.02\displaystyle\int_0^1 e^{-.02t}\, dt \approx .02$

25. $N(t) = 10,000e^{0.2t}$; $N(10) \approx 74,000$ **26.** $\displaystyle\int_1^{\infty} s(t)\, dt = \int_1^3 s(t)\, dt = \frac{1}{3}$ **27.** $r \approx 3.5\%$ **28.** $.5\displaystyle\int_2^{\infty} e^{-.5t}\, dt \approx .37$

PRACTICE TEST: CHAPTER 5

1. $(2x + 9)e^{x^2+9x}$ **2.** $1/2(x + 1)$ **3.** $\frac{1}{2}e^{x^2} + C$ **4.** $e + \ln e - 1 - \ln 1 = e + 1 - 1 - 0 = e$
5. $e^{-x}/(x + 1) - e^{-x} \ln(x + 1)$ **6.** $(x + 2)e^x + C$ **7.** $-1/6(x^2 + 3)^3 + C$ **8.** $\frac{1}{8}e^{-16}$ **9.** $(\ln x)^8/8 + C$
10. $3(\ln x)^3/x + (3/x)$ **11.** 8.7 years   **12.** $p'(x) = -30e^{-0.06x}$ **13.** $485.22; $430.35

CHAPTER 6   EXERCISE 6-1

1. 10   **3.** 1   **5.** 0   **7.** 1   **9.** 6   **11.** 150   **13.** 16π **15.** 791   **17.** 0.19   **19.** 118
21. $100e^{0.8} \approx 222.55$ **23.** $2x + \Delta x$ **25.** $2y^2$ **27.** $E(0, 0, 3)$, $F(2, 0, 3)$ **29.** \$4,400; \$6,000; \$7,100
31. (A) \$12 million   (B) \$10 million   (C) \$3 million loss
33. $T(70, 47) \approx 29$ minutes; $T(60, 27) \approx 33$ minutes   **35.** $C(6, 8) = 75$; $C(8.1, 9) = 90$
37. $I(12, 10) = 120$; $I(10, 12) \approx 83$

EXERCISE 6-2

1. 3   **3.** 2   **5.** $-4xy$ **7.** -6 **9.** $10xy^3$ **11.** 60   **13.** $2x - 2y + 6$ **15.** 6   **17.** -2
19. 2   **21.** $2e^{2x+3y}$ **23.** $6e^{2x+3y}$ **25.** $6e^2$ **27.** $4e^3$
29. $f_x(x, y) = 6x(x^2 - y^3)^2$; $f_y(x, y) = -9y^2(x^2 - y^3)^2$ **31.** $f_x(x, y) = 24xy(3x^2y - 1)^3$; $f_y(x, y) = 12x^2(3x^2y - 1)^3$
33. $f_x(x, y) = 2x/(x^2 + y^2)$; $f_y(x, y) = 2y/(x^2 + y^2)$ **35.** $f_x(x, y) = y^4e^{xy^2}$; $f_y(x, y) = 2xy^3e^{xy^2} + 2ye^{xy^2}$
37. $f_x(x, y) = 4xy^2/(x^2 + y^2)^2$; $f_y(x, y) = -4x^2y/(x^2 + y^2)^2$ **39.** $x = 2$ and $y = 4$ **41.** (A) $2x$ (B) $4y$
43. $C_x(20, 10) = \$70$, which means production costs increase \$70 for an increase in production of one standard board (production of competition boards held fixed) when $x = 20$ and $y = 10$; $C_y(20, 10) = \$100$, which means production costs increase \$100 for an increase in production of one competition board (production of standard boards held fixed) when $x = 20$ and $y = 10$
45. $P_x(1, 1) = -4$, which means that profit will decrease \$4 million per 1,000 increase in production of type A computer at the $(1, 1)$ output level; $P_y(1, 1) = 10$, which means that profit will increase \$10 million per 1,000 increase in production of type B computer at the $(1, 1)$ output level
47. $T_1(70, 47) \approx 0.41$ minute per unit increase in volume of air when $V = 70$ cubic feet and $x = 47$ feet; $T_x(70, 47) \approx -0.36$ minute per unit increase in depth when $V = 70$ cubic feet and $x = 47$ feet
49. $C_w(6, 8) = 12.5$, which means the index increases 12.5 units for 1 inch increase in width of the head (length held fixed) when $W = 6$ and $L = 8$; $C_L(6, 8) = -9.38$, which means the index decreases 9.38 units for 1 inch increase in length (width held fixed) when $W = 6$ and $L = 8$
51. $Q_M(12, 10) = 10$, which means that the IQ increases 10 points for 1 year increase of mental age (chronological age held fixed) when $M = 12$ and $C = 10$; $Q_C(12, 10) = -12$, which means that the IQ decreases 12 points for 1 year increase in chronological age (mental age held fixed) when $M = 12$ and $C = 10$

EXERCISE 6-3

1. $2x\,dx + 2y\,dy$ **3.** $4x^3y^3\,dx + 3x^4y^2\,dy$ **5.** $\frac{1}{2}x^{-1/2}\,dx - \frac{5}{2}y^{-3/2}\,dy$ **7.** $3x^2\,dx + 3y^2\,dy + 3z^2\,dz$
9. $(y + 2z)\,dx + (x + 3z)\,dy + (2x + 3y)\,dz$ **11.** $dz = -0.4, \Delta z = -0.39$ **13.** $dz = 15, \Delta z = 13.635364$
15. $dw = 1.7, \Delta w = 1.73$ **17.** $dw = -0.5, \Delta w = -0.490196$ **19.** 4.98 inches   **21.** $15\pi \approx 47.12$ cubic inches
23. 30 cubic centimeters   **25.** $dz = e^{x^2+y^2}[(y + 2x^2y)\,dx + (x + 2xy^2)\,dy]$
27. $dw = (1 + xyz)e^{xyz}(yz\,dx + xz\,dy + xy\,dz)$ **29.** \$10,460   **31.** \$5.40   **33.** $1.1k$ **35.** 14.896

EXERCISE 6-4

1. $f(-2, 0) = 10$ is a local maximum   **3.** $f(-1, 3) = 4$ is a local minimum   **5.** f has a saddle point at $(3, -2)$
7. $f(3, 2) = 33$ is a local maximum   **9.** $f(2, 2) = 8$ is a local minimum   **11.** f has a saddle point at $(0, 0)$
13. 2,000 type A and 4,000 type B; Max $P = P(2, 4) = \$15$ million
15. \$3 million on research and development and \$2 million on advertising; Max $P = P(3, 2) = \$30$ million
17. 20 inches by 20 inches by 40 inches

EXERCISE 6-5

1. Max $f(x, y) = f(3, 3) = 18$ **3.** Min $f(x, y) = f(3, 4) = 25$
5. Max $f(x, y) = f(3, 3) = f(-3, -3) = 18$; Min $f(x, y) = f(3, -3) = f(-3, 3) = -18$
7. Maximum product is 25 when each number is 5   **9.** Min $f(x, y, z) = f(-4, 2, -6) = 56$
11. 60 of model A and 30 of model B will yield a minimum cost of \$32,400 per week
13. A maximum volume of 16,000 cubic inches occurs for a 40 by 20 by 20 inch box
15. 16 kilograms of type M and 8 kilograms of type N; minimum cost is \$32 per week

EXERCISE 6-6

1.

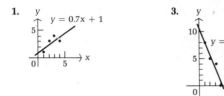

$y = 0.7x + 1$

3.

$y = -2.5x + 10.5$

5.

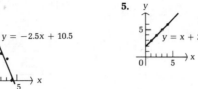

$y = x + 2$

7. $y = 2.12x + 10.8$, $y = 63.8$ when $x = 25$ **9.** $y = -1.2x + 12.6$, $y = 10.2$ when $x = 2$
11. $y = 1.53x + 26.67$, $y = 14.4$ when $x = 8$

13. $y = 0.75x^2 - 3.45x + 4.75$

15. (A) $y = 0.382x + 1.265$ (B) $10,815
17. (A) $y = 0.7x + 112$ (B) Demand, 140,000 units; revenue, $56,000
19. (A) $y = 11.9x + 69.2$ (B) 69.2 parts per million   **21.** (A) $y = 10.1x + 10.7$ (B) 9 weeks

EXERCISE 6-7

1. (A) $3x^2y^4 + C(x)$ (B) $3x^2$ **3.** (A) $2x^2 + 6xy + 5x + E(y)$ (B) $35 + 30y$
5. (A) $(y + y^2)/\sqrt{x} + C(x)$ (B) $6/\sqrt{x}$ **7.** (A) $\sqrt{y + x^2} + E(y)$ (B) $\sqrt{y + 4} - \sqrt{y}$
9. (A) $e^{x-y} + E(y)$ (B) $e^{1-y} - e^{-1-y}$ **11.** 9   **13.** 330   **15.** 24   **17.** $(56 - 20\sqrt{5})/3$
19. $e + e^{-3} - 2e^{-1}$ **21.** 16   **23.** 49   **25.** $\ln 5$ **27.** 12   **29.** $4/3$ **31.** $32/3$

33. $\displaystyle\int_0^1 \int_1^2 xe^{xy}\, dy\, dx = \frac{1}{2} + \frac{1}{2}e^2 - e$ **35.** $\displaystyle\int_0^1 \int_{-1}^1 \frac{2y + 3xy^2}{1 + x^2}\, dy\, dx = \ln 2$

EXERCISE 6-8

1. $R = \{(x, y) | 0 \le y \le 4 - x^2, \; 0 \le x \le 2\}$ **3.** $R = \{(x, y) | x^3 \le y \le 12 - 2x, \; 0 \le x \le 2\}$
     $= \{(x, y) | 0 \le x \le \sqrt{4 - y}, \; 0 \le y \le 4\}$

$y = 4 - x^2$

R is both a
regular x-region
and a regular
y-region

$y = 12 - 2x$

$y = x^3$

R is a regular
x-region

5. $R = \{(x, y) | \frac{1}{2}y^2 \le x \le y + 4, \; -2 \le y \le 4\}$

$x = \frac{1}{2}y^2$

$x = y + 4$

R is a regular
y-region

7. $\frac{1}{2}$ **9.** $39/70$ **11.** $56/3$ **13.** $-3/4$ **15.** $\frac{1}{2}e^4 - \frac{5}{2}$

17. $R = \{(x, y)|0 \leq y \leq x + 1, \;\; 0 \leq x \leq 1\}$

$$\int_0^1 \int_0^{x+1} \sqrt{1 + x + y} \, dy \, dx = (68 - 24\sqrt{2})/15$$

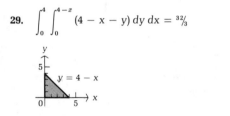

19. $R = \{(x, y)|0 \leq y \leq 4x - x^2, \;\; 0 \leq x \leq 4\}$

$$\int_0^4 \int_0^{4x-x^2} \sqrt{y + x^2} \, dy \, dx = {}^{128}\!/_5$$

21. $R = \{(x, y)|1 - \sqrt{x} \leq y \leq 1 + \sqrt{x}, \;\; 0 \leq x \leq 4\}$

$$\int_0^4 \int_{1-\sqrt{x}}^{1+\sqrt{x}} x(y - 1)^2 \, dy \, dx = \frac{512}{21}$$

23. $\displaystyle\int_0^3 \int_0^{3-y} (x + 2y) \, dx \, dy = {}^{27}\!/_2$

25. $\displaystyle\int_0^1 \int_0^{\sqrt{1-y}} x\sqrt{y} \, dx \, dy = {}^2\!/_{15}$

27. $\displaystyle\int_0^1 \int_{4y^2}^{4y} x \, dx \, dy = {}^{16}\!/_{15}$

29. $\displaystyle\int_0^4 \int_0^{4-x} (4 - x - y) \, dy \, dx = {}^{32}\!/_3$

31. $\displaystyle\int_0^1 \int_0^{1-x^2} 4 \, dy \, dx = {}^8\!/_3$

33. $\displaystyle\int_0^4 \int_0^{\sqrt{y}} \frac{4x}{1 + y^2} \, dx \, dy = \ln 17$ **35.** $\displaystyle\int_0^1 \int_0^{\sqrt{x}} 4ye^{x^2} \, dy \, dx = e - 1$

EXERCISE 6-9 CHAPTER REVIEW

1. $f(5, 10) = 2,900; f_x(x, y) = 40; f_y(x, y) = 70$ **2.** $\partial^2 z/\partial x^2 = 6xy^2; \partial^2 z/\partial x \, \partial y = 6x^2 y$ **3.** $dz = 2 \, dx + 3 \, dy$
4. $dz = 4x^3 y^3 \, dx + 3x^4 y^2 \, dy$ **5.** $2xy^3 + 2y^2 + C(x)$ **6.** $3x^2 y^2 + 4xy + E(y)$ **7.** 1 **8.** $\frac{1}{2}$
9. $f(2, 3) = 7; f_y(x, y) = -2x + 2y + 3; f_y(2, 3) = 5$ **10.** $(-8)(-6) - 4^2 = 32$
11. $\Delta z = f(1.1, 2.2) - f(1, 2) = 7.8897; dz = 4(1)^3(0.1) + 4(2)^3(0.2) = 6.8$

12. $y = -1.5x + 15.5; y = 0.5$ when $x = 10$ **13.** 18 **14.** $\displaystyle\int_0^3 \int_0^{y+1} (x + y)^3 \, dx \, dy = 408$

15. $f_x(x, y) = 2xe^{x^2+2y}; f_y(x, y) = 2e^{x^2+2y}; f_{xy}(x, y) = 4xe^{x^2+2y}$
16. $f_x(x, y) = 10x(x^2 + y^2)^4; f_{xy}(x, y) = 80xy(x^2 + y^2)^3$ **17.** 25.076 inches **18.** $y = {}^{116}\!/_{165}x + {}^{100}\!/_3$

19. $\int_0^1 \int_0^{\sqrt{1-x^2}} (x+y)\, dy\, dx = \frac{2}{3}$ **20.** $\int_0^1 \int_0^{\sqrt{1-y^2}} y\, dx\, dy = \frac{1}{3}$

21. (A) $P_x(1, 3) = 8$; profit will increase $8,000 for 100 units increase in product A if production of product B is held fixed at an output level of (1, 3)   (B) For 200 units of A and 300 units of B, $P(2, 3) = \$100$ thousand is a local maximum

22. $y = 0.63x + 1.33$; profit in sixth year is $5.11 million

23. $T_x(70, 17)$ or -0.92 minute per foot increase in depth when $V = 70$ cubic feet and $x = 17$ feet

24. $65.6k$ **25.** 50,000   **26.** $y = \frac{1}{2}x + 48$; $y = 68$ when $x = 40$

PRACTICE TEST: CHAPTER 6

1. (A) 7   (B) 8   **2.** $f_x(x, y) = 4(x^2y^3 - 2x)^3(2xy^3 - 2)$; $f_y(x, y) = 12x^2y^2(x^2y^3 - 2x)^3$

3. $dz = 3x^2y^4\, dx + 4x^3y^3\, dy$ **4.** $\Delta z = 0.85$; $dz = 0.8$ **5.** $-4x$ **6.** $2x^3ye^{x^2y} + 2xe^{x^2y}$ **7.** 72

8. $R_x(30, 20) = \$1,100$; $R_y(20, 30) = -\$600$ **9.** $f(0, 2) = -4$ is a local minimum

10. $\int_0^1 \int_0^{1-x} (1 - x - y)\, dy\, dx = \frac{1}{6}$

11. $y = 1.08x + 18.3$; cost of producing 40 units is $61,500

CHAPTER 7   EXERCISE 7-2

1. $\frac{7}{3}$ radians   **3.** $\frac{7}{4}$ radian   **5.** $\frac{7}{6}$ radian   **7.** II   **9.** IV   **11.** I   **13.** 1   **15.** 0   **17.** -1

19. 60°   **21.** 45°   **23.** 30°   **25.** $\frac{\sqrt{3}}{2}$ **27.** $\frac{1}{2}$ **29.** $-\frac{\sqrt{3}}{2}$ **31.** 0.1411   **33.** 0.6840   **35.** 0.7970

37. $\frac{3\pi}{20}$ radian   **39.** 15°   **41.** 1   **43.** 2   **45.** $\frac{1}{\sqrt{3}}$ or $\frac{\sqrt{3}}{3}$

49. (A) $P(13) = 5$, $P(26) = 10$, $P(39) = 5$, $P(52) = 0$ (B) $P(30) \approx 9.43$, $P(100) \approx 0.57$ **51.** (A) $-5.6°$ (B) $-4.7°$

EXERCISE 7-3

1. $-\sin t$ **3.** $3x^2 \cos x^3$ **5.** $t \cos t + \sin t$ **7.** $(\cos x)^2 - (\sin x)^2$ **9.** $5(\sin x)^4 \cos x$

11. $(\cos x)/2\sqrt{\sin x}$ **13.** $-\dfrac{x^{-1/2}}{2}\sin \sqrt{x} = \dfrac{-\sin \sqrt{x}}{2\sqrt{x}}$ **15.** $f'(\frac{\pi}{6}) = \cos \frac{\pi}{6} = \frac{\sqrt{3}}{2}$

17. $\dfrac{(\cos x)^2 + (\sin x)^2}{(\cos x)^2} = \dfrac{1}{(\cos x)^2} = (\sec x)^2$ **19.** $\dfrac{x \cos \sqrt{x^2 - 1}}{\sqrt{x^2 - 1}}$

21.

Δx	-0.5	-0.05	-0.005	0.005	0.05	0.5
$\dfrac{\cos \Delta x - 1}{\Delta x}$	0.2448	0.0250	0.0025	-0.0025	-0.0250	-0.2448

23. (A) $x(0) = 6$, $x(\frac{1}{40}) = \sqrt{24} \approx 4.9$, $x(\frac{1}{20}) = 4$, $x(\frac{1}{10}) = 6$ (B) $x'(t) = -20\pi \sin 20\pi t - \dfrac{20\pi \sin 20\pi t \cos 20\pi t}{\sqrt{25 - (\sin 20\pi t)^2}}$

(C) $x'(\frac{1}{40}) = -20\pi$ inches/second, $x'(\frac{1}{20}) = 0$ inches/second

EXERCISE 7-4

1. $-\cos t + C$ **3.** $\frac{1}{3}\sin 3x + C$ **5.** $\frac{1}{13}(\sin x)^{13} + C$ **7.** $-\frac{3}{4}(\cos x)^{4/3} + C$ **9.** $\frac{1}{3}\sin x^3 + C$

11. 1   **13.** 1   **15.** $\frac{\sqrt{3}}{2} - \frac{1}{2} \approx 0.366$ **17.** 1.4161   **19.** 0.0678   **21.** $e^{\sin x} + C$ **23.** $\ln|\sin x| + C$

25. $-\ln|\cos x| + C$ **27.** (A) $\int_0^{104} \left(5 - 5\cos\frac{\pi}{26}t\right) dt$ (B) 520 hundred dollars = $52,000

EXERCISE 7-5

1. $(\sec t)^2$ **3.** $\tan u \sec u$ **5.** $2(\sec 2x)^2$ **7.** $3u^2 \tan u^3 \sec u^3$ **9.** $(\tan x)^2 \sec x + (\sec x)^3$

11. $2\tan x(\sec x)^2$ **13.** $e^x(\sec e^x)^2$ **15.** $\tan t + C$ **17.** $-2\cot \frac{1}{2}x + C$ **19.** $-\ln|\cos w| + C$

21. $\dfrac{1}{2\pi}\ln|\sin 2\pi x| + C$ **23.** $-\frac{1}{3}\csc x^3 + C$ **25.** $\frac{2}{3}(\tan x)^{3/2} + C$ **27.** $-\ln|\cos(1 + e^x)| + C$ **29.** 2

31. 0 **33.** $-\ln(\sqrt{3}/2)$ **35.** $\ln(\sqrt{2} + 1)$ **37.** 0.5463 **39.** 9.1731 **41.** $\tan x - x + C$

EXERCISE 7-6   CHAPTER REVIEW

1. (A) $\pi/6$ (B) $\pi/4$ (C) $\pi/3$ (D) $\pi/2$ **2.** (A) -1 (B) 0 (C) 1 **3.** $-\sin m$
4. $\cos u$ **5.** $(\sec z)^2$ **6.** $-(\csc y)^2$ **7.** $\tan m \sec m$ **8.** $-\cot w \csc w$ **9.** $(2x - 2)\cos(x^2 - 2x + 1)$
10. $3t^2 \sin(1 - t^3)$ **11.** $2(\sec 2z)^2$ **12.** $-\pi(\cot \pi w)(\csc \pi w)$ **13.** $(-1/x^2)\tan(1/x)\sec(1/x)$
14. $(-1/2\sqrt{t})(\csc \sqrt{t})^2$ **15.** $-\frac{1}{3}\cos 3t + C$ **16.** $\frac{1}{2}\sin 2x + C$ **17.** $-2\ln|\cos \frac{1}{2}y| + C$
18. $3\ln|\sin(w/3)| + C$ **19.** (A) $30°$ (B) $45°$ (C) $60°$ (D) $90°$
20. (A) $\frac{1}{2}$ (B) $\sqrt{2}/2$ (C) $\sqrt{3}/2$ **21.** (A) -0.7147 (B) -0.0431 (C) -6.7997
22. (A) 1.8898 (B) 7.0862 (C) 0.4577 **23.** $(x^2 - 1)\cos x + 2x \sin x$ **24.** $6(\sin x)^5 \cos x$
25. $\frac{1}{3}(\sin x)^{-2/3}\cos x$ **26.** $\sec 2x + 2x \tan 2x \sec 2x$ **27.** $-\frac{1}{2}(\cot x)^{-1/2}(\csc x)^2$ **28.** $u(\sec u)^2 + \tan u$
29. $\frac{1}{2}\sin(t^2 - 1) + C$ **30.** 2 **31.** $\sqrt{3}/2$ **32.** $2\tan\sqrt{x} + C$ **33.** $-\frac{1}{2}\csc(w^2 + 1) + C$
34. $\ln(2 + \sqrt{3})$ **35.** $\sqrt{2} - 1$ **36.** -0.243 **37.** 0.7327 **38.** $-\sqrt{2}/2$ **39.** $\sqrt{2}$ **40.** $\pi/12$
41. (A) -1 (B) $-\sqrt{3}/2$ (C) $-\frac{1}{2}$ **42.** $-2x(\sin x^2)e^{\cos x^2}$ **43.** $\sin u(\sec u)^2 + \tan u \cos u$
44. $e^{\sin x} + C$ **45.** $e^{\tan x} + C$ **46.** $\frac{3}{4}(\tan x)^{4/3} + C$ **47.** $\frac{1}{5}\sec^5 x + C$ **48.** $2\sqrt{3}/3$ **49.** 15.2128
50. $-\ln|2 + \csc x| + C$

PRACTICE TEST: CHAPTER 7

1. $(2x + 1)\cos(x^2 + x)$ **2.** $2xe^{x^2}\cos e^{x^2}$ **3.** $-2xe^x \sin x^2 + e^x \cos x^2$ **4.** $\frac{2}{3}(\sin x)^{-1/3}\cos x$

5. $2\tan x(\sec x)^2$ **6.** $\sin x \tan x \sec x + \sec x \cos x$ **7.** $-\cot x(\csc x)^2 + \dfrac{\cos x}{\sin x}$ **8.** $\frac{1}{2}\sin(x^2 - 2x) + C$

9. $-e^{\cos x} + C$ **10.** $\ln|\sin x| + C$ **11.** 1 **12.** $-\cos e^x + C$ **13.** $\frac{1}{2}\ln|\sin x^2| + C$
14. $\frac{3}{7}(\tan x)^{7/3} + C$ **15.** $-\ln|1/\sqrt{3}| = \ln\sqrt{3}$

CHAPTER 8   EXERCISE 8-1

1. General solution: $y = x + C$; particular solution: $y = x + 2$
3. General solution: $y = \ln|x| + C$; particular solution: $y = \ln|x| - 2$
5. General solution: $y = Ce^x$; particular solution: $y = 10e^x$
7. General solution: $y = 25 - Ce^{-x}$; particular solution: $y = 25 - 20e^{-x}$
9. General solution: $y = Cx$; particular solution: $y = 5x$
11. General solution: $y = (3x + C)^{1/3}$; particular solution: $y = (3x + 24)^{1/3}$
13. General solution: $y = Ce^{e^x}$; particular solution: $y = 3e^{e^x}$
15. General solution: $y = \ln(e^x + C)$; particular solution: $y = \ln(e^x + 1)$
17. General solution: $y = Ce^{x^2/2} - 1$; particular solution: $y = 3e^{x^2/2} - 1$
19. General solution: $y = 2 - 1/(e^x + C)$; particular solution: $y = 2 - e^{-x}$
21. $y + \frac{1}{3}y^3 = x + \frac{1}{3}x^3 + C$ **23.** $\frac{1}{2}\ln|1 + y^2| = \ln|x| + (x^2/2) + C$ or $\ln(1 + y^2) = \ln(x^2) + x^2 + C$
25. $\ln(1 + e^y) = \frac{1}{2}x^2 + C$ **27.** $y = \sqrt{1 + (\ln x)^2}$ **29.** $y = \frac{1}{4}[x + \ln(x^2) + 3]^2$ **31.** $y = \sqrt{\ln(x^2 + 1)}$

33. $5,000e^{1.2} \approx \$16,600$ **35.** $\dfrac{\ln(0.5)}{-0.03188} \approx 22$ days **37.** $\dfrac{\ln(0.1)}{-0.02310} \approx 100$ days **39.** $\dfrac{\ln(0.2)}{-0.05579} \approx 29$ years

41. (A) $100e^{5(0.3365)} \approx 538$ bacteria (B) $\dfrac{\ln(10)}{0.3365} \approx 6.8$ hours

43. (A) $\dfrac{50,000}{1 + 499e^{-0.1617(20)}} \approx 2,422$ (B) $\dfrac{\ln(499)}{-0.1617} \approx 38.4$ days **45.** $I = As^k$

47. (A) $\dfrac{1,000}{1 + 199e^{-0.6982(7)}} \approx 400$ people (B) $\dfrac{\ln(3/3,383)}{-0.6982} \approx 10$ days

EXERCISE 8-2

1. $xy = C$ **3.** $x^2 + y^2 = C$ **5.** $2x - 3xy + 4y = C$ **7.** $x^2y^2 = C, x^2y^2 = 4$
9. $y^2\sqrt{x^2 + 9} = C, y^2\sqrt{x^2 + 9} = 20$ **11.** $x^3 + x^2y^2 + y^4 = C$ **13.** $x^2 + \ln|x + y| + y^3 = C$
15. $xye^x + e^y = C$ **17.** $xe^y + ye^x = C, xe^y + ye^x = 2e$ **19.** $(1 + y)^2/(2 - x) = C, (1 + y)^2/(2 - x) = 16$
21. $(x^2/2) - xy + y^2 = 1$ or $x^2 - 2xy + 2y^2 = 2$ **23.** $xy - \frac{1}{3}y^3 = \frac{2}{3}$ or $3xy - y^3 = 2$

25. $x^2y + xy^2 = 6$ **27.** $y = \dfrac{3 + x}{3 + 2x}$ **29.** $y = \dfrac{1 + \frac{1}{2}x^2}{1 + x}$ **31.** $y = \sqrt{6 - \dfrac{1}{x} - x}$ **33.** $\dfrac{100^2}{110} \approx \$90.91$

35. $\sqrt{33} - 5 \approx \$0.74$ **37.** $y = \dfrac{x^3 + \sqrt{x^6 + 40x}}{x}$

EXERCISE 8-3

1. $I(x) = e^{2x}, y = 2 + Ce^{-2x}$ **3.** $I(x) = e^x, y = -e^{-2x} + Ce^{-x}$ **5.** $I(x) = e^{-x}, y = 2xe^x + Ce^{-x}$
7. $I(x) = e^x, y = 3x^3e^{-x} + Ce^{-x}$ **9.** $I(x) = x, y = x + (C/x)$ **11.** $I(x) = x^2, y = 2x^3 + (C/x^2)$
13. $I(x) = e^{x^2/2}, y = 5 + Ce^{-x^2/2}$ **15.** $I(x) = e^{-2x}, y = -2x - 1 + Ce^{2x}$ **17.** $I(x) = x, y = e^x - (e^x/x) + (C/x)$
19. $I(x) = x, y = \frac{1}{2}x \ln x - \frac{1}{4}x + (C/x)$ **21.** $y = 1 + (C/x)$ **23.** $y = (x^2 + C)/(1 + x)$
25. $y = C(1 + x^2) - 1$ **27.** $60 + 40e^{-0.5} \approx \$84.3$ million   **29.** $W(t) = te^{-t} + 100e^{-t}$

EXERCISE 8-4

1. $y = C_1e^{-x} + C_2e^{-2x}$ **3.** $y = C_1e^{-5x} + C_2e^{3x}$ **5.** $y = C_1 + C_2e^{-6x}$ **7.** $y = C_1e^{2x} + C_2xe^{2x}$
9. $y = 2e^x + e^{-x}$ **11.** $y = \frac{3}{2}e^{x/3} - \frac{1}{2}e^{3x}$ **13.** $y = 2e^{-x} + 6xe^{-x}$ **15.** $y = C_1e^{4x} + C_2e^{-x} - 3$
17. $y = 2e^x + e^{-2x} - 3$ **19.** $y = C_1 \cos x + C_2 \sin x$ **21.** $y = e^{2x}(C_1 \cos 3x + C_2 \sin 3x)$
23. $p_e(t) = -50e^{-0.1t} + 100e^{-0.2t} + 25; \lim_{t \to \infty} p_e(t) - 25$ **25.** $y = 2 - e^{-t}; 2$

EXERCISE 8-5

1. $x = C_1e^t + C_2e^{-2t}, y = 2C_1e^t - C_2e^{-2t}$ **3.** $x = C_1e^t + C_2e^{-t}, y = C_1e^t + 3C_2e^{-t}; x = 2e^t - e^{-t}, y = 2e^t - 3e^{-t}$
5. $x = C_1e^{3t} + C_2, y = C_1e^{3t} - 2C_2; x = e^{3t} + 1, y = e^{3t} - 2$ **7.** $x = C_1e^t + C_2e^{-t}, y = 3C_1e^t + C_2e^{-t}$
9. $x = C_1e^{2t} + C_2e^{-t} + 2, v = C_1e^{2t} + 2C_2e^{-t} + 4$
11. $p = 5e^{-2t} + 10e^{-3t} + 85, q = 10e^{-2t} + 10e^{-3t} + 80; p$ decreases to the limiting value of 85; q decreases to the limiting value of 80
13. $x = 25e^{-4t} + 25e^{-6t}, y = 25e^{-4t} - 25e^{-6t}$; 30.5 units in compartment 1 and 3.0 units in compartment 2 after 6 minutes; 4.6 units in compartment 1 and 2.1 units in compartment 2 after 30 minutes; 0.5 unit in compartment 1 and 0.4 unit in compartment 2 after 1 hour

EXERCISE 8-6   CHAPTER REVIEW

1. $y = \dfrac{C}{x^4}$ **2.** $y = \frac{1}{6}x^2 + \dfrac{C}{x^4}$ **3.** $y = \sqrt[3]{C - x^2}$ **4.** $y = \dfrac{-1}{x^3 + C}$ **5.** $y = e^x + Ce^{2x}$

6. $3xy^3 - 2x - 2y^3 = C$ or $y = \sqrt[3]{\dfrac{2x + C}{3x - 2}}$ **7.** $y = \frac{1}{2}x^7 + Cx^5$ **8.** $y = C(2 + x) - 3$ **9.** $y = C_1e^{7x} + C_2e^{-3x}$

10. $y = C_1e^{-6x} + C_2xe^{-6x}$ **11.** $y = 10 - 10e^{-x}$ **12.** $y = x - 1 + e^{-x}$ **13.** $y = e^2e^{-2e^{-x}}$
14. $y = \dfrac{x^2 + 4}{x + 4}$ **15.** $y = \dfrac{-x + \sqrt{9x^2 + 16}}{4}$ **16.** $y = \frac{1}{3}x \ln x - \dfrac{x}{9} + \dfrac{19}{9x^2}$ **17.** $y = \sqrt{1 + 2x^2}$
18. $y = \dfrac{\sqrt{1 + 8x} - 1}{2x}$ **19.** $y = \frac{1}{2}e^{-4x} + \frac{1}{2}$ **20.** $y = e^{4x} + e^{-4x} - 1$ **21.** $x = 1 - t, y = 1 - 2t$
22. $x = e^{2t} - e^{-4t} + 2, y = 2e^{2t} + e^{-4t} - 1$ **23.** $100{,}000e^{-2.4} \approx \$9{,}071.80$
24. $p = -25e^{-3t} - 25e^{-4t} + 100, q = 25e^{-3t} + 50e^{-4t} + 75; p$ increases to a limiting value of 100; q decreases to a limiting value of 75   **25.** $y = 100 + te^{-t} - 100e^{-t}$
26. $x = 50e^{0.01t} + 25e^{0.04t}, y = 100e^{0.01t} - 25e^{0.04t}$; the first species increases without bound; the second species dies out after approximately 46.2 years

27. (A) $200 - 199e^{-0.02314(5)} \approx 23$ people   (B) $\dfrac{\ln(100/199)}{-0.02314} \approx 30$ days

---

PRACTICE TEST: CHAPTER 8

1. $y = Ce^{x^4}$ **2.** $y = \frac{1}{5}x^2 + \dfrac{C}{x^3}$ **3.** $y = C_1 e^{7x} + C_2 e^{-7x}$ **4.** $y = e^{2x} - e^x$ **5.** $y = x + \sqrt{4 - x^2}$

6. $y = 2e^x - 2e^{3x}$ **7.** $x = e^{-t} + 2e^{-2t},\ y = 2e^{-t} + 2e^{-2t}$ **8.** $\dfrac{\ln(0.25)}{-0.2877} \approx 5$ years

9. $\dfrac{\ln(0.15)}{-0.000121} \approx 15{,}700$ years   **10.** $p_e = 50 + 15e^{-t} + 10e^{-2t}$

CHAPTER 9   EXERCISE 9-1

1. $1 - x + \frac{1}{2}x^2 - \frac{1}{6}x^3 + \frac{1}{24}x^4$ **3.** $1 - 2x + 2x^2 - \frac{4}{3}x^3$ **5.** $2x - 2x^2 + \frac{8}{3}x^3$ **7.** $4 + \frac{1}{8}x - \frac{1}{512}x^2$
9. $1 + \frac{1}{3}x - \frac{1}{9}x^2 + \frac{5}{81}x^3$ **11.** 0.9048375 **13.** 0.60416667 **15.** 0.18266667 **17.** 4.1230469
19. 1.0099017 **21.** $1 + 2x + \dfrac{2^2}{2!}x^2 + \dfrac{2^3}{3!}x^3 + \cdots + \dfrac{2^n}{n!}x^n$ **23.** $1 + x + x^2 + x^3 + \cdots + x^n$

25. $2 - 2^2 x + 2^3 x^2 - 2^4 x^3 + \cdots + (-1)^n 2^{n+1} x^n$ **27.** $\ln 3 + \frac{4}{3}x - \frac{1}{2}(\frac{4}{3})^2 x^2 + \frac{1}{3}(\frac{4}{3})^3 x^3 - \cdots + \dfrac{(-1)^{n+1}}{n}(\frac{4}{3})^n x^n$

29. $1 + 2x + 3x^2 + 4x^3 + \cdots + (n+1)x^n$ **31.** $x + x^2 + \dfrac{1}{2!}x^3 + \dfrac{1}{3!}x^4$ **33.** $1 + x^2 + \frac{1}{2}x^4$

35. $1 - x^2 + x^4$ **37.** $1 + tx + \dfrac{t(t-1)}{2}x^2$ **39.** $1 + tx + \frac{1}{2}t^2 x^2$

41. (A) $p_0(x) = 1,\ p_1(x) = 1 + 4x,\ p_2(x) = 1 + 4x + 6x^2,\ p_3(x) = 1 + 4x + 6x^2 + 4x^3,$
$p_4(x) = 1 + 4x + 6x^2 + 4x^3 + x^4,\ p_5(x) = 1 + 4x + 6x^2 + 4x^3 + x^4$ (B) $f = p_4,\ p_5 = p_4,\ p_n = p_4$ for $n > 4$
43. $p_2(x) = 15 + \frac{1}{20}x^2,\ C(10) \approx p_2(10) = \$20$ **45.** $p_2(x) = 1{,}000 - 100x + 5x^2,\ D(1) \approx p_2(1) = 905$ pounds
47. $p_2(x) = 100 - x + \frac{1}{200}x^2,\ D(10) \approx p_2(10) = 90.5$ milligrams
49. $p_2(t) = 10t - \frac{1}{2}t^2,\ N(2) \approx p_2(2) = 18$ words per minute

---

EXERCISE 9-2

1. $1 + (x - 1) + \frac{1}{2}(x - 1)^2 + \frac{1}{6}(x - 1)^3 + \frac{1}{24}(x - 1)^4$ **3.** $e^2 - e^2(x + 2) + \frac{1}{2}e^2(x + 2)^2 - \frac{1}{6}e^2(x + 2)^3$
5. $3(x - \frac{1}{3}) - \frac{9}{2}(x - \frac{1}{3})^2 + 9(x - \frac{1}{3})^3$ **7.** $1 + \frac{1}{2}(x - 1) - \frac{1}{8}(x - 1)^2 + \frac{1}{16}(x - 1)^3 - \frac{5}{128}(x - 1)^4$
9. $-1 + \frac{1}{3}(x + 1) + \frac{1}{9}(x + 1)^2 + \frac{5}{81}(x + 1)^3 + \frac{10}{243}(x + 1)^4$ **11.** 1.0954375 **13.** -1.1440329

15. $e^3 + e^3(x - 3) + \dfrac{e^3}{2!}(x - 3)^2 + \dfrac{e^3}{3!}(x - 3)^3 + \cdots + \dfrac{e^3}{n!}(x - 3)^n$

17. $\ln(\frac{1}{2}) + 2(x - \frac{1}{2}) - \dfrac{2^2}{2}(x - \frac{1}{2})^2 + \dfrac{2^3}{3}(x - \frac{1}{2})^3 - \cdots + \dfrac{(-1)^{n+1}}{n}2^n(x - \frac{1}{2})^n$

19. $1 + 2(x - 1) + 3(x - 1)^2 + 4(x - 1)^3 + \cdots + (n + 1)(x - 1)^n$

21. $e^{-10} - e^{-10}(x - 10) + \dfrac{e^{-10}}{2!}(x - 10)^2 - \dfrac{e^{-10}}{3!}(x - 10)^3 + \cdots + \dfrac{e^{-10}(-1)^n}{n!}(x - 10)^n$ **23.** $1 + 2(x - 1) + (x - 1)^2$

25. $-1 + 5x - 10x^2 + 10x^3 - 5x^4 + x^5$ **27.** $e^a[1 + (x - a) + \dfrac{1}{2!}(x - a)^2 + \dfrac{1}{3!}(x - a)^3 + \cdots + \dfrac{1}{n!}(x - a)^n]$

29. $\dfrac{1}{a} - \dfrac{1}{a^2}(x - a) + \dfrac{1}{a^3}(x - a)^2 - \dfrac{1}{a^4}(x - a)^3 + \cdots + \dfrac{(-1)^n}{a^{n+1}}(x - a)^n$
31. $p_2(x) = 20 + (x - 10) - 0.05(x - 10)^2,\ R(11) \approx p_2(11) = 20.95$
33. $p_2(x) = \frac{295}{4} - \frac{295}{512}(x - 64) + \frac{885}{131{,}072}(x - 64)^2,\ y(70) \approx p_2(70) = 70.5$
35. $p_3(k) = 100\left[1 - 0.11(k - 1) + \dfrac{(0.11)^2}{2}(k - 1)^2 - \dfrac{(0.11)^3}{6}(k - 1)^3\right],\ N(3) \approx p_3(3) \approx 80$

EXERCISE 9-3

1. $1 + 2x + \frac{1}{2}x^2 + \frac{4}{3}x^3 + \cdots + \left(\frac{n + (-1)^{n+1}}{n}\right)x^n + \cdots, \quad -1 < x < 1$

3. $x^3 + x^4 + x^5 + x^6 + \cdots + x^{n+3} + \cdots, \quad -1 < x < 1$

5. $1 + \frac{1}{2}x^2 + \frac{2}{3}x^3 + \frac{3}{4}x^4 + \cdots + \frac{n-1}{n}x^n + \cdots, \quad -1 < x < 1$

7. $1 + x^2 + x^4 + x^6 + \cdots + x^{2n} + \cdots, \quad -1 < x < 1$

9. $1 + \frac{1}{2!}x^2 + \frac{1}{4!}x^4 + \frac{1}{6!}x^6 + \cdots + \frac{1}{(2n)!}x^{2n} + \cdots, \quad -\infty < x < \infty$

11. $x + x^3 + \frac{1}{2!}x^5 + \frac{1}{3!}x^7 + \cdots + \frac{1}{n!}x^{2n+1} + \cdots, \quad -\infty < x < \infty$

13. $x^2 - x^6 + x^{10} - x^{14} + \cdots + (-1)^n x^{4n+2} + \cdots, \quad -1 < x < 1$

15. $1 + 2^3 x^3 + 2^6 x^6 + 2^9 x^9 + \cdots + 2^{3n} x^{3n} + \cdots, \quad -\frac{1}{2} < x < \frac{1}{2}$

17. $e^{-2} + 2e^{-2}(x+1) + \frac{2^2 e^{-2}}{2!}(x+1)^2 + \frac{2^3 e^{-2}}{3!}(x+1)^3 + \cdots + \frac{2^n e^{-2}}{n!}(x+1)^n + \cdots, \quad -\infty < x < \infty$

19. $\ln 2 + \frac{1}{2}(x-2) - \frac{1}{2(2^2)}(x-2)^2 + \frac{1}{3(2^3)}(x-2)^3 - \cdots + \frac{(-1)^{n+1}}{n(2^n)}(x-2)^n + \cdots, \quad 0 < x < 4$

21. $1 - (x-1) + (x-1)^2 - (x-1)^3 + \cdots + (-1)^n (x-1)^n + \cdots, \quad 0 < x < 2$

23. $1 + 3(x-1) + 3^2(x-1)^2 + 3^3(x-1)^3 + \cdots + 3^n(x-1)^n + \cdots, \quad \frac{2}{3} < x < \frac{4}{3}$

25. $1 + 2x + 2x^2 + 2x^3 + \cdots + 2x^n + \cdots, \quad -1 < x < 1$

27. $x + \frac{1}{3}x^3 + \frac{1}{5}x^5 + \frac{1}{7}x^7 + \cdots + \frac{1}{2n+1}x^{2n+1} + \cdots, \quad -1 < x < 1$

EXERCISE 9-4

1. 0.998, 2 terms   **3.** 0.819, 4 terms   **5.** 0.001, 1 term   **7.** 0.095, 2 terms

9. (A) $1 + 2x + 3x^2 + 4x^3 + \cdots + nx^{n-1} + \cdots, \quad -1 < x < 1$

(B) $1 + 3x + 6x^2 + 10x^3 + \cdots + \frac{n(n-1)}{2}x^{n-2} + \cdots, \quad -1 < x < 1$

11. (A) $\frac{1}{2} + \frac{1}{2^2}(x+1) + \frac{1}{2^3}(x+1)^2 + \frac{1}{2^4}(x+1)^3 + \cdots + \frac{1}{2^{n+1}}(x+1)^n + \cdots, \quad -3 < x < 1$

(B) $\frac{1}{2^2} + \frac{2}{2^3}(x+1) + \frac{3}{2^4}(x+1)^2 + \frac{4}{2^5}(x+1)^3 + \cdots + \frac{n}{2^{n+1}}(x+1)^{n-1} + \cdots, \quad -3 < x < 1$

(C) $\frac{2}{2^4} + \frac{3(2)}{2^5}(x+1) + \frac{4(3)}{2^6}(x+1)^2 + \frac{5(4)}{2^7}(x+1)^3 + \cdots + \frac{n(n-1)}{2^{n+2}}(x+1)^{n-2} + \cdots, \quad -3 < x < 1$

13. (A) $(x-1) - \frac{1}{2}(x-1)^2 + \frac{1}{3}(x-1)^3 - \frac{1}{4}(x-1)^4 + \cdots + \frac{(-1)^{n+1}}{n}(x-1)^n + \cdots, \quad 0 < x < 2$

(B) $\frac{1}{2}(x-1)^2 - \frac{1}{3(2)}(x-1)^3 + \frac{1}{4(3)}(x-1)^4 - \frac{1}{5(4)}(x-1)^5 + \cdots + \frac{(-1)^{n+1}}{(n+1)n}(x-1)^{n+1} + \cdots, \quad 0 < x < 2$

15. 0.197   **17.** 0.066   **19.** 0.461   **21.** $y = 1 + \frac{1}{2!}x^2 + \frac{1}{3!}x^3 + \frac{1}{4!}x^4 + \cdots$

23. $y = -x + \frac{1}{2!}x^2 - \frac{1}{5!}x^5 + \frac{1}{6!}x^6 + \cdots$ **25.** $y = 1 + x - \frac{4}{3!}x^3 - \frac{7}{4!}x^4 + \cdots$

27. (A) $-1 + \frac{1}{2!}x - \frac{1}{3!}x^2 + \frac{1}{4!}x^3 - \cdots + \frac{(-1)^n}{n!}x^{n-1} + \cdots, \quad -\infty < x < \infty$ (B) -1 (C) -0.09755

29. (A) $e^2 + \frac{e^2}{2!}(x-2) + \frac{e^2}{3!}(x-2)^2 + \frac{e^2}{4!}(x-2)^3 + \cdots + \frac{e^2}{n!}(x-2)^{n-1} + \cdots, \quad -\infty < x < \infty$ (B) e^2

31. 5; 8   **33.** 9   **35.** $\int_0^1 \frac{36\pi}{4 + x^2}\, dx \approx 26.22$ **37.** 18.83°C   **39.** $\int_0^5 \frac{500}{100 + t^2}\, dt \approx 23$ items

EXERCISE 9-5

1. ½   **3.** 1   **5.** 2   **7.** 3   **9.** 1   **11.** 0   **13.** Does not exist   **15.** ⅓   **17.** $-\frac{1}{6}$ **19.** 1
21. 0   **23.** Does not exist   **25.** $1/(1 - e)$ **27.** -1 **29.** 0   **31.** Does not exist   **33.** ½

EXERCISE 9-6   CHAPTER REVIEW

1. $1 + \frac{1}{2}x - \frac{1}{8}x^2 + \frac{1}{16}x^3$ **2.** $2 + \frac{1}{4}(x - 3) - \frac{1}{64}(x - 3)^2 + \frac{1}{512}(x - 3)^3$ **3.** $1 + \frac{1}{3}x - \frac{1}{9}x^2 + \frac{5}{81}x^3$

4. $2 + \frac{1}{12}(x - 7) - \frac{1}{288}(x - 7)^2 + \frac{5}{20,736}(x - 7)^3$ **5.** $3 + \frac{1}{6}x^2$ **6.** $5 + \frac{4}{5}(x - 4) + \frac{9}{250}(x - 4)^2 - \frac{18}{3.125}(x - 4)^3$

7. 1.0049876   **8.** 1.9748418   **9.** 1.0033223   **10.** 1.9831925   **11.** 3.0016667   **12.** 5.0803542

13. $\frac{1}{4} + \frac{1}{4^2}x + \frac{1}{4^3}x^2 + \frac{1}{4^4}x^3 + \cdots + \frac{1}{4^{n+1}}x^n + \cdots, \quad -4 < x < 4$

14. $\frac{1}{2} + \frac{1}{2^2}(x - 2) + \frac{1}{2^3}(x - 2)^2 + \frac{1}{2^4}(x - 2)^3 + \cdots + \frac{1}{2^{n+1}}(x - 2)^n + \cdots, \quad 0 < x < 4$

15. $\frac{1}{8} + \frac{1}{8^2}(x + 4) + \frac{1}{8^3}(x + 4)^2 + \frac{1}{8^4}(x + 4)^3 + \cdots + \frac{1}{8^{n+1}}(x + 4)^n + \cdots, \quad -12 < x < 4$

16. $\frac{1}{4}x^2 + \frac{1}{4^2}x^4 + \frac{1}{4^3}x^6 + \frac{1}{4^4}x^8 + \cdots + \frac{1}{4^{n+1}}x^{2n+2} + \cdots, \quad -2 < x < 2$

17. $x^2 + 3x^3 + \frac{3^2}{2!}x^4 + \frac{3^3}{3!}x^5 + \cdots + \frac{3^n}{n!}x^{n+2} + \cdots, \quad -\infty < x < \infty$

18. $e^4 + 2e^4(x - 2) + \frac{2^2}{2!}e^4(x - 2)^2 + \frac{2^3}{3!}e^4(x - 2)^3 + \cdots + \frac{2^n}{n!}e^4(x - 2)^n + \cdots, \quad -\infty < x < \infty$

19. $x + \frac{1}{e}x^2 - \frac{1}{2e^2}x^3 + \frac{1}{3e^3}x^4 - \cdots + \frac{(-1)^{n+1}}{ne^n}x^{n+1} + \cdots, \quad -e < x < e$

20. $\ln 2 + (x - 2) - \frac{1}{2}(x - 2)^2 + \frac{1}{3}(x - 2)^3 - \cdots + \frac{(-1)^{n+1}}{n}(x - 2)^n + \cdots, \quad 1 < x < 3$

21. $\dfrac{1}{x} = \frac{1}{2} - \frac{1}{2^2}(x - 2) + \frac{1}{2^3}(x - 2)^2 - \frac{1}{2^4}(x - 2)^3 + \cdots + \frac{(-1)^n}{2^{n+1}}(x - 2)^n + \cdots, \quad 0 < x < 4;$

$\dfrac{1}{x^2} = \frac{1}{2^2} - \frac{2}{2^3}(x - 2) + \frac{3}{2^4}(x - 2)^2 - \frac{4}{2^5}(x - 2)^3 + \cdots + \frac{n(-1)^{n+1}}{2^{n+1}}(x - 2)^{n-1} + \cdots, \quad 0 < x < 4;$

$\dfrac{1}{x^3} = \frac{2(1)}{2^4} - \frac{3(2)}{2^5}(x - 2) + \frac{4(3)}{2^6}(x - 2)^2 - \frac{5(4)}{2^7}(x - 2)^3 + \cdots + \frac{n(n - 1)(-1)^n}{2^{n+2}}(x - 2)^{n-2} + \cdots, \quad 0 < x < 4$

22. $\dfrac{1}{x} = -1 - (x + 1) - (x + 1)^2 - (x + 1)^3 - \cdots - (x + 1)^n - \cdots, \quad -2 < x < 0;$

$\dfrac{1}{x^2} = 1 + 2(x + 1) + 3(x + 1)^2 + 4(x + 1)^3 + \cdots + n(x + 1)^{n-1} + \cdots, \quad -2 < x < 0;$

$\dfrac{1}{x^3} = -\frac{2(1)}{2} - \frac{3(2)}{2}(x + 1) - \frac{4(3)}{2}(x + 1)^2 - \frac{5(4)}{2}(x + 1)^3 - \cdots - \frac{n(n - 1)}{2}(x + 1)^{n-2} - \cdots, \quad -2 < x < 0$

23. $\frac{1}{3(9)}x^3 - \frac{1}{5(9^2)}x^5 + \frac{1}{7(9^3)}x^7 - \frac{1}{9(9^4)}x^9 + \cdots + \frac{(-1)^n}{(2n + 3)9^{n+1}}x^{2n+3} + \cdots, \quad -3 < x < 3$

24. $\frac{1}{5(16)}x^5 + \frac{1}{7(16^2)}x^7 + \frac{1}{9(16^3)}x^9 + \frac{1}{11(16^4)}x^{11} + \cdots + \frac{1}{(2n + 5)16^{n+1}}x^{2n+5} + \cdots, \quad -4 < x < 4$

25. 3   **26.** $-\frac{1}{5}$ **27.** Does not exist   **28.** 0   **29.** Does not exist   **30.** 1   **31.** 0.0612   **32.** 0.3140

33. 0.0359   **34.** 0.0976   **35.** $y = x + \frac{2}{3!}x^3 + \frac{4}{5!}x^5 + \frac{8}{7!}x^7 + \cdots$ **36.** $y = 1 + \frac{1}{2!}x^2 + \frac{2}{4!}x^4 + \frac{12}{6!}x^6 + \cdots$

37. 5 years; $25,208   **38.** $0.25   **39.** $\frac{1}{5}\int_0^5 \frac{5,000t^2}{10,000 + t^4}\, dt \approx 4.06$ **40.** $\int_3^5 \left\{ 5 - 18 \ln\left[1 + \left(\frac{t - 4}{2}\right)^2\right] \right\} dt \approx 7.20$

41. $\dfrac{1}{5}\displaystyle\int_0^5 (10 + 2t - 5e^{-0.01t^2})\, dt \approx 10.4$

PRACTICE TEST: CHAPTER 9

1. $1 - \frac{1}{2}(x - 1) + \frac{3}{8}(x - 1)^2 - \frac{5}{16}(x - 1)^3$

2. (A) $\ln 4 + \frac{1}{4}x - \dfrac{1}{2(4^2)}x^2 + \dfrac{1}{3(4^3)}x^3 - \cdots + \dfrac{(-1)^{n+1}}{n(4^n)}x^n + \cdots, \quad -4 < x < 4$

    (B) $x + \dfrac{1}{3!}x^3 + \dfrac{1}{5!}x^5 + \dfrac{1}{7!}x^7 + \cdots + \dfrac{1}{(2n + 1)!}x^{2n+1} + \cdots, \quad -\infty < x < \infty$

3. 0.7408; 5 terms   **4.** 0.0068   **5.** (A) 0   (B) -2 (C) Does not exist

6. $\dfrac{1}{1 + x} = \frac{1}{3} - \dfrac{1}{3^2}(x - 2) + \dfrac{1}{3^3}(x - 2)^2 - \dfrac{1}{3^4}(x - 2)^3 + \cdots + \dfrac{(-1)^n}{3^{n+1}}(x - 2)^n + \cdots, \quad -1 < x < 5;$

    $\dfrac{1}{(1 + x)^2} = \dfrac{1}{3^2} - \dfrac{2}{3^3}(x - 2) + \dfrac{3}{3^4}(x - 2)^2 - \dfrac{4}{3^5}(x - 2)^3 + \cdots + \dfrac{n(-1)^{n+1}}{3^{n+1}}(x - 2)^{n-1} + \cdots, \quad -1 < x < 5$

7. $1 + \dfrac{3}{2!}x^2 + \dfrac{1}{3!}x^3 + \dfrac{9}{4!}x^4 + \cdots$ **8.** (150, 147.8)

CHAPTER 10   EXERCISE 10-1

1. $x_n = \dfrac{x_{n-1}^2 + 4}{2x_{n-1}}, x_5 = 2.0000001$ **3.** $x_n = \dfrac{2x_{n-1}^3 + 8}{3x_{n-1}^2}, x_5 = 2.0000002$

5. $x_n = \dfrac{e^{x_{n-1}}(x_{n-1} - 1)}{e^{x_{n-1}} + 1}, x_5 = -0.56714329$ **7.** $x_n = \dfrac{x_{n-1}(1 - \ln x_{n-1})}{1 + x_{n-1}}, x_5 = 0.56714329$

9. $x_n = \dfrac{x_{n-1}(1 + x_{n-1}^2 - \ln x_{n-1})}{1 + 2x_{n-1}^2}, x_5 = 0.65291864$

11. 0.29843788, 6.7015621   **13.** -1.5567733 **15.** $-1.286772, 1.4436234, 11.843149$

17. 0.62870831, 5.4445699   **19.** 1.7963219   **21.** 9.8559595

23. $-1.8414057, 1.1461932$

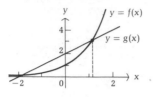

27. $x_n = \dfrac{p-1}{p}x_{n-1} + \dfrac{A}{px_{n-1}^{p-1}}$ **29.** x_n oscillates between -1 and $+1$ **31.** 2.64   **35.** 0.7 year   **37.** 9.2 hours

EXERCISE 10-2

1. $L = 1.6337994$; Absolute error $= 0.0844824$
$R = 1.8056276$; Absolute error $= 0.0873458$
$M = 1.7175661$; Absolute error $= 0.0007157$
$T = 1.7197135$; Absolute error $= 0.0014317$
$S = 1.7182819$; Absolute error $= 0.0000001$

3. 0.84529412   **5.** 3.2855456   **7.** 0.93533949   **9.** 0.7429841   **11.** 1.1692303   **13.** 1.5205957
15. 0.21439351   **17.** 0.62330691   **19.** 3.1364471   **21.** 1.3258174   **23.** 0.73716118   **25.** 4.3982297
27. 3.3300882   **29.** 11.749557   **31.** Consumers' surplus $= 4.2884055$, Producers' surplus $= 2.6739279$
33. 3 years; $3,424.78   **35.** 10.782459   **37.** Approx. 25

EXERCISE 10-3

1. Exact solution $y = x + 2$

x_n	y_n	$y(x_n)$	$y(x_n) - y_n$
0	2	2	0
0.1	2.1	2.1	0
0.2	2.2	2.2	0
0.3	2.3	2.3	0
0.4	2.4	2.4	0
0.5	2.5	2.5	0
0.6	2.6	2.6	0
0.7	2.7	2.7	0
0.8	2.8	2.8	0
0.9	2.9	2.9	0
1	3	3	0

3. Exact solution $y = x^3 + 1$

x_n	y_n	$y(x_n)$	$y(x_n) - y_n$
0	1	1	0
0.1	1	1.001	0.001
0.2	1.003	1.008	0.005
0.3	1.015	1.027	0.012
0.4	1.042	1.064	0.022
0.5	1.09	1.125	0.035
0.6	1.165	1.216	0.051
0.7	1.273	1.343	0.07
0.8	1.42	1.512	0.092
0.9	1.612	1.729	0.117
1	1.855	2	0.145

5. Exact solution $y = \ln(1 + x)$

x_n	y_n	$y(x_n)$	$y(x_n) - y_n$
0	0	0	0
0.1	0.1	0.09531018	-0.00468982
0.2	0.19090909	0.18232156	-0.00858753
0.3	0.27424242	0.26236426	-0.01187816
0.4	0.35116550	0.33647224	-0.01469326
0.5	0.42259407	0.40546511	-0.01712896
0.6	0.48926074	0.47000363	-0.01925711
0.7	0.55176074	0.53062825	-0.02113249
0.8	0.61058427	0.58778666	-0.02279761
0.9	0.66613982	0.64185389	-0.02428593
1	0.71877140	0.69314718	-0.02562422

7. Exact solution $y = (x + 1)^3$

x_n	y_n	$y(x_n)$	$y(x_n) - y_n$
0	1	1	0
0.1	1.3	1.331	0.031
0.2	1.6601818	1.728	0.0678182
0.3	2.0865303	2.197	0.1104697
0.4	2.5850326	2.744	0.1589674
0.5	3.1616778	3.375	0.2133222
0.6	3.8224563	4.096	0.2735437
0.7	4.5733599	4.913	0.3396401
0.8	5.4203810	5.832	0.411619
0.9	6.3695133	6.859	0.4894867
1	7.4267509	8	0.5732491

9. 1.2941308   **11.** -2.8139474 **13.** 1.1044663   **15.** 1.488047

17. Exact solution $y = e^{-x}$

x_n	y_n	$y(x_n)$
0	1	1
0.2	0.8	0.81873075
0.4	0.64	0.67032005
0.6	0.512	0.54881164
0.8	0.4096	0.44932896
1	0.32768	0.36787944

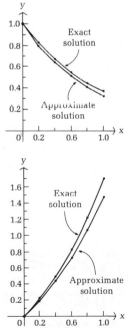

19. Exact solution $y = e^x - 1$

x_n	y_n	$y(x_n)$
0	0	0
0.2	0.2	0.22140276
0.4	0.44	0.4918247
0.6	0.728	0.8221188
0.8	1.0736	1.2255409
1	1.48832	1.7182818

21. Exact solution $x = 2e^t - e^{-t}$, $y = 2e^t - 3e^{-t}$

t_n	x_n	$x(t_n)$	y_n	$y(t_n)$
0	1	1	-1	-1
0.2	1.6	1.6240748	0	-0.01338674
0.4	2.24	2.3133294	0.96	0.97268926
0.6	2.944	3.095426	1.92	1.9978027
0.8	3.7376	4.0017529	2.9184	3.103095
1	4.64896	5.0686842	3.9936	4.3329253

23. Exact solution $x = e^{3t} + 1$, $y = e^{3t} - 2$

t_n	x_n	$x(t_n)$	y_n	$y(t_n)$
0	2	2	-1	-1
0.2	2.6	2.8221188	-0.4	-0.1778812
0.4	3.56	4.3201169	0.56	1.3201169
0.6	5.096	7.0496475	2.096	4.0496475
0.8	7.5536	12.023176	4.5536	9.0231764
1	11.48576	21.085537	8.48576	18.085537

25.

t_n	x_n	y_n
0	0	0
0.1	0.1	-0.1
0.2	0.21048374	-0.21051709
0.3	0.3344569	-0.3326607
0.4	0.47525048	-0.46746696
0.5	0.63655423	-0.61587108
0.6	0.82244983	-0.77867489
0.7	1.0374435	-0.95650928
0.8	1.2864973	-1.1497911
0.9	1.575059	-1.3586746
1	1.9090893	-1.5829965

27.

Δx	$y(0.5)$	$y(1)$
0.5	0.5	0.625
0.25	0.671875	0.75454982
0.1	0.73323389	0.80535355

29.

x_n	y_n from $y' = 2x$	y_n from $y' = 2y/x$
1	1	1
1.2	1.4	1.4
1.4	1.88	1.8666667
1.6	2.44	2.4
1.8	3.08	3
2	3.8	3.6666667

31. (A) 0.74682418   (B) 0.77781682   **33.** 42,674   **35.** $p(3) \approx 391.28$, $q(3) \approx 387.29$ **37.** 14,565
39. 26,892 foxes, 9,499 hares   **41.** 7,567

EXERCISE 10-4   CHAPTER REVIEW

1. $x_n = \dfrac{x_{n-1}^2 + 10}{2x_{n-1}}$, $x_3 = 3.1622777$ **2.** $x_n = \dfrac{x_{n-1}^2 + 1}{2x_{n-1} + 1}$, $x_4 = 0.61803399$

3. $x_n = \dfrac{-x_{n-1}\ln x_{n-1}}{1 + x_{n-1}}$, $x_4 = 0.27846454$ **4.** $x_n = \dfrac{1 + x_{n-1}}{1 + e^{x_{n-1}}}$, $x_3 = 0.56714329$

5. $L = 9.2$, Absolute error $= 0.8$; $R = 10.8$, Absolute error $= 0.8$; $M = T = S = 10$, Absolute error $= 0$

6. $L = 12.96$, Absolute error $= 3.04$; $R = 19.36$, Absolute error $= 3.36$; $M = 15.92$, Absolute error $= 0.08$; $T = 16.16$, Absolute error $= 0.16$; $S = 16$, Absolute error $= 0$

7. Exact solution $y = e^{x^2}$

x_n	y_n	$y(x_n)$	$y(x_n) - y_n$
0	1	1	0
0.1	1	1.0100502	0.0100502
0.2	1.02	1.0408108	0.0208108
0.3	1.0608	1.0941743	0.0333743
0.4	1.124448	1.1735109	0.0490629
0.5	1.2144038	1.2840254	0.0696216
0.6	1.3358442	1.4333294	0.0974852
0.7	1.4961455	1.6323162	0.1361707
0.8	1.7056059	1.8964809	0.190875
0.9	1.9785029	2.247908	0.2694051
1	2.3346334	2.7182818	0.3836484

8. Exact solution $y = (x + 1)^{-4}$

x_n	y_n	$y(x_n)$	$y(x_n) - y_n$
0	1	1	0
0.1	0.6	0.68301346	0.08301346
0.2	0.38181818	0.48225309	0.10043491
0.3	0.25454545	0.3501278	0.09558235
0.4	0.17622378	0.2603082	0.08408442
0.5	0.12587413	0.19753086	0.07165673
0.6	0.09230769	0.15258789	0.0602802
0.7	0.06923077	0.11973037	0.0504996
0.8	0.05294118	0.09525987	0.04231869
0.9	0.04117647	0.0767336	0.03555713
1	0.03250774	0.0625	0.02999226

9. Exact solution $y = e^x + e^{2x}$

x_n	y_n	$y(x_n)$	$y(x_n) - y_n$
0	2	2	0
0.1	2.3	2.3265737	0.0265737
0.2	2.6521403	2.7132275	0.0610872
0.3	3.0665368	3.1719776	0.1054408
0.4	3.5554023	3.7173656	0.1619633
0.5	4.1334967	4.3670031	0.2335064
0.6	4.8186745	5.1422357	0.3235612
0.7	5.6325537	6.0689527	0.436399
0.8	6.601329	7.1785734	0.5772444
0.9	7.7567652	8.5092506	0.7524854
1	9.1374064	10.107338	0.9699316

10. Exact solution $y = \dfrac{1}{1 + x^3}$

x_n	y_n	$y(x_n)$	$y(x_n) - y_n$
0	1	1	0
0.1	1	0.999001	-0.000999
0.2	0.997	0.99206349	-0.00493651
0.3	0.98507189	0.97370983	-0.01136206
0.4	0.95887199	0.93984962	-0.01902237
0.5	0.91473909	0.88888889	-0.0258502
0.6	0.85198302	0.82236842	-0.0296146
0.7	0.77358851	0.74460164	-0.02898687
0.8	0.68561795	0.66137566	-0.02424229
0.9	0.59536413	0.578369	-0.01699513
1	0.50923073	0.5	-0.00923073

11. $-3.5829187, 0.25132286, 3.3315958$

12. -1.1329332

13. 0.77288296

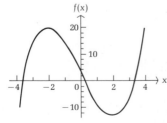

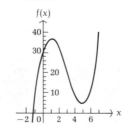

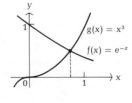

14. $-1.3802776, 0.81917251$

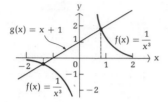

15.   0.44110398   **16.**   0.65045327   **17.**   5.3689579   **18.**   0.54557919   **19.**   0.86577378   **20.**   2.019096

21.   0.64   **22.**   3.3189676   **23.**   0.5   **24.** $x(1) \approx -0.56977534, y(1) \approx -0.88941257$ **25.** $x_n = \dfrac{ax_{n-1}^2 - c}{2ax_{n-1} + b}$

26.

x	$\Delta x = 1$	$\Delta x = 0.5$	$\Delta x = 0.25$
0	1	1	1
0.25			1.25
0.5		1.5	1.5686887
0.75			1.9803002
1	2	2.2905694	2.5096919
1.25			3.1850878
1.5		3.5402402	4.0404855
1.75			5.1179688
2	4.236068	5.4626931	6.4701916

27. $r \approx 1.2$, Area ≈ 1.0902125

28. $\displaystyle\int_1^2 y(x)\,dx \approx 1.8205765$

x_n	y_n
1	1
1.1	1.1414214
1.2	1.2911351
1.3	1.4489684
1.4	1.6147686
1.5	1.7883995
1.6	1.9697389
1.7	2.1586764
1.8	2.3551116
1.9	2.5589525
2	2.7701148

29. $x \approx 5.4916672$ **30.**   Consumers' surplus ≈ 87.970477, Producers' surplus ≈ 42.436332 **31.**   2 years; $2,284.85
32. $657,572   **33.**   0.59440558   **34.**   24,228 foxes; 51,047 hares   **35.**   6 components

PRACTICE TEST: CHAPTER 10

1. $x_n = \dfrac{x_{n-1}^2 + 3}{2x_{n-1}}$; $\sqrt{3} \approx 1.7320508$

2. 2.1200282   **3.** -8.1435335 **4.** 2.431994   **5.** 4.8507601

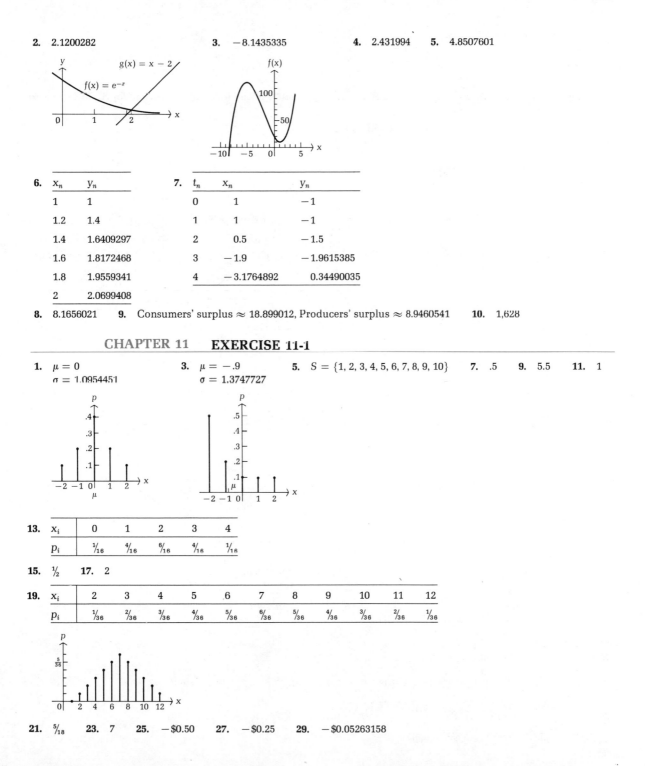

6.

x_n	y_n
1	1
1.2	1.4
1.4	1.6409297
1.6	1.8172468
1.8	1.9559341
2	2.0699408

7.

t_n	x_n	y_n
0	1	-1
1	1	-1
2	0.5	-1.5
3	-1.9	-1.9615385
4	-3.1764892	0.34490035

8. 8.1656021   **9.** Consumers' surplus $\approx$ 18.899012, Producers' surplus $\approx$ 8.9460541   **10.** 1,628

CHAPTER 11   EXERCISE 11-1

1. $\mu = 0$ **3.** $\mu = -.9$ **5.** $S = \{1, 2, 3, 4, 5, 6, 7, 8, 9, 10\}$ **7.** .5   **9.** 5.5   **11.** 1
$\sigma = 1.0954451$ $\sigma = 1.3747727$

13.

x_i	0	1	2	3	4
p_i	$\frac{1}{16}$	$\frac{4}{16}$	$\frac{6}{16}$	$\frac{4}{16}$	$\frac{1}{16}$

15. $\frac{1}{2}$ **17.** 2

19.

x_i	2	3	4	5	6	7	8	9	10	11	12
p_i	$\frac{1}{36}$	$\frac{2}{36}$	$\frac{3}{36}$	$\frac{4}{36}$	$\frac{5}{36}$	$\frac{6}{36}$	$\frac{5}{36}$	$\frac{4}{36}$	$\frac{3}{36}$	$\frac{2}{36}$	$\frac{1}{36}$

21. $\frac{5}{18}$ **23.** 7   **25.** $-\$0.50$ **27.** $-\$0.25$ **29.** $-\$0.05263158$

31.

x_i	4,850	-150
p_i	.01	.99

$E(X) = -100$

33. Site A, with $E(X) = \$3.6$ million   **35.** 1.54   **37.** A_2 is better, since for A_1, $E(X) = \$4$, and for A_2, $E(X) = \$4.80$

EXERCISE 11-2

1. $f(x)$

$f(x) \geq 0$ from graph

$\int_0^4 f(x)\,dx = 1$

3. $\int_2^3 \tfrac{1}{8}x\,dx = \tfrac{5}{16} = .3125$ **5.** $\int_3^4 \tfrac{1}{8}x\,dx = \tfrac{7}{16} = .4375$ **7.** $\int_5^\infty f(x)\,dx = 0$

9. $F(x) = \begin{cases} 0 & x < 0 \\ \tfrac{1}{16}x^2 & 0 \leq x \leq 4 \\ 1 & x > 4 \end{cases}$ **11.** $F(2) - F(0) = \tfrac{1}{4} - 0 = \tfrac{1}{4}$ **13.** $f(x)$

$f(x) \geq 0$ from graph

$\int_0^\infty \dfrac{2}{(1+x)^3}\,dx = 1$

15. $\int_3^\infty \dfrac{2}{(1+x)^3}\,dx = \tfrac{1}{16} = .0625$ **17.** $\int_1^1 \dfrac{2}{(1+x)^3}\,dx = 0$

19. $F(x) = \begin{cases} 0 & x < 0 \\ 1 - [1/(1+x)^2] & x \geq 0 \end{cases}$ **21.** $F(x) = \begin{cases} 0 & x < 0 \\ \tfrac{3}{4}x^2 - \tfrac{1}{4}x^3 & 0 \leq x \leq 2 \\ 1 & x > 2 \end{cases}$

23. $F(x) = \begin{cases} 0 & x < 1 \\ x \ln x - x + 1 & 1 \leq x \leq e \\ 1 & x > e \end{cases}$ **25.** $F(x) = \begin{cases} 1 - xe^{-x} - e^{-x} & x \geq 0 \\ 0 & \text{otherwise} \end{cases}$

$F(2) - F(1) = 2\ln 2 - 1 \approx .3863$ $1 - F(1) = 2e^{-1} \approx .7358$

27. $f(x) = \begin{cases} 2x & 0 \leq x \leq 1 \\ 0 & \text{otherwise} \end{cases}$ **29.** $f(x) = \begin{cases} 12x - 24x^2 + 12x^3 & 0 \leq x \leq 1 \\ 0 & \text{otherwise} \end{cases}$

31. $F(x) = \begin{cases} 0 & x < 0 \\ \tfrac{1}{2}x^2 & 0 \leq x < 1 \\ 2x - \tfrac{1}{2}x^2 - 1 & 1 \leq x \leq 2 \\ 1 & x > 2 \end{cases}$

33. .5555   **35.** (A) $\int_0^1 \tfrac{1}{10}e^{-x/10}\,dx = 1 - e^{-1/10} \approx .0952$ (B) $\int_4^\infty \tfrac{1}{10}e^{-x/10}\,dx = e^{-2/5} \approx .6703$

37. (A) $\int_4^{10} 0.003x\sqrt{100 - x^2}\,dx = \dfrac{(84)^{3/2}}{1,000} \approx .7699$ (B) $\int_0^8 0.003x\sqrt{100 - x^2}\,dx = .784$

(C) $\sqrt{100 - (100)^{2/3}} \approx 8,858$ pounds

39. (A) $\int_7^{10} \frac{1}{5,000}(10x^3 - x^4)\,dx = .47178$ (B) $\int_0^5 \frac{1}{5,000}(10x^3 - x^4)\,dx = \frac{3}{16} = .1875$

41. (A) $1 - F(30) = 2.5e^{-1.5} \approx .5578$ (B) $1 - F(80) = 5e^{-4} \approx .0916$

EXERCISE 11-3

1. $\int_0^2 \frac{1}{2}x^2\,dx = \frac{4}{3} \approx 1.333$ **3.** $\sqrt{2}/3 \approx .4714$ **5.** $\int_{(4-\sqrt{2})/3}^{(4+\sqrt{2})/3} f(x)\,dx = 4\sqrt{2}/9 \approx .6285$ **7.** $\sqrt{2} \approx 1.414$

9. $\int_1^\infty \frac{4}{x^4}\,dx = \frac{4}{3} \approx 1.333$ **11.** $\sqrt{2}/3 \approx .4714$ **13.** $\int_{(4-2\sqrt{2})/3}^{(4+2\sqrt{2})/3} f(x)\,dx = 1 - \left(\frac{3}{4 + 2\sqrt{2}}\right)^4 \approx .9627$

15. $2^{1/4} \approx 1.189$ **17.** $\mu = \int_2^5 \frac{1}{3}x\,dx = \frac{7}{2} = 3.5,\ V(X) = \int_2^5 \frac{1}{3}x^2\,dx - (\frac{7}{2})^2 = \frac{3}{4} = .75,\ \sigma = \sqrt{3}/2 \approx .866$

19. $\mu = \int_0^3 \frac{x}{2\sqrt{1+x}}\,dx = \frac{4}{3} \approx 1.333,\ V(X) = \int_0^3 \frac{x^2}{2\sqrt{1+x}}\,dx - (\frac{4}{3})^2 = \frac{34}{45} \approx .7556,\ \sigma = \sqrt{\frac{34}{45}} \approx .8692$

21. $e^{1/2} \approx 1.649$ **23.** 1   **25.** .61427243   **27.** 2.1555352   **29.** $x_1 = 1,\ x_2 = \sqrt{2},\ x_3 = \sqrt{3}$

31. (A) $E(X) = \frac{1}{8}\int_6^{10} (10x - x^2)\,dx = \frac{22}{3} \approx \$7.333$ thousand   (B) $x_m = 10 - 2\sqrt{2} \approx \$7.172$ thousand

33. $E(X) = \int_0^\infty [x/(1 + x^2)^{3/2}] = 1$ million gallons   **35.** $\mu = \int_0^{10} \frac{1}{5,000}(10x^4 - x^5)\,dx = \frac{20}{3} \approx 6.7$ minutes

37. $E(X) = \int_0^3 (\frac{4}{9}x^3 - \frac{4}{27}x^4)\,dx = \frac{9}{5} = 1.8$ hours

EXERCISE 11-4

1. $f(x) = \begin{cases} \frac{1}{2} & 0 \le x \le 2 \\ 0 & \text{otherwise} \end{cases}$ **3.** $\mu = 1,\ x_m = 1,\ \sigma = 1/\sqrt{3} \approx .5774$ **5.** $f(x) = \begin{cases} 20x^3(1 - x) & 0 \le x \le 1 \\ 0 & \text{otherwise} \end{cases}$

7. $\mu = \frac{2}{3},\ \sigma = \frac{1}{3}\sqrt{\frac{2}{7}} \approx .1782$ **9.** $f(x) = \begin{cases} 2e^{-2x} & x \ge 0 \\ 0 & \text{otherwise} \end{cases}$ **11.** $\mu = \frac{1}{2},\ x_m = \frac{1}{2}\ln 2 \approx .3466,\ \sigma = \frac{1}{2}$

13. $f(x) = \begin{cases} \frac{15}{4}x^{1/2}(1 - x) & 0 \le x \le 1 \\ 0 & \text{otherwise} \end{cases}$ **15.** $\mu = \frac{3}{7} \approx .4286,\ \sigma = \frac{2}{7}\sqrt{\frac{2}{3}} \approx .2333$ **17.** $\frac{1}{3}$

19. $F(x) = \begin{cases} 0 & x < 0 \\ \frac{7}{3}x^{4/3} - \frac{4}{3}x^{7/3} & 0 \le x \le 1 \\ 1 & x > 1 \end{cases}$ **21.** $\frac{2}{\ln 2} \approx 2.885$ **23.** $F(x) = \begin{cases} 1 - e^{-(x/2)(\ln 2)} & x \ge 0 \\ 0 & \text{otherwise} \end{cases}$

25. (A) $F(\lambda) - F(0) = 1 - e^{-1} \approx .6321$ (B) $F(2\lambda) - F(0) = 1 - e^{-2} \approx .8647$

     (C) $F(3\lambda) - F(0) = 1 - e^{-3} \approx .9502$ **27.** (A) $\frac{1}{b - x}$ (B) $\frac{1}{\lambda}$ **29.** $F(40) - F(25) = \frac{3}{8} = .375$

31. (A) $\beta = 1$ (B) $F(.75) - F(0) = \frac{27}{32} \approx .8438$ **33.** $F(2) - F(0) = 1 - e^{-2/3} \approx .4866$

35. (A) $E(X) = \mu = \frac{3}{8} = 37.5\%$ (B) $F(1) - F(.5) = 1 - (2.2)(.5)^{1.2} + (1.2)(.5)^{2.2} \approx .3036$

37. (A) $E(X) = \mu = -\frac{1}{\ln .7} \approx 2.8$ years   (B) $1 - F(\mu) = e^{-1} \approx .3679$

39. (A) $E(X) = \mu = .9 = 90\%$ (B) $1 - F(.95) = 1 - 19(.95)^{18} + 18(.95)^{19} \approx .2453$

EXERCISE 11-5

1. $\frac{1}{10} \int_0^{1/2} \int_0^{1/2} (x^2 + 2y)\, dy\, dx = \frac{7}{480} \approx .0146$ **3.** $\frac{1}{10} \int_0^{1/4} \int_1^3 (x^2 + 2y)\, dy\, dx = \frac{193}{960} \approx .2010$

5. $\frac{1}{10} \int_{3/4}^1 \int_0^3 (x^2 + 2y)\, dy\, dx = \frac{181}{640} \approx .2828$ **7.** $\frac{1}{10} \int_0^1 \int_x^{3x} (x^2 + 2y)\, dy\, dx = \frac{19}{60} \approx .3167$

9. $\mu_X = \frac{1}{10} = \int_0^1 \int_0^3 (x^3 + 2xy)\, dy\, dx = \frac{21}{40} = .525,\ \mu_Y = \frac{1}{10} \int_0^1 \int_0^3 (x^2y + 2y^2)\, dy\, dx = \frac{39}{20} = 1.95$

11. $\int_1^2 \int_1^3 \frac{6}{x^4y^3}\, dy\, dx = \frac{7}{9} \approx .7778$ **13.** $\int_2^3 \int_4^\infty \frac{6}{x^4y^3}\, dy\, dx = \frac{19}{3,456} \approx .0055$ **15.** $\int_1^2 \int_1^\infty \frac{6}{x^4y^3}\, dy\, dx = \frac{7}{8} = .8750$

17. $\int_1^\infty \int_1^x \frac{6}{x^4y^3}\, dy\, dx = \frac{2}{5} = .4$ **19.** $\mu_X = \int_1^\infty \int_1^\infty \frac{6}{x^3y^3}\, dy\, dx = \frac{3}{2} = 1.5,\ \mu_Y = \int_1^\infty \int_1^\infty \frac{6}{x^4y^2}\, dy\, dx = 2$

21. $\frac{1}{4} \int_0^1 \int_0^{x^2} (x + y)\, dy\, dx = \frac{7}{80} = .0875$ **23.** $\frac{1}{4} \int_0^1 \int_y^{2-y} (x + y)\, dx\, dy = \frac{1}{3} \approx .3333$

25. $\mu_X = \frac{1}{4} \int_0^2 \int_0^x (x^2 + xy)\, dy\, dx = \frac{3}{2} = 1.5,\ \mu_Y = \frac{1}{4} \int_0^2 \int_0^x (xy + y^2)\, dy\, dx = \frac{5}{6} \approx .8333$

27. $\mu_X = \frac{1}{(b-a)(d-c)} \int_a^b \int_c^d x\, dy\, dx = \frac{a+b}{2},\ \mu_Y = \frac{1}{(b-a)(d-c)} \int_a^b \int_c^d y\, dy\, dx = \frac{c+d}{2},$

$C(X, Y) = \frac{1}{(b-a)(d-c)} \int_a^b \int_c^d xy\, dy\, dx - \left(\frac{a+b}{2}\right)\left(\frac{c+d}{2}\right) = 0$

29. (A) $\int_1^2 \int_5^\infty xe^{-xy}\, dy\, dx = \frac{1}{5}(e^{-5} - e^{-10}) \approx .0013$ (B) $\int_1^2 \int_{2/x}^\infty xe^{-xy}\, dy\, dx = e^{-2} \approx .1353$

(C) $\int_1^2 \int_0^\infty x^2 e^{-xy}\, dy\, dx = \$1.50$

31. (A) $\int_0^3 \int_0^\infty \frac{1}{4} e^{-[(x/4)+y]}\, dy\, dx = 1 - e^{-3/4} \approx .5276$ (B) $\int_0^\infty \int_0^2 \frac{1}{4} e^{-[(x/4)+y]}\, dy\, dx = 1 - e^{-2} \approx .8647$

(C) $\int_0^5 \int_0^{5-x} \frac{1}{4} e^{-[(x/4)+y]}\, dy\, dx = 1 + \frac{1}{3}e^{-5} - \frac{4}{3}e^{-5/4} \approx .6202$

33. (A) $\int_{1/2}^1 \int_{1/2}^1 30x^2y^3(1 - xy)\, dy\, dx = \frac{705}{1,024} \approx .6885$

(B) First ingredient: $\int_0^1 \int_0^1 30x^3y^3(1 - xy)\, dy\, dx = \frac{27}{40} = 67.5\%;$

second ingredient: $\int_0^1 \int_0^1 30x^2y^4(1 - xy)\, dy\, dx = \frac{3}{4} = 75\%$

35. (A) $\int_{.7}^1 \int_{.7}^1 (x^2 + 2y^2)\, dy\, dx = .1971$

(B) Math: $\int_0^1 \int_0^1 (x^3 + 2xy^2)\, dy\, dx = \frac{7}{12} \approx 58.3\%;$

language: $\int_0^1 \int_0^1 (x^2y + 2y^3)\, dy\, dx = \frac{2}{3} \approx 66.7\%$

EXERCISE 11-6   CHAPTER REVIEW

1. $S = \{1, 2, 3, 4\}$ 2. $\frac{1}{2} = .5$

x_i	1	2	3	4
p_i	$\frac{1}{8}$	$\frac{2}{8}$	$\frac{3}{8}$	$\frac{2}{8}$

3. $E(X) = \frac{11}{4} \doteq 2.75$, $V(X) = \frac{15}{16} = .9375$, $\sigma = \sqrt{15}/4 \approx .9682$ 4. $-\$0.25$ 5. $\int_0^1 (1 - \frac{1}{2}x)\,dx = \frac{3}{4} = .75$

6. $\mu = \int_0^2 (x - \frac{1}{2}x^2)\,dx = \frac{2}{3} \approx 0.6667$, $V(X) = \int_0^2 (x^2 - \frac{1}{2}x^3)\,dx - (\frac{2}{3})^2 = \frac{2}{9} \approx .2222$, $\sigma = \sqrt{2}/3 \approx .4714$

7. $F(x) = \begin{cases} 0 & x < 0 \\ x - \frac{1}{4}x^2 & 0 \le x \le 2 \\ 1 & x > 2 \end{cases}$ 8. $2 - 2\sqrt{2} \approx .5858$ 9. $\int_1^4 \frac{5}{2}x^{-7/2}\,dx = \frac{31}{32} \approx .9688$

10. $\mu = \int_1^\infty \frac{5}{2}x^{-5/2}\,dx = \frac{5}{3} \approx 1.667$, $V(X) = \int_1^\infty \frac{5}{2}x^{-3/2}\,dx - (\frac{5}{3})^2 = \frac{20}{9} \approx 2.222$, $\sigma = \frac{2}{3}\sqrt{5} \approx 1.491$

11. $F(x) = \begin{cases} 1 - x^{-5/2} & x \ge 1 \\ 0 & \text{otherwise} \end{cases}$ 12. $2^{2/5} \approx 1.32$ 13. $\frac{1}{40}\int_0^3\int_0^1 (2x + y)\,dy\,dx = \frac{21}{80} \approx .2625$

14. $\frac{1}{40}\int_0^2\int_0^2 (2x + y)\,dy\,dx = .3$ 15. $\frac{1}{40}\int_0^2\int_0^{4-y} (2x + y)\,dx\,dy = .6$

16. $\mu_X = \frac{1}{40}\int_0^4\int_0^2 (2x^2 + xy)\,dy\,dx = \frac{38}{15} \approx 2.533$, $\mu_Y = \frac{1}{40}\int_0^4\int_0^2 (2xy + y^2)\,dy\,dx = \frac{16}{15} \approx 1.067$,

   $C(X, Y) = \frac{1}{40}\int_0^4\int_0^2 (2x^2y + xy^2)\,dy\,dx - \frac{38}{15}(\frac{16}{15}) = -\frac{8}{225} \approx -.0356$

17. $f(x) = \begin{cases} 42x^5(1-x) & 0 \le x \le 1 \\ 0 & \text{otherwise} \end{cases}$

18. $F(.75) - F(.25) = 7(.75)^6 - 6(.75)^7 - 7(.25)^6 + 6(.25)^7 \approx .4436$

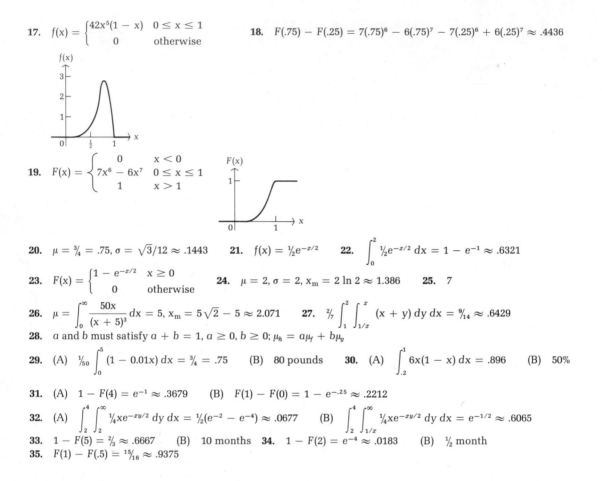

19. $F(x) = \begin{cases} 0 & x < 0 \\ 7x^6 - 6x^7 & 0 \le x \le 1 \\ 1 & x > 1 \end{cases}$

20. $\mu = \frac{3}{4} = .75, \sigma = \sqrt{3}/12 \approx .1443$ **21.** $f(x) = \frac{1}{2}e^{-x/2}$ **22.** $\int_0^2 \frac{1}{2}e^{-x/2}\, dx = 1 - e^{-1} \approx .6321$

23. $F(x) = \begin{cases} 1 - e^{-x/2} & x \ge 0 \\ 0 & \text{otherwise} \end{cases}$ **24.** $\mu = 2, \sigma = 2, x_m = 2\ln 2 \approx 1.386$ **25.** 7

26. $\mu = \int_0^\infty \frac{50x}{(x+5)^3}\, dx = 5, x_m = 5\sqrt{2} - 5 \approx 2.071$ **27.** $\frac{2}{7}\int_1^2 \int_{1/x}^x (x+y)\, dy\, dx = \frac{9}{14} \approx .6429$

28. a and b must satisfy $a + b = 1, a \ge 0, b \ge 0; \mu_h = a\mu_f + b\mu_g$

29. (A) $\frac{1}{50}\int_0^5 (1 - 0.01x)\, dx = \frac{3}{4} = .75$ (B) 80 pounds **30.** (A) $\int_{.2}^1 6x(1-x)\, dx = .896$ (B) 50%

31. (A) $1 - F(4) = e^{-1} \approx .3679$ (B) $F(1) - F(0) = 1 - e^{-.25} \approx .2212$

32. (A) $\int_2^4 \int_2^\infty \frac{1}{4}xe^{-xy/2}\, dy\, dx = \frac{1}{2}(e^{-2} - e^{-4}) \approx .0677$ (B) $\int_2^4 \int_{1/x}^\infty \frac{1}{4}xe^{-xy/2}\, dy\, dx = e^{-1/2} \approx .6065$

33. $1 - F(5) = \frac{2}{3} \approx .6667$ (B) 10 months **34.** $1 - F(2) = e^{-4} \approx .0183$ (B) $\frac{1}{2}$ month

35. $F(1) - F(.5) = \frac{15}{16} \approx .9375$

PRACTICE TEST: CHAPTER 11

1. $\mu = 3.2, V(X) = 1.96, \sigma = 1.4$

2. $\mu = \frac{1}{4}\int_0^2 (x + x^2)\, dx = \frac{7}{6} \approx 1.167, V(X) = \frac{1}{4}\int_0^2 (x^2 + x^3)\, dx - (\frac{7}{6})^2 = \frac{11}{36} \approx .3056, \sigma = \frac{1}{6}\sqrt{11} \approx .5528$

3. $\int_1^5 \frac{10}{9x^2}\, dx = \frac{8}{9} \approx .8889$

4. $F(x) = \begin{cases} 0 & x \le 1 \\ \frac{10}{9}\left(1 - \dfrac{1}{x}\right) & 1 \le x \le 10 \\ 1 & x > 10 \end{cases}$

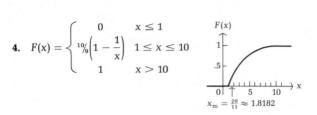

$x_m = \frac{20}{11} \approx 1.8182$

5. (A) $\frac{1}{8}\displaystyle\int_1^2 \int_0^1 (x + 4y)\, dy\, dx = \frac{7}{16} = .4375$ (B) $\frac{1}{8}\displaystyle\int_0^1 \int_y^{2-y} (x + 4y)\, dx\, dy = \frac{7}{24} \approx .2917$

6. $\displaystyle\int_5^\infty \frac{1}{5}e^{-x/5}\, dx = e^{-1} \approx .3679$ **7.** $\frac{1}{3}$

8. $E(x) = \frac{1}{8}\displaystyle\int_2^6 (6x - x^2)\, dx = \frac{10}{3}$, expected demand is 333 dozen; $x_m = 6 - 2\sqrt{2}$, median demand is approximately 317 dozen

9. $\displaystyle\int_0^4 \int_{2/x}^\infty xe^{-4xy}\, dy\, dx = e^{-8} \approx .0003$

APPENDIX A EXERCISE A-1

1. T **3.** T **5.** T **7.** T **9.** $\{1, 2, 3, 4, 5\}$ **11.** $\{3, 4\}$ **13.** $\varnothing$ **15.** $\{2\}$ **17.** $\{-7, 7\}$
19. $\{1, 3, 5, 7, 9\}$ **21.** $A' = \{1, 5\}$ **23.** 40 **25.** 60 **27.** 60 **29.** 20 **31.** 95 **33.** 40
35. (A) $\{1, 2, 3, 4, 6\}$ (B) $\{1, 2, 3, 4, 6\}$ **37.** $\{1, 2, 3, 4, 6\}$ **39.** Yes **41.** Yes **43.** Yes
45. (A) 2 (B) 4 (C) $8; 2^n$ **47.** 800 **49.** 200 **51.** 200 **53.** 800 **55.** 200 **57.** 200
59. 6 **61.** A+, AB+ **63.** A−, A+, B−, AB−, AB+, O+ **65.** O+, O− **67.** B−, B+
69. Everybody in the clique relates to each other.

EXERCISE A-3

1. $6x^{10}$ **3.** $y^6/3x^4$ **5.** 8×10^5 **7.** $1/y^8$ **9.** u^{10}/v^6 **11.** 3×10^5 **13.** 125 **15.** 125
17. x^2y **19.** $a^9/8b^4$ **21.** $y^4/9$ **23.** $n^{12}/8m^6$ **25.** $x^{2/5}$ **27.** $1/x^{1/2}$ **29.** $2y^2/3x^2$ **31.** $2mn^2\sqrt{2m}$
33. $3x\sqrt{2x}$ **35.** $4\sqrt{2x}$ **37.** 3×10^4 or 30,000 **39.** 1.12×10^4 or 11,200 **41.** $1/xy$ **43.** $3x^{1/2}$ **45.** x

EXERCISE A-4

1. $2u^4 + 6u^3$ **3.** $6m - 2n$ **5.** $-x + 8y$ **7.** $m^2 - 4m - 21$ **9.** $6x^2 - 7x - 5$ **11.** $x^2 - 9y^2$
13. $8m(1 + 2m)$ **15.** $3x(4x^2 - 2x - 1)$ **17.** $(y - 4)(y + 2)$ **19.** $(x + 3y)(x + 5y)$ **21.** $(y - 4)(y + 2)$
23. $(x + 3y)(x + 5y)$ **25.** $(x - 5)(x + 5)$ **27.** $-3y + 4$ **29.** $10m - 18$ **31.** $6x^2 - xy - 35y^2$
33. $4x^2 - 20xy + 25y^2$ **35.** $64x^2 - 9y^2$ **37.** $x^3 - 6x^2y + 10xy^2 - 3y^3$ **39.** $(2x - 3y)(x - 2y)$
41. $(3x - y)(2x + 3y)$ **43.** Not factorable **45.** $(5x - 2y)(5x + 2y)$ **47.** $2x^2(x - 2)(x - 10)$
49. $2x(x^2 + 4y^2)$ **51.** $2uv(2u + v)(u + 3v)$ **53.** $-x + 27$ **55.** $6x^4 + 2x^3 - 5x^2 + 4x - 1$
57. $-7x^2 - x - 16$ **59.** $4x^3 - 14x^2 + 8x - 6$

EXERCISE A-5

1. $3x/4y$ **3.** $4x(x - y)/3(x + y)$ **5.** y/x **7.** $a^3/48c^3$ **9.** $x/2$ **11.** $1/(x + 2)$ **13.** $(6 - x)/3x$
15. $(u^2 + uv - v^2)/v^3$ **17.** $1/(2u + 3)$ **19.** $(6 - x)/2x(x + 2)$ **21.** False **23.** False **25.** $3y/(x + 3)$
27. $1/2y$ **29.** $x(x - 4)$ **31.** $(15x^2 + 14x - 6)/36x^3$ **33.** $(5x^2 - 2x - 5)/(x + 1)(x - 1)$
35. $(2x^2 + x - 2)/2x(x - 2)(x + 2)$ **37.** $-2y/(x - y)^2(x + y)$ **39.** $(x - y)^2/y^3(x + 1)$
41. $(5x^2 + 1)/6(x + 1)^2$

EXERCISE A-6

1. $m = 5$ **3.** $x < -9$ **5.** $x \le 4$ **7.** $x < -3$ $\xrightarrow{\quad\circ\quad}^{x}_{-3}$ **9.** $-1 \le x \le 2$ $\xrightarrow{\bullet\quad\bullet}^{x}_{-1\ \ 2}$

11. $y = 8$ **13.** $x > -6$ **15.** $y = 8$ **17.** $x = 10$ **19.** $y \ge 3$ **21.** $x = 36$

23. $m < 3$ **25.** $x = 10$ **27.** $3 \le x < 7$ $\xrightarrow{\bullet\quad\circ}^{x}_{3\ \ 7}$ **29.** $-20 \le C \le 20$ $\xrightarrow{\bullet\quad\bullet}^{C}_{-20\ \ 20}$

31. $y = \frac{3}{4}x - 3$ **33.** $y = -(A/B)x + (C/B) = (-Ax + C)/B$ **35.** $C = \frac{5}{9}(F - 32)$ **37.** $B = A/(m - n)$

39. $y = -2$ **41.** $m = 3$ **43.** $y \le -14$ **45.** 3,000 \$10 tickets; 5,000 \$6 tickets

47. \$7,200 at 10%; \$4,800 at 15% **49.** (A) \$887.10 (B) \$477.67 **51.** 5,000 **53.** 12.6 years

EXERCISE A-7

1. ± 2 **3.** $\pm\sqrt{11}$ **5.** $-2, 6$ **7.** $0, 2$ **9.** $3 \pm 2\sqrt{3}$ **11.** $-2 \pm \sqrt{2}$ **13.** $0, 2$ **15.** $\pm\frac{3}{2}$

17. $\frac{1}{2}, -3$ **19.** $(-1 \pm \sqrt{5})/2$ **21.** $(3 \pm \sqrt{3})/2$ **23.** No real solutions **25.** $-4 \pm \sqrt{11}$

27. $x \le -3$ or $x \ge 4$ **29.** $-4 < x < 3$ **31.** $x < -1$ or $x > -\frac{1}{2}$ **33.** True for all real numbers

35. $r = \sqrt{A/P} - 1$ **37.** \$2 **39.** 100 miles/hour

EXERCISE A-8   APPENDIX REVIEW

1. (A) T (B) T (C) F (D) T **2.** $4y^7/3x$ **3.** $6x^2 + xy - 12y^2$ **4.** $(13 - 2x)/3(x - 2)$

5. (A) $(x - 3)(x + 6)$ (B) $(2x - 5)(2x + 5)$ **6.** $x < 4$ or $(-\infty, 4)$ $\xrightarrow{\quad\circ\quad}^{x}_{4}$ **7.** $x = 0, 5$

8. $u = 36$ **9.** (A) $\{1, 2, 3, 4\}$ (B) $\{2, 3\}$ **10.** (A) 28 (B) 5 (C) 4 (D) 10

11. (A) 90 (B) 45 **12.** $9x^6/4y^8$ **13.** $27y^9/x^6$ **14.** $2\sqrt{2x}$ **15.** $(2.2 \times 10^8)(3 \times 10^{-4}) = 6.6 \times 10^4$

16. $2x^3 - 5x^2 + 7x + 5$ **17.** $-x^2 + 17x - 11$ **18.** $x + 2$ **19.** $(x - 10)/(x - 2)^2(x + 2)$

20. $2x(3x - 1)(x + 4)$ **21.** $3x(3x^2 - 2x + 2)$ **22.** $x = 2$ **23.** $-1 < x \le 4$ **24.** $x = \pm\sqrt{7}$

25. $x = (3 \pm \sqrt{17})/4$ **26.** $x \le -1$ or $x \ge \frac{2}{3}$ **27.** $-2 < x < 5$ **28.** $y = \frac{2}{3}x - 2$ **29.** $y = 3/(x - 1)$

30. $2xy^2\sqrt[3]{2xy}$ **31.** Yes **32.** $x \ge -3$ **33.** $x = (-j \pm \sqrt{j^2 - 4k})/2$

PRACTICE TEST: APPENDIX A

1. (A) $\{2, 4, 5, 6\}$ (B) $\{5\}$ (C) $\{8\}$ (D) $\{2, 4\}$ **2.** (A) F (B) T (C) T (D) T

3. $4y^4/5x^3$ **4.** $4xy^2\sqrt{xy}$ **5.** $5x^2 - 2x - 20$ **6.** $(5 - x)/(x - 2)(x + 2)$ **7.** $m^3(m + 1)/2n(m - 1)$

8. $2 \le x < 12$ **9.** $x = \frac{4}{3}$ **10.** $x = (3 \pm \sqrt{17})/4$ **11.** $x < -3$ or $x > \frac{1}{2}$ **12.** (A) 900 (B) 350

INDEX

This book was typeset by Typothetae, and was
    printed and bound by R. R. Donnelley and Sons
Production was coordinated by Phyllis Niklas
The text was designed by Janet Bollow, and the
    cover was designed by John Williams
Interior photographs were taken by John Drooyan, and the
    cover photograph was taken by John Jensen
Technical art was drawn by Art by AYXA

BASIC DIFFERENTIATION

For C a constant:

1. $D_x C = 0$

2. $D_x u^n = n u^{n-1} D_x u$

3. $D_x(u \pm v) = D_x u \pm D_x v$

4. $D_x(Cu) = C D_x u$

5. $D_x(uv) = u D_x v + v D_x u$

6. $D_x \left(\dfrac{u}{v} \right) = \dfrac{v D_x u - u D_x v}{v^2}$

7. $D_x \ln u = \dfrac{1}{u} D_x u$

8. $D_x e^u = e^u D_x u$

9. $D_x b^u = b^u \ln b \, D_x u$

10. $D_x \sin u = \cos u \, D_x u$

11. $D_x \cos u = -\sin u \, D_x u$

12. $D_x \tan u = (\sec u)^2 \, D_x u$

13. $D_x \cot u = -(\csc u)^2 \, D_x u$

14. $D_x \sec u = \sec u \tan u \, D_x u$

15. $D_x \csc u = -\csc u \cot u \, D_x u$